Down East Books

Published by Down East Books
An imprint of Globe Pequot
Trade division of The Rowman & Littlefield Publishing Group, Inc.
4501 Forbes Blvd., Ste. 200
Lanham, MD 20706
www.rowman.com
www.downeastbooks.com

Distributed by NATIONAL BOOK NETWORK

Library of Congress Cataloging-in-Publication Data
Names: Borton, Mark C., author.
Title: Moondoggle : Franklin Roosevelt and the fight for tidal-electric
  power at Passamaquoddy Bay / Mark C. Borton.
Description: Lanham, MD : Down East Books, [2023] | Includes
  bibliographical references and index. | Summary: "This book is a
  dramatic tale about the appeal of tidal power, the difficulties in
  realizing its potential, and the engineers and U.S. presidents who tried
  to make clean and renewable tidal power a reality"— Provided by
  publisher.
Identifiers: LCCN 2022030167 (print) | LCCN 2022030168 (ebook) | ISBN
  9781608937141 (cloth) | ISBN 9781608937158 (epub)
Subjects: LCSH: Passamaquoddy Tidal Power Project. | Tidal
  power-plants—Maine—History—20th century | Passamaquoddy Bay (N.B. and
  Me.) | Roosevelt, Franklin D. (Franklin Delano), 1882–1945. | Cooper,
  Dexter P.
Classification: LCC TK1081 .B68 2023 (print) | LCC TK1081 (ebook) | DDC
  621.31/21340974142—dc23/eng/20221108
LC record available at https://lccn.loc.gov/2022030167
LC ebook record available at https://lccn.loc.gov/2022030168

Contents

Prologue
Who Killed Quoddy?

Rarely do vice presidents of the United States push the trigger.
From the comfort and safety of the Executive Office Building in Washington, D.C., John "Cactus Jack" Garner contemplated the target, 660 miles away and a few hundred yards from the US border.

The campaign had taken fifteen years. There had been battles on both sides of the aisle in Congress and of the border. One last blast and the fight should be over. He tapped his finger on the telegraph key, sending the spark racing. In milliseconds it would reach its destination and ignite 600 pounds of explosives buried in the contested territory.

An honor guard and 15,000 civilians surrounded the target.[1] At noon, the ground exploded. Everyone in Eastport, Maine, began yelling and dancing. They were celebrating their good fortune and the start of construction on the world's first tidal-hydroelectric power plant. It was a dream come true: perpetual power, pollution-free power, Depression-ending power. It was the archetype of President Franklin Roosevelt's New Deal: jobs and electric power for the American people.

With the smell of explosives and money in the air, construction of "Quoddy" was finally underway.

"Quoddy" was the Passamaquoddy Bay Tidal Power Project, a massive engineering and construction effort to dam up 150 square miles of coastal Maine and New Brunswick to produce electricity. It would use the latent energy in the eighteen-foot tides in Passamaquoddy Bay and provide more than enough power for every light bulb and electric motor in Maine.

Before a single light was lit, Quoddy was killed.

Right from the start, there were signs that Quoddy was in danger.

President Franklin Roosevelt, a summertime resident of Passamaquoddy Bay and Quoddy's national champion, *could* have triggered the celebratory explosion on July 4, 1935, from the White House or in person in Eastport.[2] He didn't.

Dexter P. Cooper, the American engineer who had conceived Quoddy and worked on it for fifteen years, *should* have ceremoniously thrown the first shovel of dirt. He didn't.

Roscoe C. Emery, Eastport's mayor and publisher of its newspaper, the *Eastport Sentinel, was* at the celebration. He had helped Eastporters understand Quoddy's economic potential and the forces—gravitational, political, and financial—at play in Passamaquoddy Bay.

These men—Franklin Roosevelt (the "Champion"), Dexter Cooper (the "Engineer"), and Roscoe Emery (the "Reporter")—are our protagonists.

There were other supporting actors, both proponents and opponents. Some founded the electric industry. Others founded nations. Some were obscure; others were the most famous people alive. Some of their roles in the Quoddy saga were unknown for years. It took a small army of federal investigators to uncover the truth.

The Quoddy saga is not just about pioneers in tidal-hydroelectric power. It is the story of efforts by the United States to define itself as it progressed through its adolescent dreams of Manifest Destiny, became shell-shocked by World War I, intoxicated itself during the Roaring Twenties, struggled through the Great Depression, fought for victory in World War II, and triggered the atomic age.

Quoddy was front-page news in the 1920s and 1930s. The *New York Times* published more than 200 articles about the project: the *Boston Herald* over 500. *Time, Life, Newsweek, Business Week, National*

Review, Saturday Evening Post, Popular Mechanics, National Geographic, and many others covered it. Many in the press took sides. The media melee made Quoddy the "storm center of one of the more violent New Deal controversies."[3]

The fight over Quoddy was so intense because it raised fundamental questions about the control of natural resources, economic justice, monopolies, free enterprise, welfare, campaign finance, states' rights, and the power of the presidency—issues that a century later are still being fought over in America's capital.

Despite the barrage of information, the *whole truth* about Quoddy has never truly been told—until now.

To fully appreciate the epic battle over Quoddy, we need to go back to when our protagonists—the Champion, the Engineer, and the Reporter—were beginning to exert their influence on the waters, nations, and economies of the world.

With new evidence presented here for the first time—from memoirs and personal letters made available after the protagonists had died, and from formerly classified government documents—we can fully appreciate the depth and breadth of the effort to kill Quoddy. And we can finally answer the questions that have dogged Downeast Maine for the last century:

Who killed Quoddy?
How? Why?

This is a detective story. With new information, we will reconstruct the crime and solve the mystery.

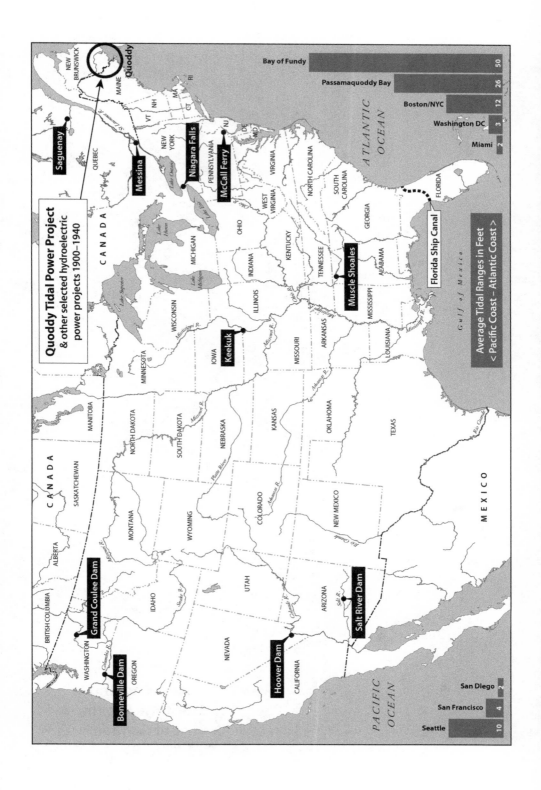

Quoddy Tidal Power Project
& other selected hydroelectric
power projects 1900–1940

Quoddy

Saguenay

Messina

Niagara Falls

McCall Ferry

Muscle Shoales

Keekuk

Florida Ship Canal

Grand Coulee Dam

Bonneville Dam

Hoover Dam

Salt River Dam

Average Tidal Ranges in Feet
< Pacific Coast – Atlantic Coast >

Bay of Fundy — 50
Passamaquoddy Bay — 26
Boston/NYC — 12
Washington DC — 3
Miami — 2

San Diego — 2
San Francisco — 4
Seattle — 10

ATLANTIC OCEAN

PACIFIC OCEAN

Gulf of Mexico

MEXICO

CANADA

NEW BRUNSWICK
MAINE
QUEBEC
VT
NH
MA
CT
RI
NEW YORK
NJ
DE
MD
PENNSYLVANIA
WEST VIRGINIA
VIRGINIA
NORTH CAROLINA
SOUTH CAROLINA
GEORGIA
FLORIDA
OHIO
KENTUCKY
TENNESSEE
ALABAMA
MISSISSIPPI
LOUISIANA
ARKANSAS
MISSOURI
INDIANA
ILLINOIS
IOWA
MICHIGAN
WISCONSIN
MINNESOTA
NORTH DAKOTA
SOUTH DAKOTA
NEBRASKA
KANSAS
OKLAHOMA
TEXAS
NEW MEXICO
COLORADO
WYOMING
MONTANA
IDAHO
UTAH
ARIZONA
NEVADA
CALIFORNIA
OREGON
WASHINGTON
BRITISH COLUMBIA
ALBERTA
SASKATCHEWAN
MANITOBA

Lake Superior
Lake Michigan
Lake Huron
Lake Erie
Lake Ontario

Mississippi R.
Missouri R.
Ohio R.
Arkansas R.
Platte River
Rio Grande
Columbia R.
Snake R.
Salt R.
St. Lawrence R.
Tennessee R.

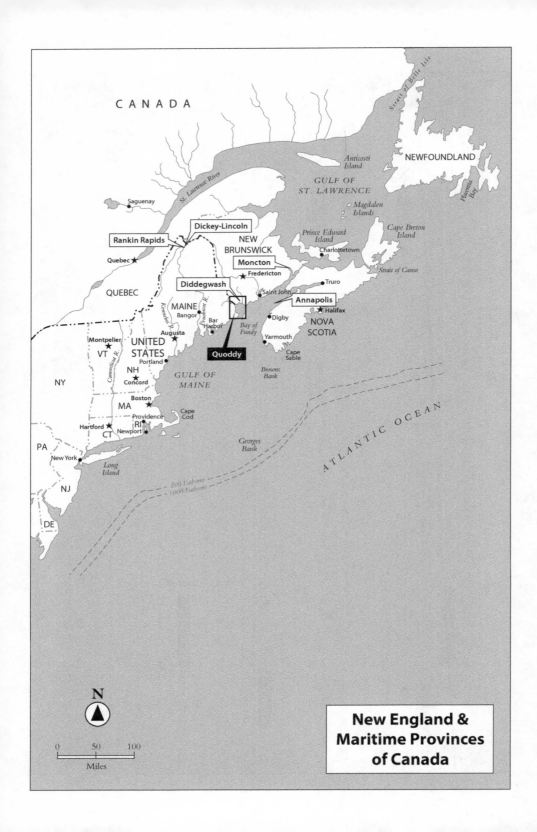

CANADA

Saguenay

St. Lawrence River

Rankin Rapids

Dickey-Lincoln

Quebec ★

QUEBEC

Kennebec R.

Diddegwash

MAINE

Bangor

Bar
Harbor

Augusta

Connecticut R.

Montpelier ★

VT

Portland

NH

Concord

Penobscot R.

Quoddy

GULF OF
MAINE

Boston

MA

Providence

*Bay of
Fundy*

Saint John

Moncton

Fredericton
★

NEW
BRUNSWICK

*Anticosti
Island*

GULF OF
ST. LAWRENCE

*Magdalen
Islands*

*Prince Edward
Island*

Charlottetown

NEWFOUNDLAND

Strait of Belle Isle

*Placentia
Bay*

*Cape Breton
Island*

Strait of Canso

Truro

Annapolis

Digby

★ Halifax

NOVA
SCOTIA

Yarmouth

*Cape
Sable*

*Browns
Bank*

ATLANTIC OCEAN

*Georges
Bank*

200 Fathoms
1000 Fathoms

Hartford ★

CT

RI

Newport

Cape
Cod

NY

PA

New York

*Long
Island*

NJ

DE

UNITED
STATES

N

0   50   100

Miles

**New England &
Maritime Provinces
of Canada**

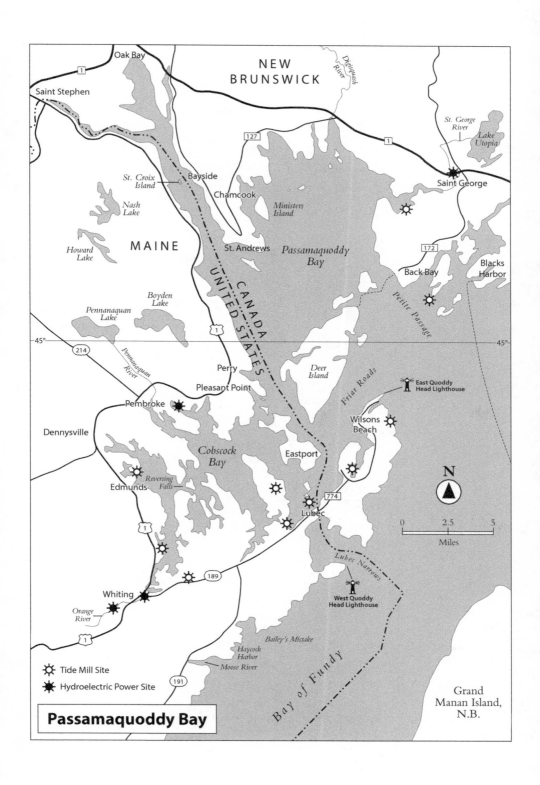

Passamaquoddy Bay

☼ Tide Mill Site

✸ Hydroelectric Power Site

President Theodore Roosevelt and John Muir at Yosemite Valley, California, 1903 ▶

PART I

MANIFEST DESTINIES

CHAPTER 1

The Protagonists

THE CHAMPION

The US Navy sent a destroyer on an errand: pick up a passenger at his vacation house, 400 miles away, and take him sightseeing.

The captain of this formidable taxicab may not have relished shuttling "the brass" while he went sightseeing but probably thought, *He gives the orders.* So Captain William Halsey steamed out from Newport, Rhode Island, headed for the easternmost tip of Maine, and throttled up.[1] *This junket will take three days—and several tons of coal.*

It was a summer morning in 1913.[2]* While the "brass" aboard would be a civilian, not a brass-buttoned naval officer, he was nonetheless powerful and important. He was the assistant secretary of the navy, Franklin Delano Roosevelt. Roosevelt was thirty-one years old, the youngest person in history to hold the lofty spot. He had been in the post for only four months and was flexing his muscles.[3]

Halsey picked up Roosevelt at his summer home, not in Maine, but just across the border on Campobello Island in New Brunswick, Canada. Roosevelt joined the captain on the ship's open-air bridge. Then they steamed seventy miles back to the southwest to inspect the fortifications around Bar Harbor.

It was a grand day. Fully satisfied with what they had seen of the ship and shore, they headed back to Campobello. When the red-and-white-banded lighthouse at West Quoddy Head came into view, Roosevelt asked Halsey whether he could pilot the ship—*Halsey's ship*—through Lubec Narrows.

Here's trouble!, Halsey may have thought. Not even the steam-powered passenger ships on their regular runs from Portland dared Lubec Narrows except at slack-high tide. Instead, they would detour twenty miles counterclockwise around Campobello to reach the protected harbors inside the bay.[4] Halsey knew regulations forbade civilian command of naval vessels but may have thought: *He's my boss.*

"All I had been told," Halsey later recalled, was that Roosevelt "had some experience in small boats, so when he asked me to transit the strait between Campobello and the mainland, and offered to pilot us himself, I gave him the conn (steering control) but stood close by. . . . The fact that a white-flannelled yachtsman can sail a catboat out to a buoy and back is no guarantee that he can handle a high-speed destroyer in narrow waters."[5]

Halsey's ship was the 294-foot *Flusser*, one of the newest and fastest ships in the US Navy. *Flusser* had three coal-fired Parsons direct-drive steam turbines that generated 10,000 horsepower (hp), a riveted steel hull, and hatchet bow.[6] It could cut through pounding seas at a breakneck 32 knots. This *wasn't* Franklin's father's 60-foot sailboat with its puny engine and 8-knot speed.

The deadly destroyer wouldn't handle like a yacht. As Halsey explained, "A destroyer's bow may point directly down the channel, yet she is not necessarily on a safe course. She pivots around a point near her bridge structure, which means that two-thirds of her length is aft of the pivot, and that her stern will swing in twice the arc of her bow."[7] In other words, a slalom course was an invitation to disaster.

Lubec Narrows is the tidal strait separating mainland United States from Campobello Island in Canada and the western entrance to Passamaquoddy Bay. The strait is not straight, but an "S" curve, with rocks, shoals, and swirling currents at every turn. Through this strait millions of tons of water rush with the twice-daily change of the tides. At high tide, the twisting channel can be 30 feet deep—but when the tide drops 20 feet, the clear channel gets very narrow. Tidal currents race through Lubec Narrows at up to eight knots,[8]* making it impossible for sailboats to sail against and tricky for powered boats because the current changes direction as the tide rises and falls, causing whirlpools to pinball from shore to rocky shore.

If Roosevelt miscalculated, and the current caught him broadsides, the power of the moving water would hurl the ship against the shore with a force roughly equal to its 902 tons being lifted into the sky and dropped.

Looking like an overgrown newsboy in his cap, Roosevelt headed the *Flusser* into this maelstrom. He was in command of a ship of war, one that was vastly larger, more powerful, and more difficult to maneuver than anything he had piloted. He was risking not only the ship but also the lives of its 89 officers and crew.

The approach to Lubec Narrows starts at the lighted whistle buoy off Sail Rock, two-tenths of a mile southwest of the West Quoddy Head Lighthouse. With his hands on the ship's wheel, a thick column of coal smoke rising from its four stacks, and the ship running at twice the speed of his father's sailboat, Roosevelt made his first turn. As he turned *Flusser* to port (left), Roosevelt looked over his shoulder and saw the purple line of Grand Manan Island[9] appear to slide into his wake. Then he refocused on the lighthouse ahead. Passing the fairway buoy, Roosevelt again turned to port and into the Narrows. The Boring Stone and Round Rock were now perilously close to *Flusser*'s starboard side. Looking over the starboard rail, he saw Liberty Point, from which he had often watched fishing boats negotiate the strait, and may have thought, *What a sight this must be!*

Roosevelt now turned to starboard to avoid Wormell Ledges, which brought the great ship into the gut—the narrowest part of the Narrows—and roared through. There were two more sets of turns before they reached Campobello's harbor, but each would be wider and easier than the last. Roosevelt's infectious grin spread quickly.

Lubec Narrows was Roosevelt's briar patch, a place he had navigated hundreds of times during a quarter century of boating on his own; with his father; and with his tutors, the local fishermen and Indian guides. As Halsey now realized, Roosevelt, *a civilian*, knew how to navigate the dangerous strait better than any uniformed officer in the US Navy.[10]

Franklin Roosevelt had salt in his veins. His grandfather, Warren Delano II, operated a fleet of clipper ships in the China trade, carrying tea, spices, opium, porcelain, and silk, in partnership with Russell & Company of Boston, the largest American trading firm in China. Other partners in the firm included the Forbes brothers and Russell Sturgis.[11] Ephraim Delano, Franklin's great-great-grandfather, was a New Bedford whaler.[12] The first De la Noye in North America sailed from Holland in 1621.[13]

While Halsey had diplomatically granted his boss his fun (and wisely stayed close to the conn), he knew what to watch for when a man takes command of a ship. "As Mr. Roosevelt made his first turn, I saw him look aft and check the swing of our stern. My worries were over; he knew his business."[14]

So began a friendship built on mutual trust, consummate skill, and iron-willed determination. Together they would win two world wars. In the latter, William "Bull" Halsey Jr. was commander of the US Navy's Third Pacific Fleet—and his friend, the "yachtsman," was his commander in chief, the president of the United States, and the undisputed leader of the free world.

It was at Passamaquoddy Bay that Franklin Roosevelt learned many of the life skills that he would draw on in times of crisis. It was at Passamaquoddy Bay that personal crisis stuck. It was to Passamaquoddy Bay that he repeatedly returned in triumph.

And it was in Passamaquoddy Bay that President Franklin Roosevelt would fight to put the power of the tides to work—for the people.

THE ENGINEERS

"It's impossible," they said, to dam the Mississippi River.[15]

It certainly seemed impossible on March 23, 1912. That winter had been unusually cold. Ice two feet thick had formed upstream of the twelve-mile-long Des Moines Rapids. A huge new dam was being built at the base of the rapids at Keokuk, Iowa. On the third day of spring, the ice broke loose and headed for the dam. Everyone in Keokuk held their breath as immeasurable tons of ice crashed into the unfinished dam.

It held.[16]

The rest of the Mississippi's spring flood roared down the rapids two weeks later. The flood waters rose higher and higher, threatening to overtop the dam. Hugh Lincoln Cooper, the world's preeminent hydroelectric engineer, had designed what he hoped would be the world's biggest dam. His younger brother, Dexter, was head of construction. On April 5, Dexter was out on the dam managing the emergency. "Men were posted every three or four feet, placing sandbags and they were on the job day and night," Hugh's secretary recalled.

> Still, the river rose, and it began to look like nature was too powerful to be overcome by the hand of man. On that critical morning I went down to the works and met Mr. Cooper, who was walking around the cofferdam encouraging the men. When I asked him what the outcome appeared to be, he replied, "It is in the lap of the gods." . . .
> As we talked, waves were lapping at the top of the sandbags. I went back up to the office with a heavy heart, sensing disaster. A few hours later word came up that the river seemed to have crested and was stationary . . . and everyone shouted for joy.[17]

US Army Lieutenant Zebulon Pike had discovered the rapids in 1805 while exploring the upper Mississippi for the US government, which had acquired the territory from France in the Louisiana Purchase of 1803.[18]* The French sold it because they couldn't control the vast and distant wilderness they had claimed based on Samuel de Champlain's explorations two centuries earlier.

In 1805, Keokuk was a seasonal encampment of Sauk Indians who caught fish in weirs built in the river. While the rapids were a hindrance to travel, the Indians could paddle their shallow-draft dugout canoes down through them and pole or drag their boats upstream.

Major Zachary Taylor and the US Army arrived seven years later and settled in.[19]* When the army prohibited troops having Indian wives, Samuel C. Muir, a surgeon, resigned his commission in favor of his Indian wife. Dozens of similar liaisons provoked the federal government to designate 119,000 acres around Keokuk as a "Half-Breed Tract" in 1824. Initially the government prohibited private land sales to protect the tract's residents from unscrupulous settlers and speculators, but the US Congress ended the provision in 1834, effectively allowing claim jumpers to steal the land.[20]

The first steamships arrived in Keokuk in 1820. Despite their power, they were too cumbersome to go up the rapids,[21] so cargo was off-loaded to small flat-bottomed boats called lighters and then poled or dragged upriver, Sauk style. Thus, the second economic resource Keokuk captured was commerce itself: the moving and trading of goods.

By 1837, the US government wanted to get more cargo through the rapids more easily and sent Lieutenant Robert E. Lee of the US Army to survey the site.[22] Lee suggested using the rapids to generate power.[23] After several false starts, a privately financed, 7.6-mile canal around the rapids was completed in 1870.[24] The following year, a railroad bridge was built across the Mississippi. With both north-south river traffic and east-west rail traffic, Keokuk became a nexus of continental commerce, a western boom town, and an exemplar of America's belief in continuous progress and its Manifest Destiny.[25]

President Theodore Roosevelt, the embodiment of American exuberance, visited Keokuk in 1903 for the centennial anniversary of the Louisiana Purchase. The same year, he vetoed a bill that would have allowed a private company to develop a soon-to-be famous hydroelectric site on the Tennessee River in Alabama.[26] Nonetheless, in 1905 he signed the legislation that allowed a private company to harness the immense energy of the nation's largest river by constructing a hydroelectric power dam across it.[27]

If it could be done, the dam would ensure Keokuk's economic future.

If it could be done, it would be huge.

Hugh Cooper had already done the huge and the impossible: he had harnessed Horseshoe Falls, the thundering torrent on the St. Lawrence River at Niagara. From that stampeding water he had corralled 200,000 hp and 91,800 kilowatts (kW) of electricity.[28]*

Hugh Cooper had no formal education beyond high school. He was "only slightly over five feet in height, of round build, and with a head Napoleonic in size." In 1903, he proposed to dig the world's largest hydraulic tunnel *under* Horseshoe Falls as the tailrace for the powerhouse to be built. Britain's Lord Kelvin, one of the most famous scientists alive, laughed at the little man whose only training was from the "university of pick and shovel."[29]*

Hugh dug the tunnel.

Under Hugh's tutelage was his much bigger but fifteen-years-younger brother, Dexter, then age 23. Their father was a millwright. Hydropower ran in their veins. Despite his success, Hugh was stung by Kelvin's criticism. Hugh sent Dexter for six years of formal training in Switzerland and Germany so he would have the world's best engineering education, as well as a world of practical experience.[30]

When Hugh Cooper heard people were trying to dam the Mississippi, he went to Keokuk to investigate. There he found a company in disarray. Intrigued, inspired, and confident in his abilities, he bought control of the company and designed what *he* thought should be built and *how*. Then he looked for the $30 million (about $900 million in 2020 dollars) he needed for construction. After five years of trying, having spent all of his own money, and with only 30 days remaining before the charter signed by President Roosevelt would expire, he found the money.[31] Hugh sent a telegram back to Keokuk on January 8, 1910: *Begin excavations immediately.* With hand shovels, workers started digging through snow and into the mighty Mississippi's left flank.[32]

Stone & Webster Corporation of Boston financed the project.[33] Wisely, Hugh made Edwin Webster the president of the Mississippi River Power Company and himself vice president and chief engineer.[34] S&W financed the project because Hugh had contracts for the purchase of the power from several communities around St. Louis, Missouri, 142 miles south of Keokuk[35]—even though that was 90 miles, or *three times*, farther than anyone had yet transmitted electricity.[36] In 1912, S&W controlled 27.8% of the hydroelectric power in the United States. General Electric controlled another 29.7%.

Hugh Cooper ran an extremely efficient operation, with up to 1,150 laborers but only 50 administrative and engineering staff. It was not, however, totally spartan. He understood the value of public relations and built viewing stands for curious tourists; hired photographers; staged events; and had a fleet of six chauffeur-driven Cadillac touring cars kept ready to carry him, his staff, or visiting dignitaries wherever needed.[37]

Philip Fleming was just a freshman at the University of Wisconsin when he got a summer job at the Keokuk dam, first as an axe man and then as a surveyor.[38] Young Philip impressed his superiors. When his childhood friend died that summer, the friend's father, US congressman Tom Hedges, was moved by Philip's letter of condolence. Hedges offered to give Philip an appointment to either the US Naval Academy at Annapolis or the US Military Academy at West Point.

Philip asked Hugh Cooper what he should do. "He immediately advised West Point—telling me that if I maintained high enough rank to keep myself in the upper portion of my class, I would get into the

Army Engineers. He thought that, as an engineering unit, they were unequaled."[39] It was the chance of a lifetime, a fateful decision. Philip Fleming joined the army.

Gertrude Sturgis thought nothing in 1911 would be different from all the other days of her twenty summers. She was at Campobello, as her family always was in July. She took the dogs for a walk in the woods. They were beautiful woods in those days, tall dark spruces, firs, and fragrant balsams arched over the road, forming a cool green tunnel with bright shafts of sunshine filtering thru. Then, suddenly out into the bright sunlight, the long crescent of Herring Beach came into view, with the blue-green Bay of Fundy beyond. As she walked, Gertrude ruminated on the bay and her future:

> It is still to me the most beautiful and fascinating body of water in the world, it's forty foot tides forming eddies, rips and whirlpools, constantly changing the color and surface of the water. But I was in no mood just then to see beauty in anything. I was bored with my life. Twenty years of doing all the things expected of us in those days—Dancing classes, "coming out," Junior League, sewing circle, Vincent Club, etc. All that is, except the one thing we were brought up to believe was the answer to all problems: Marriage.[40]

Gertrude was the daughter of Dr. and Mrs. Russell Sturgis, Boston Brahmin bluebloods. She was young, attractive, and well-to-do, though not truly wealthy. When she traveled along the Atlantic Coast, local newspapers noted her presence, along with her peers in the social stratosphere, the Astors, Stuyvesants, and Vanderbilts.[41]* She would be a catch.

> I had had numerous little attacks of what I though might turn out to be the "Real thing," even to the point of being engaged for over a year when I was seventeen, but they all ended in such boredom that as I walked along, I muttered to myself, "I'm through with men." Then I began to think, "What am I to do?" I was still turning this over in my mind when we got back to the cottage and found the mail had come, bringing a letter. For me that letter changed the whole course of my life.

As a child, Gertrude had promised her brother Russell, nine years older, that when she grew up she would keep house for him, wherever he went. Russell's letter asked if she was still interested. At the time, Russell was finishing a three-year training course with the Chicago, Burlington & Quincy railroad.[42] He got the job through an "uncle" and had progressed to being the roundhouse foreman in Keokuk.[43]

> This was my opportunity to "get away from it all," but I did not realize that the town of Keokuk would be swarming with young engineers, due to the construction of a huge dam across the Mississippi River. . . .
>
> How I had the temerity to go, knowing nothing about cooking and housekeeping, I find it hard to answer but go I did, with my *Fanny Farmer* cook book under my arm, and high hopes for adventure in my heart, to say nothing of courage born of ignorance.

Gertrude was soon swept into a whirl of gaiety when the news got around that there was a new girl in town: "I didn't have to sit at home and twiddle my thumbs many evenings. . . . It was 'out of the frying pan into the fire' as far as men were concerned. The whole town was flooded with them."[44]

Dexter Cooper met Gertrude Sturgis shortly after she arrived in Keokuk. Soon thereafter, "I changed my mind," confided Gertrude,

> about being through with men. Our friendship developed very fast, in spite of the much too anxious and obvious scheming of Hugh's wife, my "to be" sister in law who felt it was high time for her "little brother," as she called Dexter, to be thinking of settling down. If we had not been really and truly attracted to one another, our love affair never would have survived. As it was, we laughed at all the planning and went on our way. I always knew when he was coming to call, for in the morning a box of fondant covered grapes, put out by a local candy maker, would arrive. They were extremely good, and between us we would finish off the whole box in one evening. . . .
>
> Then came the day when I realized that I must be ready to make a decision. I did—I went home in February, and Dexter and I were married in Boston in April.[45] We left for a two week honeymoon the first few days a business trip for him, in Montreal, and then to my Mother's summer cottage at Campobello. It was Dexter's first sight of the Bay of Fundy, and he was enthusiastic enough to satisfy even me.

Bridging the gap was dangerous work. The mile-wide Mississippi River didn't like being squeezed and backed up behind the growing dam. The narrowing gap between the east and west sides of the dam compressed the twelve-mile-long rapids into a few feet, and down this precipice the river rushed with ever-faster speed and force.

The last gap was perhaps 75 feet wide.

Dexter, construction manager for the Illinois division, had his workmen cantilever a huge timber over the rushing water. From the other side, Horace T. Herrick, construction manager for the Iowa division, did the same. Then workmen from each side inched out onto the beams. They were not tethered with safety ropes. If a man fell, it was better to ride the roaring river and hope his natural buoyancy would pop him to the surface once he had cleared the gap and the water slowed.

Ever so gingerly, the workmen worked their way out, hammering handrails onto the cantilevered beams. Stopping just short of each other, they retreated.

Then Dexter, the newlywed, and Horace climbed onto their respective beams and met in the middle of the gap, the mighty Mississippi growling below them. They had started, two years before, nearly a mile apart. Now, at 3:47 p.m. Central Time on July 21, 1912, they shook hands, joining the continent East and West.

Horace then retreated so Dexter could walk over the water, from Illinois to Iowa, to congratulate his brother Hugh.[46]

The Mississippi was finally dammed on May 31, 1913. *The Daily Gate City* trumpeted the historic American achievement:

> **THE GREAT KEOKUK DAM IS COMPLETED**
> In the midst of tricolor bunting, waving flags, shrieking whistles, cheering men, and smiling women, the great dam across the Mississippi River was completed here yesterday with the depositing in the forms of the last concrete in spillway No. 101. . . .
> Every employee . . . was present at the simple but very impressive ceremonies, and Chief Engineer Hugh L. Cooper had as guests some of the men who were for years active in the work that finally resulted in harnessing the Mississippi. . . .
> Dexter P. Cooper, who has had direct charge of the building of the great dam, stood beside his wife as she placed in the last batch of concrete a brass box containing historical data, which is expected to remain untouched for several thousand years.[47]*

Horace T. Herrick didn't get to enjoy the celebrations. Just five days earlier, he had been electrocuted when he picked up a 1,100-volt live wire that had fallen into his workboat.[48]

Keokuk Dam was an engineering masterpiece. Seven-eighths of a mile long, it was the world's largest concrete dam and largest concrete monolith. To make the monolith—a single, solid, unseamed piece of concrete—the Cooper brothers poured fresh concrete continuously, always maintaining a "wet edge" so the next pour would fuse with the previous pour and form the single, super-strong, leak-free barrier. To keep pouring during the winter, the concrete was heated and covered.[49]

The electrical engineering was also brilliant. The dam generated 142 megawatts of electrical power and successfully transmitted it to St. Louis at a scarily high 110,000 volts, using new AC-DC transformers and transmitters designed by Hugh Cooper, which became the model for most modern high-power transmission systems.[50]*

With the "Dream of Half a Century Realized,"[51] Keokuk's boosters prophesized their city would soon rival St. Louis and Chicago.[52] Eager to see the great construction project he had authorized, Theodore Roosevelt made his third trip to Keokuk in September 1912.[53]* As many as 30,000 people have followed in his footsteps each year.

Writing six years later, Montgomery Meigs, the US Army engineer in charge of the locks, noted, "That all this intricate machinery of novel type should have operated so well is a high tribute to the chief engineer, Mr. H.L. Cooper. . . . [He] had such opportunities as fall to few men in unrestricted authority and unlimited means, the only thing asked of him was results. . . . The whole plant has operated with remarkable success."[54]

More than a century after its construction, the Keokuk Dam is still generating electricity and remains the largest privately owned dam on the Mississippi River.[55] But not for long would Keokuk remain the world's largest hydroelectric power plant. By the time it opened, Hugh was eyeing the Nile, and Dexter was thinking about the sea itself.

The engineers—Hugh Cooper, Dexter Cooper, and Philip Fleming— were like water in the atmosphere. When one project was done, they would evaporate and then precipitate somewhere else in the world where people wanted power.

THE REPORTER

Roscoe Conklin Emery's heart must have been pounding and his chest swelling on October 15, 1914. A ten-minute walk, two blocks downhill on Boynton Street and half a block south, brought him to 42 Water Street, home of the *Eastport Sentinel*, Maine's second oldest newspaper. Inserting the key, he opened the door to the three-story brick edifice built in 1886, the year he was born. Now age 28, Emery was the *Sentinel*'s new owner.

Roscoe was the youngest of eleven small fry in the Emery family.[56] His father drew his living from a herring weir. Like almost everyone else in town, his siblings either caught or processed fish or serviced the fleet and factories. While fish stock ran in Roscoe's veins, ink flowed from his fingers. Instead of schooling with the fish, he graduated Phi Beta Kappa from Colby College in Waterville, Maine.[57] His first job was as Waterville's school principal. In 1907, he moved closer to home as the principal of Lubec's all-grade, 39-student Hilltop School. He and his equally youthful faculty of three posed for a group portrait sitting on the school's front steps. They appear happy and relaxed, all looking directly into the camera.

When Hilltop School burned in 1909, Roscoe joined the weekly *Sentinel*. On July 19, 1910, he covered the visit of US president William Howard Taft.[58] In 1912, Roscoe interviewed a "somewhat erratic genius" who exhibited a model showing how to generate electric power from the tides.[59] The biggest stories that year were the attempted assassination of former president Theodore Roosevelt and the sinking of the *Titanic*, whose cut-short call for help, ***---*** (SOS), had been heard at the new wireless telegraph station at Eastport, the closest American port to the scene of the disaster.[60] Electricity was new and newsworthy. There were articles about electric ray fish and an experiment by Nobel laureate Svante Arrhenius, who surreptitiously subjected classrooms of children to high-voltage electric fields. "After six months . . . it was found those in the electrified room had improved 50 percent more in bodily and menta [*sic*] vigor."[61]*

Each week, *Sentinel* readers were entertained by nationally syndicated fiction writers of adventures, including Jack London; humor, such as O. Henry; and romance, such as Ryland Bell's "The Moon and the Ocean: Together They Exert a Peculiar Influence." Readers witnessed the dawning of the modern age in the mix and juxtaposition of the articles and advertisements. In 1912, there was an advertisement for Standard Oil's "Rayo" brand kerosene lamps, which were "easier on the eyes than any other artificial light." The paper's first automobile ad appeared in 1912, showing a $350 Liberty-Bush, open-air, two-seater *driven by a woman*. It was countered the next week by an ad announcing the expansion of Gardner's livery stables.[62]

The *Sentinel* chronicled the daily life of the area: fish prices, fires, church suppers, high school sports, celebrities, and politics. The causes of some persistent problems were becoming clearer, such as the link between contaminated drinking water and deadly diseases like typhoid, although the real solutions were still beyond reach.[63]* Easy cures of all kinds were touted in patent medicine ads. The drugstore in town sold a guaranteed cure for constipation: "A most scientific and common-sense treatment is Rexall Orderlies, which are eaten like candy. They are very pronounced, gentle and pleasant in action, and particularly agreeable in every way. They do not cause diarrhea, flatulence, gripping, or other inconveniences."[64]

When Roscoe took over the *Sentinel* in 1914, electricity was still exciting lots of attention, as were airplanes and the Panama Canal. Locally, entrepreneurs were experimenting with freezing fish and stocking salmon. The state of Maine was trying to organize a coherent highway system, now that one half of 1% of residents owned a car,[65] and to create a Public Utility Commission to regulate the proliferating water, power, and transportation companies. Nationally, Woodrow Wilson, a Democrat, was president, though few in Maine had voted for him. Internationally, the Austrian-Serbian war was rapidly engulfing Europe, and German submarines were sinking merchant ships in the Atlantic.[66]

The July 8, 1914, edition of the *Sentinel* ran several feet of effusive copy on Eastport's "biggest Independence Day celebrations east of Boston." A two-column-inch article, however, had much larger significance to the 500 members of the local Passamaquoddy Nation:

> At midnight of June 3, the Cherokee Indian nation of Oklahoma ceased to be a tribal government, thereby marking the passing of the largest tribal organization of Indians in the United States. The Cherokee will exist no more except as a citizen of the Republic. All tribal property had been converted into cash and each Cherokee had received his allotment . . . about

$15 per capita to the 41,000 members of the tribe, thereby terminating the nation and closing the books.

If 41,000 Cherokees could be cashed out of existence so cheaply, the Passamaquoddies may have wondered what small change was in hand for them.

Roscoe was writing a book of sorts, and every week he would publish a new chapter. The stories he published in the *Sentinel*, with his sharp eyes and pen, are the history of one tiny town that amounted to less than 1% of Maine's population, which was less than 1% of the US population. But the people of Passamaquoddy Bay had played, and would again play, roles in the nation's history far greater than their small numbers.

First, however, would be the changing of the guard at the *Sentinel*. On page 2, in the upper left, where the paper's editorial views were expressed, appeared the following two columns on October 14, 1914:

> The *Sentinel* has not always been brilliant, but it has been good. Its columns have been clean. In its middle age it was a "Black Republican" . . . advocating the abolition of slavery in the United States. A little later it was a "ram-rod" temperance paper and was sometimes classed locally as an organ of the "God and morality party." . . .
>
> The *Sentinel* has now been sold . . . to Roscoe C. Emery, a nephew of the owner of sixty years ago. Mr. Emery is a young man finely equipped for the work he has undertaken. . . . In his writings for publication he has developed much ability in this line, and a successful career as publisher is predicted by his friends.[67]

Roscoe Emery, the classics scholar and new proprietor of the *Sentinel*, began his first editorial—chapter 1 in his *Book of Weeks*—in Latin:

> *INGREDIOR*
> ("I enter.")
> It may not be inappropriate and is perhaps to be expected that the new management of the *Sentinel* on taking charge of its affairs should make some brief mention of its aims and policies.
> These in short will be to serve the interests of Eastport as a whole and its surrounding territory with all the completeness and efficiency we can command.
> The temptation is always present in assuming such an office to promise too much, in the way both of bettering the paper itself and of making its

influence more effective for good. Without yielding unduly to this tempta-
tion, we can only say that while we hope to accomplish both of these purposes, whatever progress is made in that direction must be made gradually,
and in order to be solid and substantial, must have the support of a majority
of the people of this community.

Those who look for startling changes of policy on the part of the new
management will therefore be disappointed. There will be no rabid politics,
no personalities, no crusade,—at least for the present. Whatever changes
are made in the paper itself will be made quietly, as experience suggests or
necessity dictates.[68]

Emery's anonymity was an affectation. If anyone in Eastport didn't
already know him, they would soon, for he would become a personality.
Roscoe was also planning to run for elected office, at a time when the
politics would become rabid. And while Roscoe didn't know it, he had
already reported on the subject that would become his crusade.

When Roscoe took the *Sentinel*'s reins, he posed for a formal portrait. This time he is alone, sitting stiffly upright, hair trimmed, with a
serious expression on his face as he stares off to the side. Roscoe's lush
locks of dark hair would be cut short and turn grey before his crusade
was over.

As reporter and politician, Roscoe Emery wrote thousands of articles
and letters to constituents and bureaucrats. His writings and his assidu-
ous collecting of articles about the Passamaquoddy Bay Tidal Power
Project give us unique insight into the facts, the fictions, the intrigue,
and the human tragedy of Quoddy.

CHAPTER 2

One Bay–Three Islands–Three Nations

PASSAMAQUODDY BAY

Earth holds the Moon at a distance. The Moon pulls longingly at the Earth, but only its mantle of water is moved. This is the tide.

As the Moon gyrates around the Earth, coming teasingly close, then shying away, her gravitational attractiveness increases and diminishes with her distance—and the Earth's waters rise and fall in response. Tilting her body, she exposes a new rhythm, further exciting her audience. This is her captivating power.

The Moon has been doing this dance for so long that her maneuvers are known to those who watch her carefully. She is predictable. Thus, tides also are predictable—they will continue to rise and fall as long as the Moon rises and the Sun sets.

Unlike manpower, horsepower, or fossil fuels, Moonpower is inexhaustible.

Astronomy creates tides.[1]* The huge tides in the Bay of Fundy are the product of astronomy multiplied by geology.

Glaciers weighed heavy on the land and depressed it. Ten thousand years ago, as the two-mile-thick ice moved south, it clear-cut the forests and strip-mined the soil, leaving rocky hills. As the planet warmed, the ice receded like an old man's hairline. The land, relieved of its load, rebounded like a mattress after slumber. The melting glaciers exposed the land but raised the sea, flooding the lingering depressions. Hills became islands; valleys, bays.

The 600-mile-long Gulf of Maine runs northeast-southwest along
the coast of New England and Atlantic Canada. Its blunt southern end
is formed by Massachusetts Bay and Cape Cod Bay. Crescent-shaped
Cape Cod is a recessional moraine of unconsolidated sand, clay, and
boulders pushed south by the glaciers. The Gulf's northern reach, the
145-mile-long Bay of Fundy, was formed by tectonic faults. Nova Scotia is an upthrust scarp south of the fault; Fundy is a rift valley, with
mainland New Brunswick to the north.

At Fundy's funnel-shaped mouth, the St. Croix River pours into
it, creating Passamaquoddy Bay and a distinct demarcation between
the lands to the east and west. Passamaquoddy Bay and the St. Croix
River lie in a northwest-southeast fissure. Here the transverse ridges
and faulting that give Maine its deeply indented coastline of granite
headlands give way to longitudinal faults parallel to the coast that give
New Brunswick its sandstone cliffs and beaches.

After the glaciers retreated, dust and stream sediments slowly accumulated, eventually becoming deep enough to be called "soil." Wind-
blown and bird-dropped seeds took root. The plants attracted animals
that fed on this new resource growing at the margins of the barren
landscape. Small prey attracted larger predators, and so the food chain
forged its links. After perhaps six thousand years, the forest reached
maturity. Meanwhile, denizens of the deep insinuated themselves
among the bay's waters, burrowed into its mud, and cemented themselves to its rocky shores.

Into this primordial ecosystem came its first human inhabitants,
the Algonquin-speaking Wabanaki—the "people of the dawn."[2] As
nomadic hunter-gatherers, they occupied territories defined by rivers,
their primary food source and transportation routes. The Penobscots
hunted in what we would now call central Maine, the Passamaquoddy
in northeastern Maine and western New Brunswick, and the Mik'maq
around the Bay of Fundy. In their nearly lost language,[3]* Passamaquoddy may have meant "too many bears,"[4] the "place of pollock,"[5] or
the "place of the undertow people,"[6] meaning *the people of the tides*.

Tides jump suddenly at Passamaquoddy Bay. As the world turns and
the ocean's tidal "bulge" or "wave" rolls in from the deep Atlantic, it
rides up the continental slope and over the shallows of Georges Bank.
The shallow seabed creates friction that slows down the wave's bot-

tom while the top continues westward at full speed. This process compresses the ocean tide wave, making it horizontally shorter but vertically steeper, thus raising the tide. Out in the Atlantic, the ocean tide wave is less than two feet high; riding up onto Georges Bank, it grows to five.

As the ocean tide wave is deflected north into the Bay of Fundy, a second dynamic comes into play: the "funnel effect." Simply, from entrance to end, Fundy gets narrower and shallower (at a remarkably consistent four feet per mile); consequently, the incoming tide wave has less space in which to fit—so it grows vertically, getting "squeezed upward."[7] When the incoming tide's momentum is exhausted, at high tide, the water piled up in the bay runs back downhill, draining away as the ebbing current.

The Gulf of Maine is a *tidal basin*, a semi-enclosed body of water that responds to tidal influences. The Gulf amplifies the tide's height through a dynamic called *harmonic resonance*, or, more commonly, the "bathtub effect." The Gulf's *period of oscillation* is 11 hours and 24 minutes, closely coinciding with the normal 12-hour-and-25-minute tidal cycle. Put another way, the ocean tide's *wavelength* is close to the length of the Gulf. This causes the ocean tide wave to be *reflected* off the Gulf's terminal ends and bounce back in harmony—like doubling the frequency and pitch of a sound wave—and doubles the height of the tide. The phenomenon gets its popular name because it behaves like water in a bathtub: a few little rhythmic waves with your hand will set up waves big enough to slosh right out of the tub.

These three dynamics—continental shallowing, the funnel effect, and the bathtub effect—combine to push the tides higher and higher. The tide that was 2 feet high in the ocean and grew to 5 on the Banks becomes 12 feet high as it enters the Gulf. When it pushes past Grand Manan Island and into the Bay of Fundy, the tide is 18 feet high. Reaching St. John, halfway up Fundy, it's 25. When forked into the fishtails of Chignecto Bay and Minas Basin, it is 40 feet high. By the time it squeezes into Truro at Chignecto's easterly end, the tide's energy has been summoned into one great leap, 40, sometimes 50 feet high—the highest tides in the world.[8]*

Humans have an uncanny ability to exploit every conceivable resource and dynamic. Since time immemorial we have been exploiting the tides. Whether it be digging for clams at low tide or floating our loads upstream on the flood, wherever there are tides, we use them.

If we were going to try to put a bridle on the Moon, to harness her tidal power, the Bay of Fundy is the place to try. While the tides in Passamaquoddy Bay are only half as large as in Chignecto Bay, Passamaquoddy Bay's tight-mouthed and granite-girdled coves offer the best sites for corralling the tides and harnessing their power.

Samuel de Champlain surveyed the New World with his employer, the French nobleman Pierre Dugua, Sieur de Mons. They were pleased with what they discovered in Passamaquoddy Bay in 1604. They saw rich stands of timber, good harbors, and plenty of fish. The natives were friendly and happy to trade furs for European manufactures.

King Henri IV of France had granted de Mons a trading monopoly and almost unlimited executive powers as lieutenant general of Nouvelle France (New France), a vast territory stretching from present-day Cape Breton 400 miles south to Pennsylvania—and from the Atlantic 3,000 miles west to the Pacific.[9] In exchange, de Mons financed the entrepreneurial expedition himself and was directed to "establish the name, power, and authority of the King of France; to summon the natives to a knowledge of the Christian religion; to people, cultivate, and settle the said lands; to make explorations and especially to seek out mines of precious metals."[10]

Having spent the summer exploring the Baie Française (fumbled into Bay of Fundy), the Frenchmen chose Isle de Saint Croix as, they hoped, a safe place to spend the winter. It proved a poor choice. The location of their settlement was almost exactly 45° north latitude, the same as Bordeaux in southern France, so they assumed the weather would be similar. It wasn't.

It was a severe winter. Ice formed on the river, and the tides broke it into pieces and swirled them around, making the river impassible by foot or boat. The explorers were stranded, cut off from fresh water, food, and firewood on the mainland, and they quickly exhausted the little island's few resources. Starvation and scurvy killed 35 of the 79 settlers.[11] Come spring, the survivors abandoned the island in favor of the milder climate and the rich tidal marshes of the Annapolis Valley in what became known as Nova Scotia (New Scotland), where they set up the continent's first tide mill.[12]*

In 1620, the Pilgrims landed in Massachusetts and started the first permanent British settlement in North America. Then, in 1621, King

James I of England granted William Alexander a royal charter to Nova Scotia based on England's claim of discovery made by John Cabot in the 1490s. Nova Scotia then included New Brunswick, which bordered the English colony of Virginia at 45° north latitude (and overlapped the territory claimed by France).[13] But since lines of longitude are discernible only to suitably skilled navigators, the real and obvious St. Croix River was chosen as the boundary between the two English possessions.[14]

It was 130 years between Christopher Columbus's arrival in the New World and the Pilgrims' landing, and in that time Maine's Indian population plummeted from perhaps 60,000 to 15,000,[15] mostly due to European diseases. By 1761, Maine's total population had grown to 17,500, and within three years European immigrants had pushed it up to 24,000. As population pressure increased, the French and English fought each other for control of the New World's resources. The native people had little choice but to choose sides in what became known locally as the French and Indian War, part of the global Seven Years' War (1756–1763). The English won, thereby taking possession of eastern North America as far west as the Mississippi River and all of Canada, while France kept the territory between the Mississippi and the Rockies.[16]

Far from profiting from their winning alliance with the English, the war had cost the Passamaquoddy Indians dearly. In 1760, Chief Michael Neptune went to Halifax, Nova Scotia, to plead his case for relief. He told Governor Charles Lawrence that war, famine, and diseases had reduced his fighting men from 1,400 to 350. The governor sent him away with no more thanks for his losses than a single blanket and a laced hat.[17]

Islands add pepper to the salt of Passamaquoddy Bay. The two largest islands, Campobello and Deer, lie athwart the bay, forming a vast and well-protected anchorage to the north. Deep channels with fast tidal currents run between the islands and the mainland. Three islands are particularly important in the story of *Moondoggle*: Campobello Island in New Brunswick; Moose Island in Maine; and a virtual island, the Passamaquoddy tribal village of Sipayik. While they are close to each other, they are also separate from each other—and a long way from the Atlantic seaboard's population centers.

The Quoddy Tidal Power Project would end their isolation. It would wed them together . . . *for better or worse, for richer or poorer, in sickness and in health.*

CAMPOBELLO ISLAND, NEW BRUNSWICK, CANADA

His service to the British Empire in India cost him his right arm. Lieutenant William Owen, a Welsh naval officer, was a veteran of the Seven Years' War.[18] He beseeched the Lords of the Admiralty, "humbly praying that they will be pleased to direct something to be done for me, either by Gratuity, Pension, or Preferment."[19] Like most, he got nothing. But then his former commander, Lord Campbell, was appointed governor of Nova Scotia,[20] and Owen redirected his lobbying efforts to the man he hoped would be more generous. Campbell replied, "If you chuse [*sic*] to go with me I . . . shall have it in my power, I fancy, to do something for you there."[21]

Owen went west with his patron in 1766 to survey the governor's new domain. When Owen's duties were done, Campbell kept his promise. Acting as "Governor-in-Chief, Captain-General, and Vice Admiral of Nova Scotia and its Dependencies," on September 30, 1767, he named Owen "The Principal Proprietary of the Great Outer Island of Passamaquoddy."[22]

What the Passamaquoddy had known as *a-bah-quiet* (probably "island in the sea"), and what Champlain had christened Port aux Coquilles ("harbor of clams"), Owen cribbed as "*Campo bello* being, so I presume, the Spanish and Italian equivalent of the French *beauchamp* or the English *fair field*" and "partly complimentary and punning on the name of the governor."[23]

Owen hoped to live in the fashion of an English gentleman. He formed the Campobello Company, and, with £16,000 of investors' money, he shipped to Campobello a boatload of supplies and 38 indentured servants with which to establish his estate and money-making enterprises.[24]*

Campbell's grant was fraudulent. By understating the island's size as 4,000 acres instead of its actual 12,000 acres[25] (about 80% the size of Manhattan, New York),[26] Campbell gave Owen three times as much land as the law allowed for a person of such rank, and Campbell included Owen's three cousins as joint grantees, even though they had performed no service for the Crown.[27] There were also three European families already living on the island, who had no intention of abandoning the home they had carved from the wilderness to a presumptuous "Willy-come-lately." Particularly stubborn were the Wilsons. Eventually, a truce was reached: Owen wouldn't charge them rent; according to him, they "cheerily acquiesced" to his jurisdiction.

When Owen's son was born in 1771, he referred to the boy, only half-jokingly, as "the Hereditary Prince of Campobello." The mother of "Prince" Edward William Campbell Rich Owen was the proprietor's housekeeper, Sarah Haslem.[28]

His wilderness fiefdom, however, was not to his liking. In 1771, when England was again about to go to war with France, Owen rejoined the British Navy. He eventually became an admiral and died in India in 1778, leaving Campobello adrift.[29]

David Owen, one of the original grantees, arrived in 1787 as Campobello's Second Principal Proprietary. His mission to the island was spurred by squatters, debts, and dubious land sales too complex to catalog, and his own political troubles in England. The 37-year-old academic dispossessed the people who had been eking out an existence on the island. He was bitter, and they were bitter. Said one islander, "his meanness was as low as his pride was great."[30]

Life on the island was hard. David Owen tried to sell the island, without success. He died in 1830, and according to a ghoulish local legend, his body was placed in a lead coffin, filled with spirits, and shipped to England. Once at sea, the coffin was opened, the spirits liberated, the body dumped, and the alcohol drunk.[31]

William admitted he was "an orphan, a bastard, a foundling."[32] He was, in fact, William Fitz-William (WFW) Owen, the second illegitimate son of William Owen and Sarah Haslem.[33] The precocious boy was rescued from the streets of Manchester, England, and placed in a navy orphanage, and a few years later, went to sea.

In 1835, he arrived in Quoddy Bay as Captain William Fitz-William Owen, RN, the Third Principal Proprietary of Campobello.[34] There were then about 600 tenants who paid him rent, a few freeholders and a fish factory at Wilson's Beach, and a tide-powered sawmill at Harbour de L' Outré (Harbor of the Otter).[35] Captain WFW Owen settled in, calling himself "The Quoddy Hermit":

> [I]n 1831, or [after] forty-three years, during which I have served under every naval man of renown and was honored by the friendship of Nelson. From the year 1797 I have held commands and been entrusted with some important service, for the most part in remote parts of the world. My character, if I may be allowed to draw it myself, contained much of good and bad. The latter, perhaps, I contrived to veil sufficiently not to mar my reputation; but, by the grace of God, he has not left me without his spirit of self-conviction. . . . I thought myself a tolerably religious man, but knew myself to be as Reuben, unstable as water. At fifty-seven my worldly ambition was barred by corruption in high places. At sixty-one I became the "Hermit."[36]

As suited an aging gentleman of the Empire, he planted English roses in his garden, set up two cannons he had captured from Spanish pirates,[37]* and built a quarterdeck for his dry-docked self.[38]* On his imaginary ship, in full dress uniform, he relived military campaigns and debated science, nature, and religion with his imaginary friends, "Academius," "Rusticus," and "Theophilus."[39]

Hardly a hermit, Captain Owen meted out justice, had St. Anne's Anglican Church built, gave lay sermons, married young lovers, held every civil post, guarded his land and fishing rights, steadily improved his rustic retreat, and welcomed every visiting dignitary. When a sailor tried to kiss his bride-to-be before it was time, the captain roared, "Back!"—and took the first kiss for himself.[40]

Captain Owen had vision. He believed Campobello could become as great as his birthplace on Morecomb Bay, where the 35-foot tides drove several tide mills. Accordingly, in 1839 the captain incorporated the Campobello Mill and Manufacturing Company. Two thousand shares were offered at $200 apiece, each bearing 6% interest. The prospectus published in London said they would "build and employ all descriptions of mills, mill dams, fulling and carding machinery" and that there was "enough spruce and pine forest to keep four double saw mills busy for forty years."[41] Captain Owen also reasoned that Campobello's strategic location on the US-Canada border and proximity to the American-European trade routes would make an excellent site for manufacturing, and, if "constituted a free Warehousing Port," it would "have a decided [tax] advantage over any other spot in British America."[42] Unable to attract investors, the captain returned to the more agreeable company of Academius, Rusticus, and Theophilus.

Captain Owen was right about one thing: Campobello was indeed strategically located—for smuggling. Laws, labor rates, and taxes varied from one side of the border to the other, and whenever an arbitrage opportunity arose, there was sure to be someone willing to exploit it.

The captain, however, was still a navy man, and he let it be known that he yearned for a return to the sea. Eleven years after his forced retirement, he got the letter he had been hoping for: a perfect assignment for a still-sharp, 65-year-old seadog. His orders read in part:

> Whereas, the Charts of the Bay of Fundy, with its Islands, Rivers and Roadsteads, as well as those of the several parts of the adjacent Coasts of New Brunswick and Nova Scotia, are in various places exceedingly erroneous; And whereas, the extraordinary fluctuations of the tides in that Bay render an exact delineation of its banks, shoals and depth, most essentially necessary; And whereas, the population along its Coasts, and the consequent intercourse between its Ports, have much increased of late years. Therefore, a fuller acquaintance with its maritime resources is required, in order to lead to a still fuller development of its Commerce and industry: we have determined to have an elaborate Survey made of the whole of the District, and we have thought fit to appoint you to conduct this Survey; not only from the skill and experience you have shown on former Surveys but from the local knowledge you have acquired during your long residence in its vicinity.[43]

The high-tech hybrid sail-and-paddle steamer HMS *Columbia* was sent for Captain Owen's use on the thousand-mile survey, which took five years. In recognition of a job well done, he was promoted to rear admiral.[44] As a final salute to Owen at age 80, the navy promoted him again, to vice admiral, with a £150 annual pension.

Vice Admiral WFW Owen died in St. John on November 3, 1857, at age 84. His body was taken to Campobello for burial. The transportation of the deceased Principal Proprietary did not go as planned: the ship ran aground as it approached Campobello. But unlike his predecessor, the eccentric, imperious, and affable admiral was much loved by his fellow Campobelloans. They waded out into the icy water, lifted his body to their shoulders, and bore him tenderly to his final resting place.[45]

The War of 1812 had been over for 30 years, but it still wasn't clear where the US-Canada border should be. Where the St. Croix River was a river, the line was clear enough: right down the middle. But as the river convoluted into islands and coves, the question wasn't so simple. That's why the Americans sent their most brilliant lawyer, Daniel Webster,[46]* to debate the issue with Canada in 1842.

The Canadian delegation received their illustrious guest most hospitably. They feasted him and toasted him with fine wine. Then they took him on a sailing tour of the disputed area. It was a fine breezy day. The Canadians loved it.

Webster didn't. He was seasick. He didn't want to be out in this blow, bobbing about with his stomach bouncing into his throat and head spinning like the top of the mast. When the Canadians said the next leg would be to circumnavigate Campobello Island by going around the leeward (windy) Atlantic side, Webster said he had seen enough and cut the trip short.

In so doing, he also cut the United States short: Campobello Island is only 750 feet from mainland Maine but 6 miles from mainland Canada. Logically, the border should have been drawn on the island's east side, thus making Campobello part of the United States. But it wasn't. In the Webster-Ashburton Treaty of 1842, the border was drawn on the island's west side, through Lubec Narrows—thus making Campobello part of Canada.[47]*

Stepping over the line has a storied history in Quoddy Bay. For many years the boundary ran through the two-acre cliff-faced island at the north end of Lubec Narrows, now called Pope's Folly. Legend has it that when Maine passed a law against the sale of intoxicating liquor, a bold spirit named Pope built a bar straddling the line, across which he could step in either direction if threatened by taxation or arrest. Pope, imbibing too freely, stepped over the line, and off the cliff, proving the certainty of death and taxes.[48] Benedict Arnold, the most infamous side-switcher in American history, also lived for a few years on Campobello at Snug Cove, just beyond America's reach, and from which the merchant-trader doubled as a smuggler.[49]

Hereditary feudalism ended on Campobello in 1881 when Admiral Owen's granddaughter, a childless widow, sold her property for $200,000 to a *new* Campobello Company, a Boston-based real estate firm, which planned to turn the island into a summer resort.[50]

In those days, optimism and progress were the watchwords as America upshifted from a slow-paced agrarian society to an industrial powerhouse at full throttle. Taxes were low, and fortunes were made fast. For the lucky few with their hands on the controls, life was exceptionally good.[51] The Campobello Company promoted the good life at their island,

> that has every possible advantage of an island that has its own little history of pirates and hidden treasure, an island famous for the natural and extraordinary beauty of its landscape as well as for the health-restoring qualities of its salubrious climate. . . .
>
> [O]nce you visit our island, which is so remarkably charming and of such home inviting aspect, where you can hook our fish, shoot our game, sail our waters, canoe our rivers, and gain new health, breathing its sweet balsamic air, you are pretty sure to come back again as a summer resident.[52]

The sugar plum prose went on for three pages. The new Campobello was intended for the wealthy who wanted a more active and less formal experience than the elaborate summer resorts at Bar Harbor, Maine, and Newport, Rhode Island. In less than a year, the 60-room Tyn-y-Coed Hotel was built and fully booked for the season. By the end of the first year (1882), fifteen private building lots had been sold.[53]

Campobello appealed to 55-year-old James Roosevelt, Esq., vice president of the Delaware & Hudson Railway, and his wife, 29-year-old Sara Delano, of Hyde Park, New York. In 1883, they stayed at the Tyn-y-Coed Hotel, bringing with them their one-year-old son, Franklin Delano Roosevelt. They liked the island so much they bought a house on a ten-acre lot overlooking Friar's Bay[54] and moved in, in July 1885.[55]

In the 1880s, it took a long time to get to Campobello from—anywhere. The nearly 300-mile trip from Boston, Massachusetts, took about 25 hours by train or steamboat. From New York City, the 500-mile trip took more than 40 hours.[56] When you finally got to Campobello, you didn't want to leave.

Most people stayed for the whole season. When the weather was nice, it was idyllic. The temperature rarely reached 90°F, there was usually a sea breeze, and there were typically no crowds. Many guests brought their own household staff, and there were plenty of local hands available to lug your luggage, cut your firewood, build your fire, cook your meals, pack and carry your picnic basket, paddle your canoe, crew your sailboat, bait your hook, and clean your fish. Not to mention clean your dishes, clothes, house, and stable. Such was the rugged life of "city slickers" turned "rusticators." Admiral Owen, Academius, Rusticus, and Theophilus would have fit right in.

Franklin Delano Roosevelt was born into privilege. He lived in a mansion, had been to Europe several times before his 21st birthday,[57] and had spent 17 summers at Campobello. He loved it.

At Campobello, when not on his own or being tutored by his governess, Franklin could often be found playing with the Sturgis kids—Russell Jr. (two years older) and his six younger sisters—or the Porter kids, sailing their permanently beached boat, the *Merry Chanter*, across the waving grass.

In 1891, Franklin's father took delivery of a 51-foot sloop, the *Half Moon*,[58] named in honor of the ship Henry Hudson used to explore the river that bears his name. Franklin was given sailing lessons by the local fishermen, who moonlighted as captains and crew for Campobello's summer residents. The high tides, strong currents, fog, winds, and many rocky islands made Passamaquoddy Bay a challenging navigational environment. But rich kid Franklin was an eager and able student, and he took to the old salts like a fish to water. He hung around the docks, peppering them with questions. He learned what things were and how to use them. He learned how to take orders and the serious consequences when orders weren't followed or when things went wrong. Soon Franklin had convinced his teachers to allow him to go fishing with them in the big and blustery Bay of Fundy. It was not long thereafter when Captain Franklin Calder told Franklin, "You're a full-fledge[d] seaman, sardine-sized."[59]

That was not just the hired man idly complimenting his client; Captain Calder and Franklin Roosevelt liked and respected each other and became lifelong friends. Years later, when the need came for help and discretion, it was Captain Calder whom the Roosevelts trusted to carry their secret across the bay to a specially chartered private train waiting at Eastport.[60]

It was at about this time that Sara Roosevelt noticed an emerging character trait in her son: he organized the activities of his playmates and gave orders, which were followed. When this was pointed out to the would-be commander in chief, he replied, "But Mommie, if I didn't give the orders, nothing would happen."[61]

When Franklin was ten years old, James Roosevelt hired Tomah Joseph, chief of the Passamaquoddy Indians, to teach Franklin about the bay, native ways, and how to paddle a canoe. Franklin proved himself competent, and in 1898 James gave his son, now age sixteen, his own sailboat, the 21-foot knockabout *New Moon*. It became his "most precious possession."[62]* Franklin was now out on the water almost every day. If he wasn't in a boat, then he was swimming in the tidal lagoons, walking the beaches, or climbing the cliffs girding the island.

Captain Calder and Tomah Joseph taught Franklin how to navigate— with a compass, of course, but also with his eyes, ears, and nose—from the color of the water and where the clouds formed; from the sound of the surf and the calling of birds; and from the tangy smells of the pine trees and seaweed. Having waded, sailed, and paddled in and around the rugged coast, Franklin knew where the rocks were, when they would be covered or uncovered by the tide, which way the currents ran as the tide turned, and how strong those tidal currents were.

The demands of the sea—skills, knowledge, and discipline—were impressed on Franklin by his instructors and by his cousin, Theodore Roosevelt, whose two-volume book, *The Naval War of 1812*, argued for a strong US Navy, worldwide bases, and military preparedness. Theodore Roosevelt's book was published in 1882 when he was just 23 years old. Within four years it was standard issue on every US Navy ship. Fifteen years after the book's publication, Theodore Roosevelt became assistant secretary of the navy.

Following thirteen family members,[63] Franklin went to Harvard (1899–1903). Three things happened while he was there that profoundly shaped his life: he took Professor William Z. Ripley's course on the concentration of power in corporations;[64] he fell in love; and the president of the United States was assassinated while on a trip to see the immense new hydroelectric power plant at Niagara Falls,[65] causing his vice president, Theodore Roosevelt, to become the 26th president of the United States.

Standing six-feet-two-inches tall, Franklin was turning into a handsome and athletic young man. Golf was a fashionable new recreation, and Franklin helped design the course for Campobello. It had a unique hazard, grazing sheep, requiring players to "aim well over their cropping heads and hope for the best." In that happy summer of 1900, Franklin aimed well and became the club's champion.[66]

But while practicing his swing, Franklin hit a golf ball through the window of the playhouse where the six Sturgis girls—Anne, Susan, Beatrice, Gertrude, Carolyn, and Frances—were painting the floor red. Hearing the crash of broken glass, he raced to the playhouse and saw the girls covered with what he thought was blood. Running for help, he was relieved, and then embarrassed, to learn his "victims" were quite all right.[67]

With Franklin's tender emotions thus revealed, Gertrude, a precocious dark-haired beauty, saw her chance. Gertrude recruited his mother to get him to dance with her at the Saturday night soiree.[68] Sara made sure Franklin was on Gertrude's dance card. The parties were held at the hotel, with a marvelous view of the sunset. They had a lovely time, but no sparks flew. Gertrude was, after all, a bit young for Franklin. She was twelve.

James Roosevelt's yacht _Half Moon_ exploded and sank in 1900. Luckily, none of the Roosevelts were aboard at the time, and none of the crew were hurt. Franklin offered to sell his beloved _New Moon_ to help pay for a replacement. James declined his son's offer and ordered a new yacht, a 60-footer, the _Half Moon II_.

In reality, it was a father's parting gift to his son.

Having been a semi-invalid for five years following a heart attack, James died on December 8, 1900, at age 72.[69]

In the first-class car of the D&H railroad running between New York City and Hyde Park, New York, Franklin met a cousin he hadn't seen since they were toddlers. Her name was Eleanor.[70]* He arranged to see her again. Being Franklin's fourth cousin and the favorite niece of the president of the United States, nineteen-year-old Eleanor was of course welcome at Sara's summer house on Campobello.[71]

The beguiling island had its intended effect. Picnics were had, beaches were walked, sunsets were savored. To keep the budding romance safe from his mother's prying eyes, Franklin wrote in his journal in code. Had Sara not been blind to the idea of her son's affections going to another woman, she might have figured out "E is an angel" meant Franklin was in love with Eleanor.[72]

Franklin contrived to meet Eleanor at the Groton School in Marion, Massachusetts, which he had attended, and which Eleanor's younger brother was then attending. There Franklin proposed marriage, and Eleanor accepted. They did not, however, tell anyone. They were sure Sara and Eleanor's guardians would think them too young.[73]*

With their engagement a secret, "Cousin Eleanor" was invited to spend August 1904 at Campobello. When Franklin finally told his mother of his engagement, she was not pleased. She tried to change his mind and to distract him with trips to the Caribbean and job offers in London. But he was unmoved. She relented, on the condition that the engagement remain secret for a while to give the prospective couple time to consider their vows.

Franklin and Eleanor's ardor survived, and they announced their engagement in December. Three months later, the betrothed sat a few feet from Theodore Roosevelt during his second inauguration. Two weeks thereafter, on March 17, 1905, Franklin and Eleanor Roosevelt were married. Standing in for her deceased father was Eleanor's uncle, President Theodore Roosevelt. "Well, Franklin," said the president as he kissed the bride, "there's nothing like keeping the name in the family."[74]

MOOSE ISLAND, MAINE, UNITED STATES

The first horse set hoof on Moose Island in 1804. So strange was the sight that a little boy ran to tell his mother about the "man sitting on a cow that ain't got any horns!"[75] Such a late date may seem hard to believe, but why would a fisherman who lives on an island need a horse?

A map from 1815 shows the gap between what was then the District of Maine—a part of Massachusetts—and Norwood Island (later Carlow) was "Fordable at Low Water in Spring Tides only." Likewise, there was a low-tide link between Carlow and Moose Islands, which were at that time occupied by the British Navy and hence part of New Brunswick, Canada.[76]

Back on the horse: Even if you crossed the bar to the mainland, a horse wasn't much use since there were hardly any roads. Traveling any distance was easier by boat: the Passamaquoddies went by canoe; the Europeans by sail. Hence, for the first two hundred years of settlement, Eastporters relied on wind and tide power for transportation and on men and women for mechanical power rather than horsepower. Besides, horses ate a lot.

The British Navy had taken possession of Moose Island on July 11, 1814, simply by asking for it. The request was made by Sir Thomas Hardy to US Army major Perley Putnam.[77] Putnam was disinclined to acquiesce to Hardy's request but was prevailed on by the towns-people, who commended him for his bravery but pointed out he really didn't stand a chance in a fight against 2,000 British troops when he had only 59.[78]

The United States had entered the War of 1812 because the British had a nasty habit of taking possession of people, or "impressing" them, and forcing them to work in the British Navy. The enslavement of American citizens was instigated by the globe-spanning Napoleonic Wars (1803–1815), in which the French and English were again at each other's throats. The United States had tried to stay out of the fray by declaring itself neutral—and thus able to keep trading with both bel-ligerents—but England didn't see it that way. Trying another tack, the United States self-imposed the Embargo Act of 1807, banning trade with both hostile nations. This strategy also backfired, hurting many American merchants as much as the Europeans. But not all. For Pas-samaquoddy Bay merchants, on the border between the United States and British Canada, the embargo was a boon and ushered in the second great smuggling era. The Anglo-American altercation was resolved with the Treaty of Ghent, and on June 30, 1818, the British returned Moose Island to the United States.[79]*

To exercise their new freedom of speech, Moose Islanders recruited a newspaper man from Boston, Benjamin Folsom, who began publish-ing the *Eastport Sentinel* on August 9, 1818. "The name *Sentinel* was selected for the new journal on account of its location on the border." Its motto evidenced its mission: "*Here shall the Press, the Peoples' rights maintain—Unmov'd by influence, and unbridg'd by gain—Here patriot truth, its glorious precepts drane—Pledg'd to Religion, Liberty and Law.*"[80] The press on which it was printed was purported to have been owned by another Benjamin, with the last name Franklin.[81]

Fish is why you went to Moose Island. The Passamaquoddies went every spring; the Europeans stayed. They were fishermen-farmers, not hunter-gatherers. The settlers grew grain, and grain needs to be ground, and hand grinding with a stone quern is hard work.

Robert Bell, an immigrant from St. Andrews, Scotland, didn't like the monotonous manual labor. He thought it would be easier, and a good long-term investment, if he built a grist mill to grind his grain. Accordingly, he made friends with the Passamaquoddies and explained to them what he was looking for: good land for farming adjacent to a tight little saltwater cove that wouldn't freeze over in winter. Bell's friends took him to what is now Crane Mill Stream on the western side of Cobscook Bay, in Edmunds. There, in 1765, Bell acquired a grant from the Massachusetts Colony for several hundred acres and built a grist mill powered by the eighteen-foot tides.

Bell was just fifteen years old when he arrived and apparently not tall. His shortness and youth earned him the nickname *sakwahiganus'isk*, or "Little Chimes."[82] His stature notwithstanding, Bell built a good foundation for his future family. Nine generations later, his descendants still own Tide Mill Farm, where they raise organic livestock and vegetables. Although the tide mill burned down in 1900, Little Chimes's millstones remain as evidence of what a man with vision can accomplish with help from his friends and power from the tides.

You could have your fish four ways: fresh, sun dried, smoked, or salted. The first variety lasted only a few days before it started to putrefy, but the latter varieties could be kept for many months; transported to distant markets; and then soaked in water, cooked, and eaten. Fishing technology then was a small boat; some twine and fishhooks; and dry weather, firewood, or salt. As such, the trade's practitioners were many, small, and scattered. That changed in 1875 when Julius Wolff imported to Eastport the new European technique of canning sardines.

Where once the lowly sardines (young herring) were too small to be of interest, they fit nicely into cans, and there seemed to be an unlimited supply of them. The new harvesting technology was the weir, a scaled-up version of the ancient fish trap. The typical arrangement was a pole-and-net fence extending several hundred feet from shore to a corral shaped to make it easy for the fish to get in but hard for them to get out. As the tide ebbed, the fish would be concentrated into the corral, where fishermen in boats would scoop them up with nets, like fishing in a barrel.[83] Wolff's success attracted competitors.

By 1900, there were almost 800 herring weirs in New Brunswick. To process all the fish, canneries were built right on the docks, in nearly every port. By 1882, Eastport, or "Sardine City," had eighteen canneries, including Clark's, the world's largest. The canneries concentrated in Eastport at the juncture of Passamaquoddy Bay and the Bay of Fundy, which was easy for the sail-powered fishing fleet to reach quickly with their perishable catch. In turn, the large cargoes attracted the railroad, which built a trestle bridge into town in 1898,[84] took over the outbound freight, and put an end to the coastal shipping and shipbuilding businesses.

In the commercial food chain, little mom-and-pop canning shops got gobbled up by conglomerates, culminating in 1900 when the three largest merged into the Seacoast Canning Company, which tallied 38 factories in Eastport, eleven in Lubec, three in Machiasport, two in Pembroke, two in Millbridge, and one in Jonesport.

The same year, "old-style" three-part hand-soldered cans (bottom, side, top) were beaten out by machine-made two-part cans (pressed dish, top). The cans of fish stacked up to staggering heights, doubling from 44,952,244 pounds in 1899 to 87,224,524 pounds in 1904. But the huge haul depressed the prices, netting the canneries only 5% more cash.[85]

Originally the fish were deep fried in imported olive oil before packing. Later they were packed in less expensive domestic cottonseed oil and then cooked in the can, thereby decreasing the labor invested. With each technological advance, each new application of mechanical power, each increase in efficiency, each record-breaking catch, Eastport's sardine industry was advancing toward extinction and taking the town with it—for there are not an unlimited number of fish in the sea. To survive economically, Eastport would soon need a new resource to exploit. Or a new way to exploit it.

Fire incinerated Eastport's downtown three times, in 1839, 1864, and 1886. In 1888, Eastport began the switch from oil lamps to electricity when the privately owned Electric Light Company opened with a 112-hp steam-powered generator. Eastport's merchants planned to light their rebuilt downtown electrically, hoping to attract customers like moths—without flames. In this age of racing progress, civic pride was won with firsts. So when Eastport heard of Machiasport's plan to turn on its new streetlights during its Independence Day celebrations, Eastport raced ahead and had its electric christening on July 3, thus starting a grand tradition of "Night Before" celebrations. Following in the merchants' wake, the canneries started shifting from steam engines to electricity.[86]

Despite the War of 1812, three disastrous fires, an earthquake (1904), and several hurricanes (notably the Saxby Gale of October 4–5, 1869),[87] Eastport could point to the 1900 Census with pride. From timber and fish Eastport had extracted a century of growth, its inhabitants now numbering 5,311.[88] It seemed Eastport could recover from any apocalypse. Only famine and plague were yet to be tested.

Lubec was as far as you could go and still be in the United States. That characteristic was important to the Reverend Prescott F. Jernegan, who in 1897 founded the Electrolytic Marine Salts Company (E.M.S. Co.) to extract gold from seawater.

Lubec, a part of Eastport until 1811,[89] was too far from Boston for a day trip, but an overnight steamship could deliver visitors to within walking distance of Jernegan's extraction plant. The not-too-far location made it possible for would-be investors to make a quick inspection in the afternoon, have an enjoyable evening with their "wives" at an inn, travel back to Boston on the morning boat, part company, and return home.

It had been known for decades that seawater contains about a milligram of gold in a ton (233.86 gallons) of seawater. Therefore, extracting one ounce of gold required processing 662,300 gallons of water—almost exactly the volume of an Olympic swimming pool.[90] Alas, the cost of chemically treating so much seawater wasn't worth it. But electricity changed everything.

Electrolysis—the passing of an electrical current through an aqueous solution—magically made some metals appear and disappear. Jernegan claimed to have invented an electrolytic process to precipitate gold from seawater at low enough cost to make it profitable. The process was of course secret. Jernegan, however, confided that Passamaquoddy Bay's great tides were critical to the effort: they would pump great volumes of seawater through Jernegan's electrically charged, platinum-plated, mercury-filled "gold accumulators,"[91] and a generator would be hitched to a traditional tide mill to produce the needed electricity.[92]

Reverend Jernegan attended and later taught at Phillips Andover Academy in Massachusetts, received his divinity degree from Brown University in Rhode Island, and in 1892 became pastor of the Baptist church in Middletown, Connecticut. He got into an argument about his salary there and left. Shortly thereafter he began experimenting with electrolysis, the secret of which was revealed to him "in a vision."[93]

To demonstrate the effectiveness of his secret process to potential investors from his former congregation, the reverend set up the apparatus in a small shack on a pier in Rhode Island. To assure them everything was legitimate, they supplied their own materials and conducted the experiment themselves—though it had to be timed to the tides, which would be just right on a particular winter night.[94] The investors lowered Jernegan's apparatus and their chemicals through the bobbing cakes of ice and into the pregnant sea. The investors stayed all night, shivering, while Jernegan retired to the comfort of a hotel. In the morning, with numbed minds and trembling hands, the would-be investors hauled up the accumulator and took its contents to a chemist of their choosing. Low and behold, the chemist assayed out five grains of gold and ten of silver. Convinced, the investors each entrusted $5,000 to their minister[95] ($130,000+ in 2020 dollars).[96]

In ten days, Jernegan collected $250,000 from professional investors in Boston and New York.[97] With cash in hand, he set to work in earnest. Jernegan leased the Comstock tidal grist mill in Lubec,[98] installed in it 220 accumulators, and surrounded the area with a high barbed-wire fence.[99] Jernegan rechristened the mill Klondike Plant No. 1, causing the *Lubec Herald* to pant about the glittering "new El Dorado" in their midst. Hitching up the old tide mill's tub-wheel turbine to a dynamo, E.M.S. Co. generated the electricity for the electrolysis and about 20 lightbulbs, probably the first in Lubec, and installed the first telephone.[100]

Klondike Plant No. 2—with 5,000 accumulators—was to be built on the Lubec Canal, a man-made tidal strait dug in 1840 to provide power to a plaster mill. During the spring and summer of 1898, E.M.S. Co. employed 700 men to enlarge the canal; build dams and tide gates; and expand the mill to house (and keep out of view) the accumulators. Tons of lumber, hardware, tools, and dynamite were brought in. The *Lubec*

Herald tried to calculate what was being paid to workers and local merchants. Stores, houses, apartment buildings, and a recreation hall sprang up around the mills.

Then on Friday, July 29, 1898, the *Hartford (CT) Courant* headlines cried, "The Bubble Burst. Gold-from-Seawater Company Suspends. Steps taken to Apprehend Jernegan."[101] Work stopped. Investors panicked. With words that would echo through the ages, the *Boston Herald* stated:

> Everything is quiet here [in Lubec] tonight as could be expected with 700 men suddenly thrown out of employment, and the fond hopes of the last eight months for the building up of a prosperous city around the fabled gold-producing plant of the marine salts company blasted beyond hope of recovery. . . . The people have been for the past two days as though suddenly awakened from a dream, and they can hardly realize as yet that silence and desolation are to fill the valley which up to Friday was filled with noisy activity.[102]

Under an assumed name, Jernegan had bought at least $80,000 ($2+ million in 2020 funds) of bearer bonds and departed for Europe.[103] He was tracked to Le Havre, France, where he boarded a train for Paris. Telegraphing ahead by Americans had French authorities waiting for him, but he slipped off the train.[104] When a reporter finally found him, the reverend protested his innocence, saying he was on legitimate business and that the real problem was the disappearance of his partner, Mr. Fisher, who had absconded with the "indispensable records" of his electrical experiments—on which Fisher's half of the gold extraction process was based—and of which Jernegan claimed to be ignorant. His own hasty departure, said the righteous reverend, was an effort to catch the slippery Fisher.[105] Furthermore, the preacher proffered, there was no fraud: the funds he had were only his share of the royalties E.M.S. Co. paid him for his knowledge of the chemical part of the electrochemical extraction process.[106] Then he disappeared.

Fisher had also disappeared, with $100,000. Notice of his death appeared in a newspaper from Australia in 1900, though it was widely believed Fisher faked his own death to throw off his adversaries.[107]

E.M.S. Co. was a hoax, reputedly one of the largest swindles in American history. The secret to Jernegan's success was not a "vision" but Mr. Fisher in a diving suit in the dark of the night "salting" the sea

by placing gold filings into the accumulators.[108] Jernegan was successful because he ran a convincing scam and kept a hefty "reserve fund,"
as he described in a letter to his accomplices read at his trial:

> Such a fund is like a revolver: if you need it at all you need it very much.
>   . . . There is just one time when I expect to be liberal with money, and
> that is the moment when, with it and us in a safe place, I sit down and divide
> up what each man has contributed to the success of the undertaking, and
> what he needs to fulfill his ideas of life. But until that time, I want to see
> every penny saved. . . .
>   Money and lust have been the two most vexing problems of my life. I
> see how both may be solved if I can get a home where I keep my body in
> touch with nature, and have a settled, if small, income.[109]

The meaning of Jernegan's letter solicited much salacious speculation. His wife divorced him. Eventually he felt remorse and returned
some of the money.[110] He reportedly trekked to Alaska, where he
panned for real gold with little success, and then drifted to the Philippines. There, free from the bonds of faith and family, he may have
finally solved his most vexing problems.

Whatever his fate, Jernegan's tide-powered fraud still haunts Quoddy
Bay.[111]*

Gone were the good old days. By 1913, Eastport fishermen were
catching fewer fish, and vacancies were appearing on Main Street.
There were fewer people who needed to buy what the merchants had to
offer. And while the sardine factories weren't as productive as at their
peak in 1900, they were still canning a lot of fish and grinding up the
heads, tails, and guts into fertilizer. In the heat of the summer, if you
were within five miles of Eastport, you knew it.[112]*

Summer was also when it was most fun to be in Eastport. Thousands
came, at least for the day, taking advantage of discounted fares offered
by the railroad and steamships to lure visitors to Eastport's famous
Independence Day celebration. Small navy ships had been part of Eastport's celebrations for many years. The crews happily marched down
Main Street, played baseball against the hometown "Quoddies" (and,
graciously, usually lost), and showed off the size and power of American fighting ships to throngs of visitors. When given their liberty, the
sailors consumed vast quantities of ice cream and soda at the pharma-

cies, gallons of beer at the bars, and many a meal at the eateries; then they danced with the local girls and, if they were lucky, found a suitable site to enjoy the fireworks with them. Or, if they were more rambunctious than romantic, they could buy packets of flour or talcum powder, which were tossed as harmless hand grenades at anyone wearing temptingly dark clothes. When Franklin Roosevelt became assistant secretary of the navy in 1913, he upped the ante. He ordered the battleship USS *North Dakota* and its 933 men to join the party.[113]

All the visitors and bunting on those few gay days briefly hid the fact that Eastport's tide of fortune had peaked more than a decade before. Campobello was in the same boat; its three hotels had all closed by 1907.[114] When the navy ships departed, there was nothing on the horizon that would bring prosperity, nothing to reverse the economic ebb.

When the party was over, when the parade floats or "horribles" were dismantled, when the hangovers subsided, Quoddyites were left wondering: *What were they going to do?*

SIPAYIK, PASSAMAQUODDY TRIBAL NATION

Sighting along the barrel, Francis Joseph Neptune aimed his musket.[115] The target was at extreme range and moving. Holding his breath to steady the instrument, he squeezed the trigger. The spring-loaded flint snapped forward, striking the steel frizzen and priming powder in the pan, sparking a miniature explosion that forced a one-ounce ball of lead out of the gun's muzzle. The target, Admiral Cox of the British Navy, fell dead into the water.[116]

It was an extraordinarily skillful shot. It threw the now leaderless attackers into confusion and resulted in the failure of their attack on Machias, Maine, on August 14, 1777. Neptune was the chief of the Passamaquoddy Indians. He was fighting for his adopted country, the United States of America.

The Passamaquoddy were asked to fight on behalf of the American rebels by General George Washington. In 1776, Washington extended a metaphoric "chain of friendship" along with his request for the natives to switch sides, the tribe having fought on the British side during the French and Indian Wars. It was Francis's father, Chief Michael Neptune, whom the British governor of Nova Scotia had sent packing in 1760 with threadbare compensation for the Passamaquoddy lives lost in the war.[117]

A second rebel request for assistance was made by the First Continental Congress through its "superintendent of the eastern Indians," Colonel John Allan.[118] In return for continuing to fight on the American side, the Passamaquoddy hunting grounds would be safeguarded and the tribe would be "forever viewed as brothers and children under the Protection and Fatherly care of the United States." The tribe accepted Congress's offer and pledged allegiance to the American flag and its democratic ideals. Allan predicted that "no people will defend the liberty of America better."[119]

Let there be no misunderstanding: because the Passamaquoddy fought for the Americans—and *not* for the British—Maine is in the United States, not Canada.

But who would fight for the Passamaquoddy?

Having won the war, the US Congress reneged on its pledge to the Passamaquoddy tribe and revoked Allan's appointment.[120] In 1783,[121] Allan wrote to Alexander Hamilton, the first secretary of the treasury, seeking a written treaty for the Passamaquoddy tribe to protect their property rights: "The Indians to have an Exclusive Right of the Beaver Hunt, on the Rivers, which they now live on. To wit—all Eastward of machias, with the Lakes that Extend from Passamaquady, River to Penobscot . . . .[and] Some methode, to prevent Unfair dealing with the Indians, and Embezzleing, their property."[122] He received no response.

Unlike the rabble in Congress, Allan was an honorable man, and when he became a commissioner for the Commonwealth of Massachusetts, he helped negotiate the Treaty of 1794 with the Passamaquoddy tribe in which they relinquished all other claims to lands within Massachusetts in exchange for clear title to 23,000 acres that became known as Indian Township, several islands, ten acres at Pleasant Point, and fishing rights.[123] But few European Americans cared what the treaty said. Most hoped the natives would become extinct, "gradually diminishing" through death, disease and dispersal, "till there shall be none remaining."[124]

Robert Pagan was banished by Massachusetts in 1778, along with 300 others. The Banishment Act sought to "to prevent the return . . . of certain persons . . . to the United States, [who had] joined the enemies thereof." Having suffered the *illegal* destruction or seizure of his property (including land, buildings, and ships), Pagan and his family fled from their home in Portland, Maine. They joined 60,000 to 100,000 other British Loyalist refugees.[125]

In 1780, Pagan settled on the Penobscot River, believing it would become a Loyalist settlement. In 1783, learning that the US-Canada border would be established farther east, leaving him in hostile territory, he again relocated. This time he moved to the immigrant enclave at St. Andrews on Passamaquoddy Bay's eastern shore.[126]

For a third time Pagan found himself at risk of being forced from his home when, in 1796, negotiations were underway between the United States and Britain about which of the three rivers named "St. Croix" was the true international border. In 1797–1798, Pagan used copies of Champlain's maps, and one drawn for him by Francis Joseph Neptune of the river he called "Skutik,"[127] to figure out where the line should be. Then he and a surveyor conducted an archaeological excavation of Dochet Island, where they found the remains of the buildings erected by de Mons and Champlain in 1604. That ended the debate: the international boundary was drawn through Passamaquoddy Bay, leaving St. Andrews safely in Canada.

Possession*, legal pragmatists say, *is nine-tenths of the law. The same high-minded First Congress of the United States that reneged on its pledge to the Passamaquoddy passed the oddly but aptly named Indian Nonintercourse Act of 1790 to protect Indians from unscrupulous land grabbers.[128] But as legal realists would say, *the Act was honored in the breach.*

On their ten acres at Pleasant Point, 100 members of the 500-member Passamaquoddy Tribe spent the summer hunting porpoises for meat to eat and for oil to light their lamps. The plot was too small for a year-round settlement, so Father Roman, a French Catholic priest, petitioned Massachusetts for additional land. "Hunger," Father Roman said, "will force them at last to take up husbandry."[129] On March 4, 1801, Massachusetts decreed, "a tract of land . . . called Pleasant Point, containing ninety acres, the property of this Commonwealth, be and hereby is appropriated to the use and improvement of the said tribe of Indians."[130]

Within a year, the Passamaquoddies paddled their canoes to Boston to petition the state to prevent trespassers on their lands and for funds to build a church. Massachusetts appropriated $500 for the church and assigned an agent, Theodore Lincoln, to oversee its construction, settle local complaints, and dispose of timber "for the use and benefit of said tribe." Mr. Lincoln did just that: he built the church and leased tribal land to loggers—for up to 999 years.[131]

Impoverishing encroachments and restrictions crept in from every side: roads were cut, eel weirs were prohibited, sawdust suffocated the shad fish, clam beds were covered by bridges, dams were built, lands were flooded, islands were seized, and lots were sold off en masse.[132]

The District of Maine, three times bigger and separated from the rest of Massachusetts by the intruding state of New Hampshire, eventually got tired of being ruled by unjust, ignorant foreigners and was allowed to secede from Massachusetts to become the twenty-third state in the Union on March 15, 1820, as part of the Missouri Compromise. The Compromise allowed Maine to join the Union as a Free State and Missouri to join as a Slave State, thereby maintaining the balance of power between the Union's northern and southern factions. In the Compromise, the Penobscot tribe relinquished their ancestral rights to the state of Massachusetts for $30,000. The Passamaquoddy did not.[133]

Maine was not, in reality, a Free State. The Passamaquoddy were as good as slaves: they couldn't vote, they were segregated and discriminated against, and they were nearly powerless to prevent their lands from being taken by squatters and swindlers.[134]

Benjamin Folsom, the *Sentinel*'s founder, drew his readers' attention to the remarkable fact that President Andrew Jackson's annual message to Congress of December 8, 1829, had reached Eastport, 780 crow miles from Washington, in only 60 hours. The news had traveled by horse and sail at an average speed of thirteen miles per hour! Folsom noted gleefully that the Eastport-captained packet ship *Boundary* traveled the fastest and "beat the New Yorkers all hollow." But Folsom conceded, "Having a considerable part of our paper in type when the Message was received, must apologize to our readers for omitting a part of it, which will be given next week." Folsom devoted almost one-quarter of his four-page paper to Jackson's message and a few column inches

of critical commentary on it, noting a *"great deal* of cunning and adroit-
ness, and considerable amount of what we shall call *electioneering*."[135]

In the *Sentinel*'s December 23 edition was more of Jackson's message and a fine example of what Folsom was warning about:

> The continuation and ulterior destiny of the Indian tribes . . . have become
> objects of much interest and importance. It has long been the policy of
> Government to introduce among them the arts of civilization in the hope
> of gradually reclaiming them from the wandering life. . . . [We] have, at
> the same time, lost no opportunity to purchase their lands and thrust them
> further into the wilderness. By this means they have not only been kept in
> a wandering state, but been led to look upon us as unjust and indifferent to
> their fate. Thus, though lavish in its expenditures upon the subject, Govern-
> ment has constantly defeated its own policy; and the Indian in general reced-
> ing further and further to the West, have retained their savage habits. A por-
> tion, however, of the Southern tribes, having mingled with the whites, and
> made some progress in the art of civilized life, have lately attempted to erect
> an independent government within the limits of Georgia and Alabama. . . .
>
> Would the people of Maine permit the Penobscot tribe to erect an Inde-
> pendent Government within their State? And unless they did, would it not
> be the duty of the General Government to support them in resisting such a
> measure? . . .
>
> I informed the Indians inhabiting the part of Georgia and Alabama that
> their attempts to establish an independent government would not be counte-
> nanced by the Executive of the United States, and advised them to emigrate
> beyond the Mississippi, or submit to the laws of those States.
>
> Our conduct towards these people is deeply interesting to our national
> character, their present condition, contrasted to what they once were, makes
> a most powerful appeal to our sympathies. Our ancestors found them the
> uncontrolled possessors of these vast regions. By persuasion and force,
> they have been made to retire from river to river, and from mountain to
> mountain, until some of the tribes have become extinct, and others have
> left but remnants to preserve for a while, their once terrible names. . . .
> Humanity and every honor demand that every effort should be made to
> avert so great a calamity. It is too late to enquire whether it was just in the
> United States to include them and their territory within the bounds of the
> new States, whose limits they could control. That step cannot be retracted.
> A State cannot be dismembered by Congress, or restricted in the exercise of
> her Constitutional power. But the people of these States, and of every State
> actuated by feelings of justice and regard for national honor, submit to you
> the interesting question, whether something cannot be done, constructively
> with the rights of the States, to preserve this much injured race?[136]

The Jackson administration and Congress wouldn't "dismember" states but would dismember tribes.

The next week, the Indian issue got even closer to home: "By the following Message of President Jackson to the House of Representatives," the *Sentinel* reported, "it will be seen that our Indian neighbors are at work—and with a prospect of success":

> A delegation from the Passamaquoddy Indians, residents within the limits of Maine, have arrived in this city, and presented a memorial, soliciting the aid of the Government in providing them with a means of support. Recollecting that this tribe, when strong and numerous, fought with us for the liberty we now enjoy, I can not refuse to present for the consideration of Congress their supplication for a small portion of the bark and timber of the country which once belonged to them.
>
> It is represented that from individuals who own the land adjoining the present small possession of this tribe, purchases can be made sufficiently extensive to secure the objects of the memorial. . . . Should Congress deem it proper to meet them, it will be necessary for their being held in trust for the use of the tribe during its existence as such.[137]

Congress did not *deem it proper*. They did not care for the tribe's *existence as such.*[138]

Indian sovereignty stayed in the news, and on July 14, 1830, Folsom ran another article about the Cherokee crisis that foreshadowed a bigger battle among Americans that would erupt a generation later, central issues of which were "states' rights" and "liberty and justice for all":

> The Governor of Georgia has issued his Proclamation announcing that the iniquitous laws . . . took effect upon the Cherokees on the 1st of last month . . . providing for the annexation of the Cherokee country to the State of Georgia. . . . The laws, usages, and immemorial customs of the Cherokee nation are declared to be "null and void, as if they had never existed"; the expression of opinion amongst the Indians is made a felony; and the Indians are disqualified from being witness in the Courts, which they are told will protest them! . . .
>
> It has been for some time known that gold in great abundance is found on the Cherokee lands: the white have in great numbers intruded upon their territory to obtain it, and after repeated representations to the United States

Government, a military force was at length dispatched to the Cherokee nation, which drove off the intruders, and left the gold lands free for the Indians, the rightful owners. . . . But the Governor of Georgia has issued a second proclamation asserting the right of Georgia to the lands of the Cherokees . . . If injustice like this is to be tolerated in our country, let us no more talk of Turkish oppression, or Miguel's tyranny, or the barbarism of Algiers. Georgia has outdone them all in infamy.[139]

Righteous indignation was the reaction of many to Jackson's and Georgia's deplorable actions. "Amongst the many eloquent and powerful speeches in support of the cause of Justice, Humanity, and the Nation's Honor . . . that of Mr. BATES, of Massachusetts, is conspicuous, both for the force of its arguments and for the beauty of its style":

Mr. Speaker, there is not an act of Georgia . . . there is not a resolve, ordinance or law of Congress; there is not a treaty of the U.S. with the Indian tribes that does not tend to establish the fact, that the Indians are the proprietors of the lands and hunting grounds they claim, subject only to the restriction upon their right of alienation. . . . Not one act, law or treaty does not establish the fact that they are sovereign. . . . Sir, the emblems of it were sparkling in the sun, when those who now inhabit Georgia, and all who ever did, were in the loins of their European ancestry. . . .

That when the United States, in consideration of the cession of land made by the Cherokee to the government, guaranteed to them the "remainder of their country forever," you meant something of it. . . . If you have honor, it is pledged; if you have truth, it is pledged; if you have faith, it is pledged—a nation's faith, and truth, and honor! And to whom pledged?— To the weak, the defenseless, the dependent. . . . And for what pledged? By a nation in its youth—a Republic, boastful of its liberty. . . .

Sir, your decision upon this subject is not to be rolled up in the scroll of your journal and forgotten. The transaction of this day, with the events that it will give rise to, will stand out upon the canvass in all future delineations of the quarter of the globe, putting your deeds of glory in the shade. You will see it everywhere—You will meet it on the page of history, in the essay of your moralist, in the tract of the jurist. You will see it in the vision of the poet; you will feel it in the sting of the satirist; you will encounter it in the indignant frown of the friend of liberty and the rights of many, wherever despotism has not subdued to its dominion, the very look. You will meet it upon the stage; you will read it in the novel, and the eyes of your children's children throughout all generations, will gush with tears as they run over the story. . . . And, sir, you will meet it at the bar above.[140]

Despite Bates's rebuke, Congress passed the Indian Removal Act of 1830, which resulted in the Trail of Tears, the forced march of the Cherokee to the inhospitable West. On the trip, as many as half of the 16,000 refugees may have died.

Neither the expulsion nor two Supreme Court cases settled the issues of Indian rights and states' rights.[141] It would take the intervention of greater presidents to undo the damage caused by President Andrew Jackson.

Koluscap, a man-god in Abanaki origin stories, made the white birch tree and told it to take care of the native people.[142] It did as it was bid.

From the roots, the natives made a sweet tea; from the fine-grained and hard wood, they made tools; and from the bark, they made amazing things. The birch's outer bark is smooth, strong, and bright white, with black flecks, calling forth its common name: the paper birch. Small sheets of bark can be peeled off the tree with ease, and, with a sharp knife and more work, the tree's entire covering can be removed like skinning the hide from a deer. Like hide, birchbark is waterproof, with a high oil content that keeps it flexible, prevents rot, and makes it excellent tinder, even when wet. The inner bark is reddish brown, and by scraping through layers, the underlying colors show through. You can also draw on paper birch with a bit of charcoal or ink. It is nature's canvas.

The epitome of native woodworking craft was the birchbark canoe. It was in essence the recreation of an animate being in birch: a rib cage of wooden slats covered by a hide of bark. Seams in the bark were sealed with spruce gum, and gunwales and seats were lashed to it with split spruce roots. It was light, fast, and stable. It could be made to fit a single man or half a dozen. It was perfectly suited to the island-strewn waters of coastal Maine and its vast network of lakes and rivers, and easily carried, or *portaged*, from stream to stream. In 1898, A. E. Wickett copied the Penobscot design, substituting machine-made cotton canvas for birchbark, and created an American icon, the Indian Old Town Canoe.[143]

The Passamaquoddy had the paper birch but not a written language. Deeds and promises were remembered by repeatedly speaking of them.

Tomah Joseph paddled his own canoe. The blooming shad bush and the male chickadees switching to spring song told him it was time to move. As his ancestors had done for millennia, he moved from his winter home in the interior to the coast.

While his great-grandfather would have loaded his canoe with beaver and mink pelts, those animals had been hunted out. Instead, Tomah loaded his canoe with birchbark. Rather than taking furs to an agent at Eastport who would ship them to Europe for finishing into men's hats and women's coats, Tomah took his finished goods directly to the consumer: the tourists.

Paddling to Campobello, Tomah carried his possessions up the beach, the pebbles rattling underfoot, and set up camp in the woods near Welshpool. No one in town seemed to mind the summertime squatter, as he provided what might have been called "local color," demonstrated and sold birchbark crafts decorated with Wabanaki origin stories, told interesting tales, harvested and sold wild berries, and guided "sports" on hunting and fishing trips. Campobello native Franklin Calder thought highly of his occasional neighbor: "He is a man of integrity, skill and gentleness. Each visitor is eager to gain his companionship and guidance in his canoe as he paddles into nooks where one less experienced might hesitate to penetrate."[144]

Franklin Calder and Tomah Joseph shared a valuable client: James Roosevelt. James hired them to teach his son Franklin the finer points of sailing and canoeing in the bay's great tidal waters and to impart to him their experience and wisdom about the ways of nature and men. James was a keen judge of character, and ten-year-old Franklin stuck like a limpet to these older men, who showed him how to understand, adapt to, and direct the forces of nature. Tomah Joseph began mentoring Franklin in 1890 and continued until Tomah's death in 1914.

In 1907, Tomah made a birchbark canoe for Franklin. It was a beautiful craft, seventeen feet long, with plenty of rocker and flare (curved front to back and side to side) to handle blustery days on the bay. On the bow Tomah etched the image of an owl, Tomah's spirit animal. When he presented the canoe to Franklin, he said, *Mikwid hamin* ("Remember me").[145]

On July 20, 1910, Franklin, age 28, paddled his canoe into Passamaquoddy Bay. Reaching his objective, the presidential yacht USS *Mayflower*, he announced himself and climbed aboard. "The preponderous president . . . greeted young Roosevelt, the personification of youth and energy."[146]* Three-hundred-pound Taft was the third president of the United States that Franklin had met.[147]* They, too, were his role models.

CHAPTER 3

Converging Currents

WATERPOWER AND ELECTRICITY

Waterpower is older than Rome. Marcus "Vitruvius" Pollio provided the first written description of water wheels in 25 BCE. Sometime after that, someone figured out that tidal straits could be dammed to create a power pool that worked year-round, regardless of summer droughts or winter freezes.

The key differences between river-powered mills and tide mills are that tides flow in two directions, and the time of that flow happens 50 minutes later each day. To deal with the reversing tidal currents, most tide mills use *gates* made of wood that are pushed open by the incoming tide, thereby filling the mill pond, and are balanced so they automatically snap shut when the tide starts to ebb. Thereafter, the water is directed through a chute to a waterwheel, which is pushed by the rushing water, thus generating mechanical power. Typically, tide mills produce usable power for only about five hours during each of the two daily ebb tides. This could provide a workday's worth of power, but in two shifts twelve hours apart, and 50 minutes later each day—hardly a convenient schedule.

Many of the early tide mills used vertically rotating undershot waterwheels because they worked well where the water's fall, or *head*, was small. Other tide mills used more efficient spout-type *penstocks*, which funneled water at the vanes of horizontally turning wheels. Another advance was to "box in" the horizontal wheels to create *tub wheels*, much like modern turbines. Spout-tub wheels became popular in northern areas because they drew water from the bottom of the mill pond, rather than the top, and therefore had fewer ice problems in the dead of winter, when even brackish water can freeze.[1]

Waterpower is harnessing the force of gravity acting on moving water and turning it into horsepower.[2*] The higher the head, the more power can be generated. Horsepower increases with the square of the head. In other words, doubling the head yields four times the horsepower. For example, if the head is 10 feet, every square mile of a millpond's surface area can theoretically produce 7,200 hp. But if you had 20 feet of tide, you'd get a stampede of 28,800 hp. There are only fifteen places in the world where the tides are greater than 20 feet. Topping the list is the Bay of Fundy with 50-foot tides.[3*] Thus, the potential energy in the 150-square-mile Passamaquoddy Bay, "Fundy's left thumb," is enormous.

Tide mills powered the New World. They were among its first power plants, an essential technology that enabled early immigrants on the coast to produce significant mechanical power and turn out products to keep them alive and sell for a profit. Waterpower allowed the immigrants to shift from fishing to farming and then to manufacturing. With each step, the infrastructure got larger, the population denser, and natural resources more intensively exploited.

Until the invention of the first steam engine by Thomas Newcomen in 1712, there had been only four types of mechanical power: manpower, animal power, wind power, and waterpower. While humans and draft animals could walk to where the work was, windmills couldn't, and waterpower—the most powerful—was not portable. It could extend only as far as the drive shaft and belts could stretch, effectively limiting its application to within a few hundred yards of the waterwheel.[4*] Thus, industry went to where the power was.

Tide mills were built on waterways (frictionless liquid highways) where both raw materials and finished goods could be transported efficiently at a time when roads and bridges were nonexistent or barely usable with heavily loaded wagons. Tide mills provided the power to process the essentials of food and shelter: to grind the grain and saw the lumber. Mills (tidal or riverine) were the anchors around which towns formed. Everyone went to the mill: to work, for flour for their daily bread, for lumber for their houses and ships, for news, for commerce, and perhaps for a pint of brew at the tavern that often stood nearby.

In colonial North America, a mill was often the single most valuable structure in a town. Grist mills were so important that they were often regulated as monopolies to prevent abusive pricing or water management or to restrict their use to locals rather than outsiders.[5]

In 1641, the Commonwealth of Massachusetts first promulgated the Great Pond Ordinance, which established "common liberties," including free speech, freedom of emigration, and that "Every Inhabitant that is a howse [sic] holder shall have free fishing and fowling in any great ponds and Bayes, Coves and Rivers, so farre [sic] as the sea ebbs and flows."[6]* The law effectively gave the legislature the authority to dispose of its property, including water rights, "as it thinks proper."[7]

In the centuries following the pioneering de Mons–Champlain tide mill at Annapolis, Nova Scotia (1605), hundreds of tide mills were set up on tight little coves all along Maine's rocky coast. One of the largest, at South Harpswell, had three 75-hp turbines that could grind a total of 50,000 bushels of grain annually.[8] There were ten tidal sawmills on the Kennebec River at Winnegance and at least nine tide mills in Passamaquoddy Bay.[9]* When tide mills reached their peak around 1800, there were several hundred in Maine; dozens in Massachusetts, New York, and Atlantic Canada; and a few as far south as the Carolinas.[10]*

Three innovations in tidal power occurred in Boston in the 1800s. Boston was a city of shippers—builders, captains, and traders—but the Embargo of 1812–1814 drastically reduced their business. Meanwhile, the young country was turning from an agricultural economy to manufacturing, but Boston—on a peninsula in Massachusetts Bay—didn't have enough waterpower for heavy manufacturing. Consequently, people and money were leaving the city for locations with river power. Boston, however, had ten-foot tides and several small tide mills.[11]* In 1813, Uriah Cotting and Isaac Davis hit on an idea to solve tide mills' intermittent power problem and small scale. They called their idea "Perpetual Power."

Instead of having just one mill pond, they would have two: a "full basin" that would be kept as high as possible by refilling at high tide, and a "receiving basin" into which the tailrace water from the mills would drain. The receiving basin would be kept as low as possible by preventing the tide from coming in and emptying the accumulated tailrace water at low tide. Their scheme could utilize nearly the full ten feet of head, 24 hours per day, every day, regardless of the outside tide level, droughts, or freshets. Their oceanfront location made transporting goods easy and inexpensive.

Their second innovation was leasing "millpower"[12]* to manufacturers as a raw material. This was quite different from the traditional business model in which an entrepreneur built a mill to produce and sell a product (e.g., lumber) or provide a service (e.g., grind grain). Cotting and Davis sold *power* and created, in effect, the first power company, the first "utility."

Cotting and Davis petitioned the state for the right to dam off a section of the Charles River. Despite fierce opposition, the plan was approved. The company then issued a 24-page prospectus to aid in raising the $250,000 needed, which ended by saying, "ERECT THESE MILLS, AND LOWER THE PRICE OF BREAD." Revenue was optimistically projected to be $450,000 per year. Stock sold for $80 a share, and on the opening day, there was such a crowd that one investor climbed into Cotting's office by a window. The stock sold out in a day.[13]

Huge stone and earth dams were constructed, and a handful of mills were erected beginning in 1822. A toll road was constructed over the main dam, and promenades, parks, and a bathing pavilion were built along the river. But there was a problem: there wasn't enough water for the six mills that were built, and some mills were taking more than their fair share of water. A lengthy court fight led to the utility developing a millpower monitoring system, the first step toward continuous power metering. These two innovations—the power "utility" company and the power "meter"—are now used worldwide.[14]*

By 1858, steam engines were widely available and replacing tidal power. Meanwhile, the tidal basins at Boston, which had been crisscrossed by railroads and steadily filling with refuse,[15]* became more valuable as filled land for the expanding city. Thus, the world's first two-pool tidal power utility died, was buried, and was forgotten. In its place rose the handsome neighborhood of Back Bay.

Electricity changed everything. In the history of the world, humans' mastery of electricity is as important as the control of fire. Electricity could provide heat, light, and mechanical power. It was infinitely scalable, from milliwatts to megawatts. While coal and oil as *fuel* for light or power could be *transported* in trains, ships, and other vehicles, electricity could be *transmitted*—the power itself could be moved to where the light or work was wanted—without having to move the power plant.[16]*

With tiny infrastructure—thin metal wires—electricity could (eventually) be transmitted vast distances, over mountains and even under oceans. Waterpower was an excellent way to generate electric power. Thus, many water mills were converted into *hydroelectric* plants, and soon thereafter people started thinking about *tidal-hydroelectric* power.

The science of electricity began in 1600 with William Gilbert's seminal book on electromagnetism. Benjamin Franklin's famous kite-and-key experiment of 1752 demonstrated that lightning and electricity are the same thing. Soon after that, static electric *generators*—typically glass jars spun by hand cranks and brushed by wool or silk—became the favorite apparatus of experimenters and tricksters for the bright and very uncomfortable sparks they could inflict on unsuspecting volunteers. Then, in 1820, Hans Orsted discovered that a compass needle would be deflected by an electromagnetic field, and in 1831 Michael Faraday tuned Orsted's discovery into the first direct current (DC) electrical generator. It was a simple machine consisting of a copper disc twelve inches in diameter that rotated between the ends of a magnet.

Current is the flow of electricity; the speed of the current or force was originally called "pressure" or "tension" and eventually defined as volts; amps is the volume, and watts the power. An electric *generator* uses mechanical power to produce electric power. An electric *motor* uses electric power to make mechanical power. Electric generators and motors are essentially the same device, run in opposite directions.

In a typical watermill, the power is the moving water and the *load* is the grindstone, saw, or other device. Similarly, an electric generator uses power to turn a coil of copper wire inside of a magnetic field (the load), thereby generating electricity. Conversely, an electric motor uses electric power to turn a shaft (the load), which is then used for mechanical purposes.

Electricity can also produce light, and Thomas Edison invented the first practical light bulb in 1879. On September 4, 1882 (eight months after Franklin Roosevelt was born), Edison powered up the first commercial-scale electric utility company from his coal-fired Pearl Street Station in New York City, lighting up the offices of J. P. Morgan and the *New York Times* as his first customers.[17] At $0.24 per kilowatt hour (kWh)—about equal to gaslight—electricity was a luxury. Adjusted for 140 years of inflation, that would be roughly $7.20 today—when the average residential price is now $0.13 per kWh.[18] Pearl Street was the first DC-powered "central station" and began the switch from small power plants at individual manufacturing companies to ever-larger central stations serving ever-larger groups of commercial and residential customers. Twenty-six days later, the first hydroelectric power plant began operating on the Fox River in Appleton, Wisconsin.[19]*

From that day on, there was a worldwide race to generate electric power and profits. It was the Wild West all over again, a frantic scramble to control territory: hydropower sites and markets. To the victor would go the spoils.[20]

Niagara Falls is the largest waterfall in North America. Situated on the international border between the United States and Canada, the mighty Niagara River falls 185 feet in one stupendous, awe-inspiring, earth-quivering, thunderous plunge. Niagara's enormous power had been evident from the moment humans first heard its roar, felt its thunder, and saw its size. The Falls, however, were too great to grasp, too huge to harness. The first mill was built along the edge of the rapids above the falls in 1758–1759.[21] Between 1847 and 1875, three waterpower companies were formed and funded, dug canals, sold mill sites, and failed, having lost $1.5 million.[22] Nonetheless, in 1877 Jacob F. Schoellkopf told his wife, "Momma, I bought the ditch" for $71,000.[23]* Thereon, in 1881 he built Niagara's first hydroelectric generating station. With an 86-foot head, his DC generator powered a few mills and lit sixteen streetlights and part of the Falls.[24]*

In 1886, Thomas Evershed[25] proposed to sell hundreds of mill sites along a large intake canal and twelve lateral canals and produce mechanical power from 238 500-hp waterwheels,[26] all connected to a common 2.5-mile tailrace tunnel.[27] For several years he tried in vain to enlist capital for construction.[28] Then a young lawyer proposed using the tunnel to feed hydroelectric turbines. Intrigued, Evershed's partners promised the young William Birch Rankine an equal share in their enterprise if he could find the funds.[29] It took Rankine almost two years to get a meeting with J. P. Morgan, who was interested but wary. Rankine wisely asked Morgan who should lead the enterprise, to which he replied that if Rankine could get Edward Dean Adams, of *the* famous family from Boston, Morgan would back him. Rankine did, and Morgan committed the first $400,000 (about $12 million in 2020 funds). With that as a starter, in 1889 Rankine brought $26.3 million (equivalent to about $800 million today) into the Niagara Falls Power Company (NFPC), from the Rothschild, Vanderbilt, and Astor families. "Thus began one of the most audacious ventures in American business history."[30]

NFPC's power plant would open with two generators of 5,500 hp[31] each—*three times* the existing market for electric power within the two-mile radius it could be transmitted.[32] The plant, however, was being built with ten generators, or *fifteen times* more power than the existing local market needed.[33]*

To find a way to transmit the power to distant markets, NFPC convened an international commission[34]* headed by Sir William Thomson, later known as Lord Kelvin, one of the most famous scientists alive.[35] The commission offered a prize of $100,000 for anyone who could transmit electricity at least 25 miles—*twelve times* the then-current limit, and the distance to the big power market at Buffalo, New York.[36]* Without waiting for a solution to the transmission problem, NFPC began blasting out nine miles of canals and tunnels. It was an enormous gamble.[37]

To transmit the power, ropes and pulleys were proposed, but friction would grind such a system to a halt or start a fire. George Westinghouse, who made his fortune from air brakes on trains, thought power might be transmitted through pipes using compressed air.[38] Thomas Edison insisted he could command DC power to flow through a cable laid along the bottom of the river. Unconvinced, the commission awarded four prizes for worthy ideas, but no grand prize for a believable solution.[39] They were convinced, however, to investigate the alternating current (AC) systems being developed.[40]*

New York City was the biggest market in the world for electricity. Its huge population and wealth attracted all the leading lights of invention and investment. But Edison's vaulting ambition was constrained by his low-voltage DC technology. Distance was death to DC, and despite expensive tree-trunk-sized copper power lines, Edison couldn't get DC to flow farther than two miles. His luminous jewel at Pearl Street was imprisoned in its DC shell, making it impossible to conquer the country with DC.

Direct current (DC) is constant in direction and voltage—graphically, a straight line. Alternating current (AC) alternates in voltage and direction—graphically, a sine wave:

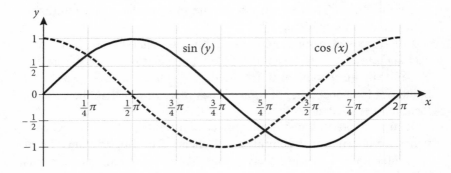

AC's advantage for the transmission of electrical power can be explained by analogy. Imagine a pipe two inches in diameter and three feet long filled with white golf balls. Paint another golf ball red. When you push the red ball into the pipe on the left end, a white ball immediately pops out the right end. With only a little effort at the left end to insert the red ball, you transmit "power" when the white ball pops out of the right end. In contrast, DC is like inserting dozens of white balls in after the red ball in an effort to push the red ball all the way through the pipe. Nikola Tesla solved the problem of alternating current (positive-negative: the red ball being pushed back and forth) by using a second circuit 90 degrees out of phase with the first (cosine: dotted line, above) which combined to create usable one-direction rotation.[41]

AC could be transmitted long distances by stepping up the voltage with transformers and then stepping it back down to make it more convenient for domestic use. AC was also more efficient for motors, gaining for it the immense manufacturing market.

George Westinghouse was one of the first Americans to comprehend AC's advantages, and in 1885 he began buying up useful patents and hiring engineers to develop the technology.[42]* In 1886, Westinghouse invented the *transformer*, which increased the voltage of the current for easier long-distance transmission, and used another transformer to decrease it for easier local distribution and use.[43] Partly in response to Westinghouse's control of the AC patents, J. P. Morgan forced the merger of Thomas Edison's many companies and his biggest competitor, Thompson-Houston, to form General Electric (GE) in 1888 and forced out Edison and his executive team.

Edison and GE saw Westinghouse and AC as threats to their DC preeminence and launched the "War of the Currents" against AC as unsafe and Westinghouse as unethical. Sparked by a few fatal accidents, the flames were fanned by the press. The war included corporate espionage[44]* as well as GE and Edison encouraging the development of AC for criminal executions by electrocution, or, as they called it, being "Westinghoused." One of Edison's accomplices, Harold P. Brown, even challenged Westinghouse to a duel: each would subject himself to higher and higher voltages of DC and AC, respectively, until one dropped out willingly—or dead.[45]

On May 1, 1893, Thompson sent Adams a cable: "Trust you avoid gigantic mistake of adoption alternating current." That same day, the Columbian Exposition opened in Chicago, celebrating the 400th anniversary of Christopher Columbus's arrival in the New World.[46] Westinghouse beat the extortionary "electrical trust" headed by GE and Chicago Edison for the contract to light the exposition, by charging one-third the price.[47] Twenty-seven million people—almost half the population of the United States—came to be dazzled by the exposition's "White City," shimmering with 180,000 AC-powered Westinghouse lights and a fantasy land of thousands of never-before-seen electrical machines and amusements: neon signs, "telautograph" picture phones, monorail trains, escalators, battery-powered ferry boats, and radio transmission.[48]* Standing in as Westinghouse's second in his duel with Brown, Tesla defied reason and his critics to prove AC's safety by subjecting his glowing self to 100,000 volts of AC power,[49] *three hundred times* the voltage Brown had used to "Westinghouse" dogs.[50]

Everyone who saw it knew it, even if they didn't understand it: the exposition was electricity's coming out party, the start of the revolution, dawn on the edge of a *New* New World. The White City convinced NFPC to award the power transmission contract to Westinghouse.[51]* In August 1895, NFPC's Plant no. 1 opened for business.[52] The very first customer was the Morgan-backed Pittsburgh Reduction Company, which moved to Niagara to take advantage of the abundant and cheap power for its patented electrolytic process.[53] The company changed its name in 1897 to Aluminum Company of America (Alcoa). Privately, Alcoa made an unusual deal with the public utility: NFPC would never sell power to any other aluminum company.[54]*

In 1896, the switch was thrown, lighting up Buffalo with AC power from Niagara.[55] Distance was destroyed. AC turned fantasy to reality. By 1897, the previously nonexistent electrochemical industry, incubated in Niagara's ultra-hot electric furnaces, was devouring more current than all other customers combined.[56]

By 1900, all ten generators at Plant no. 1 were pumping out power, and a second plant was being built with eleven more.[57] By 1905, the consumption of electric power in the United States had grown one-hundredfold.[58] But by 1907, only 8% of American homes had electricity. It was still a luxury for a few.[59]

Many were unhappy with the despoliation of Niagara Falls, so a campaign was begun in 1869 to buy the Falls for a public park. The Falls were hugely popular but also a tourist-swindling "den of iniquity" and an unsightly jumble of billboards and "sordid industry" belching noxious fumes and soot.[60] Proponents of the park included Frederick Church, painter; Frederick Law Olmsted, designer of New York City's Central Park and Chicago's fairgrounds; and Joseph Pulitzer, publisher. In 1885, both New York and Ontario finally approved the legislation for the twin parks.[61]*

Creating Canada's 35-mile-long park, complete with bridges, promenades, gardens, museums, restaurants, and "accommodations" for men and women, proved much more expensive than anticipated.[62]* To raise needed funds, the Parks Commission leased much of the waterfront land and water rights for electric power generation.[63]*

Meanwhile, popular opinion in Canada was turning against electric power monopolies and their predatory pricing practices. In the 1905 provincial elections, Conservative Party candidates ran on the platform "waterpower of Niagara should be as free as the air." They won and in 1906 appointed Adam Beck as chair of the newly created Hydro-Electric Power Commission, or "Hydro," the world's first *publicly owned* electric power utility. Then began a war of words and propaganda between the *public power* proponents and those of *private power*.[64]* Hydro won, and in 1910 it completed an AC transmission line from Niagara to Toronto to compete with the privately owned Toronto Electric Light Company. By 1914, Ontario had the lowest electric rates in the world, and Adam Beck was knighted for his services to the Commonwealth of Canada.[65]*

The fierce competition in Canada—between multiple private power companies and a public power company—drove daring and dangerous innovations.

More power was needed, so two more power plants were built on the Canadian side of Niagara Falls. The Ontario Power Company plant opened in 1905. It used huge pipes to carry water from upstream of the Falls—1.2 miles along the top of the gorge—to downstream of the Falls, where it fell through penstocks to the turbines at the base of the cliff. When investors from Toronto wanted to build the third power plant on the Canadian side, they were faced with the problem of how to get around the other power plants, canals, and tunnels.[66] A quiet little man engineered a radical solution.

Hugh Cooper engineered it and put his younger brother, Dexter, in charge of constructing it. The first thing Dexter had to build was a wing dam into the river, just above the falls, to divert the flow to the power-house intake.[67]* Dexter had trouble getting people to work on the wing dam because of the danger: if you fell into the river, you fell over Niagara Falls. To show that the work could be done safely, Dexter worked with the men. When it came time to dig the world's largest diameter hydraulic tunnel—a tunnel that would go *under* Niagara Falls—Dexter took the lead.

There were to be eleven generators, each sitting atop a 180-foot vertical shaft in one big pit excavated from the rock. At the bottom of each shaft would spin a turbine. The spillwater from the turbines would be combined into a single tailrace tunnel, 33 feet tall and 19 feet wide.[68]* The tail race would run 2,200 feet southwest and exit through the crushing curtain of water cascading over Horseshoe Falls.

Digging through the hard upper layer of limestone was not difficult for experienced miners, but the softer shale underneath was a concern. It was, after all, the softness of the shale, which eroded quicker than the limestone, that created the overhanging Falls. In other words, the roof of the tunnel was stronger than the walls holding it up.

On the work went, day after day. Eventually the builders knew they were close. Loading extra dynamite into the bore holes and connecting it to the fuse wire, they retreated into the tunnel. Pressing the plunger sent a spark down the wire, exploding the rock in a deafening blast.

When the smoke had cleared enough to breathe, Dexter led the way forward. Instead of the deathly silence of the tunnel, they heard a muffled roar. Soon the water was up to their hips. Unable to see what was under the water, they fetched a tiny rowboat, and Dexter climbed in, alone.

With a paddle in one hand, a torch in the other, and a long rope tied to the stern, Dexter inched forward. There was light at the end of the tunnel. Dim, diffuse light. Bumping and scraping forward, he reached the tunnel's end. There before him was a netherworld, not air, not water, but both: spray fuming from the wall of water that poured over the Falls and vaporized as it crashed onto the rocks. There he was, underneath the overhanging lip of limestone, under the hooves of four million horses galloping over Horseshoe Falls, striking the rock with primeval power.

Signaling with a jerk on the rope, he was pulled back into the tunnel. Trying to convey the enormity of what he had witnessed, Dexter said it was "the nearest thing to Hell I could imagine."[69]*

Only the tides could outpower Niagara, and other engineers were imagining how to generate electricity from the never-ceasing, never-freezing high tides of the Atlantic.

A two-pool tidal power project had been proposed in 1912 for Eastport's Half Moon Bay.[70] Later that same year, Wallace R. Trumbull, a propeller expert from St. John, New Brunswick, proposed building dams across the Petitcodiac and Memramcook estuaries between Hopewell and Moncton to bottle up a 40-foot tidal power head. Despite having some merit, the Trumbull plan got stuck in the mud of Moncton's "chocolate river."[71]*

In 1915, Ralph Clarkson, an engineering professor at Acadia University in Wolfville, Nova Scotia, planned to use the horizontal flow of the fast tidal currents (nine knots; ten mph) in the Minas Passage to pump water 300 feet up to a storage reservoir on Cape Split. The water would then be allowed to flow back downhill through conventional turbines to generate electricity when it was most needed and would be most valuable. Clarkson successfully tested a prototype, but with the finish line in sight, his hopes, dreams, and equipment were destroyed by fire.[72]

With the demand for electric power booming, surely someone would soon turn tidal dreams into deeds.

LESSONS FROM WORLD WAR I

Safety in numbers is often a good strategy, but during the steamy summers in the early 20th century, America's teeming cities were incubators for contagious diseases. If you could afford it, you got out of the city. The farther away, the cooler the climate, and the fewer the people, the better. Campobello was just what the doctor ordered.[73]*

For the Roosevelts, Campobello was also their escape from the worldly cares of business and where they enjoyed the "strenuous life" of vigorous outdoor recreation epitomized by their famous cousin Theodore.[74]* Campobello was also where Franklin chose to recuperate from an emergency appendectomy in 1915.[75]

The following summer, a polio epidemic swept the eastern United States, sickening many, including the daughter of the Roosevelts' coachman. Knowing it was contagious and deadly, but not what caused it or how it was transmitted, the populace was scared. There were reports of villagers turning away cars with out-of-state license plates. Worried about potential disease carriers, Franklin wrote to Eleanor at Campobello, "The infantile paralysis in New York is appalling. Please kill all the flies. . . . I think it is really important."[76]

Franklin had arranged for a flotilla of 21 navy ships to assemble in Eastport for the 1916 Independence Day celebration, but he and they had to leave in late June because of hostilities in Mexico. Eleanor and the children would stay until after the summertime disease season had passed. It was a wise decision: That summer nearly 2,500 Americans died of polio.[77]

To avoid infection on public transportation, Franklin planned to bring his family home in early September aboard a navy ship. But his boss, Secretary of War Josephus Daniels (and the owner of the Raleigh, North Carolina, *News and Observer*), advised Franklin to wait until after the midterm elections to avoid any unwanted attention from the press that such private use of government property might produce. With the elections over, Franklin chauffeured his family home aboard the 240-foot presidential yacht USS *Dolphin*.[78]

Out with the old, in with the new might have been Franklin Roosevelt's motto as the assistant secretary of the navy as he prepared for America's potential entry into the expanding war in Europe. Appointed to the Naval Consultant Board of inventors and industrialists, he got to know Thomas Edison and executives from GE, RCA, Westinghouse, and J. P. Morgan. Through it and the Lighthouse Board, Franklin became versed in radio and the other new electric technologies that would soon transform communications and war from plodding to lightning.[79]*

Franklin recognized that ship propulsion and gun technology was also changing rapidly; steam turbine power had replaced sails, and steel had long ago replaced wood. While wooden sailing ships were obsolete as fighting machines, they still had use to the navy as floating barracks. With their armaments and masts removed, they could house hundreds of sailors while in port. The accommodations were dark and dank but more cost efficient than erecting new buildings on land. Eventually, however, the leaky old tubs had to be got rid of, and Eastport was happy to help.

Although it may seem strange to tow an old hulk from Virginia to Maine just to be demolished, Eastport had a unique advantage: its eighteen-foot tides were twice those of Norfolk, New York, or Boston and could therefore get deep-draft ships out of the water at low tide. Some kindling, kerosene, and a match could then do the hard work of separating several hundred tons of valuable brass, copper, and iron from the wood. As a bonus, the blaze gave Eastport a spectacular fireworks display. The next morning there would be nearly nothing left of the hulks but hunks of metal, easily picked up from the beach.[80]* Among the spectators at the burning of the USS [Benjamin] *Franklin* on October 2, 1916, were Franklin Roosevelt and his family.[81]*

The Austrian-Serbian conflict exploded into World War I when Germany invaded its neighbors and mutual-defense treaties drew other countries, including the British Empire, into the conflict on August 4, 1914.

There were two questions for Americans: *Should we fight? On which side?* In our immigrant nation, Chicago and New York ranked as some of the largest German-speaking cities in the world. But because America's deepest cultural connections were with England, the two previous questions resolved into one: *When?*[82]

Internationalists wanted the United States to enter the war before it was too late; isolationists wanted to wait. The die was cast on April 6, 1917, when the United States entered the war on the Anglo-French side in response to the sinking of US merchant ships by German submarines and the revelation that Germany had been trying to incite Mexico to declare war on the United States.

Shortly after the United States entered the war, President Wilson asked the states to set up civilian defense councils to sniff out potential spies. Samuel Insull, an English immigrant to Chicago, focused the Illinois Council on defeating Germany rather than on harassing German immigrants. As a leader in the electric utility business, he used his executive contacts and skills to draw in leaders from industry, labor, churches, and politics to create an enormously effective patriotic force. "The brilliance of planning and organization" was "breathtaking," said historian Forrest McDonald; "the ruthless precision of its execution" was "terrifying."[83]

What did these leaders do? They created the Committee on Public Information; in other words, they created propaganda. With it they marshalled 380,000 volunteers, including 2,000 "Four-Minute Men," who gave rousing prescribed speeches at nearly every movie house, reaching 700,000 people per week. They ran scrap metal drives, collecting millions of pounds. And they sold a staggering $1.3 billion of US government war bonds.[84]*

Insull wielded so much power that when the price of coal tripled, he threatened to take over the coal mines to force the price back down. The threat to nationalize the mines was made on the spur of the moment, but within hours he had the backing of President Wilson. Instead, the Federal Fuel Administration was created, and the price of coal fell by half.[85]

The council then uncorked the bottleneck in producing armaments in the eastern states; coordinated and ramped up production in the Midwest, beyond the reach of coastal attacks and where there were fewer power shortages (and where unions were less strong); and, with its peers and the War Production Boards (which forced the interconnection of parochial utilities to increase production by more efficient use of existing capacity), overwhelmed Germany's productive capacity.[86]

When the war was won, the utility executives who had led the war charge on the home front took stock of what they had achieved and applied the lessons to business.

"Grab the children and beat it into the woods," Franklin told Eleanor in 1917, if a trigger-happy German naval officer decided to shell Campobello. Such a "fool and unexpected thing," he assured her, was "500 to 1 against the possibility," but the location of the assistant secretary of the navy's summer home was public knowledge, and a shot, regardless of what it hit, would strike a blow to the pride of both Canada and the United States. Franklin and his family were tempting targets.[87]*

When the United States entered the war in 1917, Franklin, age 31, asked President Wilson to have him transferred to active duty. Wilson refused, telling him he was more valuable to his country at his desk than on the front lines, but Franklin's yacht, *Half Moon II*, was drafted for active duty in the navy's coastal patrol.[88]

The last American serviceman killed in World War I was a Passamaquoddy Indian. Moses Neptune, son of Chief William Neptune, served in Company I, 103rd Infantry. He was killed on November 11, 1918, in the Argonne. Fellow tribesman Charles Lola, also killed in the war, was posthumously awarded the *Croix de Guerre* by the French government in recognition of his remarkable courage on the front lines defending liberty. Samuel Dana, George Stevens, Henry Sockbeson, and David Sopiel were wounded. All told, twenty-five Passamaquoddy braves volunteered to fight for the United States. None of their sacrifices were recognized by the US government, for none were recognized as US citizens.[89]

World War I demonstrated America's need for more electric power and the risk of depending on foreign sources for strategic materials like Brazilian rubber for tires or nitrates from Surinam or Chile for explosives.[90]

The National Defense Act of 1916 empowered President Wilson and the Army Corps of Engineers to construct munitions plants and their requisite electric power plants. During wartime the plants would produce explosives, and during peacetime they would produce fertilizer. This was the electrochemical equivalent of turning swords into plowshares. President Wilson selected Muscle Shoals on the Tennessee River at Florence, Alabama, as the site of the plants—a site President Theodore Roosevelt had preserved as a public resource a dozen years before. In February 1918, the army began constructing the Wilson Dam, thereby deliberately putting the federal government into the power-generating and -transmission businesses for the first time.[91]*

Without any meaningful experience in building hydroelectric dams, but with the most experienced hydroelectric engineer in the world in its command—in France—the US Army ordered its man to Alabama. He was, of course, Hugh Cooper.

Hugh Cooper had volunteered for the army at age 51 when the United States entered World War I in April 1917. He was commissioned a major in May and by July was in France designing port facilities, having been promoted to lieutenant colonel. The chief of army engineers, Maj. Gen. William M. Black (born just a few miles from McCall Ferry, Pennsylvania, where Hugh and Dexter had built another pioneering dam),[92] transferred Hugh to Muscle Shoals in March 1918. Hugh made a careful study of the problems the army was experiencing; then he redesigned the project and put Dexter in charge of construction. At 260,000 hp, it would be the world's most powerful plant, 100,000 hp (60%) bigger than Keokuk.[93]

In May, Hugh was promoted to full colonel but then transferred back to France, without explanation.[94] Apparently some in the army didn't appreciate Hugh as much as their chief did. Dexter stayed on.

Americans fought in World War I for less than two years. Construction of the Muscle Shoals Dam took seven years. The fight over the dam would last much longer.

POWER AND PROSPERITY

Roscoe was thinking about the future on New Year's Day, 1919, when he published an exhortation for the fishing industry to cooperate:

> It is the consensus of opinion that the United States faces a period of the greatest prosperity in its history. Reasons for this business optimism are based on the facts that of all the nations in the war the United States has been least drained of its resources. . . . Great things will be required of all the people and great will be their reward. The commercial world awaits our advances expectantly. Peace will bring us greater victories than ever. Every industry will feel its effects, none more so than the production and distribution of fish.[95]

Roscoe's prediction of national prosperity would come true, but by September he was turning up his nose at the smell of Eastport's rotting fishing industry: "The work of the Chamber of Commerce in securing

a new all-the-year-around industry for Eastport is such as to deserve
the unreserved commendation of the entire city. There has never been
a time when it was more needed than right now, when the sardine industry is proving once more how utterly unsuited it is to be the main
dependence of a community like this."[96]

With fish prices dropping 25%, and fishermen and canners stopping
operations, Roscoe grew concerned. His fortunes were now even more
closely tied to the fishing industry, as he had married Vera Gordon, the
daughter of a fish merchant in St. John, New Brunswick.[97] So he ruminated on the other possibilities: "[C]onsidering our command of water
routes and access to the Canadian [railroad] lines through St. Andrews,
by which we may ship to Chicago as cheaply as Boston. . . . Why not
make a thoro [sic] study of these matters, and invest as much money in
building up our industries as we have casually invested for instance, in
a single Fourth of July celebration?"[98]

Roscoe was thinking like an entrepreneur. He was proposing what
we might now call a SWOT analysis, a business development planning
process that considers strengths, weaknesses, opportunities, and threats.
He correctly identified the great harbor as one of Eastport's strengths,
but he understood its distance from major markets was a weakness.
Eastport's opportunity was still in darkness. It would soon come to
light, as would the threat.

The dawn of a new industry for Eastport glimmered in the same issue of the *Sentinel* just quoted: Lubec was planning to capture "waste
water" falling 30 feet over two dams to create electricity.[99] When Lubec's power plant opened in February 1921, Roscoe illuminated the
seminal issue:

> Profiting by its experience in municipal ownership of its water system,
> Lubec, known as one of the liveliest towns in Washington County, has
> entered the hydroelectric field under same plan. Without waiting for some
> company to come along and finance a development, the town fathers,
> backed by public sentiment, have now expended about $50,000 in har-
> nessing the Orange River at Whiting, and will, this week, see the formal
> opening of a plant capable of generating sufficient current for all lighting
> and power purposes in the town, turning motors in manufacturing plants
> and supplanting the old steam power used formerly in pumping its water
> at a savings of roughly $3,000 per year.[100]

Lubec's system generated 210 hp or 100 kW of electricity. The electrical equipment came from Westinghouse; the dam and powerhouse were built with local materials and labor. "All but a small part of the money necessary for the plant was raised by a bond issue, the bonds going almost exclusively into the hands of Lubec and Whiting people, so that the residents have a double interest in their plant—that of taxpayers, whose outlay will be reduced eventually," and as investors earning dividends.[101]

In December 1921, Eastport's *privately owned* Electric Light Company opened its second hydroelectric generating station on the Pennamaquan River.[102]* The 20-foot-high dam and GE generator doubled their output, making full-day electric service available for the first time. *What will the rates be?*, the *Sentinel* wondered, for the price of power in Maine ranged from $0.0025 to $0.12 per kWh hour.[103] In other words, some people were paying 48 times more than others.

Former President Theodore Roosevelt, who died on January 6, 1919,[104]* had predicted America was on the verge of momentous changes being sparked by hydroelectric power and its transmission to wherever it was wanted. But in his State of the Union address of 1909, he had warned that while the industry was "still in its infancy," an "astonishing consolidation" had already occurred: thirteen corporations, led by GE and Westinghouse, already controlled more than one-third of the hydroelectric power then in use. The electric industry was following the oil industry: great monopolies were developing, eradicating competition, and dictating the price citizens paid for electricity. "I esteem it my duty," President Roosevelt declared, "to use every endeavor to prevent this growing monopoly, the most threatening that has ever appeared, from being fastened upon the people of this nation."[105]

Accordingly, Teddy "the trust-buster" Roosevelt issued executive orders to restrain the monopolies by prohibiting private development of 1.5 million acres of federal land along sixteen rivers in half a dozen states in order to preserve for the people the most promising hydroelectric power sites. Roosevelt later justified his controversial executive orders, arguing that the president "is the steward of the people, and that the proper attitude for him to take is that he is bound to assume that he has the legal right to do whatever the needs of the people demand, unless the Constitution or laws explicitly forbid him to do it."[106]

In the tradition of "free land" and western expansion, until 1903 dam permits in North America had been given away on a "first-come, first-served" and *perpetual* basis by individual colonies and then states. In Maine, Percival Baxter[107] complained that "for years Maine had given away franchises, flowage rights, dam rights, eminent domain rights . . . and . . . everything else that belonged to the people of this State." In an effort to regain control of the state's assets, in 1909 Baxter and Maine's governor Bert M. Fernald[108] passed a law that restricted the giveaways. The Fernald Law was also intended to prevent the formation of large power monopolies and Maine's becoming a "power colony of Massachusetts" by prohibiting the export of hydroelectric power from Maine. The export ban, it was hoped, would attract power-hungry industries into Maine.[109]

As the Fernald Law's potentially unconstitutional restriction on interstate commerce demonstrated, electricity exponentially increased the scale and scope of riparian rights issues and created the need for the Federal Water Power Act of 1920.[110] The act gave the newly created Federal Power Commission (FPC) regulatory control over rivers and the licensing of dams (but *not* rates).[111] When considering a license, the FPC was required to give "equal consideration" to power development, energy conservation, wildlife and habitat protection, and recreational opportunities. Licenses could not exceed 50 years.[112]* In practice, the regulatory process was problematic and subject to manipulation.

An emergency appendectomy is bad enough if you are a youngster, but at age 41, the body doesn't recover as quickly. It would be a long convalescence for Dexter, a forced vacation, the first he had ever taken.[113] But being at Campobello again in 1919 with Gertrude would be like reliving their honeymoon.

The beautiful scenery and company notwithstanding, Dexter wasn't used to having time on his hands. He shifted uncomfortably in his chair; his nervous energy was restricted by his still-sore wound. As he watched Passamaquoddy Bay's great tides rise and fall, he began to wonder: *How high are those tides? How fast are those currents? How much water flows through those tidal straits?*

To satisfy his curiosity, he put a tide meter on the dock.[114] As Gertrude recalled, "We used to get up during the middle of the night to go to the dock. . . . I would hold a lantern while he read the gauge."

He made graphs of the tides and calculated their varying range and volume.[115] Then, using engineering formulas he knew by heart, he estimated their potential energy.

How many kW of electricity could the tides of Passamaquoddy Bay generate in a year—every year—forever, with nary a gallon of oil or pound of coal?

The numbers were staggeringly big. *Somewhere in the billions.*

Here was an enormous opportunity.

"As his interest grew," Gertrude recalled, "so did his conviction that, if the power created by the tides could be harnessed and utilized for electric power, it could be the salvation of that corner of New England, as well as being beneficial to their Canadian neighbors. He had been impressed by the quality of the people—as rugged as their coastline— and dismayed by the deterioration of its chief industries."

While Maine's chief industries, logging and fishing, were declining, its nascent electric industry was growing into a few local monopolies controlled by a few men.

The 1920s came in with a roar. The national economy was booming, and despite Prohibition's starting on January 1, there was still plenty of imbibing behind law-abiding facades on Main Street. Business was also up for Roscoe. As factories switched from wartime production to consumer products, Roscoe's newspaper swelled with display advertising from retailers eager to profit from the pent-up demand. His newest and biggest advertiser of the year, however, was not a consumer products company but a service company, Central Maine Power (CMP),[116]* which in a full-page ad exhorted readers:

> **DEVELOP IT.**
> **The Whole Future of Maine is Bound up in Water Power Development.**
> With plenty of developed water power, industries and businesses will come—and with them prosperity for us all.[117]

Full-page ads in the *Sentinel* were rare, and running full-page ads four weeks in a row was unprecedented.[118] CMP's ad blitz continued in 1921, with smaller ads in eighteen of the *Sentinel*'s weekly issues, adding up again to the company's being one of the largest advertisers of the year.[119]

CMP's ads were not advertising its service per se (it was a monopoly and had no need to compete for customers, and it didn't even serve Eastport), but they promoted the sale of its corporate stock to consumers. Purchasing the stock was the patriotic thing to do. The company's "7 percent Preferred stock" was "SOLID, SAFE, SECURE" and would help develop Maine. Priced at $107 per share, "you get $7 per year per share" in dividends (actually a 6.5% return—but who's counting?), and "You can buy it for as little as $10 down and $10 a month."[120]

CMP's ad campaign is a case study in consumer marketing techniques. Its ads promoted the idea that ownership of stock in a *public utility* (a private company) was synonymous with *public ownership*, when in fact the 5,000 people who had bought the stock[121] represented only one-quarter of 1% of Maine's population in 1920, and the majority of the stock was owned by only a handful of investors. CMP also promoted the idea that more hydroelectric power plants should be "developed" (i.e., construction completed) before "new industries and businesses will come" (i.e., after construction).[122]* Then it introduced scarcity: "Suppose an important industry wanted to come to Maine . . . and the power companies said, 'You'll have to go to some other state— we haven't the power.'"[123]

Then CMP countered alternatives: "Where shall I invest my money— in some other state, to build that state, or at home to build Maine?"[124] In effect, the company was asserting *the chicken must come before the golden egg* and *invest in our private company now before some other state steals Maine's eggs.* Putting this idea in wallet terms, CMP asked, "Are You Interested in Lower Taxes?" Knowing the answer, CMP informed readers, "Taxes in Maine can be brought down by attracting to the state enough new industries," and CMP stock was "[e]xempt from Taxation in Maine . . . and from the Federal Income Tax."[125]

Don't wait, CMP told potential investors, *this is a limited-time opportunity.* Maine needed to be "ready with power when the first big demand comes," and it could "get its full share of the industries that must move where water power is available, IF MAINE PEOPLE WILL INVEST NOW IN MAINE TO BUILD MAINE."

In the years to come, CMP would charge that the "build it and they will come" strategy was reckless. But for the moment, its priority was making money, not electricity.

With his paper and staff swelling, Roscoe moved the *Sentinel*, now the oldest paper in Maine, to larger offices at 93 Water Street. The expansion included a larger printing press and a Linotype automatic typesetter, and the paper completed the switch in power from hand-cranked to steam to kerosene to gasoline to electricity. Somewhat self-consciously, in his editorial of January 5, 1921, Roscoe wrote, "The improvement . . . may seem to many to be unwarranted by the conditions prevailing here," but his bold moves were motivated by his "confidence in the future of Eastport . . . [and] by doing so we shall play some part in helping to cause the future we believe in."[126]

Dexter's fascination with tidal power became an obsession. Back home in Buffalo, Dexter was no longer only working for his older brother Hugh but had set up his own company. During slack periods in the work, Dexter and his staff began drafting plans for how they would make power, and money, from the tides.[127]

In 1920, when the engineering plans were sufficiently developed, Dexter traveled to New York to get prices on machinery. He left the plans for the tidal power project with his engineering friends at the equipment company to work on the price quotes and returned to his hotel. There he received a message that the company's president wanted to see him, so he returned right away. The president inquired, "How do you plan to finance such a project?" Dexter replied, "I have not yet progressed that far."

"Here he was contradicted," Gertrude recalled. The president "assured him" that he and a couple of colleagues "would finance the investigations." The offer must have knocked Dexter onto his heels. "I believe this prompt offer was a 'first' in engineering history," said Gertrude. "Usually engineers have to beg for such financing."[128]

Once over the shock, Dexter must have been elated. He, on his own, was going to repeat—no, *beat*—his brother's entrepreneurial success at Keokuk. Dexter didn't have to spend five years and all of his own money trying to raise funding, as his brother had. Dexter was being offered funding up front! *No risk. Start now!*

Dexter could start now on *his* project, which was *five times bigger* than Keokuk. No longer would he be a hired man in second place. At Passamaquoddy Bay, Dexter would be in command. He would be the *first* to harness the power of the moon to make electric light. To power industry. To save Maine. To change the world.

Energized, Dexter finished up his work for his clients and went to Campobello.[129] What he needed next was a political champion. He didn't have to look far. The man with impeccable credentials lived next door: Gertrude's teenage heartthrob, Franklin Roosevelt, the assistant secretary of the US Navy with the presidential name.

Franklin loved Dexter's idea for generating tidal-hydroelectric power. Just as Passamaquoddy Indians had guided young Robert Bell to a suitable site for his tide mill, and just as Franklin had learned to navigate the bay from Captain Calder and Chief Tomah Joseph, now Franklin guided Dexter through the tidal waters of Passamaquoddy Bay, aided by navigational charts based in part on the work of Admiral Owen.[130]

Their emerging plan would link the formerly isolated islands—Campobello, Moose, and Sipayik—and their respective nations—the United States, Canada, and Passamaquoddy—for the benefit of all. The tidal power project would generate *economic* power for them all.

This was the beginning of the Quoddy Tidal Power Project. The currents were converging. The excitement was electrifying.

POLITICAL CALCULUS, OR DO VOLTS = VOTES?

Ohio governor James Cox was nominated for president on the 44th ballot at the Democratic National Convention in 1920. Exhausted by the ordeal, the weary delegates turned to who should be the vice-presidential candidate. When Cox chose the assistant secretary of the navy with the "magic name," the much-relieved delegates nominated Franklin Roosevelt by acclamation on July 6.[131] They hoped this Democratic, blue-blooded Roosevelt would have as much success as his recently deceased red-blooded Republican cousin, whose stellar trajectory Franklin was tracking.[132]

On July 25, 1920, Franklin took command of a navy destroyer, piloting it through Lubec Narrows to a hero's welcome in Eastport. Roscoe recorded the scene in the *Sentinel*:

Franklin D. Roosevelt, Democratic candidate for Vice President, arrived at Campobello at 7:10 p.m. Sunday night, on the destroyer *Hatfield*, after a record run from Boston . . . [at] an average of about 25 knots [28 miles] per hour in a lively sea.

The *Hatfield* . . . came in thru the Narrows at about 18 knots with Mr. Roosevelt himself at the wheel and the waves from her wake made a stirring sea for the small boats along the banks and wharves, one or two of which were thrown well out of water.[133]*

[After docking] . . . a trim 24-foot cat boat, the *Vireo*, purchased by the Assistant Secretary for his sons, was lowered over the side, greatly to the delight of the youngsters, who were on the beach, to greet their father. With them were Mrs. Roosevelt, Mrs. Sara Roosevelt, and a group of Campobello people, who left no doubt in the minds of anyone as to their pleasure at seeing Mr. Roosevelt return again after an absence of four or five years, bringing with him an international reputation as a naval administrator and with the promise of greater honors to come.[134]

The *New York Journal* cared not what Franklin did on his vacation at Campobello but how he got there. The paper complained that Franklin "had no right to use a fighting ship belonging to the American Navy and the American people . . . for commuting." It asked rhetorically, "[W]hat right has a man to burn up coal enough to heat the homes of fifty families all winter to carry his 165 pounds . . . to a summer resting place? Is a first-class ticket . . . not good enough?"[135]

The rebuke stung. Franklin resigned his post at the navy and campaigned full time for Cox. And, he realized, he would forever after need to keep a closer eye on the press.

Roscoe was conflicted about Franklin. Roscoe was running as a Republican in the Maine State Senate election, and Franklin was running as a Democrat in the national presidential election. The staunchly Republican *Sentinel* ran syndicated anti-Democratic diatribes with headlines like "REAL ROOSEVELT ABHORRED WILSON AND HIS WORKS—Yet Democratic Candidate for Vice Presidency Slanders Great Leader's Memory." Nonetheless, Roscoe, like everyone else in Eastport, wanted to

extend to him an adequate greeting . . . on a non-partisan basis, for it will be as a resident of Quoddy who has made good in the great field of national affairs and not as a partisan, that the ready welcome of this district will be extended to Mr. Roosevelt. . . . This city has seen much of him both as a boy and as a man. Respect for his ability and sterling qualities, and liking for his democratic and attractive personality are generally felt among those with whom he has come in contact. . . .

In many ways Mr. Roosevelt has been of assistance to Eastport and Passamaquoddy. The recent thoro [sic] survey of the boundary waters, the frequent calls made here by battleships and lesser craft, especially the rendezvous of the destroyer fleet here in 1916 are all traceable to his influence, and while Eastport and all the American towns around it are strongly Republican, if it were possible to vote separately for Vice-President, Mr. Roosevelt would probably outrun any candidate on either ticket in the national election.[136]

Accordingly, a contingent of Republicans, Democrats, and Passa-maquoddy chiefs called on Franklin at his home on Campobello and invited him to address the people of Eastport. Franklin accepted their invitation and let a secret out of the bag: had World War I not ended sooner than expected, Passamaquoddy Bay would have been made the home of the US Navy's Atlantic Fleet. The navy destroyer rendezvous of 1916 was to test the suitability of the bay for the fleet, and the survey work of the last two seasons was to be sure the bay hid no shallow-water surprises for deep-draft battleships. The plans were complete but not carried out. Eastport's economic salvation was still somewhere over the horizon. Roscoe ran the story at the top of page one but was forced to conclude, "Eastport . . . must content itself with thinking of what might have been."[137]

When Franklin stood on the steps of the Eastport library to address 3,000 well-wishers on July 28, 1920, he was not only the Democratic nominee for vice president of the United States but also someone who could decide Eastport's future. As the town fathers had rhetorically asked: *Wouldn't Passamaquoddy Bay be just as good for the fleet in peacetime as in war?*[138]

Franklin told the crowd he was "mighty glad" to be back after several years' absence and said, "No matter what happens next November, I shall come back every year, even if it means violating an old precedent against the Executives of the Nation going outside the United States."

Then Franklin dropped another tantalizing bit of news. It was such a casual remark that it went unreported by Roscoe and was remembered only years later: there was a plan afoot to generate electricity from the power of the tides.[139] Franklin did not say he was referring to Dexter Cooper's plan or where it would be. But anyone could have figured out that a tidal power project being promoted by the Democratic candidate for vice president of the United States would be in the United States and therefore right here in Eastport, home of America's highest tides.

Without realizing it, Roscoe Emery, the Reporter, had just experienced the first gravitational tug of what would become his "crusade."

The destinies of Franklin Roosevelt, Dexter Cooper, and Roscoe Emery—the Champion, the Engineer, and the Reporter—had converged. Their lifetimes of effort on the Quoddy Tidal Power Project had begun.

Thomas Edison in his laboratory in Menlo Park, New Jersey, in the early 1920s ▶

PART II

THE ELECTRIFYING 1920s

CHAPTER 4

International Action (1921–1924)

PUBLIC LIFE–PRIVATE PAIN

Democrats James Cox and Franklin Roosevelt lost the 1920 presidential election to Republicans Warren Harding and Calvin Coolidge. The Democrats' progressive policies were not popular following World War I; the nation was tired of "foreigners" and turned toward international isolation and racial segregation. Despite the loss, the election thrust Franklin into the national spotlight at age 38. Still four years younger than Theodore Roosevelt when he became president, Franklin's future looked bright.

Now Franklin needed a job. As assistant secretary of the navy, he had had lots of power, but not income, certainly not enough to pay for his lavish lifestyle. His mother paid for it. Luckily Franklin's friend, Van Lear Black, the owner of the *Baltimore Sun*, made Franklin vice president of a bank he owned, with a $25,000 annual salary, five times his navy pay.[1] Banking was not interesting to Franklin, but it put him in New York City with a bank roll for his political aspirations.

With characteristic energy, Franklin became involved in progressive causes and happily accepted the Boy Scouts' offer to head their New York Council. On July 28, 1921, he boarded a steamship for a Scout event at Bear Mountain, 20 miles north of Manhattan. The agenda included speeches, parades, demonstrations of knot-tying, a fried chicken dinner, and a bonfire.[2] The knot-tying particularly interested Franklin. Scouts could show off their handiwork by creating decorative neckerchief holders known as "slides," essentially tie pins. Neckerchiefs, or *bandanas*, were an exemplar of the Scout way: something they carried with them that was useful (sunshade, face mask, bandage, signal flag, etc.) and was made ornamental by printing colorful designs on it. Slides were made from all sorts of materials and ranged from simple and crude to fancy and finely crafted. One style popular with the New York troops was elaborately tied knots known as "boondoggles," a name that might harken back to Daniel Boone, who tied his tools onto his clothes so he could have his hands free to ford streams. In any case, the word's loopy letters perfectly represent the thing itself: an elaborate tangle of knots for a very simple purpose.[3]*

Franklin was given a boondoggle. Figuratively, it was tied to him for the rest of his life. At the end of the day, as Franklin boarded the steamship for the trip back to New York City, he was unaware that he carried something else that would affect the rest of his life.

Franklin could no longer requisition a navy ship to take him to his vacation home on Campobello. Instead, he traveled aboard the 141-foot yacht *Sabalo*, owned by Van Lear Black. Franklin enjoyed the cruise, much of it spent holding the ship's wheel and captivating his audience as captain and storyteller in chief.[4]

When he arrived on August 7, 1921, Eleanor and the children were there to meet him. After touring the yacht, the "chicks" were sent back to their nest while their parents were treated to an elegant dinner aboard, served by white-gloved stewards. That night the islanders performed the annual seining of the herring by torchlight, "biblical" in its beauty and symbolism. The Roosevelts and Blacks watched the firelight glimmer on the water; the stars overhead; and the red-shirted fishermen as they drew in their nets, bringing up a multitude of little fish that turned the water's surface into liquid silver and filled their dories with the seemingly inexhaustible resource that had sustained the island for centuries.[5]

Inspired by the previous night's spectacle, Franklin and his guests rose early the next morning to do some fishing of their own from the *Sabalo*'s motorized tender. Franklin, ever the energetic host, baited the hooks for his companions. It wasn't long before the warm August day, heat from the engine, and his own activity got Franklin perspiring. Then, as he dashed along the rail from bow to stern to tend to a guest, he fell overboard.

In a lifetime of boating, on all manner of boats, it was the first time Franklin had fallen overboard. It was a shock. Even in summer the water in the Bay of Fundy is rarely above 50°F, and the dousing immediately set him shivering. "I'd never felt anything as cold as that water," Franklin recalled. "I hardly went under, hardly wet my head, because I still had hold of the motor-tender, but the water was so cold it seemed paralyzing."[6]

Franklin resumed his strenuous fun the next morning. He began by taking Eleanor, Jimmy, and Elliot for a sail aboard *Vireo*. While on the bay, they discovered a brush fire on a small island. Sailing as close as they dared, they dropped anchor, splashed ashore, and for a frantic hour used evergreen boughs to beat out the flames.

Weary and begrimed, but pleased with their accomplishment, Franklin challenged Jimmy and Elliot to run with him the two miles to their favorite warm-water swimming spot, the brackish Lake Glensevern, on Campobello's southern shore. Again following tradition, Franklin and the boys then dashed across the cobbled spit and plunged into the frigid Bay of Fundy.

"I didn't feel the usual reaction," Franklin recalled, "the glow I'd expected." Despite the run back home to the big red cottage, he couldn't overcome the chill. When he reached home, he was too tired to dress. He finally made it upstairs to bed and had dinner served to him on a tray.

When he rose the next morning to go to the bathroom, his left leg buckled underneath him, and he had to drag himself across the hall to the toilet. His back and neck hurt. He had a fever of 102°F. Then his right leg refused to work.

Over the next few days, the paralysis spread to his arms. Soon he could hardly move his head. He lost nearly all muscular control. He had to have catheters inserted to pee and enemas done to remove solid waste. Eating was painful. The *only* thing he could do was talk.

After two terrifying weeks of wondering and failed remedies, Franklin's uncle, Frederic Delano, recommended Dr. Samuel Levine, a specialist from Harvard. Dr. Levine diagnosed the disease as infantile paralysis, otherwise known as poliomyelitis or polio. There was no known cause, treatment, or cure.[7]

Franklin Roosevelt left Campobello on September 15, 1921. He did not think he would ever return.

He was carried on a stretcher by five loyal friends: down from his bedroom; out the door of the big red cottage that had been his summer home for a dozen years; past his mother's house, where he had spent 27 blissful summers; down the long lawn where he and his children had played so often; to the dock from which he had launched so many pleasant excursions. There Captain Franklin Calder, the man who had taught him to sail and fish, became what he thought might be his final ferryman.

Passamaquoddy Bay's usually sparkling waters were Stygian, the crossing a torture, every wave jostling the helpless passenger. Franklin's entourage hoped he would be allowed to land. To calm officials' fears, he was accompanied by Dr. E. H. Bennett of Lubec and carried a certificate stating he was not a contagious threat to the public.[8]

To avoid the press, Eleanor arranged to deliver Franklin to a different dock in Eastport than had been planned and then secreted him to a private train. But the stretcher would not fit through the train's door. A baggage wagon was commandeered and rolled next to the train. With Franklin's bearers standing on the wagon, they tenderly lifted him through a window. With Franklin propped up on his stretcher so he could smile out the window, the press was allowed to get just a glimpse of him as the train pulled out of the station.[9]

The next day, September 16, 1921, the world learned something of Franklin Roosevelt's condition when the family issued a press release. Five days later, the *Sentinel* picked up the press release, breaking the news to Eastporters: "Roosevelt Has a Mild Paralysis."[10]

The truth was that Franklin Roosevelt, the rising star, had become a shooting star, a brief, bright light that turned to ash as it precipitated through the atmosphere.

THE RESERVATION

"CHIEF NEPTUNE STARTS RUMPUS," hooted the *Sentinel* on September 7, 1921, "SAYS WHITES SOLD LAND." Roscoe continued his clever but condescending report on what would become an important chapter in Passamaquoddy history. It was a story that Franklin

Roosevelt probably read, or had read to him, as he lay paralyzed in his bed on Campobello:

Former Governor of Pleasant Point Reservation Makes Heap Bad Medicines in Historic State Where Indians Sold Bogus Seal Noses Not So Very Long Ago

Chief William Neptune of the Passamaquoddy tribe of Indians, who is leading a contingent of braves, squaws and papooses at Plymouth, Mass., where they have been one of the leading attractions at the Tercentenary Celebrations of the landing of the Pilgrims, has been varying the monotony of selling bows and arrows to the white visitors by getting very much into the papers as will be noted from the extracts reprinted below.

It appears that the chief, who is an exceedingly intelligent and plausible redman, is a grandson of one of the former chiefs, now dead and gone, who owned a piece of land in the Indian Township . . . now owned by Sheriff Ivan Q. Tuell. Neptune says that land belongs to the Indians, and that no white man can acquire a valid title to it. To substantiate his claim he has made a thorough personal search of the records at Augusta and has now taken advantage of his trip to Massachusetts to carry his investigations to conclusion there. His method in pursuing his research, and his intelligent use of publicity show that no one need worry about the rights of the Quoddy Indian, so long as Neptune is on the job.[11]

For the tercentenary, Plymouth created grandiose monuments and an elaborate, if not historically accurate, "Pilgrim Spirit" pageant witnessed by 100,000 visitors.[12] Neptune's more important activities were described by a Boston newspaper:

Chief William Neptune of the Passamaquoddy Indians today asserted that the State of Maine has sold land granted to the tribe at Princeton, Maine, when the Pine Tree State was part of Massachusetts. Searching the records at the State House for copies of the deed of the land in 1795 to his forefathers, he declared he wanted only the return of the property and no money equivalent.

Fifty-two lots already have been taken from the tribe, Chief Neptune said, adding that they were sold to white men at $1 an acre. This seriously diminished the size of the Passamaquoddy Reservation, which now is about 8 square miles in area, he said.

At the office of Secretary of State Cook, the chief, accompanied by a number of men, women and children of his tribe, looked over the ancient parchment granting title to the property. To it were attached the names of nine leaders of his race, four of whom were members of the Neptune family. In that document it was specified that the land should be held by the Indians and their heirs forever. If he finds that the tribe has been unlawfully dispossessed of its property, legal proceedings will be started, the chief says.[13]

Roscoe called the tribe's state-appointed agent, Justin E. Gove, a storeowner in Eastport, for a statement. Gove asserted that "[n]o sale by the State of Maine of land in Indian township . . . has been made in the 29 years . . . [I have] been State agent for the tribe." Gove explained that 400 of the tribe's 500 members lived on the reservation at Perry and the rest in Indian Township, 40 miles distant. He explained that the revenues from timber sold by him on behalf of the tribe were deposited in their account at the state treasury, which had grown since he became the agent from $36,000 to $110,000. Mr. Gove "recalled that the Legislature granted many years ago a lease of a small portion of the township and that might be the basis for Chief Neptune's assertion."

The *Kennebec Journal* said the tribe's "ability to look after themselves in many matters, is unquestioned, but in many other things they are still simple and childlike and easily led. An idea may be started by idle talk or a mischief maker, and the Indian has a fancied wrong. It is probably so in this matter. Agent Gove, who has been in office for many years knows the Indians and their business better than any other person and his statement should be accepted." In other words, *trust the authorities*.

The *Boston Transcript* considered the issue more seriously, wondering, "Is there here a document in the nature of a warranty deed, under which Massachusetts would be bound morally, if not legally, to go to the assistance of the Indians were their property right invaded?" It concluded:

[T]hese Maine Indians were versed in the ways of the white man. They are wiser than a good many of the palefaces. They would not be likely, for example, to undertake to buy the South Stations as visitors from "up State" in New York have been known to purchase shares in the Grand Central Terminal. One of the Indian women who called at the White House yesterday wore a white man's spectacles to correct a defect of vision. And Chief Neptune, when he thinks the property rights of his tribe are invaded, begins a search for the title deeds just as do his white brothers in similar circumstances. Whether the Bay State is or is not under any obligation to see that the bargain is made with the Passamaquoddies many moons ago is not upset contrary to their wishes, they appear to be reasonably well able to look out for themselves, even in connection with a land deal.[14]

While the historical record is incomplete, it seems safe to assume that Neptune called at the White House in August 1921. Like Nova Scotia's governor in 1760, President Harding did not honor his country's agreement, did not recognize the Passamaquoddies' losses, and did nothing for the brave men who had fought their country's wars for them.

A greater president would.

THE INTERNATIONAL PLAN

***"What's This?*—Quoddy to Be Scene of Tremendous Power Development?"**

Roscoe reacted to his front-page story of January 30, 1924, with cautious optimism: "If this plan materializes it will solve the problem of employment about Quoddy forever and undoubtedly make this district the scene of tremendous industrial activity,—if it materializes."[15]

These were the Roaring Twenties, the *Great Gatsby*[16] days of coonskin coats, rumble seats, jazz, newly enfranchised women marching in the streets demanding social reforms, sheer-and-short flapper dresses, shockingly exuberant dancing styles, and lots of booze at Prohibition-be-damned speakeasies. Americans were shaking off the wartime austerity and drab, spending money like it was going out of style.

On Wall Street, the bulls were beating the bears into hibernation. Everything was looking up. A lucky and brave few could even enjoy the view looking down—from airplanes made of canvas, wood, and baling wire. Technology was turning dreams into realities. Commercial radio debuted in 1920 and was spreading rapidly.[17] The worldwide demand for power was insatiable. And regardless of how the power was generated, nearly everything new, in the new consumer economy—from toasters to trolleys to telephones—used electricity.[18]

In 1924, Brigadier General Gerald Tripp, chairman of Westinghouse Electric, published a petite but portentous book titled *Super-Power as* *an Aid to Progress.*[19] In it, Tripp attributed the rapid progress of civilization to increasing productivity through the control and application of power—first from the steam engine and its dissemination during the Industrial Revolution, and now through the even more profound transformation being conducted by electricity. The next wave, Tripp predicted, would be connecting everything into a "Super-Power" generation-transmission-distribution network for maximum efficiency.

Tripp extolled the benefits that would accrue to housewives, farmers, and businesses. He said the formation of this system was inevitable but could be accelerated or retarded by public policy: "constructive regulation" was needed, but the system "must be built by private capital and operated by private companies. Government ownership in any form would be fatal to their development and disastrous to the best interests of the public." He then likened Maine's Fernald Law to cutting the railroad lines at the state border, effectively ensuring its isolation.[20] With "Super-Power" available everywhere, the "population . . . tide which now flows city-ward" would be stemmed and rural communities revived. Tripp concluded his chapter "Water Power and Statesmanship" by praising George Westinghouse, who had

> inaugurated the alternating current system, by means of which large amounts of power can be transmitted over long distances; and though it was at first bitterly opposed by practically every electric interest, it is now in use all over the world. Without it, modern life would be impossible.
>
> But we now find need for someone who will build higher on the foundations laid by the great inventor. Another great organizer must now work for the power supply of future generations. He need not necessarily be an engineer, for the problems now to be solved are political, social, and economic. The technical and scientific work is done.[21]*

For Quoddy to be "the power supply of future generations," Dexter Cooper would have to do it himself, without help from Franklin Roosevelt, for Franklin was still struggling with the paralyzing effects of polio. Three years after being stricken with the disease, Franklin had recovered sufficiently to stand for short periods using his 20-pound leg braces,[22] and with his powerful upper body, built up through long and painful exercise, he could spin himself around in his "chair on wheels." But going anywhere—when many roads were unpaved, city streets had high curbs, buildings with elevators were exceptional, and trains had tiny toilets—was an ordeal. Besides the physical difficulties, Franklin would not have been his formerly charming self. He was depressed, and he retreated into his own isolationist world.

It did not appear Franklin would ever return to Campobello, the epicenter of his formerly active outdoor life, which had been ended by polio. Instead, over the next four years, Franklin, age 40, spent half of his time in Florida on an old houseboat named *Larooco* with his 25-year-old secretary and "other wife," Missy LeHand.[23] Meanwhile, Eleanor made a new and very different life for herself with her progressive female friends.

As biographer Doris Kearns Goodwin wrote, "Franklin had an almost mystical belief in the healing power of sun and water: 'Water got me into this fix,' he quipped (referring to his ice-cold dip in the Bay of Fundy), 'water will get me out again!' In swimming he found the most promising therapy of all, one allowing him to exercise his legs without the weight of gravity. . . . He floated in the warm waters lowered from the side of the boat by a contraption of his own contrivance, and bathed in the sun."[24] Eventually he would soak up enough energy to get back on his braced feet and move forward.

"Mr. Cooper . . . bought a house at Welshpool," the *Sentinel* announced on October 29, 1924, "and appears to be settling himself for an extended stay."[25]*

Dexter bought the old Flagg farmhouse as home for himself, his wife Gertrude, his daughters Nancy (age ten) and Elizabeth (seven), his son Dexter Jr. (three), a governess, and servants.[26] On this remote island the family would now live year-round so Dexter could work on his tidal power project full-time. Dexter tripled the size of his house—now called Flagstones[27]—to make it his home and corporate headquarters.

Over the following months and years on the water, Dexter developed an intimate knowledge of Passamaquoddy Bay and its tides—the knowledge that separates the local from the outsider, the amateur from the professional, the dilettante from the doer.

Like Admiral Owen, Dexter believed Campobello's geographic location and huge ice-free harbor were ideal; abundant low-cost electric power would attract industries, and surplus power could be transmitted to Boston, almost 300 miles away.[28]

Like his brother Hugh had done at Keokuk, Dexter was now living the life of an entrepreneur. He was fully committed to his hydroelectric power project. Dexter was not working for hire; he was working for equity. He wanted to own it.

Having built dams at Niagara, then Keokuk, then Muscle Shoals—each in turn the largest hydroelectric plant in the world—in 1924 Dexter Cooper filed an application with the US Federal Power Commission (FPC) for a preliminary permit to build the Quoddy Tidal Power Project. The commission's executive secretary responded, "[W]hile there is some reason to question the feasibility of such project. . . . [Quoddy's] magnitude and the public interest which is attached seem to justify its investigation." Due to the project's border-straddling location, the FPC queried the International Joint Commission before issuing a permit.[29]

The FPC permit and US patent Dexter also sought were similar to the charters France had granted to Pierre Dugua de Mons in 1604, England had granted to Captain David Owen in 1770, and Canada had granted to the Niagara Falls Power Company in 1892. They were grants of monopoly offered as an inducement for private entities to finance large and risky enterprises—whether exploring and settling the New World, trying to harness Niagara Falls, or building the world's first tidal-hydroelectric power plant.

Both patents and chartered monopolies are legal rights to exclude competitors. Patents prohibit potential competitors *from using your invention*. Competitors, however, are free to "invent a better mouse-trap"—or light bulb, motor, or transmission system—as Tesla and Westinghouse did (fluorescent lights, AC motors, AC-DC transformers). And while US patents expire within 20 years, after which time the invention goes into the "public domain," geographic monopolies *exclude all competitors—perpetually.*

That same week, the Bangor Hydro-Electric Company was organized from a group of local monopolies.[30] They also intended to generate and sell electricity—*exclusively* and *perpetually.*

Local reaction to Dexter's tidal power plan was mixed. In Roscoe's words:

[T]he people of Quoddy . . . have had it none too easy down there during the past decade, for the sardine business on which they nearly all depend had been steadily failing, and anything that promised industrial development looked good to them. But while they wished ardently to believe, they found it hard to do so. The distances to be covered by the dams was in many cases nearly a mile in length, the water was very deep, ranging to 180 feet, there is a slashing tidal current, and plenty of sweep for the wind and sea in stormy weather. It didn't look possible to build dams there, and the general attitude was one of unbelief that the thing could or would be done. There were frequent references to the "boom" that resulted from when the Rev. Mr. Jernegan erected his plant to extract gold from seawater,—a boom which collapsed when Jernegan decamped with many thousands of dollars belonging to investors.

But presently, under the friendly insistence of an Eastport newspaper man, the engineer began to talk a little for publication, and later, to deliver addresses at Eastport, Calais, Lubec, and Machias. Then, with more complete information, and closer acquaintance with the engineer, doubt began to vanish, and no western crowd of boosters ever got more enthusiastically behind a proposition than did the business men of Eastport and in fact all of Washington County, behind Cooper.[31]

The *Hartford Courant* expanded on the contrast between Jernegan and Cooper: "How different the present-day attempt! The man at the head of it today is a man of science and has seen no vision, save the vision of prosperity for Maine and the revival of New England. The process is no secret. There is nothing new about it. The same principles are applied here that have been applied since the earliest man made a wheel turn by the force of falling water. There is no belief necessary. As the scientist himself said, 'One needs only to turn a glass of water upside down to discover that it will fall.'" The *Courant* praised Cooper, saying his experience "is a guarantee of success of the present venture." With cutting humor, the *Courant* compared Cooper's inspiration with Jernegan's: "Instead of sending a vision . . . heaven sent appendicitis."[32]

Dexter Cooper expanded on the Boston Water Power Company's "perpetual power" plan, but instead of mechanical power, Dexter would sell electric power. The 100-plus square miles (70,400 acres) of Passamaquoddy Bay known as St. Andrew's Bay, most of which is in Canada, would be his "filling basin" or "upper pool." The 30-plus square miles (25,600 acres) of Cobscook Bay, which is entirely in the United States, would be his "receiving basin" or "lower pool." By opening and closing gates, he could maintain 16 to 28 feet of head. The volume of

water flowing through the turbines in a day would be roughly equal to
that flowing over Niagara Falls. The head and flow would combine to
generate 500,000 to 800,000 hp—or more than *all* the power plants at
Niagara. Dexter estimated construction cost at $75 to $100 million, and
the cost of power at three-quarters of one cent per kilowatt hour.[33]

Roscoe noted a few concerns: "The feasibility of building huge
dams without destroying the harbor in Passamaquoddy Bay is being studied by the experts. While Cooper proposes to install locks in
the dams to offset this objection, the engineers believe their success
would be doubtful due to the extremely rough weather continually
sweeping the harbor."[34]

Despite such concerns, Dexter's plan captured the fancy of the popular press, as shown in this article from the *National Geographic*:

> Because it has a short, distinctive name and an extraordinary reputation,
> Fundy is probably better known by name at least than any other bay in the
> world. . . . Twice every day a tremendous quantity of ocean water swishes
> up this 145-mile bay reaching depths of 30, 40, and 50 feet toward the
> head. Then it turns rather quickly and rushes out again. The thing has
> gripped the imagination of generations of juvenile geography students
> and has stamped "Fundy" into their memories. Now it is taking a delayed
> grip on technical imaginations, and engineers are wondering why these
> thousands of tons of rushing water cannot be made to turn power wheels.[35]

In time there would be thousands of articles in the popular press about
Quoddy, as Dexter's dream of harnessing the moon became an international story.[36]*

The significance of the story became clearer when engineers
Thomas Edison and Henry Ford, two of the most famous men alive in
1924, endorsed Dexter's plans. Edison said that harnessing the tides
and the sun was already practical and would happen when the cost
of developing the power was overtaken by the ever-increasing cost
of coal. Both Edison and Ford were rumored to be Dexter's financial
backers, though Dexter said such speculation was unfounded. But
since Ford had been trying to buy another hydroelectric power plant,
some doubted Dexter's denial.[37]

That left open the question: *Who was backing Dexter?*

CHAPTER 5

American Power (1925)

MAKING THE CASE

"Mr. Cooper is big and upstanding," Roscoe reported, "physically impressive,—a man in every sense of the word. Out-door life has left its impress, although he appears equally at home in the office, at the banquet table, or wherever his duties, social or professional, call him. He is a good 'mixer' easy to talk with, and always ready with a laugh or a yarn to match those of his associates of the moment. His outstanding characteristic, however, is sincerity, and it is this perhaps that leads other men to lend him their confidence. In fact, it has been this quality, more than any other, that has served to break down the barriers of unbelief and skepticism that first met the announcement of his plans."[1]

Dexter would need all of those smarts and charm to get all the approvals he needed to enclose 150 square miles of territory in two countries and build the world's first tidal-hydroelectric power plant. His list of "Essentials" was daunting:

1. Engineering approved by qualified reviewers
2. Preliminary funding for research and planning
3. Maine's approval: (a) House, (b) Senate, and (c) governor
4. US Federal Power Commissions preliminary permit
5. Canada's approval: (a) House, (b) Senate, and (c) governor-general
6. New Brunswick's approval: (a) Legislative Assembly and (b) premier
7. International Joint Commission approval: (a) fisheries, (b) engineering, and (c) funding

8. US Federal Power Commission construction permit
9. Financing for construction

Dexter had Essentials 1 and 2 at least partially completed, if not publicly disclosed. But to get Essential 3, Maine's approval, he would have
to "go public" with his plans and enter the world of politics.

A new electric device was being invented every day, from curling
irons to radios and refrigerators. These new appliances took advantage
of electricity's amazing capabilities: it could create light, heat, motion,
and signals. It enabled everything to go faster and with greater precision. With the flick of a switch, night turned into day, and time shifted
from dawn and dusk to hours, minutes, and seconds. Distance was destroyed as signals flashed over telegraph lines and radio waves, bringing news in a fraction of a second that would have taken days, weeks,
or months by train, ship, or foot.

Beyond the gee-whiz of new gizmos, electricity was changing
people's lives. With electricity you didn't have to fell a tree, cut it into
pieces, carry it home, and build a fire just to boil water. With electricity
you could light your house or barn without fear of the lantern's flame.

With electricity, you could run any kind of machinery, be it tiny or titanic, without heat and smoke or finger-eating belts or gears. With electricity, you could use power where it was wanted, not just where it was
produced. Electricity enabled precision. Electricity enabled control.
Electricity was liberating. It was freeing women from the servitude of
housework.[2] Electric lights chased away the shadows and the dangers
within. Electric lights improved productivity. They attracted customers.
They created demand—for more electricity.

Electricity changed our language. Excitement was electrifying. Mo-
tions were galvanized. People were bright. Ideas were light bulbs.

Electricity was the future, but as they say, "The future is here, it's
just not evenly distributed."[3] The cities got electricity first: New York,
of course, in 1882.[4] That's where the investors were who poured the
money into the companies that built the generating plants, the distribution networks, and the gizmos. And that's where the customers were in
sufficient numbers and density for the entrepreneurs to reach profitably.

Electrical wires were like rats: first they scurried from building to
building, then along walls, then inside them, until they had colonized

every conceivable space. The rat's nest of wires grew quickly in cities, but the lines crept out slowly to the country; there were few customers to feed on. In 1925, fewer than 10% of farms had electric service.[5] With fewer customers to support the distribution lines, rates were often higher in the country, and in some communities Bangor Hydro-Electric required customers to build their own pole lines and were then charged 18½ cents per kWh, "a rate even higher than their poles!"[6]

As the 1920s progressed, the divide between the big cities with bright lights and the electric-lightless sticks became as clear as day and night. In its rampant proliferation, electricity switched from a luxury to a necessity. Modern life became dependent on it. Economies of scale in power production and the need for ever more capital made it inevitable that urban electric companies expanded rapidly, gobbling up their smaller outlying competitors.[7] This situation led maverick Republican senator George Norris of Nebraska to warn in 1925, "Practically everything in the electrical world is controlled either directly or indirectly by a Gigantic Power Trust." The claim ignited a fierce debate about whether such a cabal existed and a joint resolution calling on the Federal Trade Commission to investigate the matter.[8]

Rural communities did not keep up with urban electrification and economic development. Many, like Downeast Maine, fell behind.

Real opportunities to get rich with electricity attracted brilliant and honest entrepreneurs as well as quacks and crooks. The trouble was telling them apart.

John A. Knowlton invented a method to generate electricity from the tides and had sixteen patents on the process. His promotional materials were full of facts about how much coal and oil New England consumed to generate its electric power, how much power was latent in the tides, and hyperbole. The "Knowlton Hydraulic Air Motor," he claimed, "practically will abolish the necessity of burning coal along the entire coast and for 500 miles inland." His invention was "ingeniously simple," according to the engineer he hired to critique it.

Imagine, if you will, a vertical piston that has a float in it attached to a lever, just like a reciprocating steam engine. When you fill the piston with water, it forces the float up, which lifts the lever, which turns a crank shaft. Simple. Now arrange many such contraptions along a crank shaft, add a few gears and a generator—and you get electricity.[9]*

Knowlton started selling stock in his Universal Tide Power Company to the public at $0.10 per share.[10] Thanks to wide but shallow press coverage and sales offices in seventeen cities, it climbed to $3.00 per share.[11] With these funds he built two power pools and a powerhouse at Saugus, Massachusetts, and generated electricity.

But Knowlton was more successful in monetizing moonbeams than in generating power, and in 1928 he was tried in federal court for fraud, having sold $968,679 (≈ $14.5 million in 2020 dollars) of stock to an incredible 12,000 investors. At the trial, the electricity produced was revealed to cost $1.08 per kilowatt hour (kWh)—at a time when the established wholesale rate was less than a penny. Ultimately, Knowlton was acquitted;[12] the jury decided he was a failure, not a fraud.

With Jernegan's golden seawater swindle and Knowlton's failure clouding the waters, legislators, regulators, and investors had to figure out who had the right idea. Was Dexter Cooper a fraudster, a deluded dreamer, or a true visionary with the skill to profitably harness the tides?

Convincing tidal power skeptics wasn't going to be easy. Dexter had to overcome Downeasters' legendary reticence and explain how the project would work and how it would benefit them, while cautioning them about the inherent risks and difficulties. At their weekly luncheon in early February 1925, Dexter gave the Eastport Rotary Club an overview:

> That is what is wanted here,—to start with the sympathy and cooperation of the people. Their material financial aid is not required, but it is desirable that they think about it, boost it and believe in it. . . . A proposition of this size cannot be financed by any individual,—not even Henry Ford could do it. . . . The financing in this case would probably be handled by a syndicate . . . [of] bond houses. . . . There will positively be no attempt to finance it by public subscription.
>
> No matter how practical or feasible a proposition may be, bankers who are to finance it must [be] SHOWN. For this reason, and also to supply the 15,000 to 20,000 horsepower needed during construction of the main project, it may be necessary to demonstrate it on a small scale at first. There is a chance to do this in Cobscook Bay, where a minor project similar in design to the big one can be put through at no very great expense to use as a testing and demonstration plant, with turbines, gates, etc., which can be

taken out and used in the large one after they have served their purpose. This and other preliminaries would require a year and a half or two years.

The final construction period would take four years at least and would require 5,000 men. . . . The question of housing the employees will be a serious one, as construction camps are expensive and in general detrimental to the men. It is to be hoped that Eastport and Lubec can absorb the newcomers in their vicinities. If the men and their families can come here and live it would be ideal for them and would lessen the cost and worry to the engineers in charge.[13]

Concluding his remarks, Dexter warned against speculative real estate investments: "Don't, I beg you, be premature. Many projects have fallen through when apparently on the verge of success. Don't make any move unless and until the thing is an absolute certainty."[14] Roscoe reported these words but didn't heed them. Half a dozen years before, Roscoe had started a real estate agency.[15] It was a good fit with his newspaper business because one of the largest expenses for a real estate agency is advertising, something he could do for free. To begin with, ads for properties listed by the Sentinel Agency were small and descriptive, but only three months after Dexter's warning, Roscoe was telling readers to "**Buy Eastport Real Estate Now.** This is the Time to Protect Yourself Against the Higher Prices and Higher Rents that are Coming Sooner than Anyone Supposes."[16]

Maine's Republican governor Ralph Brewster was impressed by Dexter and his plans and indicated an exception might be made to the Fernald Law prohibiting the export of hydroelectric power because it was tidal rather than "fluvial" (i.e., rivers), and the sea was federal territory, not state or private.[17]

Hoping to clear some obstacles, at the legislative hearing on March 13, 1925, Dexter declared he had "no desire to enter into competition" with the local power companies because he believed he could attract new industrial customers to Maine and export any surplus power. State senator Walter N. Miner thought the project would greatly benefit St. Croix River towns because it would give them deep water at their wharves all the time and beautify the scenery by keeping the tidal flats covered.[18] Senator Charles B. Carter[19] doubted Dexter could transmit power all the way to Boston.[20] Senator Frederic W. Hinckley[21] defended

Cooper's plan, saying it was "fraught with wonderful possibilities," and asked, "Why not give him a chance to try it?"[22]

To address concerns raised, amendments were added to Dexter's proposed charter, including the following: (1) No bonds could be issued without Maine Public Utilities Commission (MPUC) approval; (2) power allocation between the United States and Canada would be determined by the International Joint Commission; and (3) the power allocated to the United States could be exported from Maine only after Maine's needs had been met, as determined by MPUC.[23]

That last point should have relieved any concerns about the Fernald Law. But an open letter written by Maine's former governor and now US senator, Bert M. Fernald,[24]* identified the real issue behind the power export ban:

> The Fernald law was intended solely for fluvial power,—from lakes, streams and rivers. Many of us who supported it at one time are now beginning to see that it has accomplished no real good, and has held back rather than assisted in development of our resources. Perhaps, too, it was political rather than economic in origin and purpose. But whether it was wise or not, it has no possible application to tidal power, which in this case is so great that to limit it to this state would be simply to kill the project. Don't lose sight of that fact. Any man who opposes this change is not out to build up the state of Maine. He is out, for some motive of his own, to prevent Cooper from starting.[25]

Fernald concluded by comparing Cooper to another man who had threatened the powers that be when he crossed the Rubicon: "Cooper first came to Quoddy as a summer visitor. . . . He came, he saw, he got the idea, and we hope he will conquer."[26]

Fernald understood Maine's realpolitik and the consequences of the law that bore his name. While preventing the export of unused off-peak hydroelectric power, it inhibited the importing of industry because it kept the price of power high. While other New England states were interconnecting their electrical systems, by exerting its "home rule," Maine was turning itself into an economic island. And while the Maine Legislature was debating *fluvial versus tidal*, other, more motivated parties were boring through Maine's paper defenses and taking control of Maine's waterpower, as well as the state itself.[27]

The whirlpool sucked them down. Two of the Norwood boys were never seen again. They had been fishing and "were almost on the threshold of their Deer Island homes" when the wind died and their sloop slipped into the vortex. Only Robert survived his encounter with the Old Sow, the largest whirlpool in the Western Hemisphere.[28]*

Old Sow is penned in between Eastport and Deer Island Point. The whirlpool forms because the great tidal currents get agitated by the islands blocking Passamaquoddy Bay. Here the "funnel effect" speeds the tidal currents up to seven miles per hour as they squeeze through the narrow gap between the islands. The tidal current in the 400-foot deep Western Passage gets kicked upward by a 100-foot-high underwater hill; then it gets knocked sideways by another flood current coming in between Deer Island and Campobello, setting the water spinning clockwise like a pinwheel. No wonder the Old Sow squeals and grunts and snorts!

The width of the whirlpool, it was said, was a few hundred feet across, its surface declivity twelve feet deep! The Sow's roar could be heard at Grand Manan, thirteen miles distant.[29]

These figures may be fishermen's tales. Then again, the sea is powerful, and something may have happened once and not been seen since. Without exaggeration, Old Sow is a serious navigational hazard that has claimed about a dozen lives in Eastport's 250-year history.

In this maelstrom, Dexter proposed to build a dam.

It was April Fool's Day, 1925, when Governor Brewster called the legislature into joint convention and then condemned Cooper's charter, creating "a furor of astonishment."[30] Brewster objected to Cooper's charter because it allowed Dexter to export power and gave him ten years to complete the project. Brewster called the project "a phantom of the imagination," adding, "None of those responsible for the great electrical development within our state have considered this project as practicable." Trying to justify his apparent about-face, Brewster cited *Webster's Dictionary* to assert there was no difference between tidal-hydroelectric power and hydroelectric power. Finally, Brewster said Cooper should prove he could finance the project *before* the state approved his charter.[31]

Brewster's address showed he was listening to or being pressured by Maine's waterpower establishment, whose members didn't want Cooper to "conquer" their territory and were sowing doubt to prevent it.

[M]uch as his friends here have disliked to admit it, [Roscoe explained], he appears to have been merely dissembling his real attitude. Whether this concealment was for strategic purposes or to permit an attack to be made with greater force as a surprise, and to trick the friends of the measure into permitting dangerous delays in handling it, or whether it proceeded from uncertainty and indecision is a matter for each person to judge for himself. At any rate the Governor did himself no credit by the manner in which he treated the bill. There is nothing of his accustomed straight forwardness, courage and frankness in such a raid which savors rather of bushwhacking and sniping than of clean-cut fighting. . . .

So much for the Governor's attitude and methods. The message itself is of scarcely higher caliber, showing a reasoning force so feeble as to aston-ish anyone familiar with Mr. Brewster's usual facile ability to construct as well-developed, logical and cogent argument. . . . If it [power] is to be sold at all it must be sold where purchasers are willing to buy—in Massachu-setts. Honesty . . . would . . . not [have] attempted to camouflage it behind a rather school-boyish reference to *Webster's International Dictionary*.[32]

"A slashing attack was made" on Cooper's charter by Senator Carter in Maine's legislative debate on April 7, while "the defense was brilliantly conducted by Senator Hinckley. The vote, with two senators who favor it absent, was 19 to 9, a little better than two-thirds majority, which foreshadows success should it be necessary to pass the measure over the governor's veto." The bill was rushed to the House, where it passed 132 to 1. Then it sat on the governor's desk, making it appear he would veto it.[33]

"The situation was one which demanded a compromise," and one was proposed by Councilor Robert Peacock of Lubec, who suggested the charter be put before the people of Maine in a statewide referendum in six months (September 14, 1925). Having been absolved of respon-sibility, Governor Brewster signed the bill authorizing Cooper's charter *if* it were popularly approved in the referendum. The *Portland Sunday Telegram* laid bare Brewster's rationale: "Whatever the merits of the bill is a question that is debatable, [but] it was good politics to let the bill go along for the defeat of it would have made Washington County [turn] Democratic beyond a shadow of a doubt."[34]

VOX POPULI

Counting down to referendum day began. If the people of Maine approved Cooper's charter, then Dexter would have the exclusive right to export hydroelectric power from Maine, a popular endorsement, and positive momentum for winning Canada's approval. So off he went, with the *New York Times* following as he crisscrossed the state:

> Mr. Cooper knew less about methods of political campaigning than most politicians know about engineering. However, he set out on June 12, with an automobile and a chauffeur, and began to make speeches wherever he could collect a crowd. He travelled 6,000 miles, over all the kinds of roads there are in Maine—and some of them are not as good as they may be after Cooper's plan is carried out. Every county was visited; almost every township in the State. There was little or no organized opposition. What had to be overcome was popular ignorance or indifference.
>
> Mr. Cooper, though a man of commanding presence, is no orator. His usual method was to present his plan in simple language, understandable by any intelligent layman, and to illustrate it by drawing on a blackboard. If questions were asked, he answered them. Any who wished to stay after the meeting for a private interview was privileged to do so. He was patient, low-voiced, convincing.[35]

Washington County businessmen formed the "Cooper-Quoddy Publicity Organization" and published a flyer that described the project, its cost, and the jobs it would create, declaring, "The project has a specific interest to every business man, to every woman, to every voter in this State; it means the development of one of our resources. It does not take away, it produces, at no expense to the State or people."

Quoddy, the flyer noted, would be four times more powerful than Muscle Shoals. Hugh Cooper was quoted as saying, "In 1894 electric power in large quantity was first generated at Niagara Falls. Prior to that time no figures of either population or manufactures are given" by the Census, for "neither were especially important." But Niagara's population had grown to 51,000 by 1920, and the value of the manufactured products to $89 million. The flyer quoted former governor Percival P. Baxter endorsing Quoddy and concluded with the emphatically large and bold directive to "**BOOST MAINE**."[36]

When not publishing news on the project itself, Roscoe published features indirectly promoting it, such as the value of nitrate fertilizers, the increasing use of aluminum, and the ever-expanding uses of electricity. Hardly a week went by without his readers being immersed in Quoddy. As the *Sentinel* was the primary source of local and regional news for most Eastporters, it and its publisher (their former state representative) carried a lot of sway.

On the same day the flyer appeared in the *Sentinel*, the *New York Times* raised doubts that Quoddy would be built anytime soon due to the difficulty of construction and the resulting cost and financial risk. But, it noted, "[t]he prospect of harnessing the water has interested Mr. Cooper so much that he intends to devote the rest of his life to the plan if necessary."[37]

A Boston attorney claimed, "He Originated Tidal Power Plan." In April 1925, Jerome A. Petitti said he had conceived of the two-pool arrangement to generate hydroelectric power from the tides in the Bay of Fundy, apparently oblivious that the two-pool concept had been invented a century earlier in his own hometown. Petitti filed for patent in 1921 and said he actually began work by buying an option on a piece of land in Lubec, for $10. Then the locals upped the asking price to as much as $5,000, which, the *Sentinel* noted, "ended the option securing process."

"While Mr. Petitti does not say so," the *Sentinel* remarked, "it seems quite probable that there was another reason for the abandonment of the Lubec activities. . . . This company asked permission of the Massachusetts Public Utilities Commission to issue stock and bonds. The Commission took the position, after investigation, that the cost of producing electricity in this manner would be so much greater than by other methods that it was not commercially practicable that it would be unfair to the investing public to permit such securities to be offered for sale."[38]*

Bidding for the moral high ground, Petitti offered to waive his priority claims if somebody like Henry Ford, with money enough to complete the project, "for the good of humanity," were behind it. But "as there did not appear to be any such backing" for Dexter's project, Petitti came to the illogical, yet revealing, conclusion that he "intended to insist upon his rights."

Dexter was unconcerned, saying "it was simply a matter for the patent office to pass upon the question [and] in no way would [it] halt the promotion of the project."[39]

Petitti, the patent troll, retreated into his lair.

Dexter opened an office in Eastport on July 11, 1925, at the Acme
Theater building on Water Street, just a few feet from the ferry dock, the
Sentinel, the Western Union telegraph office, and everyone in Eastport.

Everyone in Eastport swung into action on Dexter's behalf. A Frontier National Bank ad fanned, "We're Going to Boost the Dam." Frost
the Clothier hailed, "Vote For the Quoddy Dam." Galle's followed suit,
so to speak, with "Dam! Dam! Dam! Is what we're all thinking about,
but that isn't what you'll say if you buy your clothing, shoes and men's
wear from Galle's."[40]

Downeasters were getting downright giddy at the prospect of prosperity, even dashing off dam-good doggerel:

The Dam at Quoddy Bay

When he dammed the Mississippi,
People said that he was dippy.
In fact, the wise ones said, it can't be done.
For no man can dam that flood
With its bottom oozing mud,
They thought the fight was lost ere it begun.

When he tackled Muscle Shoals
With its shifting sands and holes,
The natives held their sides and laughed with glee,
To think they any school would graduate a fool
Who dreamed of holding back the Tennessee.
. . .

There are wondrous stories told,
Of Oil and Mines of Gold,
And we hear of glorious things the West has done,
But there will come an hour
When we'll use our Tidal Power,
And take the place that's waiting, in the sun.

What a splendid thing for Maine
If this child of Cooper's brain,
Should be born and grow in stature from day to day,
For 'twill bring us wealth and fame,
And add glory to his name,
When Cooper builds his Dam at Quoddy Bay.

—Robert S. Lingley

The poet died within a week of penning these immortal lines.[41]

Voting would be statewide, so the boosters reached out to potential advocates and drew them in with a public relations extravaganza. Tuesday, July 21, 1925, became "a red letter day" when the governor and a large party of businessmen from western Maine arrived at Eastport by chartered steamship to tour the proposed project area. Dexter was onboard as tour guide in chief.[42]

During the two-day event the visitors were "royally entertained" and serenaded by bands from Lubec and the Passamaquoddy tribe, who were "in full ceremonial dress, wearing the eagle feathers of chiefs and with nothing omitted except the war paint." The tour included stops at Jernegan's gold mill and the site where the Eastport–Treat Island–Lubec dams would be built. When the visitors approached Eastport, they "received a tumultuous salute from the whistles of every plant in town and all the church bells as they drew near the wharf where a huge crowd . . . of several thousand people awaited his [Brewster's] statement with tense anxiety."[43]

Staring at the sea of faces, Governor Brewster declared that he would "not oppose the passage of the Cooper charter bill, providing the campaign . . . is not directed against the established policy of the State, which is contrary to the exportation of power." "Thus," exhaled the *Sentinel*, "is removed one of the greatest menaces to the success of the Cooper referendum, which is now practically assured."[44]

Eastporters strung a large banner across Water Street encouraging everyone to "Vote for the Cooper project on September 14." Placards were placed on cars, and the Eastport Dry Goods Store displayed maps and drawings of the project. Just to be certain, Roscoe told everyone what to do in a five-column headline:

DON'T FORGET TO VOTE FOR THE COOPER PROJECT
Monday next, is the fateful day on which Maine will vote for or against the Cooper-Quoddy power project which means so much to the people of this city and county. . . . [I]t is hoped that there will be no one so blind to their own interests or so unmindful of their duty to their home town as to fail to go to the polls on that day. Of prime importance in this connection is the necessity of registering every qualified vote.[45]

"Every qualified vote" included women, who had won the franchise in 1920, but also, for the first time, American Indians, including the Passamaquoddy, who had finally been recognized as US citizens and enfranchised in June 1924—148 years after the Founding Fathers declared, "All men are created equal." Recognizing this might be her *last* opportunity to exercise her right, "one woman, 100 years old, had herself carried to [the] polls in order that she might cast a vote for Cooper."[46]

Election day finally dawned and the votes were cast and counted:

COOPER-QUODDY PROJECT OVERWHELMINGLY APPROVED
Sweeping on, in a current of popular approval seldom if ever equaled . . . the Cooper Charter bill was carried in Monday's special election by a majority that will exceed ten to one, winning by great margins in every section of the State. Washington County's vote was almost unanimous, representing a majority of ten thousand or more, and in several towns, notably Jonesport, Machiasport, and Trescott, it was actually unanimous, not a single vote being thrown in opposition. In Eastport, only 16 opposing votes were thrown out of total vote-fall of 1,552, and about the same ratio obtained in Lubec . . . and in Calais.[47]

Dexter and his supporters had won a spectacular victory. It was time to celebrate: "A collection was taken up among local business men for fireworks, and for about three hours there was a display in the Post Office quadrangle. . . . At 9 p.m. Mr. Cooper appeared and . . . expressed his appreciation." The "ringing of bells" Roscoe recorded, "and other such expressions of joy continued until midnight."[48]

The *New York Times* intoned as well in a half-page article on September 27, which included photos of Eastport's waterfront at high and low tides and a map of Passamaquoddy Bay. Therein the reporter belittled the city but praised the big man and his plan:

It is easy to dream, too, of a day when this most eastern tip of the United States will not be so remote in point of time—fourteen hours of slow travelling from Boston—as it is now. The little city's jerkwater and one-horse years are perhaps nearing their end. The grinding of great turbines, slowly moving and enormously powerful, may usher in a new era.

What manner of man is this potential miracle worker? If one visits him at his shack of an office on Campobello—a bare room with technical books around the walls, no telephone and the scantiest and simplest of furniture—one is impressed first by his utter lack of pretension. Every one around Eastport bears witness to that. There is no pose or affectation about him. He is the most popular man in the vicinity, though he does not cultivate back-slapping or any of the other supposedly easy roads to popularity.

Mr. Cooper is 45 years old, six feet tall, broad-shouldered, powerful. His hair and cropped mustache are black and his eyes are mild enough to belie the squareness and firmness of his jaw. He rarely raises his voice, though there is evidence that if he did he might be heard a mile or two. He prefers, apparently, to persuade rather than to dominate. In presenting his argument he is so clear and logical as to make even the most ignorant layman think himself for the moment an engineer.[49]

Dexter now had accomplished Essentials 1, 2, and 3. The dreamer and miracle worker was one-third of his way to the moon.[50]

REACTIONS

Eastport's original electric entrepreneur threw in the towel. On the same day and page that the *Sentinel* recorded the referendum results came the competitive consequences: industry consolidation. Eastport's Pennamaquan Power Co. was sold to a holding company headed by Edward M. Graham, president of the Bangor Hydro-Electric Company.[51] More and larger consolidations rapidly followed the next week as the corporate chess board responded to the powerful new competitor on the horizon:

New England Power Companies Merged . . .
The New England Public Service Co. [NEPS] with a capitalization reckoned at $80,000,000 . . . is to be the holding company for all the Insull Public Utilities in New England, and its common stock will in turn be owned by Middle West Utilities Co. Samuel Insull is chairman of the board of directors of the Middle West Utilities Co., and Martin J. Insull is its president.

The Middle West Utilities Co. has owned the Twin State Gas and Electric Co., doing business in New Hampshire and Vermont for many years, and recently acquired other public service companies located in New England. All of these companies will be transferred immediately to the New England Public Service Co. which will then own a very large part of the common stock of the following companies: Central Maine Power Co., Manchester Traction Light and Power Co., National Light, Heat & Power Co, which controls: The Twin State Gas & Electric Co., The Berwick and Salmon Falls Electric Co., The Vermont Hydro-Electric Corporation, Inc., Rutland Railway Light and Power Co., The Pittsford Power Co.[52]

If anyone was confused by the surfeit of companies and layers of ownership, they weren't alone. NEPS, which controlled two-thirds of the electric power production in Maine,[53] was valued at $80 million. Compare that to the $75–$100 million cost to build Quoddy; it was about the same, *but* Quoddy would produce four to eight times more power than *all* the power companies in Maine, or more than 20 times NEPS's output. How could NEPS "reckon" themselves to be so valuable? *What was going on?*

Could Dexter find the money for Quoddy's construction, *and from whom?* The first well-informed expert to opine on the subject was the Murray & Flood engineering firm, which had been retained by an unnamed potential investor in Quoddy. Mr. Murray said:

> Dexter P. Cooper has presented a tremendously interesting possibility, of power development in his Passamaquoddy tidal project. It has provoked, and rightly, an investigation to determine its economic value. The investigation is being made and if the project is determined to be economical, we cannot afford to see it undeveloped. Under these conditions it is safe to say there will be no difficulty in raising the necessary money. Murray and Flood have gone sufficiently far in their investigation to be much interested in the continuance of it, and so far as we have been able to inform ourselves, the project is sound.
>
> This will be the first time that tidal power has ever been developed, although a good deal of study has been given to the subject for a good many years. So far as I know there is no other place in the world where there is the same favorable combination of high tides and land configuration.[54]

Murray went on to explain that Quoddy's 500,000-plus hp could not be consumed locally and thus would accelerate the formation of the "super-power" transmission grid covering New England that Guy Tripp of Westinghouse had predicted. Murray, who had conducted the "super-power" survey for the federal government in 1920–1921,[55]* was reported by the *New York Times* to have "endorsed the [Quoddy] project as one he believed to be both practical and economically advantageous." Note, however, Murray's *equivocations*: "*[I]f* the project is determined to be economical . . . much interested in *the continuance* of it [the *study*], and *so far as we have been able to* inform ourselves." He wasn't done with his analysis.[56]

Murray's ultimate determination would depend on the answer Dexter gave him to one final, shocking question.[57]

Who was Murray working for? was potentially answered by the *Sentinel* on November 18, 1925:

> Possibilities that the Cooper-Quoddy project may have the backing of such powerful interests as those of the Insulls and the General Electric Co. were given more than speculative confirmation at the Worcester Conference last Thursday, where speakers representing all three factors discussed the power needs of New England, and two of them emphasized the desirability of a "power pool" for that group of states, one referring specifically . . . to the Quoddy enterprise.
>
> The rapid advance of the Insull interests into Maine, marked by the acquisition of the Central Maine and Cumberland County Power & Light Companies, and the probability that the Bangor Railway and its subsidiaries, including the Eastport Electric Light Co., may be taken over by them at the proper time, thus completing their line almost without a break, to the huge power reservoir that is to be tapped at Eastport. . . .
>
> Should all these conjectures prove correct, it means that the marketing and therefore the financing of the Quoddy power is practically an accomplished fact, contingent only on the ability of the engineers to construct the dams at a cost low enough to make the project practical.[58]

The *Sentinel* continued with a long quote by Owen Young, president of General Electric, which can be summarized thus: (1) increasing wages depends on increasing productivity, which depends on increasing use of power tools; (2) waterpower, as New England's only source of power, is vital to its economic success; (3) all sources of power, including tidal, should be developed and interconnected through a grid; and (4) the power export ban is illogical. Or, in his words,

> [T]he extent to which you permit artificial barriers, like antiquated laws and petty jealousies, to interfere with such a great program, you are false to the interest of your capital investment—you fail in the advancement of your labor, and you lack the leadership which New England has always given to the Nation as a whole. . . . If the state has to choose between leaving its powers undeveloped or having the energy from those powers go out of state there can be but one economic answer, whether judged by the selfish interests of the state itself or on the higher grounds of a moral obligations to the nation as a whole.[59]

Putting aside Young's patriotic rhetoric, it's worth considering who was "selfish," who was honoring their "moral obligations," and what was "false." But sorting out what was right, what was true, and where to trust your funds was getting harder because the banks and newspapers were falling under the control of the holding companies.[60]

Samuel Insull was "present at the creation" of the electric industry.[61] He started out in 1881 as Thomas Edison's personal secretary, earned his trust and power of attorney, and acted as his chief operating officer. Edison said Insull was "one of the greatest businessmen in the USA" and as "tireless as the tides."[62]* Insull was the first to recognize seven profoundly important and related ideas, and he then ran with them for all they were worth.

The first idea was to build BIG. He and Ford learned from Edison's manufacturing operations that the bigger the plant, the lower the unit cost of production, which enabled lower prices, which attracted more customers.[63]* But unlike the one-time sale of a tangible product, Insull would sell a service—to the same customer, every month, forever.

The second idea was to connect power companies to form "super-power" networks with transmission lines linking multiple generating stations and customers, thereby enabling whatever power was generated to flow to wherever it was needed. This made more reliable power available at lower cost to more people; it converted otherwise wasted surplus energy into cash and reduced the expensive infrastructure needed to service peak demand periods. Despite his allegiance to Edison personally, in the "battle of the currents" Insull wisely chose the Westinghouse-Tesla AC system rather than Edison's DC system.[64]

The third idea was to promote state-level regulation because it would (1) remove the industry from the clutches of notoriously corrupt local politicians who often demanded bribes; (2) encourage technical standards to facilitate interconnections; (3) reassure consumers that rates were "fair" while guaranteeing a "fair" profit for the utilities, thus getting the states to effectively endorse the sale of utility stock and bonds to the public; and (4) enshrine "public" utilities as "natural monopolies," free from competitors.[65]*

The fourth idea was the power of public relations. It was Samuel Insull who ran the Illinois Council of Defense during World War I, and it was he who conceived the Committee on Public Information and its incredibly effective propaganda machine. With the war over, Insull simply converted it into the Committee on Public *Utility* Information and extended it through Edison affiliates in every state.[66]

The fifth idea Insull coined was mining his customers and employees as investors, thus conjuring the "public" as "customer-owners." If people owned the stock and earned dividends, then they had a vested interest in seeing the company do well. Selling tiny pieces of the pie got "everyone yelling for us,"[67] as Insull put it, and kept politicians and regulators at bay. Selling stock directly to consumers opened up a new source of capital and also cut out the middlemen—stockbrokers—who had previously had a monopoly on utility financing.[68]

The sixth idea Insull grasped was the power of the holding company. Like a trust, this new corporate construct could control an unlimited number of companies, could be controlled by a few board members, and was only loosely regulated by individual states, even if it operated across state lines.[69]*

Never pay cash was Insull's ultimate mantra. As the Archimedes of American capitalism, he moved mountains of cash with leverage. Mastering these concepts, Insull exerted his power across two-thirds of the country and personally controlled 9% of the electric utility industry in the United States.[70]

Massachusetts was the market for Maine's surplus hydropower. While Quoddy would benefit consumers, to the existing producers, Quoddy was an existential threat.

On November 25, the *Sentinel* chronicled the big moves on the competitive chessboard as it expanded from local to national:

INSULLS MAY HAVE TO FIGHT TO MARKET POWER
. . . To all appearances, the hidden hands of some of the Country's largest public utility companies are directing the maneuvering behind the sales and mergers. . . . Probably the most repeated story is that associated with the Insull's New England operations, particularly their activities in Maine. The Pine Tree State waterpower policy which forbids exporting any surplus has been regarded here as something as durable as the state itself, but already there are rumors that the next legislative session will see a change in the present act, even its complete abolition.
. . . [A]ny power they [the Insulls] export will have to be marketed under highly competitive conditions, for the much-sought Eastern field is being honeycombed with transmission lines of other large companies. . . .

. . . Insull, so it is said, bought the Central Maine Power Company planning to link his other New England companies into a small but compact system. Less than a month later, the formation of a huge super-power group, including the New England Company, the New England Power Corporation, Stone & Webster, Inc., and the International Paper Company, was announced. Present output of this system is reckoned at one billion kilowatt hours per year, with the expectation of gradually increasing this to a figure five times greater.

If Insull succeeds in gaining permission to export power from Maine, he will have to produce it in such quantities and at such a price as will meet conditions in the area which can be served by the rival group. Even with his proved reputation for generating power at a minimum cost, his Eastern properties are considered now as presenting conditions that he has not met before. . . .

The topic of discussion at New York is whether Insull has arrived too late and Maine's China Wall policy on waterpower has thrown her greatest natural resource into a speculative stage. . . .

Figuring prominently in the situation is the huge development at Passamaquoddy Bay, where power may be produced so cheaply and in such volume as to prove a determining factor in any such competition for control of the New England field.[71]

Samuel Insull countered the questioning of his business strategy by making a show of recruiting new industries to Maine. Just as Admiral Owen had done, Insull promoted Maine's unique combination of hydroelectric power convenient to deep harbors and foreign markets. And like GE's president, Insull wrapped his pitch in patriotic prose. Roscoe the real estate agent couldn't resist speculating on Insull's alluring words: "If this is correct, the possibilities for a boom for Eastport are greater than have been imagined."[72]

Optimism was in the air as 1925 came to a close. Business was booming; the Thanksgiving feast was bountiful; Christmas was coming; and Roscoe was selling lots of advertising, thanks in part to the utility industry, a practice he happily highlighted in an article from one of the wire services:

"Liberal use of advertising has contributed to make low rates possible and has shortened by one half the time necessary to attain the present stage of electrical development in the United States," says W. H. Hodge, manager of advertising department of the Byllesby engineering and management corporation of Chicago.[73]*

Collectively the third largest industry in the country, the utilities should be large advertisers. It is estimated that they are spending today an amount equal to about one half of 1 per cent of their gross revenues in this way. From 65 to 75 percent of their total advertising expenditures are in the newspapers.

About one third of the utility advertising dollar is spent for selling appliances and is charged to sale price of merchandise. One third goes for securities on customer-ownership and is an investment expense. One third is used to build up service output and explain the business and its affairs and is charged to operating expense. This last third is equivalent to about 4 mills on average monthly bill for the electric service and is so small that it cannot possibly add to rates.

Advertising keeps down and reduces rates by creating market value which in turn permits the economies of mass production and lower distribution expenses.[74]

Roscoe was aware of his self-interest in selling ads. He wasn't aware he was being used.

Dexter also had a prosperous 1925 and, with fresh investment capital in hand, looked forward to 1926.[75] He also was being used.

CHAPTER 6

Crossing the Border (1926–1929)

COURTING THE CANADIANS

To win Canada's approval, Dexter would have to do it all again: all the meetings with politicians and bureaucrats, businessmen and fishermen, private citizens and the press.

In April and May 1925, Cooper conferred with New Brunswick officials in Fredericton and went to Canada's capital in Ottawa, Ontario.[1] While many in the American press were early Quoddy advocates, the Canadian press were skeptics. The *St. Croix Courier* in St. Stephen feared the loss of the town's "Nature endowed harbor" and asserted that "not in all time, past or present, have the people of Charlotte County been faced with a problem of such moment."[2]

Debate focused on what the tidal dams would do to the coastal fishing industry. The Canadian fishing industry was one-sixth the size of the American, but the Canadians relied on it just as heavily. While most sardine canneries were on the American side, most fishermen and weirs were in Canada. Altogether they employed some 2,000 Canadians and 4,000 Americans.[3]*

One Canadian critic argued, "The dam scheme would not give any benefit after it was built to the fishing inhabitants of Charlotte County . . . the greatest benefit going to the Americans."[4] Others believed Quoddy's boosters that New Brunswick would become the "Florida of the North," and the tidal dams would turn Passamaquoddy Bay into the greatest salmon pool in the world. Looking through salmon-colored glasses, many Americans believed most Canadians were in favor of the project.[5] Bolstering this bullishness was Geoffrey Stead, a Canadian government engineer whom the *St. John Telegraph* quoted as saying he "had been most favorably impressed with the scheme which he believed would be a very good thing for New Brunswick generally and Charlotte County in particular. He admitted that fisheries in that part of the bay cut off by the dam would be adversely affected . . . [but] [t]he advantages, Mr. Stead thought, considerably outweighed the disadvantages."[6]

Dexter's prospects in Canada got a boost on May 12 when the US Federal Power Commission (FPC) granted him a preliminary permit (Essential 4).[7] As Roscoe reported, the permit would be good for three years, and he added, "It is probable that long before that period is ended a final and unrestricted license to build will be issued." The following week, the *Sentinel* added, "The committee is expected to report the bill favorably to the House today, and its immediate passage through the House and Senate is confidently expected by well-informed Canadians, since no valid argument has been advanced against it; since Canada, like Maine, has everything to gain and nothing to lose by it, and since sentiment in Charlotte County is practically unanimous in its favor."[8] Roscoe was partly right: Cooper's charter passed in the Canadian House of Commons.[9]

Debate in the Canadian Senate was surprisingly intense. Nonetheless, on May 31, 1926, Dominion approval was granted, and Dexter's charter was signed by the governor-general two weeks later. Dexter now had Essentials 1 through 5—though he still needed approval from the relevant agencies (navigation, power, fisheries).[10]

In June, Dexter turned his attention to New Brunswick's approval, beginning with St. Stephen's town council. To a packed house, he explained the project and then offered to pay damages to the fisheries. That got people's attention. As other problems were raised, Dexter replied with solutions: *If the tide no longer ebbs, won't St. Stephen's sewers, which empty into the St. Croix River, cease to function?* If that happened, he would fix them.[11]

An endorsement of a sort came in June from another country: Soviet Russia. The Soviets hired Dexter's brother Hugh to consult on a huge hydroelectric dam they wanted to build.[12]

In October, Roscoe wrote to the Canadian premier, John B. M. Baxter, to inquire where sentiment stood on the project. The premier replied, "My own attitude—and I am speaking for the Government as well—is distinctly favorable towards the realization of the project."[13]

The still cross *Courier*, however, opined that Dexter had not come to New Brunswick early or frequently enough and declaimed, "We know what we have and would like to know just what we are to get before a swap with a stranger who seems to seek the background until [it's] 'all over.' The rocks that he proposes to plant in our channels for all time will not float away and will not be easily removed."[14] The paper's antagonistic attitude continued in December:

> Recent utterances in the down river press and in papers elsewhere indicate that Dexter P. Cooper is going ahead with his project willy-nilly and whether the people of New Brunswick like it or not. Mr. Cooper still keeps himself carefully concealed from an interested public on this side of the line but his press across the line continues to talk of great industries to be developed following the construction of his dams and the closing of our harbor. Much as the people of Charlotte would like to be given information that is essential [but] the oracle says nothing . . . with the powerhouse safely located in Maine new industries in Canada will be like hen's teeth, few and far between.[15]

The "oracle" didn't reply, but Roscoe did:

> We do not object in the least to that ["down-river press"] designation, but we do take exception to being classified as "his (Cooper's) press." . . . He has never at any time tried to make "use" of the press. He has extended all possible courtesies to its representatives, but never at any time has he sought to exploit it or to spread what is known as propaganda through its agency. On the contrary, as the time for starting construction draws nearer, he has become more reticent than suits the news writers.[16]

Roscoe was not insensitive to Canadian concerns. He, like many Eastporters, was married to a Canadian. What happened in Canada mattered and would soon impact Quoddy, Eastport, and every citizen in the United States.

Canada's Niagara power plants became political flash points. While an unquestioned engineering success, the economic model was now in question. Murray & Flood (M&F) ignited the controversy with the 232-page report *Government Owned and Controlled Compared* *with Privately Owned and Regulated Electric Utilities in Canada and the United States*, which was highly critical of the government-ownership model. Picking up the story, the *Sentinel* wondered, "WHAT IS THE ANSWER?"[17]

Apparently Ontario was purchasing electric power from a private company for $15 per hp, transmitting it 230 miles, and reselling it for $22 per hp. But according to M&F, Ontario was selling power generated in its publicly owned (and untaxed) plant for $36.10 per hp after transmitting it only 60 miles. Ontario's publicly owned utilities were reputed to produce the cheapest electric rates in the world,[18] but M&F said it wasn't so. Sir Adam Beck, head of Ontario's Hydro-Electric Power Commission, shot back with a scathing 80-page rebuttal to what he said were M&F's many errors, distortions, and lies.[19]

The anti-government attitude was echoed and amplified by privately owned "public" utilities. A characteristic address was delivered by Bernard J. Mullaney, vice president of the People's Gas, Light & Coke Company (PGL&C, an Insull company) of Chicago on August 19, 1927. Mullaney was the marketing wizard behind Illinois's World War I Council on Public Information.[20] Mullaney framed the public versus private ownership debate this way:

> Conversion of natural resources to usefulness—putting them into usable form and making them available for use—is the natural function of business. As this conversion is impeded or facilitated by government, so business lags or grows, and the tide of prosperity rises or falls. . . .
>
> Prosperity, unparalleled in distribution as well as in degree, abounds in the United States not only because of our country's natural resources but also and chiefly because business has had here so little interference from government. . . .
>
> None but the blind, deaf, and dumb are justified in doubting the existence of efforts, insidious and definitely promoted, to change the intrinsic nature of our government, and eventually its form. . . .
>
> The subversionists would first of all reverse the so successful American policy of keeping the government out of business and would put the government more and more into business. Their efforts wear various masks; at one time it is a legislative proposal to "regulate" or "control" an industry; at another time it seeks to make some legitimate governmental enterprise an excuse for incremental government ownership and operation of a business. . . .

It is a troublesome, not to say a dangerous, minority because of its composition and the insidious way it works. Some of this minority are communists of the deepest Russian "red": others are a little socialist "red," shading into "parlor pink"; still others come in delicate mauve tints and call themselves "progressives."[21]

Business is good—regulation is evil was the corporate incantation. Indeed, business was good and there wasn't much regulation. If anything, government was at the beck and call of business, granting privileges of all sorts, and a huge customer. At the local level, Roscoe could see the positive results the pro-business culture was bringing to him. Picking up what was probably sent to him by a wire service,[22] Roscoe re-tooted this brassy encomium:

SAFEGUARDING AMERICA'S INDUSTRIAL STRUCTURE
A notable change may be seen in the trend of advertising nowadays. Not so many years ago, the really big advertisers were patent medicines and baking powders.[23]* Today, financial advertising is featured because the public has more money to invest. . . . The old day of individual ownership of every business is passing on. Men buy and sell collectively; they invest collectively, to get collective service that no individual capital could provide. They buy as corporations where the volume of the investment and its creed of universal service, makes it more safe from radical forays of every kind.

The corporations of today are "our" corporations; we,—everybody— finance their building and buy their products from ourselves as owners. "Public ownership" is here in its ideal form. . . .

America was never as truly American as today, with almost every family owning some form of industrial security that pays a profit. This is a safe and sane condition.[24]

This idea to "invest collectively" was really catching on, though saying "almost every family" owned an "industrial security" (meaning corporate stocks and bonds) was a gross exaggeration, and the industry's "creed of universal service" was fiction to nearly every farmer. Conversely, collective *action*, in the form of labor unions, was bitterly opposed by those promoting collective *buying*. And of course hardpressed Soviet Russia was experimenting with collective *ownership* and *management* of farms and industry—a repulsive idea to those in charge of the booming American economy, though in fact it was the way many early New England towns operated, with "commons" on which crops were grown and animals herded collectively.[25] Concluding conditions were "safe and sane" was . . . premature.

"An 'eleven-month' survey," Roscoe reported, "confirms the obvious fact that all previous prosperity records for this country have been eclipsed by the year now coming to a close . . . even the fearers-for-the-worst are smiling over prosperity prospects for next year."[26]

Another record fell in December 1926. Electric power was experimentally transmitted 1,000 miles—from Chicago to Boston—thus shorting out Senator Carter's challenge from the previous year and proving "the oracle" right. Quadrupling transmission distance was fantastic itself, but "[s]pecial significance is attached to it here, because the prime mover in the experiment . . . is an Insull concern, and Insull has been steadily extending his lines in this direction since the Quoddy project was launched."[27]

Dexter started drilling test holes in Eastport on December 28, 1926. There would be hundreds of these holes, both on land and in the water, wherever Dexter was thinking about placing a dam, tidal gates, navigation locks, or the powerhouse. Soil drills were 2½ inches in diameter and encased in steel pipes. They would go down 75 feet or until they hit rock, after which a diamond-tipped drill would be used to go farther into the rock to make sure it wasn't just a boulder.[28]

In the *Sentinel*'s last issue of the year, Walter S. Wyman, president of the Central Maine Power Company, took out a one-third page ad. It was not advertising washing machines or corporate bonds; it was raising a question:

> **MAINE'S WATER POWER QUESTION**
>
> Water power is one of Maine's great natural resources. Like all other wealth which Nature has given us it is of no value unless and until it is developed and used. Unlike all other natural resources it does not become exhausted with use but renews itself from year to year forever.
>
> All of the other forty-eight states allow surplus power to flow freely across their borders. Those states that are exporting the most power are the ones that are making the greatest growth. Maine alone forbids its export by law.
>
> This law should be modified or repealed so as to permit surplus power to be exported. . . .
>
> We ask the most careful consideration by all the people of the State of this important question.[29]

Note how Wyman begins by stating waterpower is one of *"Maine's"* natural resources, "which Nature has given to *us*," meaning *all* of the people of Maine—but the rights to develop it and export it would go to the "public" utility—owned by himself, Samuel Insull, and a few others—*"forever."*

Three days after Wyman's "advertorial" was published, Bangor Hydro-Electric bought three more electric companies.[30] "Us" got smaller.

AMERICAN TECHNOLOGY

Eastport could become the "Second Niagara,"[31] the next high-tech economic marvel, and the next international hydroelectric power-house. Nearly every week in 1927, Roscoe published something about how electricity was changing the world and how people were profiting from it.

To explain and promote Quoddy's life-changing implications, Roscoe hired out-of-town experts, including Joseph F. Preston, who had been involved in the early days of power development at Niagara.[32] Preston's first article, "THE ROMANCE OF ELECTRICITY," published on February 2, became the model.[33] It introduced names and concepts that would become familiar to Downeasters as Quoddy progressed: William Rankine (Canadian entrepreneur), Stone & Webster (Boston financiers and engineers), Sir Adam Beck (Canadian public power advocate), and Owen Young (GE's chairman). Featured prominently in a sidebar on page 1, Roscoe placed Eastport's own Dexter Cooper with the Titans of industry, regaling his audience with the story of his hellish descent under Niagara Falls. By the time Roscoe's readers got to page 2, they would have been wondering what their town—their lives—would be like with cheap and limitless tidal power at their disposal.

For the time being, however, electricity was limited and expensive. An indication of the real cost of power appeared in the *Sentinel* in July: "CALAIS TO GET CHEAPER LIGHT: Their Proposed Rates Make Ours Look High." Roscoe explained a possible cause of Eastport's predicament: "Bangor Hydro-Electric Co . . . employees were very active in placing the present city government in office."[34]

Meanwhile, another effort was underway that would impact Eastport's fortunes: to identify, quantify, and promote alternatives to Quoddy. The resulting report, prepared by a nationally known engineering firm, was quietly placed in the hands of influential people.[35]

The superiority of American technology and the bigger-is-better business model was axiomatic. Every day brought proof.

Communist Russia confirmed the axiom in January 1927 when the Soviets hired an American capitalist, Hugh Cooper, to oversee construction of the world's largest hydroelectric power plant. Roscoe intimated that Russia "added confidence and prestige to the Quoddy project, for it is well understood, that in working out his plans, Dexter P. Cooper has frequently consulted with his brother, who fully approves of all that he has done. Col. Cooper has also been here and looked the project over personally."[36]

Picking up another lead from the newswires, Roscoe ran an article about the "WORLD'S GREATEST POWER POOL" being formed in Pennsylvania: "Without vast capital and the social instinct of cooperation, there would be limping little local plants eking out a troubled existence and misserving [*sic*] their patrons over all this vast industrial district. Mere size is no public menace; wisely directed as this interconnection is, it is a great public benefit."[37]*

Another editorial was about Governor Brewster's veto of a bill for a New England power pool that would have allowed the export of surplus waterpower from Maine. "Had the bill become law," Roscoe wrote, "the Insull interests, which now control most of the hydroelectric resources of Maine, would have been in a position to enter the Massachusetts market, which is one of the objects of the Cooper-Quoddy project. . . . As matters stand now, Cooper can reach the Massachusetts market, while Insull cannot . . . [thus] the Governor has done Eastern Washington County a very considerable favor."[38] Roscoe was right in his analysis, *but out of date.*[39]

Governor Brewster had good reason to favor Dexter Cooper and his brother Hugh over Samuel Insull and his brother Martin. The Insulls were involved in a scandal in Chicago in 1926–1927 in which William McKinley (no relation to the late US president) became the Insulls' enemy after refusing to sell his utility company to them, thus blocking the Insulls' expansion plans. McKinley then ran for the US Senate against Frank L. Smith, the chairman of Illinois's utility commission. Samuel Insull told his brother to do whatever it took to defeat McKinley. Instead of the usual cash in an envelope under the table offered to politicians everywhere, Martin Insull wrote checks totaling $500,000 (about $7 million in 2020 dollars) for Smith's campaign.[40] It was a staggering sum to spend on a Senate race, and traceable. Hoping to head off a public uproar, one of the Insulls' cronies tried some hush money, offering Smith $550,000 to get out of the race.[41]* Smith refused. McKinley then spent more than $500,000 of his own money on his campaign, but Smith won anyway.[42]

The huge campaign contributions (at a time when Congress was considering limiting campaigns to $25,000)[43] gives an indication of the financial rewards to be had if politicians voted as the Insulls wanted. The dark money came to light through Senator James Reed's (D-MO) investigation, in which Samuel Insull was subpoenaed; testified about his senatorial contributions in a manner newspapers considered "arrogant" and "insolent"; refused to testify about his local contributions; was held in contempt of Congress; and, to avoid prosecution, testified *after* the local elections, when the recipients were incumbents, not candidates. The public was outraged. The press had a field day. And politicians piled on.[44]

Since the Insulls had attempted to "purchase" a US Senate seat for Smith, the rest of that incorruptible body refused to seat Smith and nullified the election.[45] The Reed Committee then broadened its investigation from economic power in politics to politics in electric power. The committee's agenda for February 25, 1927, included a proposal for the federal government to build a hydroelectric dam at Boulder Canyon,[46]* the government's unfinished explosives factory and power plant at Muscle Shoals, and Samuel Insull's shenanigans. As historian Forest McDonald wrote, "Legion were the events that ensued from that turbulent day in the Senate," for out of it came "some of the most far-reaching reform legislation in the history of the American Congress."[47]

The repercussions reached to faraway Maine, where they would change the course of human events for our protagonists—Franklin Roosevelt, Dexter Cooper, and Roscoe Emery—and millions more.

"The Insulls have demonstrated in Illinois what their attitude is toward clean politics," the *Sentinel* observed, "and Eastport people have been given a very positive demonstration of what an electric lighting concern will do when anyone steps between it and . . . [the] pocketbooks of the people. It is not well to open the doors too widely to unrestricted action by such interests." More broadly, Roscoe believed, "It had been evident all along that if the export people would accept the bill arranging for a power compact between Maine, New Hampshire, and Massachusetts, and also the so-called Carter bills fortifying Maine's control over her power, the Governor would sign the Smith-Wyman bill," approving the export of waterpower. "But" said Roscoe, "the export interests would have none of these bills. They wanted everything or nothing, and made their play on that basis. They lost."[48]

In the national scheme of things, Brewster's veto was a minor setback for the Insulls. They were buying control of power companies, trolley lines, banks, and newspapers at a pace that is hard to imagine. In 1927, they already controlled 2,064 companies. In Maine, the Insulls "not only dominated two thirds of our utilities," complained Brewster, "but five of our newspapers through interests closely allied with him, forty of our banking units . . . and many of the 17 major industries of our state."[49] By vetoing the power export repeal, Brewster, a "progressive" Republican, went against the Republican machine, the vainglorious Old Guard, and their backers, the "power gang,"[50] thus incurring their wrath.

Dexter would also have to contend with the Insull empire. While Quoddy was certainly not going to be a "limping little local plant," it seemed certain the tentacles of the Insulls' octopus[51]* were feeling their way around Quoddy and would soon embrace Dexter. Meanwhile, Dexter may have been mourning the death of Brigadier General Guy Tripp, chairman of Westinghouse, hero of the World War I Production Board, advocate of super-power grids, and critic of the Fernald Law. He died on June 14, 1927, following abdominal surgery.

Dexter was living the entrepreneurial dream and steadily turning his "romantic dream" of harnessing the tides into reality. He wasn't waiting for the Canadian fisheries report. While Roscoe was frustrated with the now close-mouthed Cooper, the *Sentinel*'s stringers kept an eye on everything around Quoddy, and on June 29 they reported Cooper had returned from Chicago, where he had been "called by business connected with the financing of the project," and soon thereafter started drilling at Le'Etete in Canada.[52]

Dexter's march toward success continued in September when plans were announced to utilize his tidal power technologies on the Petitcodiac River in New Brunswick. October brought word Dexter was moving to larger offices at the Seacoast Canning Company. Rumor had it he would also be leasing Eastport's wharves 4 and 5, and another warehouse, and Dexter had definitely ordered a 45-foot scow for his drilling operation. In November came news of two engineers added to Dexter's staff, and of Dexter paying the Eastport-Campobello ferry to increase its daily trips from two to three, as there were no telephones on Campobello, and he needed faster communication between his offices.[53]

Someone, or some group, was providing the funds for Dexter's expanded activities. Those funds came with strings attached. As always, *the devil is in the details*—and the hereafter.[54]

Meanwhile, Hugh Cooper was applying to the FPC for a license to develop a hydroelectric power plant on the Columbia River. He would have a long wait. The FPC had no staff. Instead, the three-headed Hydra borrowed staff from its constituent departments (Interior, Agriculture, and War), and the commissioners themselves met for only thirty minutes per month. *The corporately controlled, Republican-controlled Congress didn't want the Federal Power Commission to have any power.*[55]*

No one owns a river, according to the legal concepts of common law and the public trust doctrine. The doctrine, which can be traced back to Roman times, posits that the most fundamental "natural" law requires that the things on which we all depend—air, fresh water, and the sea—cannot be owned as private property but are "common to all" citizens and should be *held in trust for the public* by the government.[56]

Sometimes these "common pool resources" can be used by many people relatively easily, such as by walking into a public park like Boston Common, the city's former cattle grazing grounds. But to harness a river for power production, one traditionally had to dam it. Damming it requires land on both sides of the river for constructing the dam and more land upstream for the area to be flooded, often including whole towns, all of which must be acquired, even though some owners may not want to sell. People upstream and downstream are also affected by the altered levels and timing of the river's flow and any pollutants from agricultural, manufacturing, and domestic uses of the water.

To manage all this, a series of *riparian rights* evolved to exploit water resources while protecting adjoining property owners and the public. Governments also have the right of *eminent domain*, or the right to "take" or *expropriate* private property for public purposes, such as supplying drinking water or electricity to a community, so long as fair compensation is provided in return. In the United States, these rights and laws evolved based on peculiar geographic, hydrographic, and political considerations in each state, *not* federally.[57] Since all large rivers flow through multiple states, and sometimes flow through or divide multiple countries (like the St. Croix and St. Lawrence Rivers), legal turbulence is the result.

In New England, the Colonial Ordinance of 1641–1647 ceded the control of rivers to individual colonies (later states), whereas in the Louisiana Territory purchased from France in 1803, the riparian rights were held by the French Crown and transferred to the US federal government. These differences would have long-lasting implications for hydroelectric power and economic development in the United States.

State and local governments often lack the financial resources and expertise to build big and expensive infrastructure projects like hydroelectric power dams, so they often lease or grant monopoly rights (sometimes for 99 to 999 years,[58] but more appropriately for dozens of years) to private enterprises as incentive for the companies to build (and then operate) the infrastructure. Often these grants come with the right to expropriate private property. While logical and equitable in concept, in practice this approach almost always creates conflict.[59]* Thus, while no one owns the river, someone always owns the dam sites and water rights.

Samuel Insull was one of the first to fully comprehend the importance of "diversity of demand." Almost every electricity consumer uses it for only a part of the day, such as lighting in the evening, but not at other times. If customers were to build their own stand-alone power plants, as many did, it would need to be sized to cover their maximum demand—let's say 100 kW, even though there might only be one hour per day when this maximum was needed. At other times there might not be any demand, so let's say the average demand is 25 kW. Thus, another customer could be added to the stand-alone system at very little cost if they used power at a different time, such as for running a machine during business hours. The percentage of time the full capacity of the generating equipment is used is called the "load factor." The widest diversity of demand, coupled with very large generating capacity, yields the highest load factor and the highest profits. Switching from a stand-alone system to being a customer of a large network could result in theoretical cost savings, in our example, of 75%.[60] Power companies used a fraction of those savings to attract new customers but kept the lion's share for themselves.[61]

Given the varying amounts of power from hydroelectric generators and the varying demands for power, it is most efficient to link them to other types of generators, such as fossil fuel–fired plants. This "diversity of generation" allows operators to take advantage of each type of power source's benefits when available: rivers when they are flowing; tides when they are high; and oil, coal, or gas when they are needed or cheap.

Linking diversity of generation with *privately owned* transmission lines for diversity of demand creates "vertically integrated," "natural monopolies" that control *everything*.[62]

The "regulatory bargain" is the idea that in exchange for governments granting monopoly rights to a private company, the company will submit to government regulation, and the regulators will be "surrogates for competition." In theory, the regulators would set rates *as if* there were a competitive marketplace. But regulators were soon convinced to represent *not* the interests of consumers (whose interests were competitively low rates) but the "public interest," which included stockholders of "public" companies (whose interests were high returns on their investments). Thus, regulation in the public interest has built into it a conflict of interest in favor of the regulated industry over consumers—one magnified by the astronomically different scales of individuals versus industries.[63]

Electricity was and still is one of the most important elements in economic development. The availability and price of power affects what is produced; where, and how, and where it is shipped; and ultimately what is consumed. Thus, a few power producers—particularly trusts, combinations, and conglomerates with interlocking ownership and control—can have an extraordinary influence on an economy. That is exactly the situation the United States found itself in in the late 1920s.[64] The private utility companies and closely allied monopolies in mining, metals, fossil fuels, and railroads, or what Theodore Roosevelt called "the invisible government,"[65] could—and did—control most of the US economy.

Based on its investigation (which consisted of nothing more than questionnaires mailed to power companies), the Federal Trade Commission (FTC) concluded in 1927 that the "power trust" cabal didn't actually exist. Since the FTC also failed to investigate the industry's propaganda and lobbying activities as it had been directed to do by the US Senate, critics called the report a "whitewash" and accused the FTC of subservience to those it was supposed to regulate.[66]

Some feared "public" utilities stocks were "watered," and the real public—*both* rate payers and individual investors—were getting soaked. "Watering the stock," a technique invented in Chicago's meatpacking industry, means overvaluing a company's assets or inflating its costs.[67]* Since publicly chartered monopolies are typically granted a fixed percentage for profit or return-on-assets, the higher the costs or the higher the value of the assets, the more money the monopoly owners make.

Senator Thomas J. Walsh (D-MT) was concerned about the phenomenal growth of the power industry and its consolidation under a few holding companies: "Experience [has] led the long-suffering public to surmise that in such consolidations the stocks of the . . . properties merged were, generally speaking, acquired at inflated values, affording an unsafe basis for the securities issued against them, or the rates exacted of the consumers." To address his concerns, in 1927 and again in 1928, Walsh called for an investigation into potential financial manipulation and "whether, and to what extent, such corporations . . . through . . . avenues of publicity have made any, and what, effort to influence or control public opinion on account of municipal or public ownership in the electric power industry."[68]

Walsh's proposal met with fierce resistance. But due to popular support, efforts to defeat it yielded to attempts to bury it in committee and then to render it innocuous. "The most powerful lobby in the history of the nation," as Senator John J. Blaine (R-WI) called the National Electric Light Association (NELA) and similar organizations representing water and transportation utilities, descended on Washington. The lobbies expended so much effort and largess against the reforms that Walsh pessimistically concluded, "[T]his resolution will not be adopted by the Senate, because neither party wants it adopted, because it will dry up the source of campaign funds for the next election."[69]

Opponents claimed a "mania for investigation" and warned of "great damage . . . to honest companies," while proponents remarked "you would have supposed that the investigators were the real culprits, and the persons proposed to be investigated the innocent victims of partisan malice." The investigation was finally approved, but instead of sending it to the FPC, it was sent to the FTC, which had also been "captured" by industry, as further documented by Franklin Roosevelt's professor at Harvard, William Ripley, in *Wall Street and Main Street* (1927).[70]*

"Regulatory capture" occurs when a political entity, policy maker, or regulatory agency is co-opted to serve a minor constituency, such as a particular geographic area, special interest group, or industry, as opposed to the broader public. This often occurs because the specialized knowledge required of the regulators is most often gained from past employment by the industry or special interest group, potential future employment by them (often at higher pay), and the disproportionately large financial interest and resources of industries compared to individuals. The net result is "captured" regulators unconsciously or deliberately favoring the special interests. Sorting out what is bias versus what is criminal depends on what the definition of "criminal" is—or should be.

TROUBLED WATERS

The first sign of the storm came in January 1928, when Dr. A. G. Huntsman, director of the Atlantic Biological Station at St. Andrews, New Brunswick, testified before the Royal Commission on Maritime Fisheries about the effect the Quoddy Tidal Power Project might have on the region. Huntsman predicted Passamaquoddy Bay's sardine, cod, pollack, haddock, and clam fisheries would be "wiped out," as would the sardine fisheries along the whole coast. He also predicted the climate would change: there would be less fog in summer, the bay would freeze over every winter, and the tides outside of the dams would increase.[71]

Given Charlotte County's economic dependence on fishing, the news soured provincial appetites for tidal power. Huntsman was basing his dire predictions on the radically new idea that the waters in Passamaquoddy Bay were not all the same, but different temperatures at different depths, and certain types of fish only bred and lived in waters of certain temperatures at certain times. Furthermore, the tidal currents were essential to stirring up the water and carrying nutrients from the bottom sediments to the upper waters, where sun and plants turned them into fish food. The tidal power dams would interfere with the natural flow of the currents and hence the food and fish. Huntsman admitted this was just a "theory" but warned that building the dams would be an "enormous experiment" without precedent,[72] and irreversible.

Nonetheless, New Brunswick's lieutenant governor W. F. Todd declared himself in favor of the project because he believed it would benefit the province as a whole. Todd told the *Sentinel*, "About the only ones to be affected by the proposed development would be inshore fishermen, who had about 78 weirs on the Canadian side, with two men to a weir. Sardines, despite the development, would still continue to come to their feeding grounds, and deeper nets would have to be employed, that was all." Roscoe dumped Huntsman's testimony in the "gutter" (the inside-left column in newspaper parlance) on page 7 of his eight-page paper. In the prime position (the top right corner), Bangor Hydro-Electric advertised toasters, on sale for $4.25 (regular price $8.00).[73]

Most New Brunswick legislators were more interested in what the Canadian papers were saying when they took up the debate about Quoddy in March 1928.[74] According to St. John's *Evening Times Globe*, "it was the most contentious piece of legislation to come before the House" and monopolized the Corporations Committee for three weeks.[75] Both the proponents and the opponents came out in force. Lined up against Quoddy were Canadian Cottons, Maritime Electric Company, Connors Brothers of Black's Harbor ("the largest sardine cannery in the British Empire"), the towns of St. Stephen and St. Andrews, and the Canadian Pacific Railroad (CPR).[76] CPR's flagship hotel, the 86-room Algonquin House in St. Andrews, was the first major seaside resort built in Canada, and unlike Campobello's isolated hotels, which had all failed, the railroad-fed Algonquin was going strong. "After . . . one of the most spectacular legislative battles ever fought in New Brunswick, the Cooper Dam Bill was passed . . . by a vote of 24 to 17."[77] When the governor signed the charter on March 30, Dexter, Roscoe, and almost everyone in Eastport breathed a sigh of relief.

Cooper's New Brunswick Charter was not carte blanche. He had to (1) maintain open navigation in the St. Croix River at all times, (2) pay damages to businesses injured by the project, (3) deposit a $2 million bond in a Canadian bank as a guarantee, (4) build a bridge to Ministers Island,[78]* (5) begin construction within three years, (6) complete it within six years, and (7) use Canadian labor in Canada.[79]*

Dexter also still needed Essential 7: International Joint Commission (IJC) approval. The IJC was only supposed to determine *if* the project could be financed and built and, *if so*, what the allocation of the power between the two nations should be. Commenting on the IJC's second criterion, Roscoe averred, "It is certain, in any event, that when he [Dexter] does go before the Commission, his engineering plans will be complete to the last minute detail, and that he will have proof of abundant financial backing."[80]

Sensing a sure thing, Maine's governor Brewster said he "had no doubt, if Mr. Cooper solved his engineering problem, that money would be forthcoming in abundance to finance the tidal power project; and that he for one had always believed that the project would come to pass."[81]

Their affairs in New Brunswick apparently in good stead, Dexter and Gertrude traveled to Montreal and Ottawa for meetings with Canada's national government; then they boarded a steamer for England for a well-earned month-long vacation (and a visit to a proposed tidal-hydro-electric dam on the Severn River). After returning to Maine, on October 22, 1928, Dexter filed his application with the FPC for a permit to operate the Quoddy Tidal Power Plant (Essential 8), to produce 464,000 hp initially and ultimately 1,087,000 hp.[82]

Fisheries concerns wouldn't go away just because they were punted for further study (Essential 7-A). Getting annoyed, Roscoe called Dr. Huntsman a "'Mixing Bowl' Scientist [who] Won't Give Up Fight" and complained that articles in the Canadian press were "evidently written to conform to the view of the opponents" and "an attempt to arouse opposition in the United States." Assuring himself and his readers, Roscoe concluded, "as the practical sardine men here are not at all deceived by this propaganda, and have always viewed the project with indifference so far as its effect on their business was concerned, the proposed investigation is not taken very seriously here."[83]

The fisheries commission went to Campobello to investigate the issue in person. There, Dexter pointed out "the [tide] gates had been so located that nothing can obstruct the free flow of water to and from them, and that water would flow through them at about 24 feet per second. . . . More water, he asserted, will pass through the gates than goes through the open passages now, in the same time. . . . This flow is estimated at 2,400,000 cubic feet per second, equivalent to ten times that of the St. Lawrence River. . . . All this flow will prevent the Bay from freezing over, for the water there will be kept in circulation." As for raising the tides in the Bay of Fundy, "Mr. [Moses] Pike [Dexter's chief assistant engineer] demonstrated to the satisfaction of the committee that it would be very slight indeed,—a matter of inches at most."[84]

Roscoe thought Dexter and his team had won the day, thanks in part to the strong support of fishing and cannery men from Campobello, Lubec, and Eastport. As an example, Robert J. Peacock, the man who had proposed Maine's statewide referendum on the Cooper Charter and president of the Seacoast Canning Company of Maine, said:

> I don't suppose anyone can tell whether or not this would be the effect. But even if this were true, it is my opinion that it should not be permitted to interfere with the project, which will mean so much to Washington County. We get only 4 percent of the fish we use in our canneries from this bay, so the loss would not be irreparable if none were obtained there. Of course I don't suppose anyone knows, but I have always said that if the fish did not come up the Passamaquoddy they would go somewhere else in the vicinity, so that the canning business would not be injured in any event.[85]

Ill winds were blowing in the American Midwest. Five generations of settlers had been clear-cutting forests, plowing through the prairie, and otherwise taking the bounty of the land in quantities unprecedented in the history of the world. The huge harvest fed huge numbers of immigrants as the United States grew tenfold, from a population of 12.9 million in 1830 to 122.8 million in 1930. But silviculture and agriculture practices were extractive, not sustainable, and the topsoil was losing its fertility or being lost to erosion. Drought made it worse. The Great Plains—America's breadbasket—would soon shrivel into the Dust Bowl. Heat radiating from the parched land raised winds that would sweep up the dust and carry it in gargantuan black clouds as far east as New York City, where it darkened the day like walking through a smokestack.[86]

Modern technology was solving some of the problems. A tractor could plow more land with less labor. *But then what were the workers to do?* Chemical fertilizers could mask the disappearing topsoil. *But what if fertilizer was unaffordable?* The *Sentinel*, of course, flagged the issue:

> **FARMERS DEMAND EARLY OPERATION OF MUSCLE SHOALS NITRATE FERTILIZER PLANT**
> The biggest cyanamid plant in the world—and the only one of any size not operating—stills stands idle at Muscle Shoals, Alabama . . . . At a conservative estimate this cyanamid plant—nitrate plant Number 2—is worth today $15,000,000 to $20,000,000 . . . . Farm leaders and engineers who visit the Muscle Shoals works are amazed to think that anyone would suggest the scrapping of this magnificent plant, yet that is virtually what the electric power interests propose in case they get possession.[87]

The ill winds were on their way to Maine.

Strong lights cast dark shadows. Such could be said of Samuel and Martin Insull, the brilliant industrialists from Chicago. Despite Roscoe's previously expressed concerns, he saw the Insulls' presence at Campobello on July 3, 1928, as reason for optimism:

> **INSULL LOOKS COOPER DAM OVER With Group of Hydro-Electric and Financial Leaders; Gets First-Hand Information About Quoddy Project.**
> . . . The party included Martin J. Insull, recognized as perhaps the leading figure in the electrical development field in the north central and eastern states, through the control exercised by him and by his brother, Samuel Insull, over a chain of public service companies dealing in electric power extending from Chicago to Waterville, Me.; Walter S. Wyman, president of the Central Maine Power Company, the largest Insull unit in Maine, which is said to be about to absorb the Bangor Hydro-Electric Co., thus completing the line to Eastport; Charles L. Edgar, president of the Edison Electric Illuminating Company of Boston, an important unit in the Insull group; and Gerald F. Swope, president of the General Electric Company.[88]

In the entrepreneurial game, getting any one of these potential investors would be a home run. Getting all four was a grand slam. That they had all agreed to travel together to the ends of the earth—Downeast Maine—made it the World Series.

They had all studied Dexter's plans and found the engineering sound, the financial proposition compelling. They weren't fools. They knew what they were doing and had good reason to believe Dexter's plans were truly visionary—pioneering and farsighted—not quackery or quixotic. They had the technical and financial expertise to understand what had to be done, what could be done, and what could be achieved as a result.[89] They had the money. They just had to decide.

There was a problem with Dexter's World Series strategy: all his eggs were in one private rail car. And while they tallied four, at least one was in the fifth column.

While Roscoe the reporter speculated who might fund Dexter's plans, Roscoe the real estate agent began speculating in land and promoting his interests in a half-page ad headlined "We offer for Sale the following items of Real Estate. . . . Remember these are all located within the district that will be boomed by the Cooper Development."

Two months later, Dexter began buying options for dam landing sites.[90]* On October 19, 1928, Dexter submitted his application to the FPC for a license to operate the Passamaquoddy Tidal Power Project.[91]

Prosperity, Eastport's *huckleberry friend*, was just *'round the bend*.[92]*

Franklin Roosevelt got back in fighting form. The languorous days on *Larooca* were not just idylls. Florida's mythical waters may not have returned to Franklin his youth or the use of his legs, but they helped him regain his optimism and energy. He got back on T. R.'s trail.[93]*

When Franklin won the Democratic nomination for governor in September 1928, Roscoe, the red-blooded Republican who had known and admired Franklin for decades, did not stint in his praise: "From the time when, as editor of *The Crimson,* he fought for and obtained adequate fire escapes for the Harvard dormitories, he has always been fighting, generally for someone else." Roscoe even characterized Franklin's run as James Cox's vice presidential candidate as "fighting not so much for himself as for someone else." Roscoe continued: "Finally, he won his fight with death, but was left in a paralyzed condition from the waist down, a tragic fate for a crack tennis player, outdoor man and swimmer. His spirit was not broken and he continued his law work on crutches. On those crutches he was helped to Madison Square Garden, and as many who had known him in full health wept, unashamed, placed in nomination for the presidency his friend, Governor Smith."[94]*

Continuing his march to T. R.'s greatness, Franklin Roosevelt was elected governor of New York in 1928, in part by building on Al Smith's decade-old proposal for public power plants on the St. Lawrence River and attracting traditionally Republican rural voters to Franklin's plan for rural electrification.[95]* With his focus now on the Empire State, Franklin did not forget about the Pine Tree State, or Campobello, but it had been seven years since he was carried out of the big red cottage, and it would be several more before he piloted himself back to Passamaquoddy Bay. In the meantime, he had a powerful foe to fight.

Power wasn't just in hydroelectric plants; it was also in drinking water. In the decade following World War I, 1,200 mergers swallowed up more than 6,000 previously independent companies, and by 1929 some 200 companies controlled almost half of American industry.[96] The consolidations were squeezing Eastport, and Roscoe, now mayor, fought them.

Eastport led the opposition to the merger plans of the Maine State Water & Electric Companies (MSW&EC) at the Maine Public Utility Commission (MPUC) hearings in February 1929. MSW&EC was owned by a Boston-based holding company and represented by attorney William B. Skelton, MPUC's former chairman.[97] Eastport was joined by the towns of Sangerville, Guildford, Hartland, Old Town, Greenville, Norway, and Presque Isle, and was represented by attorney Charles E. Gurney, another former MPUC chairman.

The testimony showed MSW&EC had inflated the value of the fifteen companies by $1.1 million based on appraisals by their consultants. "It was contended by the opponents that this represented 'water' in the stock, on which the people in the towns concerned would later be called to pay a return, estimated at between $66,000 and $88,000 annually in the form of increased water rates." MSW&EC had purchased Eastport's water company for $181,000 but appraised it at $375,000, thus inflating the value by more than 100%.[98]

Under cross-examination, MSW&EC's appraiser was forced to admit "he had made an appraisal without visiting the town; that his appraisal was superficial; that he would neither invest his own money in bonds based on such an appraisal nor advise anyone else to do so; and that he didn't know what part of the plant was used or useful at the present time." This despite knowing the appraisal would be used to sell $2 million of securities to the public.[99]

In a stunning rebuke of Skelton, its former chairman, MPUC refused to allow the merger, saying, "We cannot see any sound economic reason to support the plan proposed"; MSW&EC would not be allowed to issue stock for sale to the public, and MSW&EC's actions were judged "utterly inconsistent with the spirit of the law if not an evasion."[100]

There were three outcomes of the water fight. First, MPUC prohibited MSW&EC from basing its rates on the inflated valuations. Second, and with a nod to Lubec's fine example, Eastport planned to acquire by eminent domain the local water company MSW&EC was going to swallow, but at a price determined by three honest appraisers. And third, Roscoe was renominated as the Republican candidate for mayor, without opposition.[101]

Dexter's deadline was fast approaching. His Canadian national charter had given him three years in which to begin construction. All the bureaucratic hoops he had to jump through had eaten up his precious time. Whereas in Keokuk his brother Hugh had gone from first visit to finished dam in under five years, Dexter had been working on Quoddy for a total of nine years, and full-time for five years, but not a drop of concrete had been poured.

In January 1929, Dexter's prospects dimmed when the fisheries committee stated, "1) In its opinion, if the proposed construction is carried out, the weir fisheries for herring inside the dams will be almost wholly eliminated; and 2) It is recognized that the effects on the fishers outside of the dams, predicted in the report on the subject by Doctor Huntsman, may follow, but the committee as a whole was not prepared to forecast whether these result will or will not follow, believing that a fuller investigation was necessary."[102] A "fuller investigation" would take time. Instead of clear sailing, Dexter would now have to tack back to the Canadian Parliament and ask for his charter to be extended. He asked for two years so it would conform with his New Brunswick charter.[103]

Having already spent most of his capital, on March 13, 1929, Dexter had to lay off most of his staff.[104] In the little island towns of Eastport and Campobello, the layoffs highlighted the financial impact Quoddy had on the people and heightened the already tense mood: "While the news of the hearing will be awaited here with interest amounting to anxiety, it is difficult to see how the Dominion parliament can do otherwise than to grant the petition."[105]

Ottawa, Canada's capital, is 500 miles west of Eastport. Dexter Cooper and his supporters from the St. Stephen Board of Trade, the St. George Town Council, and individuals from Grand Manan and Campobello, as well as Maine's former lieutenant governor, boarded the train together at Eastport on April 8, 1929, and headed west.[106] They made the trek to personally present their case for renewal of Dexter's Canadian charter, which would otherwise expire on June 15.[107] Opponents also headed west.

At the hearing, Mr. MacLean told the parliamentary committee that the St. Andrews tourist business would be endangered if the project were built. He also raised concern about who would be in charge if push ever came to shove, since the all-important powerhouse would be in the United States, not Canada. MacLean made sure the committee members understood the human cost of the project if it were to be the end of Black Harbor's herring fishery, where he personally employed 300 to 400 people, and where, *he said*, the industry as a whole furnished a living for 8,000 to 10,000 Canadian citizens.[108]* Dr. Huntsman stated definitively that fishing within the dams would be eliminated if they were built, though he admitted he was uncertain as to the effects outside the dams. E. P. Flintloft said the Canadian Pacific Railroad favored anything that would aid economic development, but if Huntsman was right, he would oppose the charter.

Dexter was put on the defensive by the strong opposition. A New Brunswick legislator made the point for him that cheap and abundant power could attract new industries to the region and reminded his colleagues that the charter required reimbursement for fishery losses.[109]

After three days of hearings, on April 16, 1929, it was time for the vote: 10 voted in favor and 25 were opposed. Extension denied.[110]

Dexter's Canadian national charter was not renewed. After nine years of work, after winning all state, provincial, federal, and Dominion approvals, after lining up the preliminary financing, and with construction financing in the works . . . it was over.

That was the end of Dexter Cooper's dream. What should have been a simple bureaucratic extension to allow the bureaucratically required processes to run their course became a way for a few strategically positioned politicians to kill Quoddy. "Killing of the Cooper Canadian

Charter Seems to Have Been a Pleasant Party Affair," Roscoe noted with scorn, "arranged for in advance" by the Liberals in the opposition party.[111]

But why would they kill a popularly supported $100 million, privately funded economic development project in favor of the relatively tiny and financially protected fishing industry? The *Bangor Commercial* suspected a "red herring." The *Lubec Herald* thought the reason was more sinister: "We do not care to echo the allegation that is being made, but we all remember the recent Aroostook railroad holdup; and history, especially the history of high finance, has a way of repeating itself."[112]

Dexter didn't accept the Dominion's decision. He protested to the US minister in Ottawa and the US State Department.[113] Then he vented his frustrations in a letter to the *Sentinel*: "We . . . have spent over $300,000 on our work here, and have not received approval of our plans or a charter extension. I consider this procedure confiscatory."[114]

On the Canadian side of the border, the *St. Croix Courier* was sympathetic toward Cooper but critical of its counterpart in Eastport: "[T]his paper has never waxed over enthusiastic over the benefits to be derived from a realization of Mr. Cooper's dream. The harm that it would cause to these communities was real, while the benefits to be derived, while wonderful to think about, were problematical. . . . [O]ne cannot but feel regret for the man whose greatest dream, and probably most important life work, has come to such a sudden ending."[115]

Canada's rejection may have been the end of what became known as the International Plan, but it wasn't the end of the Quoddy Tidal Power Project.

Canada's Liberal politicians and fishermen may have thought *they* killed Quoddy.

Roscoe was acerbic about Cooper's defeat in Canada. He began his commentary by saying, "Not many here realized the full import of the Ottawa hearing. So many other barriers had been surmounted, and so substantial a belief had grown up that up that the project must go on, that the possibility of its being blocked at the Dominion capital was not very seriously considered." Then he considered the consequences:

[T]he weir fishermen of Quoddy, who have been steadily growing poorer still have the inestimable privilege of keeping right on with their fishing . . . [and] keep right on exporting their most valuable product—namely their young men and young women,—who will more and more go to the regions governed with sufficient intelligence to welcome industrial development, with its factories and payrolls, leaving behind them, more and more, vacant homes and abandoned farms. . . . [While] Dr. Huntsman can still maintain his pleasant summer residence at St. Andrews, where congenial company abounds, and where are located the best golf courses in the Maritimes . . . and those others who for reasons peculiar to themselves have contributed to this result, are at liberty to derive whatever of profit or satisfaction obstructionists may feel in the evil work they have accomplished.[116]

The failure was "[t]o a very large degree . . . due to ineptness, over-confidence and lack of generalship in presenting the matter to them. Plenty of assistance could have been had in handling the matter,—such assistance as was ninety percent responsible for the success of the project in Maine,—but it was curtly, even brusquely ignored." Among Dexter's mistakes was hiring a legal team politically opposed to the government they were petitioning. Dexter was thrown on the defensive and failed to drive home the project's economic benefits; thus, "The reply to Dr. Huntsman was feeble, to say the least. That eminent scientist, whose characterization of St. Andrews Bay as a 'mixing bowl' savors more of the kitchen than of the laboratory, was allowed to get completely away with a line of what amounted . . . to nothing more or less than learned nonsense. . . . [H]is assertions, unchallenged and uncontroverted, undeniably had great weight as expert testimony with the Canadian officials, most of whom didn't know herring from shrimp."[117]

As the *St. Croix Courier* pointed out, no harm could have been done by granting Cooper the extension because he could not proceed with construction until he had approvals from three more government agencies, and there was plenty of time to address any issues they raised. Thus, said Roscoe, the preemptive cancellation "amounts to drowning the dog to get rid of the fleas." His anger dissipated, Roscoe returned to being an optimist:

The Project is of course not dead. The power is here, and Cooper's activities have so well advertised its presence that sometime, perhaps even earlier than we suppose, it will be developed. Niagara's power history is much the same as that of Quoddy's thus far. It had to pass through many discouragements and trials before development was actually achieved. It will be so here. Cooper himself says he is not through. Whether he means to try again for Dominion approval, or whether he will build on a smaller scale in Cobscook Bay, he has not yet made plain.[118]

Killing the Quoddy Tidal Power Project was to some just a tempest in Maine's teacup. But the ripples were evidence of seismic events and, like the tiny mid-ocean waves of a tsunami racing toward shore, were soon to rear up into a tidal wave that would engulf the entire country.

CHAPTER 7

Back in the USA (1929–1931)

THE ALL-AMERICAN PLAN

"All-American" was the promotional spin put on Dexter's fallback plan after his Canadian charter wasn't extended. An All-American project would be simpler in some ways, and the American economy was red hot, with the stock market up 20% in the first half of the year. Dexter's All-American Plan required about two-thirds of the International Plan's infrastructure but would generate only about half as much power—a still enormous 300,000 hp, and still economically "practicable."[1] Both the upper and the lower power pools would be within Cobscook Bay.

In addition, Dexter proposed to construct a 16.5-square-mile "pumped storage" reservoir and auxiliary power plant (like Ralph Clarkson had proposed for Cape Split in 1915)[2]* to ensure continuous power production. It would be located a dozen miles westward at Haycock Harbor and flood parts of Trescott, Cutler, and Whiting.[3]* While effective, the redundant infrastructure was costly.[4] The truly unique aspect of the reservoir was that it would be seawater, or, as Roscoe dubbed it, "The Great Salt Lake for the Pine Tree State."[5]

The dams, gates, and powerhouse were reconfigured for the All-American Plan, and a marine railway was added at Denbow Neck to lift boats into the upper pool. Dexter estimated the All-American Plan would cost $36 million[6] and that it could be incorporated into the full International Plan "if and when Canada gives her consent."[7] The *Sentinel* noted approvingly, "Cooper is apparently abandoning the extremely ill-advised policy of silence he had been following, and is making some effort to get his story into print."[8]

The *Sentinel* also applauded Central Maine Power's industrial survey as "a Wise Policy."[9] It was part of CMP's economic development efforts, and that little bit of free publicity primed the pump for the company's primary effort: selling stock in the company to *consumers*.

Beginning in July and continuing almost every week through September 4, CMP ran ads in the *Sentinel*. The first was for the new $10 million Wyman Dam on the Kennebec River. It would employ up to 1,200 men during construction and, at 100,000 hp, was as powerful as the combined output of CMP's 23 other hydroelectric stations. Its power would be "cheap enough to really ATTRACT the large power using industries." CMP encouraged readers to join the nearly 14,000 Maine people who were already CMP investors and to "Invest for Double Dividends: 6.12 per cent—Build Maine." In another ad they touted their bonds as "a home investment that has paid dividends for 23 years, that has a high degree of safety, fair yield and an opportunity to BUILD MAINE. It is a legal investment for Maine savings banks."[10]

But another policy was in question: In August, CMP and Bangor Hydro-Electric ran ads encouraging voters to repeal the Fernald Law power export ban in another statewide referendum. Asserting that "MAINE consumers pay for waste power," they said repeal would lead to lower electric rates and lower taxes.[11] By the time of the vote, "every telephone pole, fence post, and vacant show window [was] plastered with posters favoring export, and every grange and town hall filled with debate."[12]

Despite the plastering, Maine voters declined to repeal the ban on exporting power—though Washington County voted for it, for they had something to export.[13]

The *New York Word* rang the bell, and the *Sentinel* called the bout on September 4, 1929:

FIGHT FOR POWER RIGHTS IS ON
[T]he Cooper-Quoddy project, with its huge potential output may become one of the prizes or pawns in the battle of giant interests now in progress. . . .

Hydroelectric experts here say that the Insulls have demonstrated their cleverness in power contests by getting control of very valuable key properties in the East and they can make trouble for their rivals, and ultimately quit the territory in dispute with large profits.

The Federal Trade Commission, under the authority of the Walsh resolution to inquire into public utility activities, will begin in September a thorough investigation of the financial support of the forty or more groups which exercise control over power corporations. . . .

The Electric Bond and Share Company of New York refused to submit its books to the trade commission but the desired evidence can be brought out by interrogating witnesses under oath, and that the commission is prepared to do.[14]* . . .

Leading politicians in Washington assert that the power problem will be the principal issue in the next Presidential campaign.[15]

Among the "giants" of energy investors were Samuel Insull and Thomas Edison. Late in 1928, Edison had assured Americans, "Over every function of operation of this great public utility the Federal Government and State Governments have exercised the greatest scrutiny and control, this control being established essentially for the protection of those who use and pay for the commodity, electric energy."[16]

As financial writer John F. Wasik has explained, Insull Utilities Investments stock came on the market initially at $12 a share. Per Wasik, "Insull Utilities stock started 1929 at $25 a share, rose to $80 by the end of spring and was at more than $150 by the end of the summer. . . . Insull was personally worth more than $150 million [roughly $3 billion in 2020 dollars] by September. 'My God,' he exclaimed to an associate, 'A hundred and fifty million dollars! Do you know what I am going to do? I'm going to buy me an ocean liner!'" Instead, in October 1929, he formed another investment trust[17] (i.e., a holding company). So did a lot of entrepreneurs. At the beginning of 1929, one a day was being organized, and by September there were more than five hundred—building nothing, manufacturing nothing, selling nothing but paper shares in their paper selves.[18]

With the power of leverage, Insull grew his portfolio from 2,064 companies in 1927 to 4,045 by the end of 1929. During one fifty-day period, Insull securities appreciated at the rate of $7,000 *per minute*. "That kind of capital appreciation, not based on earnings growth, but on pure, insane optimism, brought 3 million Americans into the stock market. A third of the market players got into the market on margin. They only needed 10 to 15 percent in cash to buy securities; the remainder was lent by the brokers."[19]

Before Dexter could get funding from Thomas Edison, Samuel Insull, or other professional investors, he would need a new set of permits for his All-American Plan. Accordingly, he sent his revised application to the Federal Power Commission (FPC) on September 16, 1929. His timing was not opportune.

The Stock Market Crash started on "Black Thursday," October 24, 1929, almost exactly one year after Thomas Edison had assured the American people that public utility investments were safe. Within one week the American stock market lost more than 30% of its value. Prospects for Dexter's All-American Plan fell with every click of the ticker-tape machines as they punched holes in the value of overleveraged corporate stocks.

Trustworthy they weren't.

The Roaring Twenties were over. America collapsed into the Great Depression.

BLAME

The scale of the propaganda campaign was staggering. As documented in the Federal Trade Commission's 14,293-page report, the power trusts' propaganda campaign involved printing and distributing *one hundred million* flyers, pamphlets, and books, whose contents ranged from misleading to outright lies; millions of column inches of bogus newspaper articles and editorials—the equivalent of 25,929 full-page ads, which circulated through 12,784 newspapers with untold millions of readers in 1927 alone; and tens of thousands of paid speeches. Most galling of all is that *all* of this propaganda was conducted at public expense. The utility companies included the cost of these contortions of the truth in their operating expenses and therefore, as publicly chartered monopolies with a guaranteed rate of return, made a profit on it![20]

The campaign had been instigated in 1916 by Samuel Insull when he created the Committee on Public Utility Information (CPUI) with

Bernard Mullaney as its head.[21] The intent of this massive malediction can be understood from an editorial in *Electrical World* magazine:

> What is most to be feared from Washington this session, is the threatened investigation of the electric light and power companies, and in particular holding companies . . . not necessarily because of any shortcoming of the electric light and power industry as a whole, for it has an admirable record, but because of the detrimental effect of the publicity on the security market.[22]

As we saw in 1920–1921, CMP's advertisements in the *Sentinel* pitched its securities to "Investors who require Complete and unquestioned safety." Other companies hocked their stocks to "widows and orphans."[23]

A second objective of the propaganda was stated in 1927 by Arthur W. Stace, director of the Michigan CPUI: "[W]e are trying to 'kill' the municipal-ownership idea, consequently the information we want is information showing the failure of municipal ownership, its inefficiency compared with private operation, and the fact that municipal ownership, in the last analysis, is more expensive." To camouflage the killing, one power trust executive suggested cloaking it in "Red": "My idea would not be to try logic or reason, but to try to pin the Bolshevik idea on my opponent."[24]

The propagandists took advantage of the social media networks of the day, the Associated Press (AP) and United Press International (UPI), to disseminate fake news. This was particularly easy to do through the AP because of its federated organizational model, in which each newspaper acted as an agent for the whole system and could post features to the national network without review or vetting. The power trusts simply found a few friendly or corruptible editors to do their bidding, and then their slanted news articles were reproduced by hundreds of unsuspecting newspapers.[25]

An example of the power trusts' fudging was Murray & Flood's report on Niagara, which purported to prove private utility companies were more efficient than publicly owned utilities. In 1922, the National Electric Light Association (NELA) paid Murray & Flood $8,830 to write the slanted study and spent another $13,000 to have 10,000 copies printed and bound. This at a time when a new Ford Runabout automobile cost $265.[26]*

Another flagrant example was perpetrated by the *Boston Herald*, which on December 4, 1927, ran a full-page article titled "HERE IS THE TRUTH ABOUT ONTARIO HYDRO-ELECTRIC." In truth, it was fake news, which NELA agreed in advance to buy 10,000 copies of for $1,000, and in which were quoted such purportedly unbiased experts as James Mavor, emeritus professor of political economy at the University of Toronto; Professor A. E. Stewart, head of the Department of Agricultural Engineering at the University of Minnesota; and Dr. A. E. Kennelly, professor of electrical engineering at Harvard University—all of whom were secretly in the employ of NELA![27]

Thus, Roscoe Emery and the *Eastport Sentinel* were the innocent beneficiaries of the power trust's advertising largess and unwitting victims of its fake news, while Lubec's municipal water and power system and Eastport's efforts to create something similar were the targets of the power trusts' sabotage.

Samuel Insull was on the cover of *Time* magazine for the second time on November 4, 1929, one week after the crash. The article was about the grand opening of "his favorite plaything,"[28] the 21-story Chicago Opera building, known as "Insull's throne,"[29] over which Insull "rules as power primate."[30] *Time* likened the Opera's overwrought "design" to Insull's utility empire spreading from Illinois to Maine, that "almost blots out New Jersey and New Hampshire, parts of Pennsylvania and Kentucky." Most of *Time*, however, was about the panic on Wall Street and the fate of "margineers."

Following the lead of the famous Chicago retailer Sears-Roebuck, Insull covered the margin calls for his employee stock-buyers. When the City of Chicago couldn't meet its payroll, he helped raise $100 million to pay teachers, policemen, and firefighters. To deal with the national crisis, President Herbert Hoover summoned Insull and a group of business leaders to the White House and urged them to keep investing in their businesses, as Insull had done, to maintain employment and to declare that business conditions were "fundamentally sound."[31] They did.

Herbert Hoover had known the Insulls since at least 1925. He was then US secretary of commerce and the heir apparent to the president of the United States. Following in the laissez-faire, hands-off business tradition—which was often the *anything-is-fair* handouts to business tradition—of President McKinley and the Republican Party, Hoover enthusiastically supported Insull's vision of a privately owned, super-power electric grid stretching across the country. The grid should be free of not only government interference but also municipal competition (which was then but 3.5% of the nation's generating capacity). Having said as much in his keynote address at NELA's 1925 convention, Hoover's speech was presumably cheered and his campaign well funded.[32]*

Now, in 1930, the great friend of the power companies was president of the United States, but it was no longer a thriving country over which he presided; it was a collapsing country. Hoover appeared powerless to stop it. First he tried to ignore the crash, then wish it away, and only slowly did he realize real action was required.

One action Hoover took was appointing new members to the FPC: Chairman George Otis Smith, the former head of the Geological Survey; Colonel Marcel Garsaud, an engineer with close ties to the Electric Bond & Share Company; and Claude L. Draper, head of Wyoming's Public Service Commission. They immediately dismissed the chief accountant and the chief attorney, the FPC's two staffers best known for protecting the public interest. The FPC commissioners, the *New Republic* complained, "act[ed] precisely as they would have done if they had been taking orders from the power trust." Progressives were outraged. The Senate sued the Supreme Court to remove Smith, but the court concluded that since he had been seated by the Senate, only a president could unseat him.[33]

Again showing his friendliness to private power, President Hoover gave a helping hand to Dexter Cooper and the Quoddy Tidal Power Project. While not technically blocking Dexter's All-American Plan, fisheries concerns had remained an impediment because Dexter kept Canada's potential participation on the table. Therefore, the "fuller investigation" of the fisheries had to be done. Accordingly, on February 4, President Hoover recommended to Congress that $45,000 of public funds be expended on Dexter's behalf as the US government's share of a joint US-Canada investigation. Canada had proposed the joint investigation as a way to resolve the dispute created by its unjust refusal to extend Dexter's charter while its departmental investigations were still underway.[34]

Roscoe thought the money might be better spent investigating why the fisheries were already in decline, though many believed "mass destruction" by overfishing was the obvious culprit. Roscoe and others were also concerned the fisheries investigation was not about real herring but another "red herring" and a delay the opposition would use to make trouble.[35] Nonetheless, the US House approved the allocation in April, and the Senate in June.[36] An equal allocation was made by Canada, and a blue-ribbon panel began its two-year task.[37]

For Dexter, more encouraging news arrived on May 25, 1930, when the US Department of Commerce awarded him a patent.[38] Dexter's idea was not a Rube Goldberg–like contraption like those invented by Knowlton and others.[39] Instead, Dexter's patent was for a method of managing water flow in and out of a two-pool system with gates or pipes to increase the effective hydraulic head and thus the horsepower. In the case of Quoddy, Dexter estimated his method would increase the output by an astonishing 75,000 hp!

Like many great ideas, the secret was simple: configure the dams, head races, and tailraces such that you could draw water from whichever is higher, the high pool or the ocean, and send the tailrace water into the low pool or the ocean, whichever is lower. There is almost always a difference in height between the ocean and constricted bays because there is always a time lag in filling or emptying the bays. Thus, Dexter had realized the fundamental nature of a constriction and resulting differential—and how he could maximize it to generate power.

More good news arrived on August 6, when Congress granted Dexter P. Cooper, Inc. the right to develop and utilize tidal waters in the United States to generate power.[40]

With optimism born of progress, Moses Pike proposed to Laura Sleight. She accepted, and they were married on August 16, 1930, in West Lubec.[41]

With Canada's troublesome requirements removed from his list of Essentials, all Dexter needed now was number 9: money for construction.

Finding money in 1930 was a problem for everyone.[42] With his funds running low and the worldwide financial situation worsening, work at Cooper's office in Eastport "essentially ceased" by Thanksgiving Day. Even Miss Genevieve Roches, his secretary for the last five years, and an engineer on loan from Stone & Webster were laid off. Only Moses Pike would remain on the payroll for a few months. Roscoe reported that "cessation of work here has no special significance. Practically everything that can be done locally in the way of surveying, planning and other preliminaries has now been accomplished, and the field of activities has been transferred to New York or Boston, from which points negotiations looking toward government permits and financial support can be much more conveniently handled."[43]

Roscoe was attempting to put a positive spin on what was in reality another layoff adding to Eastport's economic woes. While the population continued to shrink, the ranks of the unemployed continued to grow.[44]

Roscoe wrote to the FPC on January 9, 1931, asking if, in consideration of Eastport's desperate condition, review and approval of Dexter's application could be sped up. The FPC replied on January 29 that they appreciated Roscoe's efforts and Eastport's plight, but the delay was not a bureaucratic issue, as Roscoe imagined: "The project appears to be mechanically practicable, but the corporation has not been able to make the necessary showing as to its ability to finance the work."[45]

THE CAMPAIGNS

As governor of New York, Franklin Roosevelt criticized the high price of electricity charged by private monopolies. In the spring of 1929, using rates from the publicly owned Ontario Hydroelectric Company for comparison, he graphically showed New Yorkers the costs of their high-priced electricity. He laid out a table with illustrations of fourteen household appliances—including lights, stoves, vacuums, and waffle irons—all of which a Toronto household could operate with $3.40 of electricity per month. In comparison, the cost for someone in Albany would be $9.90 per month, and for someone in New York City, $19.75. Extrapolating those costs, Roosevelt estimated private utilities were overcharging New Yorkers by $400 to $700 million per year, making those labor-saving electric devices the industry had been selling too expensive to use.[46]

While the economic injustice was felt keenly by working people, Republicans ridiculed Roosevelt's "waffle-iron campaign"—which got him reelected as governor and, in the words of Will Rogers, "threw him in the ring as the next Democratic candidate for President."[47]

As governor of the nation's most populous state, Franklin Roosevelt witnessed the stock market crash in October 1929 and the first three years of the Depression. With thousands of people unemployed and begging in the streets while the power barons luxuriated in Gilded Age castles, Roosevelt set out to level the field. In January 1930, he proposed developing public hydroelectric power plants on the St. Lawrence River and building competitive distribution networks, if necessary, to create fairly priced power for the people. Instead, the Republican-controlled state legislature created a commission to investigate the state's Public Service Commission.[48]

Back in 1927, Maine's governor Ralph Brewster had vetoed his own party's bill to allow hydroelectric power to be exported. Brewster was a progressive Republican who, according to *Current Events*, "refused to be bound by the dictates of the State G.O.P. machine." The "machine"—the running of the party based on relationships, favors, and money rather than public policy—didn't like rebellious Brewster. In retribution, the Republican machine, which according to the *Sentinel* was "closely identified with the Insull interests . . . decreed his political extinction."[49]

When Brewster completed his second two-year term in 1929, the machine-made candidate, William Tudor Gardiner, took over as governor. Gardiner reintroduced the power export bill, passed it, and signed it, but, sensing that anything so desperately wanted by the power trusts would probably benefit them more than the people, voters defeated it in a statewide referendum.

Meanwhile, Brewster set his sights on the next step in his political career, the US Senate, but he was defeated by the Republican machine in the 1928 primary and again in 1930. Setting his sights a bit lower, in 1932 Brewster entered the race to be US Representative from eastern Maine. Since he appeared likely to beat his four machine-made competitors, the Republican Party campaigned for the Democratic candidate, John G. Utterback. Utterback was elected to Congress by a slim margin. But before he could take his seat, hundreds of improperly cast votes were discovered, throwing the election into question. Accordingly, Governor Gardiner and his council asked the State Supreme Court whether they could count the improperly cast votes. The Supreme Court said, "No." But because it was an opinion, not a ruling, the council declared Utterback the winner. Brewster appealed but lost.[50]

Dawn was about to break on Downeast Maine, however, thanks to the Federal Trade Commission (FTC) investigation illuminating the shadowy financial dealings of the power trusts. Among the hundreds of unethical, unscrupulous, fraudulent, and illegal actions of the power trusts were bribes. Two of the three members of Maine's Executive Council who had voted against Brewster were revealed by the FTC to have been "on the secret Insull payroll in the export fight in 1929"[51]— and thus paid opposition to Brewster. In Maine, the Insulls were replaying their illicit effort to buy control of Illinois.

The moral of this story is that it's easier to make money if you control lawmakers who limit competition for you, sanction your artificially high prices, and give you a 75% discount on your property taxes.[52]*

Nationwide, the lights dimmed for one minute on October 31, 1931.[53] This was not a problem with the system; it was the electric industry's tribute on the death of Thomas Edison and a subtle reminder of how dependent the world had become on the commodity controlled by a few.

The fisheries issue remained contentious, with dueling facts and opinions appearing in publications. Overfishing was blamed for their decline, as was the consolidation of and collusion among the packing companies. Some claimed the Tidal Power Project would cause the death of the industry and local economies. Countering that view was Alexander Calder of Campobello, who pronounced in December 1931, "Our fisheries are wrecked beyond repair and will soon be submerged entirely. If our fishermen do not desert the sinking craft they will go down with it. There is only one lifeboat in sight—the Cooper project. Shall we take to that or go to the bottom with the wreck?"[54]

Two weeks later, reports circulated of an unnamed New York concern "not previously in the sardine business" that planned to build in Eastport a factory that could pack 200,000 cases annually. The prospect for such new jobs was surprisingly good news. But Sidney Stevens, a local attorney, was suspicious: "[It] appears to me to be a plain barefaced bluff." Since "the existing factories in Eastport cannot operate at over a third of their capacity," Stevens believed the rumor to be "a fraud invented as propaganda to offset the growing discontent of fishermen and to prevent them from seeking the Cooper remedy for their financial troubles." The instigator of the rumor, he believed, was "a mighty power trust," which was trying to block Cooper from entering its market and "never hesitated to use devious methods by which to accomplish its ends." Therefore, Stevens announced:

> I am prepared to back up my belief that this proposition is all a bluff by seeing that bluff, if it is such.
>
> I have a good site for a sardine factory, as good, I believe, as can be found in Eastport, and if that concern, if it exists, will give me a bond conditioned that it will erect and equip a sardine factory thereon during the year 1932 with even a capacity of 100,000 cases I will deliver to it as a free present an acre of land having over one hundred feet of shore frontage.
>
> Come on now, you mysterious concerns!
>
> Are you a "spook concern" or do you really exist?
>
> Verily, it is always my delight to see a bluff when I think I detect one.
>
> —Yours truly, SIDNEY STEVENS.[55]

No one took up Sidney Stevens's offer.

Financial "leverage" made the power trusts wildly rich. Here's one of the ways they did it. Let's say you bought ABC Electric Company for $100 million, paying 10% with cash and the rest with bank loans securitized by your other stock holdings. We'll assume ABC was making 7% annual profit, so you would get $7 million of cash at the end of one year, and in the meantime the value of ABC's capital stock appreciated 7%, so you get a "paper profit" of another $7 million, thus earning a theoretical total of $14 million on your $10 million investment, or a 140% return in one year (minus the bank interest). Pretty good. Now repeat. And repeat again. And again. Pretty soon you're a billionaire. To paraphrase Archimedes, "Give me a lever large enough and I can buy the world." To be effective, a lever must sit on a fulcrum—a pyramid. For ever-larger financial leverage, you need an ever-larger pyramid, which requires ever-increasing asset values (the 7% capital appreciation in our example).

Now imagine if, instead of 7% annual capital appreciation, you could generate 89%. Think of the leverage that would give you. This is not a hypothetical example. The Insulls did it in 1931 in the famously prosperous metropolis of Eastport.

The Insulls' holding company, the Eastern Public Service Company, bought 27 drinking water companies, mostly from the Maine State Water & Electric Company (MSW&EC), the company Eastport had successfully sued to prevent it from acquiring the Eastport Water Company at a 100% inflated value based on a fraudulent appraisal.[56]

The Insulls realized their 89% profit on MSW&EC in just three months when they sold it to—*themselves*, thus earning for themselves a "profit" of $5,825,318.09[57] *in the wink of an eye* and bolstering the value of their "watered stock" and their ability to sell more of it to more suckers.

With the "profits" from these transactions, the Insulls bought large power users, including manufacturing companies in Maine, to ensure for themselves a market for their electricity—*and to take control of more hydropower dam sites*.[58]*

Proving the pretense in the term "publicly owned" utility, in December 1931 the FTC revealed that despite the 14,000 individual investors in CMP (1.75% of Maine's 800,000 residents), 81% of CMP's stock was owned by a handful of the Insulls' managers.[59] The nominal "owners" of "public" utility stocks typically received annual dividends of perhaps 7%, whereas those tiny few who sat at the top of the pyramid and controlled the holding companies were raking in profits of 19% to 55%.[60]*

The Insulls' "leverage" turned upside down in early 1932 when the "security" the banks held as collateral for loans lost value. Accordingly, the banks made "margin calls" and demanded more security in the form of more stocks. But stocks were losing value, so the banks demanded cash, which meant that other stocks had to be liquidated, which drove down the price, which meant more stocks had to be sold, which further eroded the value . . . and so on in a vicious cycle until there was nothing left but worthless paper. Unable to stop the downward spiral, the Insulls' empire cracked, then crumbled, then disintegrated into dust.[61]*

The situation was just as bad at Insull-infected banks, like the Central Republic Bank of Chicago, where 90% of the bank's deposits had been lent to Insull companies in flagrant violation of state law, which forbade more than 10% of a bank's loans being made to a single company.[62] More corruption, in the form of half-price stock secretly sold to powerful people (who could resell it for an instant 100% profit), involved Anton Cermak, mayor of Chicago; South Trimble, clerk of the US House of Representatives; and Owen Young, president of General Electric.[63] The Insulls' financial collapse took with it the savings of 600,000 investors, ruining many of them. The $750 million in investor money lost made this the largest corporate collapse in US history.[64]*

"To keep the record straight," the *Portland Evening News* declared, and despite "the pledging of the last penny of collateral by the Central Maine Power Co. [CMP] to guarantee bond principal and interest," the Insulls' fraudulent financial dealings and liberal spreading of slush money to sway the Fernald Law repeal "should not be forgotten." Nor should they be believed that the FTC investigation caused the collapse of their empire and their brilliant, $100 million, fifteen-year program to develop 500,000 hp from the Kennebec River. (A program that would have taken at least ten years longer and been more expensive per horsepower than Dexter's International Plan.) The *News* thought it was "extremely doubtful whether this dream of a bigger, better and electrically busier Kennebec River would ever have eventuated," and then it chided Maine's power trust:

> The interests attempted to buy an election; failing, they attempt to excul-
> pate themselves for their offense against political morality by raising sweet
> visions of the stately structure they would have reared had not obstinacy,
> contumacy, *lese majeste* amongst a mulish electorate defeated them.
>   "Just for that," the interests say, "there won't be any big project."
>   Well, there wouldn't have been any anyway. . . . And all these sugges-
> tions that Maine has bitten the hand that fed it are but rubbish, camouflage,
> and cheap attempts at exculpation.[65]*

In a few years, there would be another example of CMP biting the hands that fed it.[66]

Cleaning up the rat's nest of the Insulls' empire would take more than fifteen years. The mess in Maine, according to one of the federal government's chief untanglers, was "one of the most fantastic examples of monopoly gone wild."[67]*

With American capitalism in shambles and the Depression spreading like a disease, Dexter Cooper left the United States on February 6, 1932, and went to work for the Communists.[68] Instead of using his talents to build the world's *first* tidal-hydroelectric power plant, he helped his brother Hugh build the world's *largest* hydroelectric power plant for a batch of Bolsheviks in Soviet Russia. It was a job.

On the eve of his departure,[69] Dexter told the *Sentinel* he would return to the Quoddy Project when economic conditions improved: "The present depression can not last much longer, and an increased demand for electric power will be one of the very first results of returning activity."[70]

Presidential candidate Franklin Roosevelt attacked the "Insulls and Ishmaels" as the personification of corporate evil in a speech on April 18, 1932. To Roosevelt, the Insulls' greed and lies were the root cause of the ruin of the entire American economy. Some viewed Roosevelt's vehemence as demagoguery,[71] but he knew more about electricity and the Insulls' activities than the public and the press, and maybe more than the FTC, for he may have known what Dexter knew.

Roosevelt kept attacking the power trusts and their Republican allies in a summer-long series of speeches that addressed the historic changes in the country: from a rustic colony to a rapidly growing agrarian nation of small farms that pushed America's frontier to its continental destiny. And now even more rapid industrialization was concentrating the nation's population in the cities and its wealth in a few individuals.[72]*

Through all of this, Roosevelt said, the role of government had changed, from defending personal liberty and national security to fostering the great enterprises like the railroads that brought the nation together and allowed it to fully utilize its vast resources with which to raise the standard of living. And now the vast power of the corporations—power that had been given to them through government grants of money and monopoly—needed to be regulated to serve all the people, not just the few who sat on top of the corporate pyramid.[73]

In Roosevelt's speech on September 21 at Portland, Oregon, he declared that "the question of power, of electrical development and distribution, is primarily a national problem."[74]* He then stated his philosophy that "the object of government is the welfare of the people," and "when the interests of the many are concerned[,] the interests of the few must yield."[75] He used "simple, honest terms" to describe the "principles" of "sound government." Such honesty was needed, he said, because,

> as the Federal Trade Commission has shown, a systematic, subtle, deliberate and unprincipled campaign of misinformation, of propaganda,

and, if I may use the words, of lies and falsehood. The spreading of this information has been bought and paid for by certain great private utility corporations. . . . A false public policy has been spread throughout the land, through the use of every means, from the innocent school teacher down to a certainly less innocent former chairman of the Republican National Committee itself.[76]

Roosevelt then explained the origins of government regulation of monopolies, starting with England's King James (1566–1625), who, when confronted by predatory practices of ferry boat operators, asked a judge to investigate. The judge determined ferry operators controlled the free flow of transportation and were therefore different from other business in that they were "vested with a public character" and, as such, "ought to be under public regulation." But, added Roosevelt, "the public service commissions of many states have often failed to live up to the very high purpose for which they were created. . . . In many instances their selection has been obtained by the public utility corporations themselves. And we can prove it!"

The corruption of the state regulatory agencies and the judiciary was one reason the industry needed to be regulated at the national level. Another was that utility holding companies were operating across state lines and using an absurd court ruling, dubbed the "Attleboro gap," that prohibited states from regulating the out-of-state operations of companies *and* the in-state operations of companies based out of state![77]* Thus, the industry, "this 'lusty younger child' of the United States," as Roosevelt put it, "needs to be kept very closely under the watchful eye of its parent—the people of the United States." He continued:

The Insull failure has done more to open the eyes of the American public to the truth than anything [else] that has happened. It shows us that the development of these financial monstrosities was such as to compel inevitable and ultimate ruin. . . .

As always, the public paid and paid dearly . . . [and] is beginning to understand the need for reform after the same public has been fleeced out of millions of dollars.

I have spoken on several occasions of a "new deal" for the American people. . . . It can be applied . . . my friends, to the relationship between the electric utilities on the one side, and the consumer and the investor on the other.[78]

Franklin Roosevelt then laid out his New Deal Public Power Policy in eight points, including transparency of the ownership and finances of utilities, the national regulation of the industry, restrictions on holding companies, the outlawing of deceptive propaganda and securities selling practices, and an end to selling off or giving away the public's common law rights to hydropower sites. While he was not in favor of nationalizing the electric utilities (as Insull had proposed for England),[79]* Franklin Roosevelt argued for publicly owned electric power plants, saying that such plants would

> create a yardstick . . . [that] will, in most cases, guarantee good service and low rates to its population. I might call the right of the people to own and operate their own utility something like this—a "birch rod" in the cupboard to be taken out and used only when the "child" gets beyond the point where a mere scolding doesn't do any good. . . .
>
> The history of the Federal Power Commission . . . the Muscle Shoals veto, and the closing of the White House doors to the public interest in the St. Lawrence project—all demonstrate that the policy of the present Republican leadership is dominated by private rather than public interest. . . .[80]*
>
> My friends, judge me by the enemies I have made. Yes, judge me by the selfish purposes of these utility leaders who have talked of radicalism while they were selling watered stock to the people and using our schools to deceive the coming generation.[81]*

Everyone was now *yelling for* the Insulls' heads.[82] To escape prosecution, they fled: Samuel to France, Martin to Canada.[83] Once the Insulls' great friend, Herbert Hoover (no doubt reluctantly) signed the warrants for their arrest.[84]

Far from demagoguery, Roosevelt's "Portland Address" was, according to historian Phillip Funigiello, "the most forthright and lucid statement of policy to emanate from either presidential candidate . . . a milestone in a national power policy."[85]

President Franklin Roosevelt with his dog Fala and
Ruthie Bie in Hyde Park, New York, in 1941 ▶

CHAPTER 8

Despair and Hope (1929–1934)

COMMUNISM VERSUS CAPITALISM

Vladimir Lenin defined Communism in 1920 as "Soviet government plus electrification," for "industry cannot be developed without electrification," and without electrification, Soviet Russia would remain a "small peasant economy."[1] Having fought through six years of World War I, followed by two years of civil war, which in combination killed ten to fifteen million of its citizens, Russia was economically devastated. Rather than the troops being supplied in an orderly fashion, the combatants pillaged, taking whatever foodstuffs they wanted, often including the seed grain, leaving nothing to eat and nothing to grow. A drought in 1921 caused famine and another five million deaths.[2] For Russia, the postwar 1920s were not roaring but depressing.

Hugh Cooper had been called before the US Congress to testify about the Muscle Shoals Dam in 1922. He testified that despite World War I being over, the chief of the Army Engineers had told Hugh he "owed it to the government to keep right on [working] . . . and do this work for nothing, and I have been fool enough to do it." Hugh thought this "a little strange, in my experience at least, to be the designer and responsible for the execution of plans for forty or fifty million dollars of work for $1 a year. . . . [T]hat is another reason I would like to see somebody hurry up and get this dam done."[3]*

Hugh Cooper's public testimony that he had to redesign the dam attracted a lot of attention and aggravated his already antagonistic relationship with many Army Engineers. The delay Hugh alluded to was due to the political war over whether the nitrate and power plant should be owned and operated by the public or by private enterprises. The bickering left the complex sitting for two years unfinished, unused, and apparently abandoned.[4]

America's most popular movie star, syndicated newspaper columnist, cowboy, *and* Cherokee Indian, Will Rogers, enlightened his readers about the cause: "When you see a hundred-and-fifty-million-dollar plant lying here idle, it gives you an idea of the pull in the legislation that the Power Trust exerts. They say 'If we don't get it, nobody else will.'"[5]

Henry Ford had proposed to solve the impasse at Muscle Shoals in 1921 by buying it for $5 million up front and $46,547 a year for a century.[6] He claimed his offer was sufficient for the government to recoup its investment if the income was invested at a 4% return.[7] Ford promised to develop the poor surrounding area into the "Detroit of the South."[8] Ford obtained surprising support from labor unions and powerful people, including Secretary of Commerce Herbert Hoover and Thomas Edison.[9] Despite his low offer, Ford's level of interest in the plant can be understood from his response to the question "Is there anything you would rather have than another billion dollars?" Ford replied, "Muscle Shoals."[10]

Former US secretary of war Newton D. Baker[11] was similarly enthused: "If I were greedy for power over my fellow men, I would rather control Muscle Shoals than to be continuously elected President of the United States."[12] There would soon be two men who would try to do both.

Had Ford built and paid for Muscle Shoals himself, Hugh Cooper would have cheered him on. But Ford hadn't, and Hugh was aghast at Ford's covetousness and misleading financial ploy. Hugh charged during the congressional hearing that if Ford's offer was accepted, it would cost the government $1.275 billion in lost revenue and would be "a tremendous calamity to the South if the greatest water power they have should be taken out of the field of public utility for 100 years."[13] Sometime later Hugh added, "I know selfishness is one of the strongest motives in the human heart, and I do not believe that the vast possibilities of the Tennessee River should go to anyone uncontrolled. There should be competition in waterpower as in everything else."[14]

Alabama Power Company also made a low offer for the power plant.[15] It, too, was repulsed by the Progressives, who wanted it to be operated by the government. Not wanting anyone else to get it, the private power and electrochemical companies kept the $150 million asset in a state of unfinished uselessness.[16]*

Seeing a man with means and motive, Dexter Cooper saw an opportunity and approached Ford as a potential investor in the Quoddy Tidal Power Project. Ford withdrew his offer for Muscle Shoals and made public comments about the practicality of tidal power.[17]* Lubec saw this as the town's opportunity and in 1924 offered to Ford all the land he needed for his vision—*for free.* Ford replied that he would accept Lubec's offer *if* he decided to build.[18] It wasn't really a "yes," but it added to Quoddy's excitement.

Instead, Ford set up several new factories for his single biggest customer.[19]*

Hugh Cooper, a capitalist, demanded payment in advance, not expecting his $50,000 consulting fee for four months of exploratory work would be acceptable to the Bolsheviks who had said they wanted to hire him. But in 1926 Soviet Russia wanted to build a very big hydroelectric power dam and knew it needed outside expertise.

No doubt Hugh Cooper's outrageous fee set off a fierce debate among the Communist leaders, who believed everyone should be paid the same—and whose salaries were only a tiny fraction of what Hugh was demanding. Hugh set his price and terms so that he had no risk: his efforts would not be speculative but work for hire.

The Soviets paid the fee, in advance, in gold.[20] It was the highest-priced consulting contract the Soviet Union had ever paid.[21] Hugh set to work on what would be the world's largest hydroelectric power plant. It was to be built on the Dnieper River in Ukraine and originally was to be managed by Leon Trotsky.[22]*

In 1927, the Soviets hired Hugh to manage the actual construction. In addition to the dam, Cooper began building a "garden city" with brick cottages for his American staff, six tennis courts (two concrete, four clay), and a golf course.[23] As Hugh described it, the Dnieper Dam would be "the center of a vast super-power system for the support of existing and new future industries within a radius of 275 miles. The area served will be about 200,000 square miles . . . [and] support an industrial population of approximately 16,000,000 or more than twice the present population of our six New England States."[24]

Practically all the construction equipment for the more than $100 million project was supplied from the United States. The nine enormous turbines were made by Newport News Shipbuilding, and 130 railroad carloads of equipment, including five generators, were made by GE. Hedging their bets, the Russians had four other generators built by Siemens of Germany.[25]*

Over a six-year period, Hugh Cooper spent at least one month each year in Russia, during which time he got to know a few Bolsheviks.[26]

GOLERO was the Soviets' 1928 plan for postwar economic recovery and development, its first Five-Year Plan. In Lenin's words, its objectives were "the organization of industry on the basis of modern, advanced technology, of electrification which will . . . put an end to the division between town and country . . . and to overcome, even in the most remote corners of land, backwardness, ignorance, poverty, disease, and barbarism." To quadruple the national power production to 8.8 billion kWh annually by 1932, GOLERO included plans for building 30 electric power plants. GOLERO was not designed to meet the existing demand for electricity; it was meant to create demand, to create the industries that would employ and feed the hungry masses.[27]*

In 1930, when America was falling into the Great Depression, Russia was powering out of it.[28]*

Joseph Stalin liked the mild-mannered American capitalist who sat next to him on that sunny day in 1930. They were at the Soviet leader's house in the Caucasus region of Soviet Georgia. Lunch was served outside, in the shade of a tree, at a long table with a white tablecloth. Stalin wore a white linen jacket and a tie, which accentuated his thick dark hair and mustache and his rugged vigor. The older, conservative Hugh Cooper wore a dark jacket and tie, his bald pate aglow in the sun. The wine and the relaxed setting helped to make the conversation flow. Mrs. Cooper took out her Kodak Brownie and clicked a snapshot of her husband and host. Philosophically poles apart, the communist and the capitalist nonetheless found their discussions over those two days to be interesting and insightful. They impressed each other. They would talk again.

Construction of the Dnieper Dam was going well. Hugh Cooper was witnessing, indeed leading, Russia's accelerated industrialization. As the damming of a river compresses the gravitational force and potential power into a short distance, the Soviet Five-Year Plan forced forward Russia's economic development, driving its citizenry from an illiterate peasantry to an industrialized infantry, at a double-time march.

The United States was technologically far in the lead, but the Soviets were determined to catch up fast. Helping them were Ford and GE. Ford set up tractor, truck, and car factories and is believed to have been primarily responsible for the increase in tractors in Russian from 26,000 in 1928 to 477,000 in 1936.[29] GE supplied Russia with all the equipment and expertise to set up its entire electric industry. All told, there were dozens of American companies and 1,500 technical advisers working in Russia in 1929.[30]

From 1927 to 1931, Russian purchases of American exports had grown from $65 million to $104 million, a 60% increase during the Depression when overall American exports dropped 50%. Russia became America's seventh-largest customer and the largest customer for industrial machinery.[31] The sales and technology transfers were driven by near-term business interests, but the personal relationships and technical standards shared by American and Russian engineers would prove critical to a very different cooperative effort in years to come.[32]

Hugh Cooper's conviction that capitalism was the best economic system to advance the welfare of humankind was not shaken by the stock market crash. He also believed we could learn a lot from Communist Russia, as he explained in a speech on August 1, 1930:

> [W]e cannot ignore the 150,000,000 people in Russia, who occupy one-sixth of the world's landed area and hold in their possession a wealth of undeveloped natural resources undoubtedly greater than are to be found in the balance of Europe, or deny that they represent a major problem in the world today from the standpoint of universal peace. . . .
>
> Churchill says, first, that if it had not been for Russia [in World War I], Germany would have conquered France in sixty days. . . . [T]he losses in the Russian ranks during this period were nearly as great as all the losses of all the allies. . . . Had it not been for Russia the Central Powers would have won before we in the United States could even think of preparing for war. . . . What our sacrifices and what our economic situation would be today,

if to save the world for democratic government, we had been forced to
conquer a victorious Central Powers, defies my imagination.

. . . It would seem to me that these sacrifices produced communism and
that these people have earned the right to try it out and see for themselves
whether it is really the road to Utopia. . . .

When we examine Communism, we find that, after all, it is an expression of dissatisfaction with existing economic conditions. . . . [I]n Germany
today over 2,300,000 unemployed are receiving state aid, and nearly this
same number of unemployed in England are on the dole also, [so] we
should not be greatly surprised if Moscow erroneously believes a world
revolution may be approaching. . . .

There are three special thoughts I want to leave with you:

The first is that communism means that the state shall own and control
all property and that its people shall in a very large degree live in equality
with one another . . . [and it] is designed to eradicate human selfishness on
a scale that has never been attempted before anywhere in the world.

The second thought is that these people are capable of great achievement in time. Also that they are blessed with a wealth of natural resources
that under any kind of rational development will have a formidable influence not only upon their own people but also on the economic life of the
outside world. . . .

The third. . . . Whether a permanently satisfactory industrialization can
ever be brought about with the words "commercial competition" taken
out of the formula or whether this plan will result in the breaking down
of personal initiative is a question we will all await the answer to with
deepest interest. . . .

Russia and the United States of America have within their combined
boundaries about half of the known natural resources of the world. In the
last analysis, natural resources are the foundation of all human progress,
and the degree of intelligence employed in their development determines
the health and happiness of all people.

The United States is developing its natural resources under the private
ownership plan and will continue to do so. Russia is developing its natural
resources under the communistic or state ownership plan and will continue
to do so until its leaders learn from experience that they must change. The
future economic history of the world largely depends on the outcome of this
un-avoidable competition.[33]

These issues—the intelligent development (and sustainable use) of
natural resources, and competition—would prove central to the story of
Quoddy and to the debate within the United States about who controls
resources and who benefits from them.[34]*

Communism was an "awful disease," Hugh Cooper thought, and would "erode . . . gradually from within" because the Soviet economy could not function without adequate incentives. Hugh's prediction was partially confirmed in July 1931 when Stalin announced his New Economic Plan (NEP), in which he called for material incentives.[35]

Hugh recognized the positive effect Stalin's reforms would have on American opinion, and he hailed the NEP as "revolutionary."[36] Hugh may well have had something to do with that revolutionary change in Stalin's thinking. As the *Christian Science Monitor* reported, Hugh was "one of the few foreigners who enjoy ready access to Mr. Joseph Stalin, and problems of the Soviet economic development have been the subject of more than one personal conversation between the Communist leader and the American engineer."[37]

Stalin wouldn't be the only world leader to call on Hugh Cooper— the dam builder without a college education—for advice on domestic and international affairs.

The Dnieper Dam was the crown jewel of the Soviet Union's first Five Year Plan.[38] It was completed on schedule on October 10, 1932. During the dedication, Hugh Cooper was awarded the Order of the Red Star, the first foreigner to receive the highest civilian honor bestowed by the Soviet government.[39] Two hundred feet below Hugh, as the Bolsheviks pinned their medal on his chest, was a hollow in Hugh's concrete monolith. The cavity was made intentionally, its purpose a secret.

THE COMMANDER IN CHIEF

Franklin's personal Arc de Triomphe was Lubec Narrows. That was where he chose to celebrate his personal successes and exercise his personal powers.

Back in 1913, just three months after becoming the youngest assistant secretary of the navy in US history, he had piloted the destroyer *Flusser* through the fast currents of Lubec Narrows, arriving in Eastport as the local hero. He did it again eight days after being nominated to run for vice president of the United States in 1920, with a record-breaking run from Boston. Now, on June 29, 1933, having been president of the United States for just three months, he approached Lubec Narrows as a national hero—with three ships and 1,200 fighting men[40]—intent on exorcising a demon that had haunted him for twelve years.

He may have been busy, and Campobello may have been a long way from New York, but it wasn't time or distance that prevented Roosevelt from returning to what had been his "beloved island." Surely it was the painful memories of losing the life he loved—his life of active outdoor recreation—which had been shattered by polio.[41]

But Franklin's spirit hadn't been shattered. He might be a cripple, but he was the most powerful man on the earth. Sailing through his personal Arc, President Franklin Roosevelt sailed out of purgatory and into Passamaquoddy Bay.

Eastport was ecstatic. There could hardly be anything more exciting for a small town than the local boy who made good coming home—with thousands of followers to boot! From the day the president's visit had been announced, the excitement had been building. As chronicled in the *Sentinel*, the president would be sailing with his sons aboard a friend's 46-foot yacht, the *Amberjack II*, on a two-week cruise from Marion, Massachusetts, to Campobello.

Eastport hired professional decorators to trim the town with flags, bunting, and "Welcome" signs and scheduled a parade, a baseball game, and a ballroom dance for the visitors.[42] Lean and lonely from the Depression, Lubec was agog at all the strangers in fancy cars with out-of-state license plates who arrived hoping to get a glimpse of the president.[43]

Franklin insisted on riding in an open-air touring car so people could see their president, despite having narrowly survived an assassination attempt on February 15. The attempt was foiled by Mrs. Lillian Cross, who hit the assassin's arm with her handbag, deflecting the bullet, which then killed Chicago's mayor, Anton Cermak. It was another eerie parallel with T. R. and proof that neither disease nor assassins could kill a Roosevelt president.[44]

It was going to be a grand party in Eastport, now that Prohibition was almost over, President Franklin Roosevelt having signed the Cullen-Harrison Act legalizing beer. (Prohibition was fully uncorked at year's end.)[45]

At the time of Franklin's inauguration, the *Sentinel* had said, "The hope of direct benefit from the fact that President Roosevelt is a former summer resident here is probably not a very dependable one." But now, with more than a little hope, "It has been intimated that the Administration is interested in the Quoddy Tidal Power project."[46]

Eastporters all remembered Franklin's interest in Quoddy a dozen years before, when he was just the assistant secretary of the navy and married to the president's niece. Now he was married to the president's wife!

Franklin was living proof that no matter what hardships you faced, you could overcome. If the expression "walk the talk" existed in 1932, it could not have better characterized Franklin's performance as the Democratic nominee for president. Polio had "tempered" and toughened Franklin, the child of privilege, much as the death of Theodore's wife and the danger of battle had turned "Teddy" into the lovable "Rough Rider." Polio forced on Franklin humility and empathy for people less fortunate than himself. He was no longer the dashing and athletic young cousin of the president. He was older and crippled, yet he stood on his own two feet.

Here he was, the man who could lead the nation out of the Depression.

Breaking precedent, Roosevelt flew to Chicago to accept his party's nomination in person, telling the delegates, "I pledge to you, I pledge myself, a new deal for the American people. Let all of us here assembled constitute ourselves prophets of a new order of competence and of courage. This is more than a political campaign; it is a call to arms. Give me your help, not to win votes alone, but to win in this crusade to restore American to its own greatness."[47]

In his acceptance speech, Franklin promised a vision for the future in which the government would act directly to fix the economy and support its people. Franklin's personality and plans were in stark contrast with his opponent, then president Herbert Hoover, a man who rarely stepped out of his office and did nothing to engender the people's affection.

The people judged Herbert Hoover and found him lacking. In a landslide on November 8, 1932, Franklin Delano Roosevelt was elected as the 32nd President of the United States. Roosevelt won 42 of the 48 states—but not Maine.

It would be four shockingly bad months before Franklin was sworn into office. He didn't wait to be given the power of the presidency; he took it.

President-elect Franklin Roosevelt visited Muscle Shoals in January 1933 with Progressive senator George W. Norris (R-NB). Back in 1926, Norris had warned that if the federal government didn't get into the electric power generation business at Muscle Shoals, "we will find ourselves in the grips of a privately owned, privately managed monopoly and it will be extremely difficult to shake off the shackles that will then be fastened upon us all."[48]

By 1932, and despite Theodore Roosevelt's warning, the congressional investigations, and the abuses exposed by Insull's corporate collapse, the electric industry had continued to consolidate, such that just eight holding companies controlled 71% of the industry.[49]

When Franklin saw the completed but abandoned $100-plus million dam, powerhouse, and nitrate plant at Muscle Shoals, he was tremendously impressed and enthralled by Senator Norris's vision for the Tennessee Valley, saying, "[W]e have an opportunity of setting an example of planning, not just for ourselves but for future generations to come, tying in industry and agriculture and forestry and flood prevention, tying them all into a unified whole over a distance of a thousand miles so that we can afford better opportunities and better places for living for millions of yet unborn."[50]

This was not the Tripp-Westinghouse *capitalistic* vision for future generations. It was Theodore Roosevelt's *progressive* and *conservative* vision.[51]* Franklin Roosevelt quickly turned his enthusiasm into action. Wartime imperatives had galvanized Muscle Shoals into existence. Peaceful prosperity had paralyzed it. Another national emergency, the Depression, reenergized it.

Three days after President-elect Roosevelt visited Muscle Shoals, the electric monopoly in Tennessee hired their own new president.[52]

On Inauguration Day, March 4, 1933, the American economy was in free fall. One in four workers, about 30 million people, were unemployed. Agricultural prices had fallen by 60%, bankrupting hundreds of thousands of farmers. Industrial production had fallen by half. Two million people were homeless. With mobs of desperate depositors trying to get their cash out before their local banks collapsed (5,000 already had), 32 of the 48 states closed their banking systems. The panic that started on Wall Street in 1929 was spreading nationwide. But in his inaugural address, President Franklin Roosevelt assured the American people:

[T]he only thing we have to fear is fear itself—nameless, unreasoning, unjustified terror which paralyzes needed efforts to convert retreat into advance. . . .

I am prepared under my constitutional duty to recommend the measures that a stricken nation in the midst of a stricken world may require. . . . But . . . in the event that the national emergency is still critical I shall not evade the clear course of duty that will then confront me. I shall ask Congress for the one remaining instrument to meet the crisis—broad Executive power to wage war against the emergency, as great as the power that would be given me if we were in fact invaded by a foreign foe. . . .

The people of the United States have not failed. In their need they have registered a mandate that they want direct, vigorous action. They have asked for discipline and action under leadership. They have made me the present instrument of their wishes. In the spirit of the gift, I take it.[53]

"President Roosevelt completely won the approval of his countrymen by his fine inaugural address," said Roscoe in an editorial. "If his administration succeeds in putting into effect the purposes, and remains true to the ideals therein expressed, the American people will be fortunate indeed in having such a leader in the White House during the long and probably painful process of beating back from the condition in which we now find ourselves."[54]* Roscoe predicted, "The 'New Deal' will go far toward becoming a square deal if it protects us from these . . . financial wolves and more power to the President in his efforts in this direction."[55]

Just two days after his inauguration, President Roosevelt ordered *every* financial institution in the country to take a four-day "bank holiday." By creating calm, and time for the Emergency Banking Act to be hastily written, approved—unanimously—and implemented, he prevented panic and the collapse of many financial institutions, demonstrating exactly what he had promised the American people in his inaugural address. Said the *Sentinel*:

The swift and decisive action of President Roosevelt in declaring a banking holiday for the nation was undoubtedly essential and came none too soon. . . . It has already changed the psychology of the American people. We have come to realize that we are "all in the same boat." . . . Indeed, there is something akin to the war spirit arising in the nation. People are beginning to buckle on their swords, figuratively speaking, and getting ready to go out and battle against the depression.[56]

One month after the bank holiday, Maine had the lowest percentage of banks in the country solvent enough to reopen. As the *Nation* explained:

> Its disastrous bank situation may be ascribed in no small degree to laxity of bank supervision and connivance at patent abuses by the State authorities during the four-year [Republican] administration of William Tudor Gardiner. . . . In the State whose motto is *Dirigo*, "I Direct," the directing came from the heads of an interlocking group of power companies, chain banks, newspapers, textile industries, department stores and political machines. As Donald Richberg rightly points out . . . "Economic power depends on political power." If the American people will learn as a result of their staggering losses the price of vested privilege, the depression may not be wholly in vain.[57]*

If not, then the consequences would be even more staggering. President Roosevelt resolved to prevent those consequences from materializing. He used the crisis as the opportunity for fundamental change. The bank holiday set the tone and pace for what would become the most impressive legislative accomplishment in US history: Roosevelt's "First 100 Days."

International boundaries weren't going to stop Roosevelt from getting what he wanted. When the United States asked France to extradite Samuel Insull, he moved to Italy, and then to Greece, which didn't have an extradition treaty with the United States. In Greece, Insull became a much-sought-after dinner guest and was twice tried (at the request of the United States) but acquitted, thus fending off extradition.[58]

A frustrated Roosevelt threatened Greece: if the little country didn't turn Insull over, he would embargo American trade with it. Sufficiently intimidated, Greece asked Insull to leave.[59] Obliging, Insull slipped out of the country in disguise.[60] All the while, he insisted on his innocence and even wrote to Roosevelt that he would come back voluntarily if he were tried alone (any fault was his own, he said, not his associates') and in federal court (because he feared he couldn't get a fair trial in Chicago).[61]

Samuel Insull was later found and arrested in Turkey, from which he was extradited back to the United States in March 1934 to face federal charges.[62]* Meanwhile, Martin Insull was extradited from Canada.[63] The central charges in the trial were that Insull's stock was heavily "watered" and that he had made a series of fraudulent transactions designed to milk money from investors.[64]

The devil was in the details, and the volume and complexity of the case confused nearly everyone. The prosecution's logic was simple: phantom profits created false dividends to lure more buyers of the stock.[65] Quoting from the Bible, the prosecutor warned the jury, "Beware of false prophets, which come to you in sheep's clothing, but inwardly they are ravening wolves."[66]

As the case wore on, many of the facts seemed to align with Insull's defense: that the bewildering blizzard of transactions—in which he spent all of his own money—were his desperate attempts to save his companies from bankruptcy.[67]

Insull was acquitted of all charges on November 24, 1934.[68] Nothing he had done was illegal—*yet*. There was plenty that was unethical, but in the power barons' game of thrones, "Anything Goes."[69]* Or, as the *Nation* observed, the verdict "illustrates once more the difficulty of sending a rich man to jail, no matter how flagrant his crime."[70]

THE DEAL MAKER

FDR, as President Franklin Delano Roosevelt was now called to eliminate any confusion with President Theodore Roosevelt (T. R.), continued reaching out to the people, who needed to see and hear their new leader. Barely a week in office, FDR used the newly popular radio technology to present his first "fireside chat," during which he explained to a vast audience his policies and how he would implement the New Deal.[71]

The New Deal intended to fundamentally change the US economy and social structure through "Relief, Recovery, and Reform." Relief programs were supports for the unemployed; recovery was to boost the economy back to health; and reform meant fixing what was wrong, especially the financial and utility industries. An "alphabet soup" of 42 new government agencies was created between 1933 and 1939 to implement these programs,[72] including the Federal Emergency Relief Administration (FERA), the Social Security Board (SSB), the Federal Deposit Insurance Corporation (FDIC), and the Securities and Exchange Commission (SEC). These agencies gave FDR control of the entire US

economy, and he steered it as decisively as he had steered the fastest
ships in the US Navy through treacherous tides in Lubec Narrows.

The priority was jobs. Congress funded FERA in May 1933 with
$500 million. Harry Hopkins, its head, spent $5 million during his
first day—because, he said, the people needed it *now*. As historian
T. H. Watkins relates, FDR differed from Hoover in "his belief that the
government itself could and should provide the work. . . . By the end of
October, he and his aids had devised a blueprint for a wide-scale program that . . . could employ four million people over the winter months.
Roosevelt liked the scheme and said so during a press conference.
Hopkins took this as a signal to begin. When an aide asked him later
whether the President had approved the plan, he answered, 'Approved
it, hell! He just announced it at his press conference!'"[73]

Speed and experimentation were essential. Within two weeks,
800,000 people were working. By the middle of January four million
were working, with a weekly payroll exceeding $62 million (about
$1 billion in 2020 dollars). They worked on thousands of projects,
among them roads, bridges, schools, parks, hospitals, and airports. The
program was enormously popular. When the original allocation was
gone, members of Congress, thinking of their constituents (and their
own reelection), provided another $950 million.[74]

In response to Maine's need for economic assistance, Republican
senator Frederick Hale sent a letter to the president on April 10: "If the
Administration has in mind a program of public works in New England I
think Quoddy presents an excellent opportunity."[75]* It was good timing.

FDR had already invited Hugh Cooper to the White House for a
meeting on April 14, just five weeks into his presidency,[76] to discuss
economic development strategy with him, Senator Norris, and Secretary of State Cordell Hull.[77] They had a lot to talk about: Muscle Shoals
as the centerpiece of a federally driven regional economic development
program in the Tennessee River Valley, Cooper's experience with Stalin's Five-Year Plan for economic development in the Dnieper River
Valley, the possibility of selling more American goods to Russia and
the normalization of diplomatic relations with the Soviet regime, and
the Quoddy Tidal Power Project.[78]*

Hugh Cooper recommended "outright recognition" of the Soviet Union. Given his personal relationship with Stalin, Hugh is presumed to have conveyed a personal message from Stalin to Roosevelt.[79]

Hugh Cooper traveled back to Russia in August 1933, during which he came across a German who had Soviet support to make a film about American Negroes living in Russia and that would defame America. Hugh met with Vyacheslav Molotov,[80]* then Russia's minister of foreign affairs, and, by threatening to boycott the opening ceremonies of the Dnieper Dam, convinced the Soviets to withdraw their support for the film. He reported these events to the US State Department, along with his belief that recognition of Soviet sovereignty would help to counter the Japanese expansionist threat to the United States and Russia.[81]

After a series of covert negotiations, Maxim Litvinoff,[82] Russia's other foreign minister, arrived in Washington on November 7, 1933, at the invitation of President Roosevelt. He was greeted at Union Station by State Department officials without top hats, as he did not *yet* represent a recognized foreign government.[83] Nine days later, and sixteen years after the Russian Revolution, US officials donned and doffed their hats to the Soviet Union. At the party celebrating the event at the Waldorf-Astoria Hotel in New York City, Litvinoff was the guest of honor; the hosts were Secretary of State Cordell Hull and the president of the American-Russian Chamber of Commerce, Hugh Cooper.[84]

With the help of Senator Norris, President Roosevelt moved even faster on the creation of the Tennessee Valley Authority (TVA), the two men's "experiment" in public power and centralized "social planning."[85] Despite the forceful objections of Wendell Willkie, the new president of Commonwealth & Southern (C&S), the private power monopoly in the Tennessee Valley, who said "to take our markets is to take our property,"[86] Norris pushed the legislation through Congress, and on May 18, 1933, FDR signed the act creating the TVA. Its mission was to bring about "the maximum amount of flood control; the maximum development . . . for navigation purposes; the maximum generation of electric power consistent with flood control and navigation; the proper use of marginal lands; the proper method of reforestation . . . and the economic and social well-being of the people living in said river basin."[87] The TVA would become one of the most important public works projects in American history, in part by replicating Russia's regional model, implementing Samuel Insull's marketing plan, and doing exactly what Willkie feared.[88]

According to the *Sentinel*, in the meeting on April 14,[89] Hugh Cooper told President Roosevelt "Quoddy could produce power at a cost that would permit its sale . . . as far south as Hartford . . . at a price less than any existing hydro-electric or any modern steam station in the New England area."[90] Hugh then laid before the president two propositions. The first was Dexter's International Plan, which would provide 800,000 hp and cost $100 million (about $2 billion in 2020 funds). The second proposal was the All-American Plan, costing $36 million[91] and generating 400,000 hp.[92] Hugh's promotional efforts on behalf of his brother were well received, so he arranged for Dexter's plans to be sent to Washington.[93]

Hugh Cooper's version of events is different from the *Sentinel*'s. Here's how he described his meeting with President Roosevelt in a letter to Senator Hale:

> On April 14th I had a general conference with President Roosevelt, during which he asked my opinion of the Quoddy Tidal Power proposal belonging to my brother, Dexter P. Cooper. I told the President that I had great confidence in my brother's engineering ability, and while I was not familiar with his cost and production estimates I felt off-hand that Dexter's proposal could not be looked upon as a present public necessity.
>
> Since the above interview, I have been studying the small amount of data about Quoddy we have in our files, and now have the view that if government money at around 4% was available, plus reasonably low present labor and material costs, the project after a thorough examination might be found worth careful consideration in any public works program the government might be considering.
>
> As I explained to you in your office yesterday, we cannot take an active part in the Quoddy project, but will always be glad to help Dexter personally in every possible way.[94]*

An "offhand" dismissal followed by a tepid endorsement conditioned by an "if" and two "mights" is hardly the confident promotion portrayed in the *Sentinel*, but it was an endorsement—as a public works project.

Why couldn't Hugh take an active part in the Quoddy project? Was he just busy? Or would it have jeopardized his business with the private power industry? Or didn't he believe in Quoddy? Or did he like having Dexter work for him, not the other way around?

And why were there two versions of the same event? What had Hugh told Dexter?

President Roosevelt was inspired to visit Quoddy Bay in June 1933, perhaps by Maine's offer of the Blanchard estate in Eastport as the president's official summer residence in the United States.[95] He would also be reprising his presidential campaign kickoff cruise along the New England coast, which had attracted tremendous press coverage and demonstrated convincingly that despite being unable to walk without aid, Franklin was strong and brave enough to be president.[96]

On June 28, Roscoe happily reported that FDR's "trip here is lending some color to this prediction [of reviving Quoddy] and further support is given it by the announcement . . . that experts are studying the project as a possible part of the Administration's public works program."[97] Power was indeed on FDR's mind, for on that same day the administration announced grants for construction of the Bonneville and Grand Coulee Dams.[98]

"At Campobello the program starts at 11 a.m., with water sports, motorboat races, etc.," announced the *Lubec Herald*, and "[a]t 1 p.m. the Connors Bros. big fleet of sixteen boats will parade all gaily decorated. At 2 o'clock President Roosevelt will be officially welcomed by J.F. Calder"[99]—the man who had taught the young Franklin how to sail and had carried the polio-stricken Franklin across Lubec Narrows. FDR addressed the crowd at Campobello on June 29:

> I can only address you as my old friends of Campobello—old and new. I was figuring this morning on the passage of time and I remembered that I was brought here because I was teething forty-nine years ago. I have been coming for many months almost every year until about twelve years ago when [*pause*] there has been a gap.
>
> It seems to me that memory is a very wonderful thing because this morning, when we came out of the fog at West Quoddy Head, the boys who were aboard said, "There is land, a head[land]," and I started ahead full speed because I knew it was Lubec Narrow[s]. . . .

I was thinking also, as I came through the Narrows and saw the line of
fishing boats and the people on the wharves, both here at Welch Pool [*sic*]
and also in Eastport, that this reception here is probably the finest example
of friendship between nations—permanent friendship between nations that
we can possibly have. . . .

I hope and am very confident that if peace continues in this world and
that if the other nations of the world follow the very good example of the
United States and Canada, I will be able to come back here for a holiday
during the next three years.[100]

The festivities over, the Roosevelts relaxed in the glow from the
fireplaces and kerosene lamps in the big red cottage. Eleanor wrote to
a friend, "No telephone. Absolute peace. It is a joy."[101] Franklin's visit
was brief—only two days—but it accomplished what he had come to
do. The demon had been purged, and he could again enjoy his "beloved
island" and the pleasures of Passamaquoddy Bay. But one of his oldest
friends wasn't there. She was in England.

Having again experienced the power of Quoddy's tides, in July President Roosevelt let Hugh Cooper know his interest in the tidal power
project. Hugh wired Dexter, who was still in Russia, to return at once.
Dexter sailed straight home without stopping in England to pick up Gertrude and the children, where they were living with Gertrude's sister.[102]

Dexter returned from Russia on July 13, 1933.[103] On July 24, he and
Moses Pike left for Washington to discuss with the president on July
26 including Quoddy in the public works program. They also met with
Maine's congressional delegation.[104] Representative Edward Moran's
idea was to develop it like Muscle Shoals, as "a more or less experi-
mental venture in Federal power." Moran discussed the matter with
George Otis Smith, a Maine native and chairman of the Federal Power
Commission, and asked the FPC for a report on the project's feasibil-
ity.[105] Smith agreed, so Dexter met with the FPC's director, Colonel
Glen E. Edgerton.[106]

During the meeting, Dexter noticed on Edgerton's desk a report by
Murray & Flood.[107] It had apparently been commissioned by Dexter's
financial backers as part of their investment in Quoddy to "determine
how the 'Quoddy' project would tie in with other water power resources of the state of Maine."[108] But instead of focusing on Quoddy's
tie-in, M&F promoted the development of hydroelectric power from

Maine's rivers. When asked about how the report came to be on his disk, Edgerton was evasive.[109] For Dexter, Edgerton's evasion was yet more evidence of Murray & Flood's duplicity.

President Roosevelt liked the plan Dexter presented because he wanted to see Campobello and the surrounding territory return to its former prosperity, and the Quoddy project could be relief, recovery, and reform all in one. FDR asked Dexter, "Could the Army Engineers build it?"

Dexter replied, "Any fool could. The real work's already been done."[110]

Excited by the president's interest, Dexter met with his investors in New York and Boston. He asked for their cooperation in promoting the project to the government and a little more.

Reaching home, Dexter reopened his offices in Eastport on August 7, 1933,[111] and "appeared in especially buoyant spirits,"[112] even though the FPC's preliminary report on Quoddy was "said to be partially unfavorable, on the basis that the project would not be an economical source of power and that no adequate market for it exists at present." The report, however, "did not cover the most important point of Mr. Moran's inquiry": its suitability as a public works project.[113]

Dexter had followed up his meetings in New York and Boston with a letter to all his corporate investors on August 3. He got their replies on August 9. He got part of what he wanted, but the letters from and to Dexter differ significantly in substance and tone.[114]

By August 30, Roscoe could report three concrete advances. First, three engineers were coming for "a thoro [*sic*] inspection of the dam sites and . . . on the effect of tidal currents on filler dumped into them, and of the handling of tugs and scows in the tide. While they were naturally reticent, it is understood their findings were favorable to the project, which they characterized as by no means as difficult as some other undertakings they had been employed on." Second, federal and state employment officers were on their way. And third, FPC Chairman Otis Smith himself would arrive the next week.[115]

At a dinner in his honor, Smith spoke of the need for "power to strengthen the arm of the working man." More power, he said, was the only way to increase productivity and wages, and electricity was the way to do it. Thrillingly, Smith said, "Ralph Waldo Emerson was inspired to make his famous remark, 'Hitch your wagon to a star' by watching the operation of a tide mill at Lubec . . . [meaning] mankind should supplement his own puny strength by utilizing the forces of nature. Perhaps the time has come for Eastport to 'hitch its wagon to a star.'"[116]

In the essay "Civilization" (1870), Emerson considered the sources of power and their use for the benefit of humanity. He noted the power in flowing water and marveled at the newly discovered omnipresence of electricity and man's ability to send instantaneous messages by telegraph. Then he noted:

> I admire still more than the saw-mill the skill which, on the seashore, makes the tides drive the wheels and grind corn, and which thus engages the assistance of the moon, like a hired hand, to grind, and wind, and pump, and saw, and split stone, and roll iron.
>
> Now that is the wisdom of a man, in every instance of his labor, to hitch his wagon to a star, and see his chore done by the gods themselves. That is the way we are strong, by borrowing the might of the elements. The forces of steam, gravity, galvanism, light, magnets, wind, fire, serve us day by day and cost us nothing.[117]

Thus galvanized, Dexter P. Cooper, Inc., applied to the Maine Legislature for a new charter (to be good until 1945), which would reaffirm its right to export power and to acquire land for transmission lines by eminent domain. The new charter was approved overwhelmingly by the legislature and signed by Governor Louis J. Brann on December 13, 1933.[118]

Maybe, just maybe, the sun, the moon, and the stars—the means, motive, and opportunity—were finally going to align and bring the Quoddy Tidal Power Project to life.

Downeast Maine was still going down when national employment and payrolls were stabilizing.[119] Even in the US Congress, Maine's apportionment was dropping, based on the 1930 Census, from four representatives to three.

"Read this Platform!" exhorted the pamphlet Roscoe printed for his mayoral campaign of 1933.[120] Roscoe "favor[ed] retrenchment in every practical manner" but recommended spending some money to replace "worn-out wooden and tar sidewalks with permanent cement walks, as rapidly as we can afford it." Ever the booster, Roscoe wanted more feet on those sidewalks and advocated advertising "Eastport's industrial opportunities and attractions as a summer resort." He also encouraged "capitalizing to the fullest possible extent the opportunities offered by the Cooper-Quoddy project." And, hitching his wagon to a star, he said, "I . . . believe I am well qualified by training and experience to take excellent care of Eastport's interests in connection with it."

Returning to cost cutting, Roscoe insisted on a fair street-lighting contract. While "we have already forced the Light Company to reduce their offer from $3500 to $2831," he considered it "still unfair in many respects." And as "the present rate of 15 cents per kilowatt hour for electric lighting service in private homes is much too high, and as such is a burden on our people which it is the duty of their public officials to lighten . . . I favor an appeal to the Public Utilities Commission to secure this reduction if other means fail."[121]* Roscoe's earnestness won him the election.

Roscoe didn't want to lose the Quoddy Project again, so on October 5, 1933, he offered advice to Dexter on how to run an effective campaign:

[I]f what I am about to write is out of order, just file it in the waste basket and ascribe it to a perhaps mistaken zeal for your success. . . . I would recommend that a county-wide organization, bipartisan in character, be formed at once and put into action with several definite objectives. . . .

At the present moment at least two major projects are being strongly advanced in Maine for the approval of the Public Works Administration. . . . Both . . . are now competing . . . [for] federal funds, have support in Maine, and might conceivably be used as sops to appease Maine sentiment should your project be rejected. . . .

A good deal of ground has already been lost by amateurish fiddling and rather comical jockeying for position by certain of your friends here, but this thing can be done even now, if handled properly. Murchie is not the man to do it, since he is now an attorney for the Bangor Hydro-Electric and forthwith may or may not be friendly to you. It is utterly beyond Beale, who is only a good errand boy for local matters, but Brown would be effective if he had someone tell him what to do. . . .

I have written very plainly, as it is my custom to speak and write. . . . I want nothing for myself in the way of favors, a job or special consideration otherwise. I only want the project to go through. If it does, I shall probably be able to paddle my own canoe very nicely without hollering for help. The only request I have to make is that this letter be kept strictly confidential, and my only wish is that it may prove of service to you.[122]

Roscoe may not have realized how important his warning was, for Murchie was not just Dexter's former lawyer but also one of his current stockholders and creditors.[123]*

Dexter replied on October 7: "Your suggestion . . . is appreciated. . . . Recently I did advise the Eastport C. of C. to not spend any money, having in mind the financial situation at home." He then asked for the plan to be put into action.[124] Three weeks later, Roscoe sent another letter to Dexter:

> You have been getting some fine support from Ruby Black the Washington correspondent of the *Portland Evening News*. You may be interested to know that I have just been offered control of their editorial department, but think I shall decline. I want to stick around Eastport until we know what Roosevelt is going to do for the project. . . . Everyone here is marking time, pending a decision at Washington and the current query heard a hundred times a day is "Any news from Cooper?"[125]

Roscoe might regret his decision to stay in deeply depressed Eastport. Then again, taking the job at the *Portland Evening News* would be like jumping out of the frying pan and into the fire, as Ruby Black would soon reveal to her close friend.

Most commercial banks were either closed or wary of funding a hydroelectric power project until the Roosevelt administration's public power policy was clearly delineated.[126] So Dexter P. Cooper, Inc., a two-person company, asked the federal government for a $43 million loan (about $800 million in 2020 dollars).

As outrageous as that might seem, it was not unreasonable given the circumstances. The government knew banks weren't able to make many large loans and had established programs to fill the need. The government also wanted commercial businesses to create permanent jobs to replace temporary relief programs, and cheap power would foster economic development and more hiring.[127]*

Cooper submitted his loan application to the Public Works Administration (PWA) on September 11, 1933. On October 5, George Williamson of Maine's PWA office reported that in his analysis the project could not support a normal interest rate of 5%, or even 4%. He therefore recommended 2% interest for 50 years.[128] While not glowing, the analysis was satisfactory enough for the PWA to ask the FPC to examine Dexter's application.[129] On October 31, 1933, Governor Brann sent telegrams to the president,[130] the PWA, and the FPC stating, "[I]t is the almost unanimous belief in Maine that favorable action on the development of the Passamaquoddy power project will do more to relieve unemployment presently and provide industrial prosperity eventually for Maine and New England than any other individual [relief] project."[131]

Col. Henry M. Waite, assistant administrator of the PWA, was, according to Senator Frederick Hale, "as dumb as an oyster when questioned about the chances of the Quoddy project getting approval."[132] When reminded that the project had been before the PWA for a month, Waite replied, "If you're going to pour millions of dollars into the Baltic Sea, you're going to think about it pretty carefully beforehand."[133] The oyster was dumb, not mute.

Someone who knew better and presumably cared about Maine was its native son, FPC chairman Otis Smith. But Otis was being pushed out and replaced, according to the *Portland Press-Herald*, as part of the Roosevelt administration's distribution of political patronage jobs. In doing so, the *Press-Herald* lamented, "The government is losing the services of a man better qualified for such a job than any other man in the country."[134] The pro–private power *Press-Herald* failed to mention that the US Senate had sued the Supreme Court to sack Smith.[135]

Eleanor Roosevelt wanted to know why Quoddy had experienced so many difficulties. Eleanor's interest may have been motivated by her concern for her summertime next-door neighbors, but it is more likely she was acting on Franklin's request. In either case, Eleanor asked someone she trusted who could get to the bottom of the problem and report back confidentially. Eleanor called on her journalist friend, Ruby A. Black.[136]* Ruby wrote back, with her professional affiliation on her letterhead crossed out:

November 2, 1933.
Dear Mrs. Roosevelt:
Here is the memorandum for which you asked. It may be too long, but I couldn't seem to get the essentials in less than seven double-spaced pages. . . .
Can't you persuade your husband to be more discreet? Was my face red when Mr. Kanne[137] read off to some hundred correspondents the transcript of the President's conference, in bed, with a delegation of correspondents,

in which the President said, "Ruby Black brought Mrs. Roosevelt some clippings about the Gannett in Maine. . ." etc. etc.!

I consulted Dr. Gruening[138]* and Dexter P. Cooper, after talking with you, on whether or not Insull fought Cooper, with the results in this memorandum. That, of course, is not sworn testimony or official records, as most of the rest of the stuff in the memorandum is. But I think it is reliable. Cooper told me confidentially that the Insull people approached him with the ostensible idea of financing his project and with the result of finding out about it in order to fight it at Ottawa and at Augusta. But he can tell you about that. . . .

Yours sincerely,
Ruby A. Black[139]

MEMORANDUM ON INSULL-WYMAN-GANNETT CONTROL OVER MAINE WATER POWER, BANKS, INDUSTRIES, NEWSPAPERS, AND POLITICS

How Insull Got Control

In 1925 Insull first went into Maine power and soon controlled about two-thirds of the electrical energy consumed in the state. The Middle West Utilities Company paid Walter S. Wyman and Guy P. Gannett (owner of newspapers, husband of the Republican National Committeewoman) $140 a share for stock in the Central Maine Power Company when it was selling on the market for $60 a share. The minutes of the Middle West's board meeting recording the transaction said: "The corporation (Central Maine) to a very large extent controls the water power situation in the State of Maine and the opportunities for future development of the property are very promising; and the price agreed to be paid, all things considered, is reasonable and fair." . . . Maine still believes that "all things considered" meant that the Insull people were sure they were buying the Governor and the legislature and the repeal of the Fernald anti-export law. Certainly events justified this belief.[140]*

Newspaper Control

When Insull bought Central Maine [P]ower stock from Guy P. Gannett, Gannett bought more newspapers and soon owned the major papers of the state, outside of Bangor. His papers are the *Press-Herald*,[141] *Evening Express*, and *Sunday Telegram* of Portland; the *Kennebec Journal*, Augusta; and the *Waterville Sentinel*. When the *Portland Evening News* was started there on Oct. 3, 1927, edited by Dr. Ernest Gruening, every effort was made to kill it, even to the extent of causing the Fidelity Trust Company, which Wyman and Gannett controlled, to refuse to lend money to merchants who advertised in the *Portland Evening News*. . . .[142]

Political Activities of Insull-Wyman-Gannett Group

In 1927 the Legislature passed, and Gov. Ralph O. Brewster vetoed the Smith-Wyman bill, drawn by the Insull interests, to repeal the law prohibiting the export of waterpower from Maine.

Power company support then, in 1928, obtained the election of William Tudor Gardiner, Republican, as Governor, pledged to sign the repeal of the Fernald law.

Since the people had not invariably elected the people sought by the Insull interests, the Insull politicians and the Gannett papers started a campaign to abolish the direct primary. The *Portland Evening News* waged its first campaign to save the direct primary, and the people voted on the referendum, in December, 1929, to kill the bill repealing it. . . .

In 1930, the power export question went to the people in a referendum battle which saw every telephone pole, fence post, and vacant show window plastered with posters favoring export, and every grange and town hall filled with debate. The Insull companies admitted, in the Federal Trade Commission investigation, to having spent $199,815.65 in that campaign. Of this, $55,700 went to the Gannett newspapers and publishing companies for advertising, printing, and thousands of copies of the newspapers containing favorable editorials, news, and advertising. Most of the rest, or $136,464.54, according to the Federal Trade Commission's tabulation from the company's books, went for personal services and expenses of various community leaders, both Republican and Democrat. . . . The "expenses" received by them were not itemized.

The people voted against repealing the law, despite the campaign. They had previously voted, on referendum, to grant Dexter P. Cooper a charter permitting him to export power when and if he develops it at Cobscook and Passamaquoddy bays, a project which Insull has fought both in Canada and in Maine. Wyman attempted to defeat the amendment to Cooper's charter to permit him to develop the project solely in American waters.[143]*

Control of Banks

The Insull-Wyman-Gannett combination formed a holding company, Financial Institutions, Inc., into which more than 50 banks were merged in a chain controlling the situation nearly everywhere except in Bangor. The Fidelity Trust Company at Portland was the parent bank. During the Gardiner regime, these banks practically named the bank examiners and shortly before the bank moratorium they were all examined and found "sound." Only one of these 39 banks, the Augusta Trust, opened after the banking holiday, and it was soon closed. It is said (I have no sworn testimony on this) that in the few days it was open, Blaine Viles,[144] probably the next Republican candidate for Governor, who is a director of the bank, drew out $40,000. . . .

The story of the illegal loans of the Fidelity and the illegal purchase of
bank stocks is shown in the bill in equity brought against the directors and
officers, headed by Wyman and Gannett. Clippings giving the text of these
charges have been made available to you and the complete official copy of
the bill is transmitted herewith. The "executive committee," which Wyman
and Gannett controlled, lent the depositors' money, without adequate security, to their own industrial companies, to political and social friends, and
thus wrecked the state financially. Chief Justice William R. Pattangall, Democrat, appointed by Gov. Gardiner (Rep.), named Robert Braun, a director of
the bank, as conservator, and now Braun as conservator is suing himself as
director, in this $3,500,000 bill against the officers and directors.[145]

Control of Industries

The Insull-Wyman-Gannett group organized New England Industries,
Inc. to buy paper, shoe, steel, textile and other factories, ostensibly to build
up the state and to provide a market for electric power when the people refused to permit them to sell power in Massachusetts. It is these companies,
most of which are not operating, some of which are in receivership, that the
banks in Financial Institutions, Inc. lent money without security.[146]

Attempt to Continue Control

However discredited the group is, it is not relinquishing its efforts to
control the state. Wyman and his friends bought the New England Public
[S]ervice Company in April 1933, after the Insull debacle. Every effort has
been made to have the Democrats appoint the "Insull crowd" to all public
offices. . . . The common stock is owned by New England Public Service
Company, owned by Wyman and his friends, while preferred non-voting
stock is sold to consumers in the well-known way. . . .

Likewise the fight to wreck the [*Portland Evening*] *News* and thus control the Maine press is still going on, and the Gannett *Evening Express* has
been sued for $500,000 libel in this connection.

The fight for control of the power situation goes on in the Wyman-
Gannett opposition to the Cooper project, known as the Quoddy project.[147]

Dexter met with Franklin at the White House on November 16,[148]
the same day his brother Hugh was celebrating US recognition of Soviet Russia. The same day, Roscoe the politician wrote to Dexter the
engineer with some advice:

I have just learned of the opposition you are getting at Washington from
the power interests, and want to make a suggestion. . . . The power people
don't fight as you do, fairly and openly, with fact and argument. They do
their stuff with money, and use facts only in a secondary way. So if you

run into any unreasoning or unexplainable blockading, be on your guard for it will probably be paid opposition. If it is, you will merely waste time trying to convert or refute it. Take your cause direct[ly] to the President, explain what has been done by the power group in other ways, tell him plainly what you suspect and ask his advice and help as to how to prove your case over the heads of or around whoever is doing the blocking. He knows something of power company tactics, and by such an appeal you have a chance to force his already warm sympathy into positive action. He is your one best bet. Once he gets going, he can and will have his way, at least until Congress meets.[149]

Roscoe followed up with a letter to Maine's former governor Percival P. Baxter about the forthcoming New England Council (a business coalition) negative report on Quoddy: "Indication[s] are very plain that the power group in Maine headed by Wyman, and aided we believe by [Edward] Graham of Bangor [Hydro-Electric] are running a drive against the Quoddy project, in conjunction with power interests in Massachusetts and Southern New York."[150] As Bangor Hydro-Electric (BHE) was one of the *Sentinel*'s largest advertisers, Roscoe would have to tread carefully or risk the solvency of his own paper.

On December 20, the Fisheries Commission finally published its findings. "Ho, hum!" said Roscoe. "They are as expected, constituting a report such as the *Sentinel* offered to write in advance, free of charge. The investigation took two years and cost $90,000." Said the report, "There is little probability of the dams affecting the fish along the coast of Maine . . . but would reduce the herring fishery inside Passamaquoddy and Cobscook Bays to negligible proportions." The report concluded further research was necessary. Roscoe retorted, "[T]he scientists would probably need some few years more. . . . The Algonquin is a wonderful hotel, the golf links at St. Andrews is the best on the coast, and the yachting is good on St. Andrews Bay, so why not keep the matter open?"[151]

LIQUIDITY

The year 1934 started badly for Dexter Cooper and the Quoddy Tidal Power Project. On January 3, the FPC killed Dexter's $43 million PWA loan request because, it said, it would not be financially "self-liquidating."[152]

Written by Roger B. McWhorter, the FPC's chief engineer, the report's reasons against Quoddy were that (1) it would cost about $40 million while a comparable steam-electric plant would cost only $16 million, (2) steam-electric power would be cheaper at higher load factors,[153]* (3) Quoddy could not compete with steam-power rates in export, and (4) there was no present or prospective market for Quoddy power at any price.[154] "It is my opinion," spoke McWhorter, the "tidal power project would be an unsuccessful venture,"[155] and he warned the PWA "contingencies might cause the cost to greatly exceed the estimates."[156]

Taken at face value, McWhorter's report was damning. But it could be readily refuted. McWhorter's assertion that there was not and would not be a market for Quoddy power "at any price" was hyperbole, as evidenced by the rapidly growing market for power all over the country, even during the Depression, and the development of industry anywhere cheap power became available. As Maine representative Edward Moran pointed out in a letter to President Roosevelt, "This objection could have been raised against Muscle Shoals, but was not: on the contrary, a Federal Corporation was set up to loan money to individuals to create a market. . . . And where is there absolutely known power market for any of the proposed projects under consideration? Why apply this test solely to Quoddy?"[157]

All hydroelectric power plants are more expensive to build than steam-electric plants. But steam plants require fuel, whereas hydro-power plants don't. Thus, the real question was: *What would be the avoided cost of fuel for the century or so a hydropower plant could be expected to operate?* If one assumed fuel costs would rise, instead of staying forever low, as McWhorter did, then steam power would eventually become more expensive than hydropower.

"A third objection interests me," wrote Moran, "that Maine rivers should first be developed. This idea that project 'A' should not be undertaken because of a possible project 'B' exists seems very strained logic; certainly THAT objection has not been raised to ANY Federal project undertaken by this Administration. The solicitude for the development of Maine rivers, as expressed by Engineer McWhorter, has caused many Maine people to believe they see the real reason for the disapproval of Quoddy . . . the sinister suspicion . . . as to the influence exerted by any 'special interests'"—namely, the power trusts.[158]*

McWhorter was likely basing his opinions on the duplicitous report by Murray & Flood.

Despite its apparent bias, McWhorter's report was published by the FPC and therefore trusted by the public as authoritative, fair, and definitive. *Just as its patrons wanted.*

Colonel Waite, the PWA's acting administrator, refused to consider Dexter's rebuttal and appeal, so Dexter went over his head. He asked for and received a meeting with President Roosevelt on January 18.[159] Waite resigned.[160] On February 20, Maine's governor Brann and representative Utterback also met with FDR and asked him for an impartial review.[161] Accordingly, a public hearing was scheduled with the PWA's Technical Board of Review for March 30.[162] The board's focus would not be what its name implies.

The New England Council (NEC) denounced Quoddy as economically "unsound" and urged that federal funds not be used for it. Not surprisingly, NEC's power committee consisted of power producers, not users. Roscoe reported that they "left no doubt as to their desire to kill Quoddy's chances for governmental approval, two members . . . bluntly admitting that to be their intention. . . . Maine people are thus afforded the curious spectacle of a giant Maine development which is ardently favored by at least ninety-five percent of the State's population, opposed in the name of the State, by its representatives on the New England Council."[163]*

Chastened, Roscoe wrote to his friend, the journalist Joseph F. Preston: "The Cooper project doesn't look quite so good but we haven't given up hope yet, and are not going to until we know it's no use to continue. It will be given a hearing March 30th by the Board of Technical Review of the P.W.A. but in my opinion it [is] marked for slaughter."[164]

The tide appeared to turn in Dexter's favor during the PWA technical review. "While there was plenty of opposition," reported Roscoe, "friends of the project feel that he more than held his own against it and built up a case that should justify approval by the Public Works Administration and by the President." The hearing's importance was shown by who was present. Representatives from the PWA, FPC, TVA, National Resource Planning Board, and National Academy of Science were on hand to listen or testify, as were Maine's congressional delegation and power companies.[165]

Dexter "sprang something of a surprise" when he explained the market for Quoddy's power: 60% would be used to produce aluminum and 8% for making stainless steel—in plants he proposed to build at Eastport—and the remainder would be sold at $0.01 per kWh to local utilities for distribution in northeastern Maine.[166]*

Four million dollars of Dexter's forty-three-million-dollar request was for metal production.[167]* The aluminum plant would produce twenty-five million pounds annually, at a cost of $0.13 per pound, and be sold for $0.16 per pound. The bauxite ore would be obtained from Russia, Greece, and Italy. Stainless steel could be manufactured for $42 per ton and sold for $100 to $225 per ton. As existing aluminum patents were about to expire and steel patents would not seriously impede his production, Dexter could operate independent of the existing monopolies.[168]

Then another surprise: Dexter named his investors—Boston Edison, Westinghouse, General Electric, and the Central Maine Power Company (i.e., the Insulls' Middle West Utilities), four of the top ten largest utility investors in the county.

Then another surprise: His investors had agreed to "cash out" their investments (with interest) rather than insisting on the presumably large ownership percentages they were entitled to. Three million dollars of Dexter's loan request was for such "liquidation costs" and other corporate debts not publicly disclosed, including to himself personally, as he had paid all the operating costs for the last few years.[169]

While not remarked on by Roscoe, this was remarkably good news for Dexter because he could become the sole owner of the $40–$50 million Quoddy Tidal Power Project *if* the loan was approved. It was also bad news because it *seemed to suggest* his investors had lost faith in the project.[170]

When Dexter was asked what security he would provide for the loan, he gave no clear answer. Asked to whom profits (if any) would go, he again gave no clear answer, saying he was waiting for guidance from the government as to the financial structure it wanted, including, possibly, a consumer cooperative *he* had proposed.[171]

Then came the attacks.

McWhorter, the deskbound bureaucrat, challenged the construction timeline and cost estimates of the most experienced hydroelectric construction manager in the world. McWhorter repeated his assertions that Quoddy power would be too expensive to market. Dexter rebutted this rehash with new and authoritative detail.[172]

The Aluminum Association of America had been alerted to Dexter's threatening proposal, bemoaned the drop in production from 275 million pounds in 1930 to 177 million pounds in 1932, and argued that no more capacity was needed. Queries by the board brought forth the rather awkward facts that aluminum production in North America was entirely—100%—controlled by Alcoa (and had been since 1888), and the decline in production was due to the Depression and did not reflect the light metal's bright future.[173]*

NEC sent a letter declaring Quoddy impractical because there was no existing local market and the "prohibitive" costs of transmitting it to distant markets. As Dexter later documented, the transmission system was controlled by an "oligarchy" that charged potential competitors prohibitively high rates to keep them out—thus Dexter's need to build his own 300-mile transmission line.[174]*

Questioning by the board revealed another awkward fact: Dexter had asked Bangor Hydro-Electric for 7,000 to 10,000 kW for use during construction but was told the company couldn't supply it.[175]

Dexter's supporters then had their turn.

Senator Hale said, "[T]he government could not do a better job than to take hold of this development," and it would give employment where little relief had been provided. Senator White said, "I am old-fashioned enough to believe in competition in business. The figures given here by Mr. Cooper are of tremendous significance in the metal business. . . . If Cooper's figures are correct it will take only a few years for Quoddy to pay for itself 10 times over in benefits to the consumer." Representative Utterback stressed the importance of cheap power and fertilizer for Maine's farmers. Representative Moran said it would be better to fund one great permanent construction than to spread large sums on small projects of little or doubtful value.

The "most distinguished" advocate for Quoddy, noted Roscoe, was Frederic Delano, President Roosevelt's uncle, a former US Army Engineer, a railroad executive, vice chair of the US Federal Reserve, and currently with the National Resources Planning Board.[176] Delano said, "I see no insurmountable obstacles in the Quoddy project. The big industries are looking for power at the seaports where they can make long-term contracts."[177] Stating that he was speaking as an individual since the NRPB had not officially considered Quoddy, he then submitted his own written "brief" for the board's review:

> It is no condemnation of the Quoddy Project to say that it comes in the class of pioneering enterprises which can only be put through successfully with some Government support . . . [such as] the early transcontinental railways across the Rockies, or the early reclamation done in our arid regions . . . [and] the Panama Canal. . . .
>
> The real trouble is that an important part of the opposition cannot be brushed aside by engineering analysis or mathematical figures. "It hasn't been done before," is a difficult statement to controvert. . . .
>
> Other kinds of opposition are more easily disposed of. This is of two kinds: first, those controlling other sources of power whose promotors feel that they will be hurt; and second, from those who ask us to say, "Where and to whom will you sell your power?"
>
> . . . [C]heap and abundant power will make its own market, just as it has done in many other instances. Can anyone doubt that the State of Maine would have been a populous state if a great field of coal of good quality had been discovered there? This "white coal" should do for Maine what it has done in Italy and Scandinavia. To fail to realize its great possibilities would be short sighted indeed.
>
> . . . [C]ooper, like most men of action, is not in any sense a good promoter. He has the instinct like so many men we admire, who would prefer to let their work speak for itself.
>
> Perhaps he would have done better and saved himself much time if he had come to Washington with a good spokesman, a man glib of speech and having the very qualities which he lacks. But he did not, for although this Project represents ten years of study and hard work, he would rather lay his cards on the table and let his simple unvarnished story speak for itself.
>
> He ought to succeed because he "has his stuff," and this Project ought to go through because it is basically sound and Dexter Cooper is the man who can make it succeed.[178]

Delano sent his brief to Arthur E. Morgan, chair of the TVA. He then sent Morgan's reply to President Roosevelt. Morgan believed Quoddy's principal disadvantages, compared to the TVA, were that it was not near a population center or valuable mineral deposits. Its advantages were that it would be one of the few large waterpower plants near sea lanes for efficient shipping, and there was a good workforce available. Morgan then gave the board something to contemplate: "A great surplus of power would be an interesting social-economic phenomenon—something unprecedented in the world's history."[179]

Conversely, what would happen if Maine—or America—had a shortage of power?

While President Roosevelt was thinking about generating hydro-electric power from Passamaquoddy Bay, another matter arose, of little importance, but that would have a profound impact on the people of the bay: Franklin had forgotten about his canoe, the one Passamaquoddy chief Tomah Joseph had made for him and presented with the words *Mikwid hamin* ("Remember me").

The canoe had been in storage at Greenlaw's boat shop in Eastport since 1921, when Franklin had contracted polio. Due to the "officiousness [of] someone who should have known better," a dunning notice for storage fees was sent to the president in April 1934. Missy LeHand, the president's secretary, asked Roscoe to investigate. Roscoe obliged and made arrangements so the boat could "remain there indefinitely, or we will dispose of it as Mr. Roosevelt wishes."[180] Missy replied that the president would like to donate the canoe to a local boys' camp.[181] Since there was no such entity, Roscoe proposed it be sold, "devoting the proceeds to . . . pay for plastic surgery for an Eastport boy with an unfortunate nasal disfigurement."[182] The president approved, and it was done.[183]

The timing of the dunning notice was probably just cosmic coincidence, but it may also have reminded Franklin of another debt that was owed. Maybe, by reminding Franklin of his Indian mentor and the services the Passamaquoddy had performed during the American Revolution; the unfulfilled treaties made by previous US presidents; and the injustices inflicted on tribes since, FDR was prompted to create the Indian Reorganization Act of June 1934, aka the Indian New Deal.[184]

John Collier, the newly appointed commissioner of the Bureau of Indian Affairs, made sure that Indian tribes were included in all the relief and recovery programs. Then he tackled reform, which essentially meant undoing the harm done to the tribes by the Dawes Severalty Act of 1887, the disastrous attempt to force tribes to assimilate into mainstream America. Instead, Collier designed a system to enable the Indians to exist within white society while keeping their tribal identities and culture.[185]

The Indian New Deal organized the tribes into self-governing bodies that, after they had adopted suitable constitutions, could police themselves and act as legal entities in their relationship with federal, state, and local governments; allowed such tribes to control land sales and leases on the reservations; and stipulated that the Indians themselves had to vote on their proposed constitutions by secret ballot before they would become effective.[186]

Simply stated, the Indian New Deal reestablished Indian sovereignty. It was the most positive change for American Indians in a century. It would, however, take at least fifty years to be substantially realized. Once again, it would be Passamaquoddies who would lead the fight.

Dexter's PWA loan application was "slaughtered," as Roscoe had feared. On May 1, the Board of Technical Review decreed Quoddy was "economically and financially unsound." Its members believed the bureaucrat, not the builder. For the board, fifteen years of work wasn't enough, and it called for a "more detailed study of the most suitable dam design under the adverse conditions that will be encountered."

Switching to Dexter's plan to use Quoddy's power for metal manufacturing, the Public Works Administration's Board of Technical Review, meant to represent the *public interest*, caved to the cartels:

> The applicant states that he has had no previous experience in the manufacture of these metals. The plants and the distribution of their output are a business that must be built up in the face of strenuous competition from some of the largest organizations in this country, who have spent many years and large sums of money in research and in perfecting their manufacturing methods, who now have a large and successful sales organizations and who might be unusually jealous of and antagonistic to such competition. . . . Without these metallurgical plants the entire project is admittedly not economically justified.[187]*

Congressman Moran analyzed the Board's public policy thusly: "The PWA says the outlook for success is not bright for this aluminum venture because there is an aluminum trust which would be hostile and would fight Cooper. In other words, we should lie down, do nothing, and suffer at its hands."[188]

The PWA Board of Technical Review was no truer to its *technical* mission when considering the relative costs of river power and steam power. Citing "detailed reports" (probably the Murray & Flood report) indicating the Kennebec and Penobscot Rivers could be developed for power more efficiently, the board dismissed the tidal power project. Furthermore, it asserted that steam power was even more efficient than river power, assuming "present costs of fuel"—a huge assumption that most unbiased analysts would not accept given the continued consumption of the limited supply of fossil fuels. The assertion that steam power was the most efficient also flew in the face of Maine's private power companies' publicly stated intention to expand their production 500%,[189] mostly with river power, not to mention the hundreds of millions of dollars the federal government was investing in river power projects on the Tennessee, Columbia, and Colorado Rivers. "However," the board conceded, "if costs of fuel increase . . . the time may arrive when the development of this project will be economically sound."

On June 7, 1934, "Honest" Harold Ickes,[190]* the secretary of the interior (PWA's parent agency), told Dexter his loan request was "finally and formally" rejected.[191] Thus ended Dexter's *third attempt* to get the Quoddy Tidal Power Project funded and built. Thus far, he had poured fifteen years of work into the project but not a drop of concrete. The dream of harnessing the tides was still as far out of reach as the moon itself.

Maine's Democratic leaders swung into action on Dexter's behalf (and their own). Congressman Utterback wrote to the president, "I do not believe there is a community in the United States in a more forlorn and helpless situation than Eastport. Her plight is desperate. . . . [I]s there not some plan or method by which the Quoddy project may be kept alive? Otherwise, the reaction in September will be most emphatic. . . . I am enclosing a few of many recent editorials, all of similar vein, attacking the Administration, and as we have no Democratic press in the State we are helpless in attempting to refute these attacks."[192]

The president told his secretary he wanted to see Dexter before leaving for a fishing trip.[193] To Dexter's surprise, when called to the White House on June 29, he was met not by the president but by his political adviser, Louis Howe, and by Governor Brann. Dexter was told that getting a federal loan was hopeless, but there might be a way to get what they all wanted.[194]

Dexter capitulated. On June 29, 1934, he proposed to President Roosevelt to sell the Passamaquoddy Bay Tidal Power Project to the US government "for the benefit of the people of the State of Maine." He proposed the following:

1. The US government would take an option for six months to purchase all rights, charters, engineering plans, patents, and property of Dexter P. Cooper, Inc.
2. The government would investigate the proposal at once with the help of Dexter P. Cooper, Inc., which would be paid $5,000 monthly beginning July 1, 1934.
3. The purchase price of the Quoddy project would be $1,000,000, which sum represented the actual cash expenditures to date of approximately $500,000, plus $375,000 to Dexter personally for his equity in the project, and the balance was to pay certain indebtedness.
4. Dexter would be awarded the engineering and direction of the project on a fee basis and would operate the plant for a period after construction, on terms to be decided later.[195]

If the government paid Dexter what he proposed, he would be comfortably wealthy, but not as flamboyantly rich as if he had been able to keep his equity in the project.

If the government did as Dexter proposed, he would no longer be an entrepreneur but a *hired hand*, and Quoddy would no longer be "capitalism" at work but something else. Some would say it would be "public," others "socialist" or even "communist."

Roscoe capitulated. He was unable to carry the heavy load of publisher, printer, real estate agent, insurance agent, fishing weir owner, and mayor, not to mention husband and father of two. His zeal for Quoddy was mistaken. He gave up. Citing his health and the press of other responsibilities, he put the *Eastport Sentinel* up for sale on July 18, 1934.[196]

Finding a buyer for a newspaper in a tiny town that was getting tinier wasn't going to be easy. It might be the end for Maine's oldest newspaper. Without its paper to hold the community together, could Eastport survive?

Even the most optimistic, the most talented, and most energetic were being crushed by the Depression.

OUT OF COMMISSION

"All Maine . . . was on the alert Monday night," July 23, 1934, to hear Governor Brann read over live radio a letter from President Roosevelt:

> I have been interested in what is known as the "Quoddy project" for a long time, and it has been my hope that eventually the State of Maine would become not only a great industrial center of the nation but that its agricultural population would be among the first to enjoy the manifest advantages of cheap electric power on the farm. . . .
>
> There is for the time being a considerable amount of Government funds available for public works, some of which, of course, could be diverted to this purpose should it be found, first, that the creation of the power is practical and economical, and second, that it can be made available for certain definite uses, and will not lie idle for many years to come. . . .
>
> I think it is important that any definite plan be based upon the securing of the title to the power site to the citizens and its use and development under the control of the State. No corporation or small group of corporations should be given a monopoly in the use of this power or a title to what is the heritage of every citizen.[197]

Acting on the president's instructions to appoint a commission to study Quoddy whose members' "reputation for judgement and ability are of the highest order and the disinterestedness of whose motives cannot be questioned," Brann appointed Kenneth Sills, president of Bowdoin College (a Democrat, but a New Deal critic) and *four* Republicans: E. S. French, president of the Maine Central and Boston & Maine railroads; Harry B. Crawford, master of the State Grange; Wingate F. Cram, treasurer of the Bangor & Aroostook Railroad; and William N. Campbell, executive of Sanford Mills.[198]

The president requested that "a time limitation be put on the preparation of this report in order to take advantage of such public funds as are now at the disposal for purposes such as this. . . . I will await the report of such a commission with the greatest interest and can assure you that you will have a sympathetic ear in Washington for whatever recommendation may result."

This was encouraging because it took the evaluation, development, and operation of Quoddy out of the clutches of "captured" federal agencies and put them in the hands of Maine.

Pushed by the president, Secretary of the Interior Harold Ickes went to Eastport for a fact-finding tour, to meet with the Sills Commission, and to be subjected to Maine's lobbying effort.[199] Ickes brought with him a major in the US Army Corps of Engineers and two others.[200]* Raymond Moley stayed with Moses Pike in Lubec, while the secretary and the major were invited to stay with the Coopers on Campobello.[201]

It was August 21, but it looked like July 4 in Eastport, with decorative buntings, a parade, and speeches. Reminiscent of FDR's triumphant homecoming a year earlier and Governor Brewster's pre-referendum visit in 1925, Eastport's Library Park was again swarming with two thousand Quoddy supporters.[202] Impressed, but watching his words, Ickes said he thought Quoddy was "a wonderful thing," which would be "one of the greatest engineering proposals in history, if and when it is developed"— to which someone in the audience shouted, "How about that 'if'?"—and he replied, "I'll leave out the 'if' and just say when it is developed."[203]

"Another reason I would like to see the Quoddy project go through," said Ickes, "is that it would reward the dream of the great engineer. It is not often a man can dream of a great plan and live to see it realized, but I hope for Mr. Cooper that he will live to see it built. . . . I urge you not to give up your fight for Quoddy. You won't anyway."[204]

Writing in his private diary on August 22, 1934 (which was published posthumously in 1954), Ickes wrote: "I praised the project as a project which I could do with propriety. Then I told them that their likelihood of being able to finance it at any time in the future depended upon their ability to develop a market for the power. . . . What I said seemed to be very well received."[205]

Ickes's recollections might not be reliable, for in his diary he wrote of his cohosts, "The Trouts, by the way, are a prominent family in those parts and they are delightful people and hospitable."[206]

There were other important things Ickes knew and should have said but didn't. Nor did he say what he was thinking and later wrote in his diary: "There is no doubt about the interest in this project in that section of Maine. They think that their economic life depends on it. Certainly, unless something happens, Eastport and other communities like it will die a death of slow economic strangulation."[207]

Ickes's public statements were taken as a sure sign of success. Another encouraging step occurred four days later when Ickes appointed the army major as an adviser to the Sills Commission. *Who was he? Why him?*

The major had helped Ickes draft the legislation creating the PWA. He had worked in the Philippines and on the Panama Canal.[208] He had been manager of athletics at the US Military Academy at West Point, New York.[209] While a student there, the affable cadet frequently visited his "Uncle Fred," a railroad executive who had lived in Burlington, Iowa, but now lived a stone's throw away in Newburg, New York. On one such visit the young cadet met the man for whom Frederic Delano was a *real* uncle, Franklin Delano Roosevelt.[210] The cadet received his commission to West Point from Congressman Tom Hedges of Iowa and on the recommendation of Hugh L. Cooper. The major was the young axman and surveyor who had worked for Dexter Cooper on the Keokuk Dam back in 1910: Philip B. Fleming.

As fate would have it, 24 years later, Fleming was now a guest in Dexter's home and in a position to pass judgment on his former employer. With Fleming's benefactors all being Quoddy advocates, his appointment augured well for Quoddy's approval.

On the second evening, Major Fleming arrived at the Coopers' house for dinner two and a half hours late, too drunk to stand.[211]

Was he an alcoholic? Or was something troubling him?

The key players met the next day. Included were Governor Brann; the Sills Commission; Dexter Cooper; Major Fleming; a PWA financial analyst; and K. F. Wingfield, a PWA engineer.[212]* The primary subject was the potential market for power. Aluminum, stainless steel, and fertilizers were the obvious possibilities.[213]* Another subject was whether the project should be privately owned or publicly owned; if the latter, whether by the state or the federal government; and the certainty that existing "public" utilities would fight any truly public power project. During deliberations, Commissioner William N. Campbell, who was hostile to the idea of a government loan, left the closed-door meeting to confer with Walter S. Wyman, president of CMP.[214]

With rumors swirling that Quoddy was being considered more for
its political value than for its economic value, Sills postponed further
activities until after Maine's midterm elections on September 10.[215]
Dexter returned home from the commission meetings "visibly tired and
. . . discouraged."[216]

Major Fleming raised red flags on August 30 when he revealed that
the PWA didn't have enough money for the Quoddy project, even if it
were approved.[217] Much more ominously, he said:

> The PWA . . . cannot act blindly or upon altruistic impulse. . . . Dexter
> Cooper has done a genuinely wonderful job. . . . But after awarding full
> tribute to his ability, we are now confronted by the practical questions: Is
> there a market for the power? . . . Without detracting in any way from Mr.
> Cooper's achievement as an engineer, it must be said that he will not be
> able to produce cheap power.
>
> In fact, if this project were completed in accordance with present plans,
> the cost of installing power would be a little over $700 per kilowatt. Com-
> pare this with the projects at Bonneville, or Grand Coulee, or Boulder Dam,
> where the cost per kilowatt was between $200 and $250!
>
> But—and it is an important "but"—that is not the whole story. Coal,
> in western industrial centers, can be obtained for $1.50 to $2 a ton. In
> Maine, due to the cost of transportation, it is approximately $5. And so, if
> a sufficient market for power actually does exist in Maine, the differential
> between producing this power by coal and producing it by electricity gener-
> ated through the Cooper project may make the latter feasible.[218]*

Major Fleming, who was promoted to deputy administrator of the
PWA the following week, was also quoted as saying that federal power
projects like Bonneville were cheaper to finance than private enterprises because the government could obtain money at lower rates and
afford to wait longer for returns.[219] Whereas a private project would
have to carry a 5% interest on a loan, a federal project would be charged
only 2.5%.[220] For a $40 million project financed for fifty years, the sav-
ings could be $40 million.

As Dexter later pointed out in a letter to White House staffers,
"The Boulder Dam development is the outstanding exception but
even here it was necessary for the Federal Government to finance a
$250,000,000 community enterprise in order that a power customer
for 35 percent of the output might be created."[221]* Switching to fed-

erally funded irrigation projects, Dexter noted that $200 million had been provided to western states *interest free for 40 years* (a $100 million subsidy). Drawing the two examples together, Dexter concluded, "[P]ower will be for the New England district what irrigation has been to the West"[222]*—the foundational investment that made the desert West habitable and economically viable.

This begged the question: *When would the federal government take over Quoddy?*

The answer was becoming more complicated because the Edison Electric Institute had arranged for a group of Alabama Power Company shareholders to sue the federal government over the constitutionality of the TVA's ownership of electrical transmission lines, in what became known as the *Ashwander* case.[223]* The Sills Commission also bogged down because it didn't have any money with which to hire staff, and thus appealed to the state for $25,000 on October 2.[224] But the Republican-controlled Executive Council dragged its feet until early November[225] and only allotted $5,000[226]—making it obvious that its members listened with a "sympathetic ear" to their political patrons, the power trusts.[227]

Maine became the national political battleground, and Quoddy the hot issue. Both the Republican and the Democratic national parties sent operatives and money to Maine to swing the midterm election.[228] Republicans didn't want Democratic governor Brann being reelected. Also up for grabs were one US Senate seat and all three House seats. The Republican concern was that the adage "As Maine goes, so goes the nation" might be true—when the Democrats were in the lead. As Maine's elections were two months earlier than most of the country's, the results were going to be leveraged as a referendum for or against the New Deal.[229]

The anti-Quoddy barrage began in the *Boston Herald* on August 16, when it asked rhetorically of Ickes, "Why this tentative repudiation of the engineers' advice and reversal of his own judgement?" The *Herald* then offered "a plausible explanation":

> The Democrats wish, naturally, to repeat their gubernatorial and congressional victories of 1932, and the outlook is not what the politicians call promising. So why not josh the Down East boys a little? . . . $42,000,000 is loose change for Uncle Sam as thing are going now.

So Maine and Gov. Brann and the Democratic congressmen and the Quoddy folks will therefore entertain next week the gentleman who was against grab-bagging and pork-barreling a year ago but is in a reconsidering mood now. The old rejection is one thing. The coming election is another.[230]

More pulp for the political paper mill came out on August 23 when the *Bangor Daily News* announced that the Maine State Planning Board's report "will state that electricity can be generated much more cheaply by the rivers of Maine than by any series of tidal dams or reservoir that might be constructed in Quoddy Bay." *That* was news to the Maine State Planning Board—for it was fake news concocted by the paper.[231]

Nonetheless, Republican newspapers echoed and amplified the charges, accusing the Roosevelt administration of trying to "bribe"[232] the people of Maine into supporting the New Deal by using the Quoddy project as "bait."[233] The Republican National Committee charged Quoddy's revival was "fake," a "gesture of despair," a "bluff."[234] Congressman Bolton (R-OH) said, "The Maine people should not be misled by veiled promises and clever gestures. . . . Undoubtedly, the Democratic party will be making great preparations to build the Quoddy Project up to September 10 (election day). After that date it will be found that the building of the Quoddy project will stop."[235]

In the midst of all this, Gertrude wrote a letter to Franklin, telling him:

The people of Maine are in a quandary. If they vote for Brann, they vote for the "Power gang" who have overplayed their hand again, and been too generous, and the people argue, if Brann is double-crossing them, he is also double-crossing you. They are very loyal to you. . . . On the other hand if they vote for [Republican candidate] Ames, they give the impression to the world at large that they are slapping you in the face. . . . When the Governor read your letter over the radio, the people were filled with hope. The Commission was formed, did nothing, and hope began to die. Then came Mr. Ickes and filled them with hope once more. A few days later the newspapers published statements, which they attributed to Major Fleming, condemning Quoddy. . . . Whether he ever made the statements (and from the gross errors it looks unlikely that he did) they have not been contradicted and now the poor people don't know what to think. Starvation & freezing are hard to face for one's children & winter is coming.

Yours as ever, Gertrude S. Cooper.[236]

On election day, President Roosevelt replied:

> I realize well the difficulties of Maine politics and I am dictating this while the vote is in progress. I fear that it will take a long time before Maine throws off the grip of the old-line utilities, for they cut into both political parties. However, I have by no means abandoned the hope of the Quoddy project itself. I am trying to get President Sills' Board to report quickly. I know how hard it is for you and Dexter but all I can tell you is to stick it out a little longer. . . .
> Affectionately, Franklin.[237]

Democrats in Maine fared well in the 1934 elections. Governor Brann was reelected, and Simon Hamlin and Edward Moran were elected to the US House. Of Quoddy's Republican supporters, Frederick Hale was reelected (by 1%), and Ralph Brewster ousted Democrat John Utterback from his House seat.[238] Since the results appeared to endorse the New Deal,[239] and the rest of the country would not go to the polls until November, Quoddy remained a national issue. The "most sensational" charge was led by Colonel Theodore Roosevelt Jr., who claimed Democratic victories in Maine were "bought and paid for" with "millions of dollars of taxpayers' money,"[240] a bit of a rhetorical overhang since no funds for Quoddy had yet been spent or even authorized.

"With his characteristic vigor,"[241] Ickes denied the allegation that he had been "Quoddyizing" (i.e., vote buying),[242] calling it "fantastic" and something "[t]he Partisan press and those who sought to win office on any terms, chose to deliberately misrepresent." Ickes also pointed out that "fewer projects were approved for Maine than for any other state in the country."[243]

Dexter was discouraged. The summer residents were gone and with them that season's gay activity. Fall was when the remaining inhabitants of the far north focused on survival. Unlike so many others who had moved away from the cold and the unemployment of Downeast Maine, Dexter had moved into it. The year had been hard; there had been many disappointments. Nearing the end of his rope, on September 20 Dexter wrote to Franklin:

Dear Mr. President:

Eleven years of personal expenditures on Quoddy have exhausted my financial resources and I have reached a point where I must make a very definite decision as to the future.

This decision will be complicated by factors other than personal ones for the people of Maine and particularly northeastern Maine have been and still are looking to me for help. Then too, you have shown great sympathy and understanding for Quoddy and I am unwilling to abandon it without first discussing it with you.

If you can arrange for a conference at an early date I will greatly appreciate it.[244]

He did not get an immediate reply. When on October 4, 1934,[245] Dexter received a letter from Chairman Sills with news of yet another delay, he snapped.[246]

On October 9, Roscoe wrote to Ralph Brewster: Dexter was "in bed with a form of breakdown due to strain."[247] For Dexter, Depression was now a medical condition.

Dr. Bennett of Lubec sent a telegram to President Roosevelt saying that Dexter "needed something to live for."[248] The situation was so serious that President Roosevelt's son and official secretary, James Roosevelt, flew up from Washington. Dexter's brother Hugh also rushed to Dexter's side, bringing with him a medical specialist.[249]

Frederic Delano was worried about Dexter's precarious condition. He had known Hugh and Dexter since he ran the railroad running through Keokuk and had later served with Hugh in the US Army Corps of Engineers in France during World War I.[250] Now on the National Resources Planning Board, Delano had a public policy interest in Quoddy and asked Hugh about it. Hugh replied that he had "never been associated in any way with the Quoddy project but believes that it is sound from an engineering standpoint and is just as important to the State of Maine as Grand Coulee is to the northwest section." Regarding public policy, Hugh said, "The question is: Is it better to try to develop a new country or to try to revive an old country in which large sums of money . . . have been invested in homes, schools, highways, etc.?"[251]

From a political perspective, Hugh noted, "There is one point of earning power in this case of Quoddy that does not exist in many other cases. . . . Quoddy would attract an immense amount of attention in a populous region. As you know, the most important crop of the State of

Maine is the tourist crop, and I feel sure that the Quoddy project would attract a steady flow of visitors, coming by rail and motor to see one of the real wonders of the world. It is hard to estimate what this might be, but I am certain it would be an important item."[252]

On October 10, Frederic Delano sent a memorandum to President Franklin Delano Roosevelt, outlining his concerns and what he thought should be done:

1. Dexter Cooper has had a breakdown and his condition is quite serious. He not only used up his own funds, but is also badly discouraged.
2. After careful study of the Quoddy project, I believe it would compare favorably with the large Government projects elsewhere. Furthermore, the proximity of an international harbor makes it an ideal place for the chemical industry. Talks with a number of men convince me that there is less real hazard in this plan than is usual in these large plants.
3. I would suggest that you consult some outstanding manufacturer of electrical equipment, like Gerald Swope, or someone familiar with the chemical industry, like Bell, President of the Cyanamid Company.
4. If it is decided to advance money to build the project, I believe I can suggest a plan of procedure which will result in the work being done on a very low cost basis.
5. I convinced myself that the fundamental opposition to this enterprise arises from a group of power plants which sold out to Insull and received Insull certificates.[253]

Franklin Roosevelt immediately sent a telegram to Dexter inviting him to the White House.[254] Dexter didn't go. On November 1, Franklin wrote again to his ailing friend:

Dear Dexter . . . I have not lost my interest in Quoddy. . . . I am doing everything possible to hasten matters. I know that you will realize that the project presents a great many difficulties that have to be straightened out and that these difficulties do not lie wholly with the Federal government but relate also to the State of Maine, the Charter, and to a fair certainty as to use of at least part of the power before government funds can be made available. . . .

In the meantime, I hope much that you and Gertrude will give serious thought to your working temporarily on some other project. As you know, we are doing much studying of many projects—not only T.V.A., but many others. These fall under Public Works, Army Engineers, Reclamation Service and Natural Resources Board. I can fit you in temporarily until such a time as there is definite word on Quoddy. . . .

[Y]ou and Gertrude and the children ought to have a little change of scene from Campobello Island! It would do you both a world of good. I hope that you will write me and let me know if you can come to Washington. Give me a little notice in order that I can be locating different kinds of suggestions for you. Meantime, keep up your good spirits and remember my interest.
     Always sincerely, *Franklin*.[255]

Dexter didn't go to visit Franklin at the White House. He didn't take a temporary job offered by the president. He didn't leave Campobello.[256] Gertrude did.[257]

"The temporizing tactics of the Quoddy [Sills] Commission," the *Sentinel* complained, "which has wasted nearly four months without accomplishing a single thing of value or making any real effort, have apparently not been pleasing to Mr. Roosevelt, who, with his accustomed vigor and directness is now undoubtedly about to make an independent investigation to be followed by such executive action as the situation thus disclosed, may warrant."[258]

That action took place on November 14 when Secretary Ickes appointed *another* commission. Officially the Passamaquoddy Bay Tidal Power Commission, but known as the PWA-Hunt Commission, it was composed of Henry T. Hunt, PWA's general counsel; H. T. Cory, an engineer who had worked on the Colorado and Columbia River power projects; the stalwart Moses Pike; and Dexter Cooper.[259] The PWA-Hunt Commission would evaluate the cost to build and produce power at Quoddy, the Maine State Planning Board would examine the project's effects on Maine's economy, and the Sills Commission would try to come to an overall policy recommendation. Collectively, the three reports were to answer the question: *Was Quoddy a good investment of public funds given the circumstances in Maine?* They were *not*, as the industry and FPC wanted, limiting the question to: *Will Quoddy pay for itself in X years as a stand-alone for-profit project?*

"These new developments," Roscoe reported, "followed closely after contact was made with President Roosevelt himself by a personal representative of Mr. Cooper,"[260] at which time "Mr. Roosevelt was placed in possession of information not previously available to him."[261] The *Sentinel* didn't say who this personal representative was or what new information had prompted the president to take action.

The Sills and PWA-Hunt commissions met on November 27. Senator White sent a letter communicating advice from Washington insiders that the project might be better off if it were pushed on grounds of public welfare, like the TVA, rather than on the same basis as when the PWA rejected Dexter's loan request for Quoddy as a commercial enterprise.[262]

Promoting Quoddy as another TVA was a risky strategy. The power trusts were likely to challenge the constitutionality of the TVA in the Supreme Court. Writing to FDR, Huston Thompson, former head of the Federal Trade Commission (FTC) when it had investigated power trust propaganda, proposed a test case, predicting the president "will go down in history with great praise if he can put this over."[263]*

Sills answered White's letter by saying, "So far we have been maintaining a judicial attitude leaning toward sympathy."[264] But by December 12, Roscoe reported the Sills Commission had "turned over $4,200 of this $5,000 to the Planning Board, which is now handling the investigation practically under the direction of Messrs. Cooper, Pike, Corey and Hunt."[265]

"You can best serve the party by resigning," Roscoe wrote to Maine's Republican Party chairman, Arthur E. Sewall, on November 13.[266] He meant "no animus" but felt it was "logical and necessary to Republican welfare." Roscoe felt Republicans should

> quit bleating about over-expenditure of government funds and bellowing loftily about abstract principles that many voters cannot afford to consider under present conditions. The thing most men with families are thinking of is jobs and something for their children and wives to eat and wear. The Democrats are coming across with practical aid in making necessities available while we are offering nothing. . . . Republicans practically ignore this issue, and attack the social and humanitarian features of the New Deal. Thereby they won the recent election for him.[267]

Rather than being off base, Roscoe was on point, as evidenced by Sewall's reply: "Your suggestions . . . [are] somewhat in line with my own thoughts as in fact immediately following the September election, I . . . [offered] to resign."[268]

At the heart of their exchange of letters, Roscoe and Sewall were debating the role of government in business and the support of its people.

What system was most likely to provide prosperity for the majority of its people? Sewell continued:

> In trying to appraise the present situation, I have tried to visualize my own past to . . . figure out whether or not I myself would have been better off under a progressive regime or under a conservative regime. . . . I have never depended upon someone else to pull me along. . . . The result is that I am not greatly in sympathy with these millions who are willing to sit on a wall and whittle in the belief that the government will take care of them. I am a thorough believer in a capitalistic state and I see no hope of progress in a socialistic state.
>
> A study of history indicates that every country that has attempted planned economy has in time gone to ruin and if we are to judge the future by the past, that is just exactly what we are headed for at the present time.[269]

Sewall then discussed the depression in his hometown of York. The natural beauty attracted visitors, who built houses and occupied hotels, making tourism Maine's largest industry. But mansions built for $125,000 now sat unsold at $25,000. This "destruction of capital . . . simply means that the houses . . . will go into ruin. . . . Hundreds of people who have been employed in keeping these properties up will be out of employment." The Republican Party

> tried at least to point out to the electorate that the only road back to prosperity was to stop the tremendous orgy of spending and make it possible for private industry to get back once more upon its feet because it is private industry and private industry alone that is going to take up this slack in employment. . . .
>
> On the other hand, the whole Democratic campaign was carried on with promises to coin out money from the federal treasury. No one seemed to think and no one seems to care now where all this money is coming from, so long as it continues to pour out but you and I know . . . there has to come a day of reckoning. . . .
>
> You ask me in your letter to state who it was that told me that the Quoddy project was impracticable. This I cannot do, but I will say it was not Mr. Campbell [of the Sills Commission]. If you will analyze the project yourself, you will realize. . . . Quoddy power will not be cheap power. It can be manufactured for nearly one-half price by steam . . . [and] there is absolutely no need for the tremendous amount of power. . . . Furthermore . . . I do not believe that the State of Maine would ever vote to obligate itself in any such project involving nearly fifty millions.[270]

While Maine had voted overwhelmingly for the Cooper-Quoddy Charter in the 1925 referendum, it did so as a privately funded project, not a publicly funded project. This was certainly not a case of bait and switch, but it was certainly different. While Sewall's assertions were debatable,[271]* one wonders: *Why couldn't he say where he was getting his information?*

It "struck the industry like a lightening bolt" on November 17, 1934, when President Roosevelt announced he intended to build multiple "little TVA" projects all across the country. In response, the private power industry mounted an "aggressive self-defense," with the Edison Electric Institute declaring it would challenge the constitutionality of each and every project.[272]

Christmas 1934 was four days away. Overnight temperatures in Eastport were dipping into the single digits, with days hovering around freezing. In a month the mercury would pool in the bulb at 20 degrees below zero Fahrenheit.

Roscoe's mission as mayor was to keep his flock from freezing or starving. For the last four years Eastport's population had been dwindling, each week losing a person or family that cranked up the jalopy or hitched a ride south, hoping to find a job or at least someplace where in their homelessness they wouldn't freeze. "Hooverville" shanty towns sprang up outside cities; the farther south, the more of them. Up in Eastport it was said, "You could stand on Water Street, which is the only real street in town, and throw stones all day without hitting anyone."[273]

Eastport's population, which had hit a high in 1900 of 5,311, had dropped to 2,800, of whom 1,800 (64%) were now on relief.[274] Nationwide, 30 million people (24%) were on the dole.

On a December day, Roscoe drafted a letter to the Federal Emergency Relief Administration (FERA). He skipped his usual pleasantries and got right to the point: "I have to advise that our [weekly] hour allocation to individual workers is not sufficient to meet their needs under winter conditions. . . . The allocation [is] as follows: one in a family 9 hours; two 12 hours; three 16 hours; four 20 hours; five 24 hours; six 28 hours; seven 30 hours."[275]

FERA would pay only 35 cents per hour. Thus, a single person's earnings for a week—during which they would be able to work for only nine hours—would amount to $3.15. For a family of four, the head of household could earn $7.00 with which to feed, clothe, and shelter their family. It was cold comfort that women would, for once, be paid as much as men.

Roscoe did not succeed in increasing the hourly wage or the allotments. He did, however, succeed in getting funding for a project sewing towels for eight weeks—*if* he could find 27 sewing machines.[276] Roscoe called on the women of Eastport to lend the town their machines "to keep 50 families from want this winter."

Mrs. Robert Lodge answered the call, as did Mrs. John H. Stark, who added to her reply, "Would like a job as none of us working." Blanche Wilson was also willing to lend her sewing machine, if she got the job.[277] The women of Eastport came through with the needed machines and stitched together a better Christmas for their families than they had feared.

Stitching together such a marginal existence was not a long-term solution to Eastport's unemployment problems. It needed a new industry.

CHAPTER 9

Green Lights (1935)

ANTICIPATION

The Maine State Planning Board issued its report on December 22, 1934. Tourism in Washington County was worth less than $1 million annually, but neighboring Hancock County registered 20 times as much revenue. Based on the experience at the Boulder Dam, which attracted 225,000 tourists per year despite being more distant from population centers than Eastport, the board predicted Quoddy would attract up to 500,000 tourists annually, who would spend $10 million a year and consume $300,000 of electricity.[1]

The case for tourism as a *better* economic development strategy than heavy industry (or tidal power) was more picturesquely framed by Lubec native Mrs. Anita Kimball, who wrote to President Roosevelt in February 1935:

> Instead of marring the natural beauty of this region, why not procure all of the available land in Washington County, and make a fish and game preserve of it, a national park for the Eastern Section. . . . Perhaps you will recall that between Lubec and Machias there are fifteen miles of forest that stretches across the whole peninsula and through that region there are several lakes. Former President Cleveland used to come here for hunting. Can you imagine the joy of the Passamaquoddy Tribe, to become once more guides and pilots through the woods and across the treacherous waters of Passamaquoddy? Picture how beautiful this region will be with the unsightly sardine factories removed and the piers used for docking pleasure craft.[2]

Returning to the prosaic, the board saw value in diversifying the state's generating resources, noting that Bangor Hydro-Electric (BHE) had to purchase power from other sources during shortages. "In contrast with . . . rivers . . . Quoddy . . . is not affected by droughts or floods but can be depended upon." Though the board thought "the hazard from floods in Maine is not great," it nonetheless said Quoddy's predictability put it "in a class by itself."[3]

Surprisingly, the most important facts in the report were not reported by the *Sentinel*. On page 150 was a table comparing existing *private* hydroelectric dams in Maine with *public* hydro projects including Quoddy (Dexter's All-American Plan at $42 million), Muscle Shoals, Boulder, Grand Coulee, and Bonneville. The report indicated the costs of producing electricity at Quoddy would be *cheaper* than *all* the private power plants in Maine, including Central Maine Power's (CMP) new Wyman Dam on the Kennebec River. While Quoddy was projected to be three times as expensive to build as the Wyman Dam, it would produce two to fourteen times as much power at an average cost of $630 per kW, versus the Wyman Dam's cost of $870.[4]

While there was an excess of electric generating capacity in southwest Maine, there was a shortage in northeast Maine. Residential use of power was expected to increase steadily but slowly, and wherever cheap power became available, new industries developed. Assuming "reasonable rates" for Quoddy's power, the board found multiple potential markets for it, including electrically under-served communities in Aroostook County; the paper mill on the St. Croix River at Houlton; small-scale manufacturing; chemical and metal manufacturing; electrifying railroads; and, of course, BHE. The board estimated aluminum, stainless steel, and fertilizer manufacturing plants could all be built at Quoddy for a total of $8 million, and if operated at one-quarter to one-half capacity, they would be profitable enough to pay off the necessary bonds in thirty years and would consume *all* of Quoddy's power output.[5]*

Comparison of Costs of Relevant Hydroelectric Power Dams in MAINE				
	Head & Flow	kW Capacity	Total Cost	Cost per kW
Quoddy Tidal Power Project, Maine—(All-American 2-pool project)				
Initial installation*	13 to 20 feet	232,000	$42 million	$182
Average output		67,000		$630
Prime power		139,000		$296
Wyman Dam on the Kennebec River, Bingham, Maine				
Present installation	135 feet	46,800[1]	$13 million	$ 224
Average output		27,500		$ 870
Prime power	5,500 cfs	10,500		$1,230
Central Maine Power Company—all plants				
Present installation	Various outputs	103,400	$26 million	$242
Average output	Head 14 feet to 135 feet	67,000		$387
Prime power	60 to 5,950 cfs	31,000		$840
FEDERAL Public Power Projects				
Muscle Shoals, Alabama: Mississippi River, 1918–1926				
Initial installation	137 feet	184,000	$37 million	$200
Prime power		50,000		$740
Boulder [Hoover] Dam, Arizona: Colorado River, 1931–1936[2]				
Initial installation	550 to 600 feet[3]	164,000	$126.5 million	$770
Grand Coulee Dam, Washington: Columbia River, 1933–1942[4]				
Initial installation	145 feet[5]	103,000	$63 million	$615
Bonneville Dam, Washington-Oregon, 1934–1937–1943[6]				
Initial installation	64 feet[7]	43,000	$35 million[8]	$810

Prime or primary power is based on the amount of continuous power developed from the water available 90% of the time. *Average output* is based on the total amount of power developed from the average flow over 50% of the time for undeveloped powers and to the capacity of the present wheels on developed powers. *Secondary power* is that available in excess of primary power. *Surplus power* is the excess, at any given time, of either primary or secondary power over the immediate demand.

[1] Listed as 68,000 hp and 53,334 kVa on *MSPB-34*, 159.
[2] Wikipedia, s.v. "Hoover Dam," accessed 7 October 2020, https://en.wikipedia.org/wiki/Hoover _Dam.
[3] The finished height of the Boulder (aka Hoover) Dam is 726.4 feet.
[4] Wikipedia, s.v. "Grand Coulee Dam," accessed 7 October 2020, https://en.wikipedia.org/wiki /Grand_Coulee_Dam.
[5] The finished height of the Grand Coulee Dam is 550 feet.
[6] Wikipedia, s.v. "Bonneville Dam," accessed 7 October 2020, https://en.wikipedia.org/wiki/Bonne ville_Dam.
[7] The finished height of the Bonneville Dam is 197 feet.
[8] The cost of the Bonneville Dam reported as $83 million at William F. Willingham, "Bonneville Dam," *Oregon Encyclopedia*, https://oregonencyclopedia.org/articles/bonneville_dam/#.XvdlyC2z3_8.

Source: MSPB-34, 125, 150–51, with additional information as noted above.

The Public Works Administration (PWA) used broader criteria than the Federal Power Commission (FPC) in determining the economic *justification* for a proposed public works project as opposed to the stand-alone economic *feasibility* (i.e., *profitability*) of a proposed commercial power plant. The PWA-Hunt Commission submitted its report on 17 January 1935. The two most important points were that (1) it recommended Quoddy be built as a work-relief project, and (2) it was kept secret.[6]

Commissioner Sills sent his draft report to the governor. The governor sent it back by State Police courier. He didn't want any leaks. He did, however, want it changed. Just one sentence. It wasn't a big change, but it was an important one.

The governor wanted the wording "It is our opinion that Quoddy power would not be very cheap power but that it would be cheap enough to act as a yardstick and to be an important factor if properly developed in the whole problem of power in New England" changed to "It is our opinion that Quoddy power would prove to be an important factor if properly developed in the whole problem of power in New England."[7] This was perfect politician speak—positively innocuous— and excised the negatively charged, yard-long lightning rod and the equivocation critics would have surely seized on. Sills asked his co-commissioners for their approval; all agreed.[8]

After six months of deliberations, Maine's finest, most educated, most informed, and most independent men who could be found—Commissioners Sills, a Democrat; and *four* Republicans: French, Campbell, Cram, and Crawford—found in favor of Quoddy.

They favored Quoddy as a matter of *public policy*: "We would argue for federal support for this project largely on the grounds of social desirability . . . without definite assurance that the project would be within a specified number of years self-liquidating."[9] In other words, *Quoddy may not make money, but Maine needs jobs now.*

The commissioners made their recommendation based not just on near-term profitability but also on a longer-term and wider view of the project, as was the case with most large federal water projects in the South and West, where flood control, drinking water, irrigation, power development, economic development, and recreation were all part of the calculus, and where predictable markets for power rarely existed.[10]*

The long-winded commission opined, "If we judge the future by the past and recall how markets have often come to power sites, such, for example, as Niagara, it seems clear that industry will still seek cheap power, and it is quite within the bounds of reasonable possibility that the manufacturing of stainless steel, aluminum, fertilizer, and other similar industries, might well be developed in the Quoddy region which is through climate and proximity to Europe, particularly well adapted to such enterprises." (*That's 77 words, 10 commas, and 6 equivocations in one sentence!*)

"If the eastern part of the state could be, in a liberal way, industrialized," the commission imagined, "it would stir the life of the whole state and send a now rather sluggish blood of commerce coursing more rapidly through the veins of trade." To conservatives, "liberal" sounded "lavish."

In another colorful but politically poor choice of phrase, the commission noted, "It would, however, require a prophet of the first order to predict the possible use of power a few years hence." Rather than emphasizing the uncertainties inherent in all development efforts, the "prophet" line brought to mind charlatans and soothsayers.

Since Maine had supported with its congressional votes and tax dollars the many huge federal power and irrigation projects in the South and West, the commission felt other states should reciprocate and give Maine its "fair share." "Maine alone" said the commission, "cannot prosper cut off from the central energizing power of the government. Maine has always been in an eddy of the national stream little affected by the tides of adversity and prosperity until the past few years." Through no fault of its own, macroeconomic events had turned against Maine and it was now "one of the most hard hit regions of the whole country." These assertions, unsubstantiated by data, were cause for the casting of aspersions about Maine's "selfishness." And the unfortunate phrases "fair share" and "social desirability"—rather than perhaps "proportionate allocation" and "benefits to the public welfare"—put the project in the category of "wants" rather than "needs," exposing it to criticism as political pork.[11]

While the Sills report was at times eloquent, Governor Brann should have exercised his politically sharp red pen more often.

Dexter got a job with the PWA as a consulting engineer on January 16, 1935,[12] and some desperately needed money. More good news arrived on January 18 in an invitation for Dexter to meet with Franklin at the White House.[13] Three days later, Maine's political leadership formally delivered to President Roosevelt the three favorable reports on Quoddy. "Following the interview," reported the *Sentinel*, "apparently by agreement, optimistic but rather guarded statements almost identical in text were issued."[14]

The **"best news Maine ... heard in many years"** broke on January 23. In a seven-column headline, the *Sentinel* announced, "COOPER-QUODDY TIDAL PROJECT APPROVAL NOW SEEMS ASSURED."[15]

The PWA recommended $30 million for construction of Quoddy as a federal relief program, and for the state to establish a Quoddy Authority to operate it. Roscoe speculated the reduction from $47 to $30 million was based on the elimination of the metallurgical plants and the delaying of expenses for electrical equipment until the next biennial budget, when the dams would be finished and a purchaser found for the power.[16]

The federal plan was to lease the power plant to the Quoddy Authority, which would issue bonds based on future income. Maine would thus pay back the federal government for Quoddy, and eventually own it, *without any financial risk*.[17] An act establishing the Quoddy Authority would have to be passed by the Maine Legislature, the text of which was being written by the US district attorney for Maine in cooperation with the PWA. It would be presented to the legislature in a special session to be called by Governor Brann, who would ask for its immediate approval so construction could begin by April 1.[18] Funds for the project would be included in the $4.8 billion relief bill soon to be voted on by Congress.

"[I]t is impossible from this distance to gauge sentiment in Congress," the *Sentinel* acknowledged, "but there is every indication that the Relief bill will be passed substantially as submitted. As to the Maine Legislature, it is unthinkable that any serious objection should be made."[19] Turning toward home, the *Sentinel* said:

> Sentiment among Quoddy residents is one of intense, but quiet, satisfaction and relief. The depression in this section has not been one of three or four years duration. It has been acute for six years, and for nine years before that there was a descending scale of prosperity, and of population in all this region. The imminent approval of Quoddy means to thousands

of families here a renewal of hope and to thousands of others a restoration to normal American standards of living. The long fight for the development, with its many periods of keen expectation followed by disappointment has left them determined not to be premature again and there will be no celebration until not a shred of uncertainty remains. When the time comes, then, there will be such a celebration as will leave no doubt in the minds of anyone in Maine as to how they feel. All the repression and restraint of the last years will be cast aside in a demonstration of joy and exultation that will long be remembered.[20]

Roscoe, *still* the *Sentinel*'s publisher, Quoddy advocate, and civic leader, then cautioned, "While the battle seems now to have been won, it would be a mistake to relax the vigilance. . . . The public interest is so obviously on the side of the development that opposition should serve to brand the opponents as unworthy of public trust, and it should not be tolerated. . . . [T]he golden stream of wealth and business that must proceed from the dams will not stop at the borders of Washington County. . . . Isn't it about time for Augusta to stop shouting 'me first'?"[21]

But jealousy rules the human heart, as the *Portland Evening News* noted: "The suggestion of Mayor Payne of Augusta that the $37,000,000 which Quoddy tidal development would cost be used instead to develop Kennebec [County] industries has made him as popular as a skunk at a lawn party."[22]

Payne wasn't the only skunk. Individualism and regionalism ran deep in Maine. They manifested locally in home rule of towns versus state authority; in states' rights versus the federal government; and in the power export ban, which was intended to increase manufacturing in Maine but failed to recognize the reciprocal fact: Maine's manufacturers needed out-of-state markets.[23]

President Roosevelt was willing to negotiate with the enemy: the private power trusts.[24]* But before doing so on January 11, 1935,[25] he wanted to establish the basis for the discussion. Here is his starting position:

AGREED FACTS:
   FIRST: That the government by giving Public Utilities a practicable monopoly in their territory and protecting them from ruinous competition

has not only the right but the duty to see that the rates charged are as low as can be fixed with a reasonable profit.

TWO: That the interest rate on any investment has always been in every country in the world made dependent upon the security of the investment. The safer the investment the lower the rate.

THREE: That the statistics during the recent depression and all other depressions before show that the returns and profits of the Utility Companies in general have been least affected of all other forms of investment, except Government Bonds.[26]

If the power trust agreed with these points, then FDR would discuss ways to resolve the conflicts between the administration and the industry, particularly:

1. Rural electrification by the companies and the extent to which the Government will finance same.
2. An arrangement for some kind of rate fixing body or bodies, national or state, whose decisions will not be constantly appealed to the courts.
3. A proper basis on which a determination of investment and expense, on which can be made a proper charge for service.
4. Ways and means by which Holding Companies can either be modified or extinguished without receivership proceedings in a way that is fair to all parties.
5. On what basis power can be exchanged between government owned power plants and private power plants.
6. A basis for the sale and distribution of electrical appliances for rural electrification.[27]

For four decades, the low level of rural electrification (in 1935 only 11% of farms were electrified) had highlighted the divide between the haves and the have-nots. The forays into farm country made by power companies tended to follow the main roads, lighting up those farms within easy reach but leaving the hinterland in the dark. This inequality, and the failure or refusal of the monopolies to serve everyone in their service area, caused these cream-skimming circuits to be dubbed "snake lines."[28] FDR, who had been pushing for rural power for a decade, was now in a position to force it forward.

The enemies met on January 24. Representing the administration were President Roosevelt, Frank R. McNinch and Basil Manly of the FPC, and David E. Lilienthal of the Tennessee Valley Authority (TVA). Representing the power trust were Harvey Couch, president of a holding company; Alex J. Groesbeck of GE's Electric Bond & Share Company and former Republican governor of Michigan; and Wendell L. Willkie, president of Commonwealth & Southern.[29]

As President Roosevelt read through a list of holding company abuses, Lilienthal saw Willkie "getting hotter and hotter." Leaning over and pointing his glasses like a weapon at the president, Willkie argued against breaking up the holding companies. His manner shocked even Grosebeck and Couch, who recalled that it was "as if Willkie had pulled out a gun and started shooting." From there the discussion degenerated and ended decisively when Willkie asked, "Do I understand then that any further effort to avoid breaking up of utility holding companies is futile?" The president gave him one look and said, "It's futile."[30]

Would you like to receive 3% of the income of every person in your state? If you were already receiving that much revenue—as Maine's electric power monopolies were—would you want to give it up?[31]* The established electric monopolies didn't, and so they attacked Quoddy on January 31 at the New England Council, asserting that "the arguments advanced in favor of the project are all found to derive from selfish motives rather than from social."[32]

Roscoe's rebuke was in grand rhetorical style: "If it is 'selfish' to promote the expenditure of thirty millions of dollars in Maine that would otherwise be spent somewhere else, then Quoddy supporters must plead guilty. . . . If it is 'selfish' to bring to this State a development that offers employment at decent wages to the breadwinners of five thousand families now on direct or work-relief, and to thousands of other families just on the verge of applying for aid, then Quoddy is highly selfish."[33]

Roscoe addressed what he believed underlaid the ad hominem-like attack by the power trust, saying, "Maine people not on the power pay roll will find it hard to agree" with the charge of selfishness. Then, referring to the Insull debacle, Roscoe's incisive pen touched Maine's still raw wound: "To the category of unintended humor may be assigned Mr. [Edward E.] Chase's[34] description of Quoddy as a piece of 'political plunder.' Representing, as many will believe, the viewpoint of the most dangerous plunderbund Maine has ever known,—a gang of racketeers in politics, power and finance whose operations Chief Justice Pattangall said cost Maine investors some thirty-seven millions of dollars. . . . Mr. Chase's use of the word 'plunder' in connection with Quoddy is nothing less than rich."[35]

The *Boston Herald* raised the rhetoric the following week. Citing the FPC and a "Maine advisory board" report, it condemned Quoddy as "not economically justified." Then it snidely allowed that "nobody begrudges Maine a $30,000,000 or $47,000,000[36] loan or gift. All the states are grabbing for everything in sight, and our Maine brethren would not be traditional Downeasters if they did not try to get their share." "But," warned the *Herald*,

> Quoddy has become a symbol. If the President should authorize the work, it would be taken as a symbol of government competition all along the line. A great utility project, initiated against the best of expert advice, would be interpreted as a warning by every electric light, water, and gas company in the country that it might have a federal rival sooner or later. As a result, expansion would end. Replacements would be deferred. Improvements would not be made. The heavy industries, from which the government expects so much, would suffer. Uncertainty would be magnified, and the return of prosperity would be delayed again. . . . Quoddy has been a threat to the county since last fall. Is it not time that we should have done with it, once for all?[37]

Forget about the unemployed. Quoddy was now a "symbol." It was not just "a threat" to Maine's power barons, or to the private power industry, or even to legions of small investors in "public" utilities of all types—Quoddy could cause the entire economy to shut down. Quoddy, the *Herald* cried, was a *"threat to the country"!* (Emphasis added, though hardly needed.)

With that, Quoddy left the realm of reason and became a religious war.

Mayor Emery was less sanguine in private than publisher Emery was in public. He wrote to his friend Harold Dubord[38] on February 9 "in strict confidence." Roscoe was worried about the negotiations between the PWA and the Maine Legislature, saying, "[W]e are very anxious that a situation shall not be created in which the President shall feel it is better to step quietly in and kill Quoddy in order to avoid embarrassment to the Governor or others . . . we want Quoddy on any terms and are prepared to go to bat with the power outfit on any conditions the PWA may impose. . . . Anything you can do to help straighten all this out through a compromise or otherwise, will be deeply appreciated and well remembered."[39]

Downeasters' anxieties were heightened by the *New York Times* on February 10. While the *Times* reasoned, "no Legislature would dream of turning down a $30,000,000 grant," it then headlined the next section of the article in bold type—"**Engineer Disapproved**"—and said, "Two reports, neither of which have been published in full and . . . the contents of which are almost unknown . . . [portray] Quoddy as an uneconomic undertaking," and "A board of review named by Secretary Ickes held long hearings and rendered an adverse report."[40]

Picking apart the *Times*'s reporting, we find a few problems. First, the boldfaced headline confuses the issues of engineering and financial review. Second, the two reports are not identified but seem likely to have been the FPC report requested by Congressman Moran and the PWA-Hunt report. Third, the review criteria for the reports vary—commercial enterprise versus work relief—and thus the conclusion may or may not be fully relevant. Fourth, the assumptions used are enormously important and subject to manipulation. Finally, the *Times* was headlining negatively what it admitted it knew almost nothing about!

"Senator Hale is for the dole instead of work relief," gasped the *Portland Evening News* in February. Maine's Republican senator would "rather sacrifice the $30,000,000 Quoddy project with its great possibilities for employment and future development than do violence to his conviction that the dole is a better aid to recovery than work relief." Hale declared, "I cannot . . . sacrifice my convictions and vote to spend four billion dollars of the money of the people of the United States for an uncertain and in my opinion wasteful program [the Relief Bill] of public works construction which I do not believe will bring back prosperity to the Country."

Even the Republican megaphone, the *Boston Herald*, thought the Relief Bill was worthy, saying, "The President's plan as outlined in his annual message is admirable. He would steer a safe middle course between the cheap, outright dole and costly unemployment projects. He would avoid the demoralizing breakdown in morale which dogs the dole system and would avert the national bankruptcy which over lavish appropriations invite." Hale assured his confused constituents that despite voting against the necessary funding, he would still do whatever he could to protect Quoddy. The *Portland Evening News* wondered "if Senator Hale would be as stout-hearted in his convictions for the worthiness of the dole—at the expense of Quoddy and of the morale of the unemployed—if his re-election had not been settled, although by a scanty margin, in 1934, instead of being slated for 1936."

To some, Hale's convictions seemed more like calculations, more show than substance, a desire to have his cake and eat it too. Challenging him, Roscoe wrote, "Quoddy proponents over the entire state see no distinction between direct and work-relief sufficient to justify your position. Quoddy will admittedly do much to restore prosperity in Maine, which should be your first consideration."[41]

Concern about the Quoddy Authority Act grew when two changes in its wording were discovered after it left Washington and before its presentation to the Maine Legislature.[42] PWA secretary Ickes considered the changes destructive to the project and began investigating who was responsible.

The first change was the elimination of the sentence, "The State will not limit or alter the rights vested in the Authority, until the said obligations with interest are fully met and discharged." Roscoe reported a PWA official as saying, "[T]he inference drawn [is] that it might be possible for the Legislature, if this paragraph is not included in the law, to amend the Authority's Act after the great Quoddy dam is built, to prevent the Authority from even distributing power. The PWA official said privately, 'It sounds like sabotage.'"[43]

In the original version of the act, the Authority's directors were to be "appointed by the Governor." But the bill now said they should be "nominated by the Governor with the advice and consent of the Council." The council, you will recall, was controlled by the power trust.

There was good reason for concern about these not so little or innocent changes. And as the *Sentinel* alerted its readers, the shadowy opposition was again fomenting factional jealousies: "Rumors have already reached Washington of the whispering campaign being conducted against Quoddy. Word is being carried to all supporters of other Maine projects that if Quoddy is approved, there is little or no chance for other Maine projects."

"Insidious, isn't it," said Secretary Ickes.[44]

Maine Supreme Court chief justice William R. Pattangall attacked the draft Quoddy Authority Act in an open letter to the *Bangor Daily News*.[45] His attack was redolent of Andrew Jackson's sleight of hand condemning the Cherokee to oblivion on the pretense of states' rights. This, too, was about keeping possession of power, and it stunk like a power trust skunk.

A typical corporate charter empowers a company to undertake almost any legal activity, but usually doesn't require it to do anything other than follow state laws and regulations. Pattangall took this corporate liberty to its illogical extreme. His first complaint was that the "rates charged for its services are to be determined by the corporation itself and do not depend on the value of its property": an attempt to reinstate what had caused the problem in the "watered stock" scandals. While the Insull holding companies had collapsed, the operating companies still had a firm grip on their local monopolies and public utility commissions. Quoddy's rates would have to be *cheaper* than those granted to the existing power companies to attract customers. The existing monopoly could, of course, lower its rates in response. That was, of course, the intent of the public power "yardstick."

Pattangall's hue and cry was classic big government fearmongering: "[T]his corporation is given the authority to take over any property in the State of Maine, and more especially the property of any existing power company," implying a pervasive Communist-like federal takeover was inevitable. "It is not bound by the limitations of the Fernald Act in respect to the exportation of power" (which Quoddy had already been granted by statewide referendum), and "[i]t contains not a single word on the part of anybody to build Quoddy Dam."[46]

Roscoe refuted the judge's flailings, gaining himself some statewide notoriety:

> The dangers you cite here are wholly imaginary and suppositional—mere stuffing for a legal scarecrow. . . .
>
> [I]t would have been much more helpful if you had sought rectification of any alleged defects in the bill by quiet consultation with the Governor, the Attorney-General or others who have the matter in charge. Your resort to newspaper publicity does not savor of the judicial or constructive method. . . .
>
> Quoddy supporters are entitled to the right to present their case without the active opposition of the Chief Justice or any other Justice of the

Supreme Court, a part of whose salaries they pay and whose absolute and unquestionable impartiality they have a right to demand.[47]

On the same page on which Roscoe printed his reply, he reported that he had been popularly reelected as Eastport's mayor for the sixth time.[48]* While there was plenty of reason to suspect the chief justice's motives (as a political appointee with gubernatorial aspirations),[49]* there was no question about Roscoe's motives. He was clearly for Quoddy and the people of Maine.

The PWA-Hunt report "was kept in the deepest secrecy," complained Elizabeth May Craig, the Washington correspondent for the *Portland Press Herald*, on March 8, 1933, "and the PWA persistently refused to give it to reporters or to members of Congress."[50]

Two days later, the president held a meeting at the White House. There was a short discussion, and then the president scratched onto a slip of paper "Quoddy Project," two names, blanks for two more names, and a dollar figure. Then he handed the postcard-sized piece of paper to General Edward M. Markham, head of the Army Corps of Engineers. As informal as it was, it was an order.

The following day, Franklin met with Dexter.[51]

Anxiety turned to anticipation when the president's $4.8 billion Relief Bill passed on March 23. At last, Maine's salvation was within reach.[52] Said Senator Hale, "The government having adopted a policy of developing water power projects and a very large sum of money having been appropriated over my vote for such developments, I am glad indeed that Maine was not left out and that it has been decided to go ahead with the Quoddy project."[53]

Hale could almost taste that cake.

Two days later came word from Washington that no Quoddy Authority would be asked of the Maine Legislature. As Roscoe admitted, "it did constitute something of a hazard" and was now "happily removed."[54]

Meanwhile, Eastport couldn't pay its bills, so it issued scrip—*funny money*—for debts as small as 50 cents.[55]

The train was too slow to get the news to Maine, so Governor Brann, who learned the news on April 4 as he got on the train in Washington, got off the train at Baltimore to send a telegram to the *Sentinel*: Quoddy would be built as a public power project under federal control.[56]

That was the president's plan—but not yet officially announced or funded.

When the $4.88 billion Relief Bill passed on April 8, the *Boston Herald* editorialized against **"SHODDY QUODDY,"** saying "the rush for the spoils is well underway and Quoddy seems to be in the van." Then the paper lambasted the Sills Commission for considering "social desirability" instead of "straight and simple business grounds," and lamented that "while everybody else is getting cash from Uncle Sam, Maine may as well 'get hers.'"[57]

An adjoining editorial titled **"BOON-DOGGLING"** criticized "the study of semantics and boon-doggling" (the science of word meanings and the weaving of belts, hammocks, etc., from twine or rope, respectively) as examples of useless instruction being paid for by the government through work-relief programs employing academics and artists for their expertise instead of something useful like ditch-digging.[58]

Thus, knot-tying got tangled up with "boon" (a political favor or spoil) to define wasteful government spending—and the *Boston Herald* made semantic history!

The handmaiden of Propaganda, *Miss-Information*, slyly insinuated herself between Quoddy and the public. Typical of the problem was a *New York Times* article that *misleadingly* cited cost estimates ranging from $30 million to $100 million for the All-American Plan—when the higher estimates were for the International Plan. This sloppy reporting was followed with another article on April 11 pejoratively titled **"MONEY FLOWS INTO THE SEA,"** which *misaligned* "technical" with "economic" opinion and then opined:

> Pertinacity has prevailed over the best technical and economic opinion. . . .
>
> The concession made to the Quoddy project is a triumph for Governor Brann. After the board of review had voted down the proposal to let the tides of Fundy make aluminum and steel, light houses, drive trolley cars and milk cows, the Governor invited Ickes to Maine and high pressured him into taking a different view. . . .
>
> There is nothing technically wrong with the indefatigable Mr. Cooper's vision of making the moon and the sun work for us through the tides. Neither is there anything technically wrong with making bathroom fixtures of platinum.[59]

Money for Quoddy came teasingly closer on May 1: "INITIAL ALLOCATION OF $20,000,000 FOR QUODDY MAY SOON BE MADE."[60] Success seemed imminent on May 15 when the *Sentinel* reported that Quoddy's allocation was on the president's agenda for approval the very next day:

> **Victory For Project Now At Hand; All Obstacles Cleared Away and Army Engineers are Set To Go**
> Press dispatches . . . all confirm what appears to be authoritative information from a source close to the White House to the effect that Quoddy is definitely slated for approval and allocations at the meeting of the Public Works Board which takes place Thursday afternoon in the President's offices.[61]

Some may have wondered, *Who was Roscoe's "source close to the White House"?*

The tantalizing news brought the patent trolls out from under their rocks. The attorneys for Jerome Petitti and John Knowlton each claimed the idea for the Quoddy Tidal Power Project had been "stolen" from their respective clients and threatened to sue. They intended to press their claims not with the financially exhausted Dexter but with President Roosevelt, the man who was about to spend $30 million— *and, they must have hoped, a little more.*

Petitti's lawyer admitted his client had failed to secure patent rights to the project but said he was "morally" entitled to have the project named for him and to be appointed chair of the Quoddy Authority, to which Cooper, he said, "should not object."[62]*

For Dexter's investors, time was running out.

There was a hitch. Federal attorneys were still concerned that construction should not be started until after the Quoddy Authority Act had been approved by the Maine Legislature—something Governor Brann thought had been dropped as unnecessary and in any case was not going to happen in the current session given the overwhelming Republican opposition. But waiting for the next session (in January) would delay the start of the project and the desperately needed work-relief funding.[63] Searching for a solution, Representative Brewster called three Maine attorneys to Washington "at his own expense" to help convince the federal attorneys to drop their objections. The PWA also needed assurances that Maine's power companies wouldn't file nuisance suits to tie up Quoddy, like those that had plagued Muscle Shoals, and that regulatory reform bills would be approved. The discussion lasted until morning.[64]

Then it happened. On May 16, 1935, the PWA allocated $10 million to immediately start work on Quoddy,[65] with a total budget of $36,284,000.[66] After fifteen years of work, after having been killed four times, the Quoddy Tidal Power Project came to life. Dexter could open his eyes—it wasn't a dream.

Time had *not* run out.

CELEBRATIONS

"EASTPORT GOES WILD," cheered the *Sentinel* to the news Quoddy would finally be built. Streets were filled with crowds, and fire alarms, church bells, and whistles added to the joyous noise on that glorious day, Thursday, May 16, 1935.[67]

In Roscoe's reporting about the biggest thing ever to happen in eastern Maine, his relief and joy—*everyone's relief and joy*—came gushing out. The overwhelming tide of emotion, long repressed by opposing currents, broke like a tidal bore[68]* and roared into Passamaquoddy Bay:

> Eastport blew off the lid last Thursday night with the receipt of the good news from Washington that an allotment of $10,000,000 had been made for the first year's construction of the Quoddy tidal power project.
>
> A celebration which began within a few minutes after the announcement by wire and radio continued into the early morning hours. . . .
>
> Over in Lubec the riot call was sounded on the fire alarm, bringing out hundreds, who soon learned the reason therefor. . . .
>
> Had Dexter P. Cooper been at his Campobello home, instead of in Washington, undoubtedly there would have been a pilgrimage to that place to serenade him.[69]

Not just the adults *Wahooed!* Every kid in the area—all 800 of them—took Friday off and strutted in their own parade, waving American flags, blowing horns, and carrying miniature picks and shovels.[70] Lubec was elated and reprised its spontaneous revelry on Thursday night with an officially sanctioned celebration on Saturday. Free beer, provided by an anonymous benefactor, was just one of the delights that made Lubec's celebration one to remember. From miles around, everyone who could come, came, and the usually quiet village

> entertained more visitors and produced more racket than it has for the last decade. . . . Because of the absence of the leader, the band did not play but possibly nobody could have heard the music above the other din, a good part of which was produced by banging on sections of tin roofing.
>
> Lifting the ban on fireworks, the Lubec town fathers told the people to go to it—and they did! Throughout the evening skyrockets, Roman candles and red fire were being set off, factory whistles were blowing and the bells were ringing.
>
> Oratory was conspicuous by its absence. Those who could orate were too happy to bother with it. . . .
>
> Far into the night the celebration lasted, and the residents of Lubec and outlying towns returned to their homes to dream of the day soon to come when the actual work on Quoddy would start.[71]

Roscoe was happy to orate on paper. He realized the historic import of the events and his roles in making them happen and documenting them. On May 22, 1935, the *Eastport Sentinel*, the newspaper of record for the world's first tidal-hydroelectric generating plant, devoted its entire front page and sixteen articles to the Quoddy Tidal Power Project. In his formal statement as mayor, Roscoe C. Emery said graciously:

> We unite in rendering our thanks to our great President, who has perceived the possibilities for good in this epoch-making development and has refused to be turned aside from his determination to make of it a reality; to Dexter P. Cooper for his genius as an engineer and the steadfast courage he has shown under difficulties that would have overwhelmed a man of lesser stature. . . .
>
> No news more encouraging or more significant has come to Maine since the Armistice was declared. That event ended a great war; this will end the depression so far as this State is concerned.[72]

In his editorial, Roscoe's emotions welled up, betraying his sympathy with Dexter and the latent fear of failure that haunted everyone in Eastport:

> Dexter P. Cooper's penetrating eye perceived the light beyond the clouds, saw the dawning of a better day; and such was his zeal, his vast power of

comprehension, his sterling incorruptible integrity, and the clearness of his vision, that he brought the near-forgotten Quoddy project up on the summit of world-wide attention. It seems as if Divine Providence called him suddenly into existence that he might assist to stop a depression. . . .

Eastport, Washington County, Maine and the nation owe Dexter P. Cooper a debt that can never be repaid. . . . Amidst the rush and acclaim in next week's formal opening of the project's construction, do not forget whose brain conceived the plan, whose indomitable will bestowed upon it the winning touch. Let us remember and thank God for Dexter P. Cooper.[73]

The *Sentinel* also published a full-page ad, subscribed to by 55 citizens and businesses, thanking Dexter Cooper and President Roosevelt.[74] But the *Portland Evening News* intoned, "Let us be fair, when assigning . . . Credit For Quoddy." "First," the editor said, is the "peculiar coast formation and the tremendous rise and fall of the tides in that southeastern corner of Maine, a gift of God made the project possible." The "honor roll . . . of Maine's progress," the *News* insisted, should also be inscribed with the names of the other civic and business leaders "who have been energetic, tireless and hopeful, if not always confident, of ultimate victory." The *News* then mused, "Now that Quoddy is actually to be built it will be interesting to observe if there is not a change of front on the part of some powerful interests and individuals in Maine that have opposed the undertaking."[75]

The next day, Governor Brann sent a telegram to Louis Howe, the president's political adviser: "Republicans attempting to appropriate credit for Quoddy, attempting to kidnap our child." Brann proposed to "neutralize this" by having Major Fleming "report to Governor."[76]

Secretary Ickes agreed with the suggestion but responded that he had "no right to give orders" to the army.[77]

Information poured from the pages of the *Sentinel* on May 22, 1935: The first spadeful of earth would be turned by Secretary of War George H. Dern. Preferences would be given to local labor, and Washington was "hoping that there will be no 'gold rush' of workers from outside the state."[78] Fifty-two percent of the budget was for labor; 600 people were expected to be employed within one month, and ultimately 7,000. Major Fleming—*not* Dexter Cooper—would be in charge.[79]

The *Sentinel* also published a letter, "ADVICE ON BUSINESS BOOM AT QUODDY," which cautioned the hordes of hopefuls that "there are a few hundred persons who are set upon engaging in . . . the lunch-beer business; whereas only a small part of this number will be able to secure permits." The writer also advised against buying property without seeing it and against coming without reservations, as all the hotels and guest houses would be booked.

Another interesting, if ominous, item appeared in that celebratory paper: the FPC had forecast a shortage of power in the event of war.

Ralph Brewster hosted a celebratory dinner party on May 23 in Washington. Among the 38 guests were Dexter and Gertrude Cooper, Governor Brann, Maine's congressional delegation, Mayor Emery, Major Fleming, PWA staffers Hunt and Cory, and a few special friends.[80] Brewster introduced Amelia Earhart, the pioneering aviator, to Gertrude, saying, "This is the wife of the man who should have all of the credit for Quoddy." Colonel Fleming winked at a friend. The "wife" who had introduced her husband to Quoddy and her Campobello neighbors wondered what that wink was about.[81] Three days later, the "wife" and her husband had dinner with her neighbors at their other house, the White House.[82]

Two days after that, President Franklin Roosevelt signed Quoddy's $10 million initial allocation.[83] The news was "received here with quiet satisfaction," sighed the *Sentinel*. "There had been no doubt as to his intentions, but various newspapers had been laying increasing stress on the fact that he had not given the project formal approval and the removal of this last barrier was naturally most gratifying to the thousands of Maine people who had been watching the progress of the project . . . with almost desperate anxiety."[84]

The anxiety was heightened when the Coopers were in Washington. There Gertrude reconnected with another old friend, Alice Mayo Hof, who summered on Grand Manan Island, just seven miles from Campobello. Alice invited Gertrude to lunch. It wasn't just a friendly visit.

Alice had been sent by her husband, Major General Samuel Hof, to tell Gertrude what he could not say himself. "He was convinced that the Army would ruin the project and he did not want such an outstanding plan to be destroyed or to have his beloved Army take part in such a scheme."

Why? What had Dexter done to incur the army's wrath?

Dexter didn't believe Hof.[85]

Everything started happening in Eastport. A seaplane winged the bluebird sky, and when it alighted on the sapphire sea, many were sure it was an omen. Sure enough, out stepped Captain Harry Jones, the supervisor of airports for the Federal Emergency Recovery Administration (FERA). The dashing captain had three passengers with him, but he refused to divulge their identities to reporters (who did not divulge their gender). Captain Jones was happy to identify the "best site on the coast" for an amphibious airport was at Emery's Beach. More important than praise, he recommended its construction as a FERA project, noting it could be operational within weeks.[86]

The impetus for an airport gained ground when the *Portland Sunday Telegram* reported on May 19, "Ralph Brewster has been in conference today with representatives of the Department of Commerce and the Boston and Maine Airways, regarding plane service to the Quoddy area. Passenger service to Bar Harbor will be established on June 15. Service to Calais, Lubec, and Eastport is under discussion . . . [and] Amelia Earhart is due here next week."[87]

Perhaps riffing on Dexter's proposal from a decade before, rumors circulated that the navy's 400-bed hospital ship *Mercy* would be used for housing workers; the retired ocean liner *Leviathan* was also floated. The army, however, said it planned to build 250 houses and a dormitory somewhere near Eastport to house the professional staff and their families. The idea was to construct a "model community which will blend harmoniously with other Maine towns." The houses would be the Cape Cod or Dutch Colonial type, costing an average $1,500 each, and would be rented, not sold.[88]

Enlarging Eastport to accommodate thousands of new residents and visitors was going to be a challenge. Accordingly, Representative Brewster, Mayor Emery, and Oscar Brown of Eastport's Chamber of Commerce literally flew to Washington to meet with PWA staff about improvements to the city's water, sewers, roads, and schools.[89] It was probably the Eastporters' first airplane ride. They were glad to hear the city could get the money it needed as 45% federal grants and the rest as loans.[90]

More critical than housing were wages, and it was good news to all that the government would be paying well for the work. The monthly earnings were to be unskilled labor at $40 to $55, skilled at $55 to $85, and professional and technical at $61 to $95.[91] That was a *dam site* better than the $12.60 to $28.00 per month the government had been paying for sewing towels!

On June 26, it was announced that Quoddy's opening ceremonies and Eastport's annual Independence Day bash would be combined into one big party, Eastport's "Celebration of the Century."[92] As usual, the party would start with the "Night Before" reunion of "old timers" and former residents. There would be music, dancing, and masquerading, and the city would be lavishly decorated with the national colors and pictures of President Roosevelt and Dexter Cooper. A seaplane would be available for sightseers.[93] There would be more than the usual fireworks; as the *Sentinel* explained, "The big noise will come . . . when the first blast will be exploded to remove materials in readiness for construction of the dam. It is hoped this same blast will explode the suggestion that Quoddy will never be built."[94]

The style of the army's accommodations was set by the arrival on June 27 of its 148-foot yacht *Sea King*.[95] The yacht had a crew of eighteen and accommodations for guests in eight staterooms, with three baths, a dining room, and a salon, as well as an awninged area at the stern. As the *Sentinel* savored, "The dining room saloon, with the sunlight filtering through the big skylight is a place for relaxation. With its dark wood finish, lounge furnishings, mammoth silk cushions, radio, reading lights, easy chairs, it could easily pass for the library of a bachelor's estate ashore."[96] Everyone wanted to see the army's yacht and the men who would live on it—the men who would harness the tides.

"All the engineers of the Army are crazy to get a chance to work on Quoddy," the *Portland Press Herald* proclaimed, because it was a "unique project and bound to be the center of engineering attention the world over."[97] All were graduates of the US Military Academy at West Point. Each was profiled in the *Sentinel* with a photograph.

Major Philip B. Fleming, the most senior in age (48) and rank, had broad experience in the administration of flood control and navigation projects.[98]

Second in command was Captain Samuel D. Sturgis Jr. (37), a fourth-generation West Pointer.[99] He had been responsible for civilian-style construction projects at army bases in the Philippines and Kansas and had graduated from officer training school in June. Captain Sturgis (no relation to Gertrude Sturgis) was to be in charge of Quoddy's Camp and Land Division.[100]

Lieutenant Donald J. Leehey (38) would head the Administration Division. In the preceding decade, he had been responsible for the postwar dismantling of Fort Dix in New Jersey, dredging and diking the Mississippi River, and five years of teaching at West Point.[101]

Lieutenant Hugh J. Casey (38) would lead the Engineering Department. He had most recently been working on dams and locks on the Ohio River near Pittsburgh.[102]

Lieutenant Royal D. Lord (35) was to be Quoddy's director of operations. He had worked on harbor projects in California and Washington, designed several athletic facilities at West Point, and most recently worked on locks and dams on the Mississippi.[103]

The résumés were impressive, but something important was missing.

Gertrude was given another warning in June while having lunch with her daughter at the Lincolnshire Hotel in Boston. There she heard a gentleman behind her talking excitedly about Quoddy. He seemed to be criticizing the government's plans. Not one to leave a challenge unanswered, Gertrude called over the headwaiter and sent the critic her card with a message on it: "Mr. Cooper has not yet sold his project to the government."

"He jumped up and came over to our table, asking if I was Mrs. Cooper," Gertrude recalled. "Then he said he had just returned from Washington and stated that he was sure the Army engineers were out to wreck the project. He asked me to tell this to my husband, but again Dexter, wanting above all else to see the successful completion of his plan—for he had for years devoted all his time and energy to it—failed to heed this admonition."[104]

Validation for Quoddy came on June 12 from Congressman John Rankin (D-MS), who noted Maine's electric consumers could expect to save five to six million dollars per year for their "juice" if Quoddy resulted in electric rates similar to those of other publicly owned power plants. "No wonder," he said, "the power trust is opposed to the development of the great Quoddy project."[105]

Remote parts of Houlton, Maine, were soon to get "juice" for the first time. Their $58,000 request to the Federal Rural Electrification

Administration (created in May 1935) had been approved, effectively
starting the first electric "co-op" in New England.[106]*

The engineering details of the tidal project were announced in the
June 17 issue of the *Sentinel*, which noted, "[T]he most significant is
the cryptic sentence in the very first paragraph,—'Extensions of the
project are contemplated.'" No explanation was given by the army, but
many assumed it meant the eventual construction of Dexter's International Plan, as the project underway was "practically identical with the
American half of the larger plan."[107]

The army's plans called for two and a half miles of dams to be built
in the sea at depths from 30 to 140 feet. Of particular interest were the
turbines, "which are to be the largest in the world." Also unique would
be the pumped storage facility and its 12.8-square-mile saltwater lake,
which would make Lubec "practically an Island, but will give it the
tremendous asset of possessing the most remarkable body of water of
its kind in the Western Hemisphere, a fact which will draw tourists
and students of marine life from far and wide."[108] (The *Sentinel* didn't
say, but presumably the lake would be *contra-tidal*—high tide in the
lake when it was low tide in the sea—a confusing situation for sea
creatures in the lake.)

The army subdivided Quoddy into ten smaller projects:

1. Dam: Pleasant Point to Carlow Island; 33' high × 2,700' long;
   $150,000.
2. Dam: Carlow Island to Moose Island; 33' × 1,500'; $100,000.
3. Dam: Carrying Place Cove to Mathews Island; 30' × 2,000';
   $150,000.
4. Main Powerhouse: Between Mathews and Moose islands; 22-unit
   capacity, 10 initially installed, each of 16,667 kVA, turning at 40–45
   rpms; head race 100' wide, 50' deep[109] across Carrying Place Cove;
   $12,000,000.
5. Dam: Eastport to Treat Island; 140' × 3,400'; $2,500,000.
6. Emptying Gates: 25 vertical lift, 30' × 30' gates between Treat and
   Dudley islands; 550,000 cfs; $3,248,000.
7. Dam: Lubec to Dudley Island; 90' × 3,800'; $1,100,000.
8. Navigation Lock: 45' × 200' in Lubec Dam into Cobscook Bay;
   $300,000.

9. Transmission Line: Eastport to Haycock powerhouses; $1,500,000.
10. Haycock Harbor pumped storage facility; 8,170 acres at 120' to
    130' above sea level; main dam 155' × 4,000'; 9 pumps of 20,000
    hp each; 2 generators at 22,400 kW and 30,000 hp; $9,077,000.[110]

The total cost of the ten items was $30.125 million: *not* the $36.284
million reported by the *Boston Herald* on May 19,[111] *not* the $42 million
used in the Maine Planning Board report,[112] *not* the $47 million to $50
million asserted by others,[113] and *not* the $100 million high ball lobbed
by the *New York Times*.[114]

At the end of the long, fact-filled *Sentinel* article was a bit of
opinion, expressed by someone with some knowledge of the situation—Dexter P. Cooper: "These figures are low though possible if
achievement of the work is carefully and masterfully handled. The
figures for hydraulic and electric equipment likewise seem low and
the allowance for overhead and contingencies below the customary
amount. To be safe I believe 15% or $4,500,0000 should be added
to the estimate."[115] In other words, the 55-year-old Dexter P. Cooper
was suggesting to the young engineers—*none of whom had ever built
a hydroelectric power plant*[116]*—that a more realistic budget—*for
them*—would be $35.6 million.

Curiously, the "worker's village," soon to be known as Quoddy
Village, with an initial price tag of $700,000, was not included in the
army's list of ten construction projects, even though it was to be the first
thing constructed.[117]

Something else was missing from the army's menu: "pork."

The third warning came from Major Philip Fleming. He told Gertrude what may have driven him to drink on that day when he was an
overnight guest in the Coopers' home. He said the army was planning
to "rook" Dexter, and the amount it was planning to offer him for his
rights in the Quoddy Tidal Power Project was ridiculously low.[118]

On June 19, Gertrude wrote to James Roosevelt, the president's son,
his personal secretary, and her summertime next door neighbor, and
told "Jimmy" what Fleming had said.

To rook means to defraud, overcharge, or swindle. A rook is a chess
piece in the shape of a crenelated defensive tower. A rook is also the
emblem of the US Army Corp of Engineers.

On June 26, Dexter Cooper and the US Army Corps of Engineers signed a contract for the sale of Dexter's rights in the Quoddy Tidal Power Project to the US government.[119] Dexter's new job as "consulting engineer" would *not* be to help engineer the dams, locks, or power plant. Congressman Brewster tried to explain the situation: "The President felt that the entire Quoddy question should be settled before work began. He wanted no loose ends. He wanted a complete settlement with Engineer Cooper for his work accomplished, and obligations he assumed before the government became interested. He wanted plans for construction understood to the last detail. Finally, he wanted a definite basis for the future—because, if the great dam can't be practically utilized when it is completed, it would be a monument to folly."[120]

A skeptical reporter then asked Brewster, "Just what DO they expect to do with Quoddy when the work of construction is done?" "The idea," said Brewster, choosing his words carefully,

> is that Dexter Cooper shall devote his time . . . to an intensive study of how the power can be utilized. . . . [H]e will not be required—or supposed to—bother with the details of construction. His big task, in the next two or three years, will be to contact a market, so there will be an immediate outlet for the power when construction is ended. It will be the big job of Dexter Cooper, operating under the National Power Policy Committee, to contact these industries, interest them, "sign them on the dotted line," to use a colloquial expression. . . . And that is a bigger job, I think, than is the building of the dam—for upon it depends a large share of the future prosperity of Eastern Maine.[121]

To complete the "big job," Dexter (or someone) had to be empowered to "sign them on the dotted line."

While Relief could keep Downeasters alive, Quoddy would change their lives. Everyone in Campobello, Lubec, and Eastport knew Dexter; knew the ten years of effort he had invested in Quoddy; knew some of what he had been up against; knew the heartache he had gone through; knew this was not a selfish effort, but one he truly believed would benefit the whole community. When their hero returned from Washington on June 28, *everyone* went to greet him:

Dexter P. Cooper was given a tremendous ovation here at 7:00 Friday afternoon, and, included in the welcome were Major Philip B. Fleming, of the
Corps of U.S. Army Engineers, who has been designated to handle the job
of building the dams, and Congressman Ralph O. Brewster. . . .

More than 300 automobiles, some of them gaily decorated, and all filled
with joyous residents of the Quoddy area, thousands of people on foot, and
two bands were in the parade that moved through the streets of the old town
to Library Park where the program of address was held. . . . Bells rang,
whistles blew, horns tooted, cheers rang out and faces were wreathed in
smiles as the heroes of the day drove by. . . .

Mr. Cooper . . . was visibly affected as he paused to allow the wild
applause that greeted him to subside. "I don't know what to tell you," he
said, "except that I am glad to be home and that I have brought back the
bacon." . . .

Major Fleming was introduced as the last speaker. He won his audience
instantly by sheer forces of personality, as in a friendly, informal way he
told of his visit here last year, the way in which the undertaking had captured his interest then, and of his plans to carry it to successful completion.
. . .

This closed the most remarkable demonstration given two individuals
here within the memory of anyone now living. It was the more notable
because it was almost wholly spontaneous.[122]

The army didn't offer what Dexter thought he was due—not the
$1 million for the equity, nor the contract to build the project,[123] nor
the right to operate the power plant for a number of years. Instead, the
army argued that Dexter's company had failed, and, as such, it was
worthless; and, since Army Engineers knew better how to build such
big infrastructure projects, they didn't need all those studies or plans
Dexter had so diligently developed; and no, Dexter wasn't needed to
operate the power plant once the power was sold, so the army would
just give him the honorary title of consulting engineer and pay him $50
per day for consulting—but *only* for the days they actually called on
him. For all of this—for more than a decade of work, for his pioneering and patented designs, for ruining his health, for ruining his personal
finances, and in recognition of the public's heartfelt gratitude—on May
24, 1935, the US Army Corps of Engineers considered it fair to pay
Dexter P. Cooper $973.76.[124]

Perhaps it was unwise of the army to mistreat a friend of their commander in chief. Four days later, Dexter and Gertrude had dinner at the White House with the president of the United States. As it would have been inappropriate for Franklin to decide what the government should pay Dexter, an "impartial person" was chosen to arbitrate. The arbiter was an engineer, a financier, an army colonel, and a member of the National Resources Planning Board, and he had known all the parties involved for decades: He was Franklin's uncle, Frederic Delano.[125]

Neither the army nor Dexter got anything close to what they wanted. Delano recommended a sales price of $60,000 (approximately $1.1 million in 2020 funds). It was roughly 1/100 of what Dexter was hoping for—and 60 times what the army had hoped to get away with. According to Gertrude, the proposed sales price would leave Dexter $25,000 in debt.[126] Why this was so, or why Frederic Delano would think that this was fair, isn't clear, but the ramifications would be.[127]

With the life nearly crushed out of him; with the project so near to becoming a reality that it would break his heart if he became an obstacle to its realization; with the people in Downeast Maine in such desperate need of salvation; with the knowledge that *he* wouldn't get another chance; with the hope that somehow all would work out; with the naïve belief in the goodwill of men—on June 26, 1935, Dexter accepted Delano's decision.[128]

Apparently the army didn't, for the engineer on whose work "depends a large share of the future prosperity of Eastern Maine," and to whom was due a "debt that can never be repaid," wasn't paid.

The US Army Corps of Engineers changed the deal.[129]

Then the Corps dragged its feet.[130]

Then it stiffed him.[131]

But Dexter wouldn't know that for months. For now, at least publicly, it was all smiles.

The army's stiff arm was also extended at Dexter's investors—GE, Westinghouse, Boston Edison, and Middle West Utilities—four of the largest electric utility investors in the country.[132] The army did so despite the following clause in the Five-Party Agreement between Dexter and his corporate investors, dated August 31, 1927:

> This agreement shall be in effect from the date hereof until June 1, 1935, and in the event the financing or construction of said project shall be

actually commenced and underway on said date, shall continue in force and effect until the completion of the project. In the event the financing or construction of the project shall not have been commenced on or before said date (unless prevented or delayed by the refusal or failure of Cooper to perform his undertakings herein) this agreement shall terminate on said date without any obligation on the part of Cooper with respect to the funds advance by the Companies.[133]

At this time, Quoddy's corporate investors had provided $375,000 in cash, paid another $50,000 or so for the two reports by Murray & Flood, and were due about $75,000 in interest, for a total of about $500,000 (or about $9.4 million in 2020 funds).[134]

Given President Roosevelt's initial allocation on May 16, 1935, of $10 million to begin construction of Quoddy, *how would you interpret this clause?*

Another warning was given to Gertrude while she was in England visiting her sister and escaping the "strain" and "countless small harassments and annoyances" to which Dexter was subjected while the project started up. The warning was from an aptly but ironically named Mr. Beavers, an engineer at the Royal Academy, whom Gertrude had sought out for advice. Based on his experience trying to build a tidal power dam across the Severn River, which had been vehemently opposed by the entrenched power companies; his high regard for Dexter's plans; and his concern that Dexter *wasn't* being consulted, he advised Gertrude to "go home and beg Dexter to move away from the site and pull out completely." Beavers believed the army would "not only wreck the project but attempt to pull Dexter down with it."

The unblinking gaze of Clio, the muse of history, looked down on Statuary Hall in the US Capitol in Washington, D.C. As she sat in her marble chariot with wheels that turned with the hands of time, holding a stylus and tablet, she noted a meeting that began at 1:58 p.m. on July 1, 1935. The meeting had been hastily called between a member of Congress and a member of the administration. The heated exchange was witnessed by Clio and her silent colleagues, including inert effigies of George Washington, Andrew Jackson, and Daniel Webster, and by a very much alive and reliable witness. Not many times would the chariot wheels turn before their plan unraveled.

Governor Brann sent another warning to President Roosevelt on July 1: "So far Quoddy entirely in Republican hands and under Republican officers. A continuation of this policy will cause loss of Maine by fifty thousand in thirty-six."[135]* Brann sent another telegram to the Democratic National Committee: "Situation is alarming Maine Democrats greatly. Is in control of Republicans. Fleming's entry into Maine under Republican auspices solely. If positions are filled by Republicans it will be difficult to support. Matter should be handled immediately."[136]

Henry Morgenthau, secretary of the treasury, also sent a telegram to the president: "Brewster very shrewdly is arrogating to himself all the credit for Quoddy project. Suggest you appoint Governor Brann to represent you at dedication exercises on July fourth."[137] FDR didn't.

Celebration Day was a day without equal. Eastport, population 2,800, exploded with 10,000 visitors.[138] There is no better way to describe it than to quote from the people who witnessed it and recorded the events in the *Sentinel*:

> The greatest occasion, the greatest celebration, the finest weather, and the largest attendance on record here combined last Thursday to make July Fourth, 1935, the most significant and outstanding day in the history of Eastport, and all of Eastern Maine. Starting with the riotously joyful "Night Before," coming to a climax when the Vice President of the United States participated by exploding the first blast in the construction, and ending with the last strains of "Home Sweet Home" waltz of the official ball early Friday morning, it was a colorful and brilliant observation of both the National birthday and of the rebirth of prosperity for a whole commonwealth. . . .
>
> The "Night Before" was one of the liveliest and best attended of these unique affairs in the recent history of the town. Water Street was crowded with a good-natured, jostling throng, bent on making the most of the time-honored relaxation of ordinary rules of conduct, and the carnival spirit reigned supreme. Confetti, but no other material, was permitted, and fell in showers everywhere. Fireworks rendered the night noisy but brilliant. . . .
>
> Dexter P. Cooper, originator of the project, was not present. He had been called to Washington in the early morning to confer about important details, it was said. . . .
>
> Vice President Garner, sitting in the Capitol at Washington, with Senator Wallace H. White and several newspaper men present, touched the button that exploded the series of six charges of dynamite . . . and amid a crashing roar, a huge curtain of gravel, stone and debris rose skyward. Work on Quoddy was thus formally and officially opened. . . .

With Major Fleming on the job for the next two years the dam is as good as built [Brewster said.] . . . In conclusion he predicted the eventual international development with an expenditure of more than $100,000,000 as our neighbors in Canada awaken to the possibilities latent in the tides. Major Philip B. Fleming spoke in his accustomed friendly and intimate manner, expressing deep satisfaction at being in charge of a project so epoch-making. . . . Amid cheers, and the tooting of myriad horns he then stepped from the platform and literally "made the dirt fly" as he turned the first shovelful of earth in the project. . . .

Major Fleming, in opening his remarks, set forth some humorous protests, among others mentioning the absence of pie from the menu, that being a requisite to any New England banquet. At this point, Mrs. Hutchinson arrived from somewhere "out back" with a beautiful example of the pie-makers' art and set it before the speaker. "There's your pie, Major," she said while a storm of applause and laughter swept the room. It was one of those spontaneous, unforeseen incidents that sometimes happen and constitute a delicious bit of humor. The Major responded very nicely and later bore the trophy off to the *Sea King*. He said a little later, that should any investigation of Quoddy occur, nothing would be found to discredit the Government, and that the work would go on.[139]

Fleming would have to eat those words. No one in Eastport thought it delicious or funny.

It was true that Dexter had been called to Washington on July 4. But that wasn't why he wasn't at the celebration. As Gertrude Cooper explained in her memoirs:

Quite an elaborate celebration was planned; invitations were sent to important people, who had to present them at the gate in order to be admitted. None was sent to Dexter; I recall his pacing up and down in my bedroom that day. When we heard the first explosion and saw the dirt fly into the air, he turned to me and said, "That's my job, and I'm not even there." This was only one of a series of humiliating experiences; their excuse at that time was that they had not known he was in town![140]

CONSTRUCTION

"Abolition of the evil *features* of holding companies," President Roosevelt said in his State of the Union address on January 4, 1935, was "a measure of national importance."[141]* To that end, he pushed for passage of the Wheeler-Rayburn Act. Title 1 thereof was the Public Utility Holding Company Act (PUHCA), which would give the Securities and Exchange Commission (SEC) authority to regulate and break up utility holding companies. Title 2 would give the FPC authority over interstate power sales, thus closing the "Attleboro gap," through which 20% of the nation's electricity flowed.[142] The effort to enact PUHCA became one of the New Deal's most bitter battles. Downeasters found themselves in the middle of the melee, in another nationwide power struggle and more scandals.

Having weathered the storm from the 1927–1929 congressional investigations of their propaganda practices thanks to co-conspirators in the press who smothered it, the power trusts launched another offensive against PUHCA and the SEC. In a bid for sympathy, Wendell Willkie branded PUHCA's breakup clause the "death sentence" for holding companies, implying that it would cause financial ruin for millions of mom-and-pop investors.[143]

Pressuring the politicians, the power trust flooded the capital with lobbyists, letters, and telegrams. Nonetheless, PUHCA was approved by the Senate 56 to 32.[144] But on July 1 the House voted against it, causing Republican newspapers to gleefully pronounce the "death sentence" dead.[145] To the surprise of some, Maine's Progressive Republican representative, Ralph Brewster, a critic of the power trusts, voted against the "death sentence." Brewster said his vote was to protest a "threat" from Thomas Corcoran of the Reconstruction Finance Corporation (RFC) that if he didn't vote for PUHCA, the Roosevelt administration would stop work on Quoddy.[146]

The charge of such a drastic political reprisal by the administration was immediately dismissed by Maine's Democratic representative, Edward Moran. But Brewster drew applause from many in Congress for bringing into the open the excessive lobbying which, in the words of Representative Duncan (D-MI), would "disgust anyone."[147] Senator Rankin (D-MS) accused the power trust of tapping his telephone, and said that Brewster had sold out to the power trust.[148] Resolutions to conduct investigations into the lobbying activities of both the power trust and the administration were then passed unanimously by the House and Senate. Senator Hugo Black[149]* (D-AL) and Representative John J.

O'Connor[150] (D-NY) were assigned the task, and a hearing about Brewster's charge was scheduled for immediately after the July 4 holiday.[151]

Newspapers the next morning were full of headlines about Brewster's charge and Corcoran's denial. The *New York Times* hit the nail on the head:

> Consider how often that the President has directed there shall be no trace of politics in the expenditures for work-relief, and how often the Republicans have insisted his subordinates won't keep politics out, and you can readily see how important the charge could become.
>
> Particularly dear as a goat to the Republican heart is the works project to harness the tides of the Bay of Fundy. . . . To the critics of the whole spending program, and to those suspicious of the claim that no politics has been or will be mixed with relief by any one close to the President, the Quoddy scheme, from the beginning, offered the best promise of grief for the administration. If Mr. Brewster can prove, or reasonably establish, that anyone nearly so important in the New Deal as Thomas F. Corcoran of RFC used the Quoddy as a political bait or threat, what Republicans have hoped for since the Maine campaign of 1934 will have come true. . . .
>
> "This ends the depression in Maine," was the remark ascribed to [Mayor Emery]. . . . The Republicans were naturally irritated. After all, Maine is their traditional country. . . . They are particularly anxious to spread the impression that their attitude was based on high economic and engineering grounds, while the New Deal is animated by politics only.
>
> That is why they took today's development as a possible windfall.[152]

The *Times* then described Corcoran in glowing terms that would make anyone but a politician blush.[153]* The paper made it clear Corcoran was one of the most formidable attorneys in Washington, not one to be attacked without consequences.

"You're a liar," shouted Brewster during Corcoran's testimony in the Senate on July 9.[154] Replying "with cold, lucid, driving fury," Thomas Corcoran "tore Ralph Brewster's tale to shameful shreds."[155] "Tommy the Cork"[156] then countercharged that Brewster had committed to voting for the "death sentence" clause, as well as the overall act; to speak for it during the time allocated to Republicans; and to bring along 25 Republican votes for the bill, but had reneged at the last minute because of a "delicate political situation in Maine." Corcoran also said Brewster had agreed to keep politics out of Quoddy and *not* to attend Quoddy's opening ceremonies on July 4 *unless* accompanied by Democratic representative Moran. But Brewster went without Moran, was memorialized in photographs as the highest-ranking political official at Quoddy's opening ceremonies, and thus captured much of the credit.

Corcoran had a witness: Brewster's friend, the former publisher of the *Portland Evening News* and now a member of the Roosevelt administration, Ernest Gruening.[157]* "[S]turdy, self-possessed Dr. Gruening"[158] corroborated Corcoran's testimony of their "three-cornered" conversation on July 1 in Statuary Hall.

Said Corcoran, "If, as you say, you are not a free man politically and must take power company support into your calculations, then you will understand perfectly that from now on you can't expect me to trust you to protect Quoddy or trust your assurances that we will get the Maine Power Authority out of the Maine Legislature."

But, said Brewster, "I felt exactly the same as I did when Insull tried to put the screws to me up in Maine. At that time I was defeated for the Senate and for Congress and went through a veritable hell because I fought Insull. But when it comes to a fundamental issue of that kind and someone tries to use a club on me, there is only one answer—to tell them to go to hell and do the opposite. . . . I had been fighting the Insull crowd in Maine and came to Washington after the terrific effects of the Insull collapse on the State. I was impressed by the complete helplessness of the State against such a huge concentration of wealth and came here with a view that concentrated economic power had to be fought with concentrated political power. So I was sympathetic toward the bill. . . . But there was the other side of the argument in the losses to be sustained by innocent investors, and lining them all up against the wall to be shot. . . . [W]hen people in my district are starving I am ready to sacrifice everything, perhaps even my reputation, to save this project."[159]

Time magazine thought the Brewster-Corcoran debacle "simmered down to a question of veracity, in which the weight of the evidence, if not conclusive, was on the side of the administration."[160] A typical reaction to the imbroglio was that of a laborer in Eastport, who said, "Brewster shot off his mouth and he better have kept still, but that's politics; he can't stop Quoddy."[161] Others were more vehement. At Lubec, Brewster was hung in effigy with a banner bearing the words "Our double crossing Congressman" and "Quoddy Traitor." Stones and other "missiles" were thrown[162] at it by a "jeering punching crowd" of "several hundred people" including "town officials."[163] At least that was what the out-of-town papers splashed sensationally across their front pages on July 10–11. After all, bad news sells well, and bad news for Quoddy was good news for the powerful in Boston. The *Sentinel* reported a very different story in a small article a week later:

> A group of half-grown boys or very young men riding on a truck belonging to a man named Quick were the actors in the little drama, no adults being involved with the possible exception of a newspaper man who was on hand with a camera. The outfit drove up and down Water Street once or twice as it was chased from point to point by Night Officer John L. McCurdy. There was no disorder, no crowd, no fighting, and nothing of a sort to justify the lurid and apparently inspired stories that appeared in the Boston papers. . . . There is some evidence that the whole pitiable performance was planned and stage-managed for purposes of publicity by perhaps two individuals, one of whom is not a resident of Lubec.[164]

While the "little drama" in Lubec was of little consequence, the investigation Brewster ignited would explode with consequences.

Vindication for Roosevelt's public power policy came from a surprising source: the Edison Electric Institute (EEI), the successor to the disgraced and disbanded National Electric Light Association. The winner of EEI's annual prize was the Tennessee Electric Power Company (TEPCo), which had "established one of the most, if not the most, remarkable sales increases . . . in the history of the Electrical Industry." TEPCo was smack in the middle of the TVA territory. Rather than wither on the TVA's undercharged vine, TEPCo blossomed.

On the front page of the *Sentinel* that recorded Quoddy's opening ceremonies was the story of TEPCo and commentary by TVA's director, David E. Lilienthal:[165]* "This shows how unfounded are the claims of private power interests that the TVA will destroy their business and impair their investment. . . . The fact of the matter is that the Electric Industry has fallen into a state of self-pity," with electric utility executives "seeking to destroy the TVA" as an "unjust interference" in their "exclusive dominion."[166] TEPCo proved the power trust's claims weren't true. Lilienthal then asked: *Wasn't this proof of enormous latent demand for more power—if it was cheap?*[167]

Roscoe didn't realize or report who was running TEPCo: Wendell Willkie.[168]

More sensational headlines erupted from the Senate's investigation on August 8 when Senator Black went on nationwide radio to describe the $5 million ($90-plus million in 2020 funds) propaganda war mounted against PUHCA.[169] Just as in the 1920s, the 1930s propaganda was characterized as operating expenses—so the public paid for it. The thrust of the power trust assault was to create a "conspiracy neurosis" that PUHCA threatened the entire free enterprise system: insurance companies' assets would be seized to cover the national debt, mom-and-pop utility investors would be wiped out, freedom of the press would be throttled, Communism and dictators would take over, and the president was physically sick and insane.[170] The attack backfired.

Using controversial "drag-net subpoenas," Black exposed the depth to which the power trust had sunk to scare the electorate and subvert the democratic process.[171] Along with much else, Black proved the giant holding company Associated Gas & Electric (AG&E) and its president, Howard C. Hopson, had sent 250,000 fake telegrams using names they had taken from phone books. Black also documented AG&E's attempts to cover up the fraud by destroying the evidence.[172]* Hopson was subpoenaed but didn't show up, leading to a twelve-day headline-grabbing manhunt. Alluding to a scene from Harriet Beecher Stowe's *Uncle Tom's Cabin*, one reporter said, "Eliza crossing the ice, hotly pursued by bloodhounds, had nothing on Howard C. Hopson. Through the corridors of leading Washington hotels, over the rolling meadows of nearby Virginia, through the highways and byways of adjacent Maryland, troops of Congressional agents, G men, police, and assorted sleuths sniffed excitedly on the trail of the elusive utility baron."[173]

Thanks to the chairman of the House investigation, Representative O'Connor, and his brother, Basil, an attorney for AG&E, Hopson nearly got away again.[174]* Apparently their scheme was to place Hopson under "House arrest" and then release him, thus effectively giving him immunity from investigation by the Senate. The scandal proved Roosevelt was right about the corrupting evils of holding companies and gave him the leverage to force the House to revote on PUHCA with the "death sentence" included; the House "stampede[d] for cover" and voted for it.[175] With more of Hopson's crimes revealed through his federal tax returns, which President Roosevelt ordered released, Hopson was convicted of fraud and jailed.[176]

Back in Maine, one subscriber wrote to the *Sentinel* congratulating Quoddy's supporters for what they had so far overcome, but cautioning them:

> There are already established profit producing sources of power not desirous of meeting competition. This is not a large group, but strong and those in command, [are] subtle, to say the least and quite conscienceless. . . . [They] would have a halt called [to Quoddy], delight to see what has already been done be undone, the workers sent back home and a tomb of stone, recording the demise of the whole matter set up. . . . There are snakes lurking in the grass nearly everywhere. . . . [Y]ou folk of Quoddy, who have built your hopes high . . . be on your guard, be watchful, see to it, before you sink back contented with promises made, with what has been done or is designed to be done, that you have swept the ground clear.[177]

Every conceivable space in Eastport was soon rented. Every empty fish factory, store, and shack was taken over for warehousing, the soil lab, the cement lab, the drafting group, the purchasing office, the employment office, or some other function. Every half-empty or quarter-empty church and fraternal hall was filled. Rents doubled.[178] Still more space was needed. As the workforce grew, they outgrew their offices, and a city-wide game of musical chairs began, with signs appearing on doors redirecting would-be job seekers and new employees to other locations.

Hotels and rooming houses were booked solid, indefinitely. Houses and unheated summer cottages that had for years sat empty as the town had shrunk with the fishing industry were sold or rehabbed and rented.[179] In conjunction with the Federal Housing Authority (FHA), local women organized a housing drive. They called on every homeowner to ask about plans for improving property; handed out flyers on low-interest FHA loans; and collected information on potential improvements and then posted it for contractors to bid on. In addition to the FHA's national objectives of spurring spending and jobs, in Eastport there were other objectives. "The first of these," as Mayor Emery explained, "is to extend to the utmost possible limit the facilities of the community for housing employees on the Quoddy project, by making every house and every room in town that can be adapted to this purpose available for it. . . . Another objective is to increase the income of as many households as practical by . . . [fitting] up houses and rooms for renting. . . . A third objective is to improve the appearance of town."[180]

Building and business permit requests piled up at City Hall. Mayor Emery saw the need for zoning to control the rampant activity, so he contacted cities that had experienced rapid expansions and then appointed a commission, headed by Dexter Cooper and Philip Fleming. They sought citizen input in adapting the experts' advice to local conditions. Their plan was implemented with strong support. Eastport would now have residential, commercial, industrial, and civic zones, as well as transportation thoroughfares and hubs for pedestrians, cars, trucks, trains, ships, and airplanes. Eastporters were going to see a new city grow up around them. Surrounded by stimulus, everyone in Eastport could imagine themselves in this new land of prosperity.

Harder for everyone to get used to was the traffic. The once-quiet town was now packed with people, cars, and trucks, most of which were "from away." The out-of-towners didn't know where they were going. No longer could you park anywhere you wanted because you wouldn't be blocking anyone else. Now people were parking everywhere. The helter-skelter caused havoc. You couldn't run into town without risk of being run into. Many visitors weren't *driving like they lived here*. There were too many close calls, too many accidents. Something had to be done.[181]

Mayor Emery, who must have been at a civic meeting nearly every night, organized a traffic commission. It recommended speed limits (*20 mph was still too fast—Make it 15!*); stop signs (*How novel!*); parking restricted to one side of the street in downtown—parallel to the curb here, diagonal facing out there (*How confusing!*); taxis (*We have taxis?*) only at designated stands; trucks only in specified loading zones at certain times (*When? Where?*); and then (*No way!*) one-way streets in the business district.[182] The blinking lights at the new railroad crossings, however, seemed like a good idea to everyone.

The original plan, as Eastporters understood it, was for the workers' village to be built convenient to the railroad station, at "Sodom," the town's rundown south end. The government would then lease the necessary land from its current owners for $15 per acre per year, promising to abandon to the landowners the leasehold improvements thereon when the leases expired. The few dozen landowners potentially involved readily agreed to this opportunity for a windfall. But then Captain Sturgis, head of Quoddy's Camp and Land Division, arrived on the scene and announced the government could not make such a deal but would buy property outright. The unmet expectations left hard feelings.[183]

Nonetheless, for Downeasters there was still much to be excited about and many reasons to go into Eastport, especially since many were now earning at least a little more money and new merchandise was filling the stores. One delicious new offering was made by Miss Christine Miles, who installed an automatic doughnut machine in the Eastport Café's front window. There she turned out ten dozen hot crisp doughnuts each hour. Miss Miles priced the doughnuts at only 25 cents per dozen, and her marvelous machine stirred and stamped and fried and pulled in the customers for fifteen hours per day.[184]

One thing missing from town, in the eyes of Captain Sturgis, was a tart of a different sort. He asked that a "red-light district" be allowed to develop since he had "four thousand men to take care of." Roscoe refused.[185]

In the heat of August, Mayor Emery found himself immersed in the "public" water supply debate. Back in 1929 Eastport had successfully sued to prevent "watered" utility stocks from artificially driving up utility rates, but residents were still at the mercy of a private utility holding company that controlled the city's drinking water. With the expected tripling of Eastport's population, the city was going to need more water. The water company said a temporary solution could be done for $50,000 per year, but a permanent solution would cost $175,000 to $300,000. The company felt it "would not be justified in making [such expenditures] without definite guaranties that the revenue would be sufficiently large and permanent to warrant it."[186] John Wise, the Eastport Water Company's general manager, was reported to have said:

> [W]hile the [Quoddy] Project is approved and actually under construction there is nothing to show it will ever be completed or that if it is completed, industries will be found to locate here to use the power. He also maintained that the company has been steadily losing money and that it would be unable to borrow several hundred thousand dollars from any bank on its record or prospects. . . . He thought the city might be able to borrow the money to buy the Company, through the RFC or the PWA although in his experience he had found that municipally-owned water companies ordinarily were not successful.[187]

Wise's words about the failures of municipally owned utilities were
quite ironic, as he had just admitted that he, a professional manager
of a private utility, was unable to run the company profitably and thus
willing to sell it to the public—and of course Lubec knew how to run
a *real* public utility.[188]*

Wise's words were also a warning. The utility holding companies had
not given up their fight against Quoddy.

Selfishness affects everyone, as Hugh Cooper had noted, not just rich
out-of-towners. That truth was demonstrated on July 19 when Major
Fleming charged local merchants and landowners with price gouging.
As an out-of-town newspaper approvingly remarked:

> Army discipline was shown here today for the second time, when Major
> Philip B. Fleming fired 100 workers and ordered a halt on the Quoddy
> seaboard power project. Major Fleming said work would not continue
> here unless the owners agree to sell their land for reasonable prices.
> . . . Today's action came as a result of difficulty in land dealings. Owners
> asked as much as $4,000 an acre for untilled land, the tax assessed value
> of which is $7.50 an acre. Fleming cited that, on the other hand, one man
> sold 40 acres to the government for $1 an acre. He declared he would take
> his "model village," for which he is purchasing land, to another town,
> unless he is met halfway.[189]

According to the *Sentinel*, "Major Fleming had on hand Friday afternoon but two offers which he termed reasonable, those of Mayor
Roscoe C. Emery and Sidney Stevens. Major Fleming stated that
condemnation proceedings by the right of eminent domain would be
invoked . . . if reasonable terms cannot be negotiated immediately."[190]
The *Boston American* sensationalized (with a double banner headline)
and dramatically distorted the dispute and layoff to imply the *entire*
Quoddy Project was halted.[191] Fleming's bluff had its desired effect.
That very day, a civic-minded citizen stepped forward. With tears in his
eyes, he patriotically offered his land to the government despite having
previously turned down a higher offer from speculative bidders.[192] He
instantly became a local hero.

The purchase was a good thing for Quoddy, as the *New York Times*
explained: "If the government starts condemnation proceedings . . .
the power interests may challenge the constitutionality of the law. The

Bangor Hydro-Electric Company and the Central Maine Power Company have adopted a watchful-waiting policy and will neglect no step to protect their alleged rights. When the 1937 Legislature undertakes the creation of an authority to sell power these private companies also foresee an opportunity to contest the legality of the plan."[193]

The day when the army took possession of the land for Quoddy Village was recorded by Sidney Stevens. He was the attorney who had called the power trust's bluff about a mysterious (and fictional) company rumored to be interested in building a fish-packing plant in Eastport. That was not the first time Stevens had put his money where his mouth was. Many *Sentinel* readers would have missed it, buried as it was in an article about astronomy and tides, but therein Stevens revealed that he had been the first private citizen to invest in the Quoddy Tidal Power Project, back in 1925.[194] Now Stevens encouraged the *Sentinel*'s readers:

> **VISUALIZES NEW EASTPORT TO BE:**
> . . . As the nucleus of this new city, the government will erect here an administration building of stone or brick three hundred and thirty two feet in length, to be one of the finest permanent public buildings in Maine; also other permanent public buildings, and twelve permanent residences for its officials. . . . The intention of the government from this appears, to this writer, to be that of making New Eastport both a far eastern military and naval base for all time. . . .
>
> With the view it commands, no lovelier spot on the face of this earth can be found than just back of Farmer Rice's little yellow house on the hillside to the right as the tourist drives through the site of New Eastport. . . .
>
> [F]rom that point alone nearly all the works to be, upon the dams, the great power house, the thousand-foot reach of the ornate bridge to be constructed over the "Carrying Place" and the work in cutting that "Carrying Place" away will be spread out before the observer as a panorama extending around him on every side.
>
> George Rice is a widower eighty years old, a life-long farmer, and he has loved those fifty acres of his best land which he has so generously given to his country at its own price thus breaking the deadlock. . . . George Rice, however, and long may his name be honored, has chosen to make the sacrifice. The writer was present with him on Monday last, July 22, at 9:20 a.m. when the government moved in and took possession. . . . The old man was much affected by what he knew was about to take place. I succeeded in comforting him somewhat by the assurance that he had earned a warm

place in the hearts of the people, and especially in the hearts of those who would come to occupy this new city of New Eastport, the little city that would come to be the gem of the Far East. He would be as a father, reverenced by all those people.

Then as he glanced past me thru the open door he saw them coming and exclaimed, "The army is moving in." . . . At his suggestion we stepped outside the house to view the government's formal occupation. . . .

As the old man at my side saw them move onto his beloved acres he broke down; tears streamed from his eyes and he trembled and showed such signs of an approaching collapse that I felt compelled to grasp him firmly by the arm to steady him. George Rice loved those lands if ever man loved that dearest treasure which Nature and God have bestowed upon mortals.

I will leave George Rice to your imagination standing there. I leave thus to you, and to posterity, a scene never to be forgotten and fit for an immortal painting by the brush of a master artist.[195]

Sidney Stevens visualized what *should* be. To complete the vision, the army needed to acquire twelve properties between Lubec and Pleasant Point on which the dam abutments were to land. One man owned two of them, and his family owned three more: the local real estate speculator, Roscoe Emery.[196]

It would be a "fight to the finish," the *Boston Herald* declared,

between the New Deal and the thousands of investors who have placed their funds in Maine power companies. . . .

Officials of the Bangor Hydro-Electric Company and the Central Maine Power Company were frank in stating that the utility companies who have adopted a policy of watchful waiting for the present, would take every possible step to protect their rights and the rights of their shareholders. It was definitely said that this policy would be pursued until the dam was completed, if it ever was. . . . They believe that if Quoddy is given enough rope it will hang itself.[197]

One of those power company presidents stated, "on the authority of our engineers" (*who had never built anything even one-tenth as big*), Quoddy would "take twice that time and could cost nearer to $100,000,000 than $36,000,000," and then he went on to *misstate* lots of facts, which Richard O. Boyer, the reporter, did not refute.

But Boyer's article is also rich with real facts and insights, as revealed by his interview with Dexter Cooper, who when asked if his plan had been abandoned, said, "I do not know. I have not recently talked to Maj. Fleming, who is in charge." Yet, in point of fact, as everyone knows here, Cooper, large, wistful, silent, has been severely hurt by the fact that his project, his ideal, his mania and abiding preoccupation, has been taken from him, while he has been relegated to a comparatively minor position. He has given his fortune to the idea that harnessing the Bay of Fundy would reclaim Maine and be a milestone in the history of the entire country. He has staked his professional reputation upon this conception, slaved for it for ten years and now that it approached actuality, he finds himself with no part in the actual construction of his brain child. But, as we shall presently see, he does not complain.[198]

Boyer also interviewed workers on the project and local residents and came away with strong impressions of the people, the project, and the town:

> [T]here is not a soul in this maritime community of 3200 against the project. One might as well expect a Christian to be against heaven as count on a citizen here to be against "Quoddy" as it is familiarly and affectionately called. "Quoddy" is regarded as the millennium, as Utopia. This town until the New Deal touched it and made all milk and honey and gold, was as down and out as any bum in a gutter. . . . Its citizens have known the depression as few have, for the economic emergency threatened to wipe it out. Homes were sold for back taxes. Merchants were going bankrupt through continual extension of credit to impoverished customers. Relief rolls were mounting and there was but one ray of hope on the horizon.
>
> That emanated from large, slow, deliberate, Dexter Cooper. For ten years he had been striving to bring his plan into actuality. . . .
>
> But through the years, Cooper's plan gave hope to the town's citizens, instead of saying, "When my ship comes in," they said, "When Dexter's plan goes through." . . . They know if they charge too much for the land that it will impede the construction of Quoddy. They know if Quoddy is abandoned the town once more will sink towards extinction.[199]

Boyer's characterization of the town and its inhabitants was harsh but not unfair. The same might be said of his next article, titled "SEEK 'KILLING' ON QUODDY LAND."[200] Therein, Boyer recounted Major Fleming's negotiations with local property owners, focusing on David MacNichol, who was said to have been offered $5,000 for his waterfront property on the south edge of town but wanted $50,000—

causing Fleming to shift the location of Quoddy Village to Rice Farm,
three miles from town.[201]

Then Boyer jabbed his pen at Roscoe: "Mr. Emery, straw hat on the
back of his head and resembling a rural city slicker, said, 'I'll get the
best price I can, but I won't sell to anyone but the government. Why I
told the Major yesterday, if you need my land and we can't get together
on price, I'll give it to you. No, sir, no one's going to stand in the way
of Quoddy around here. Of course there's a little difference as to what
the land's worth.'"[202]

The stab in the back that took a *pound of flesh* and *a drop of blood*
came in the *Boston Herald*'s unsigned editorial:

> **A GAME OF GRAB**
>
> A citizen of Eastport is quoted in the news columns as having said that
> "of course there's a little difference of opinion as to what the land's worth."
> As an example of understatement, that seems hard to beat. There is not
> simply a little difference, there is an enormous difference as to what land
> needed for Quoddy purposes is worth. . . .
>
> Eastport has received perhaps the greatest single gift ever made by Uncle
> Sam to a small town in all our history. Population was dwindling, values
> were crumbling, and nothing much except extinction loomed ahead. Yet
> no Shylockian speculator in corner lots in our cities ever displayed a more
> mercenary disposition and a more unreasonable unwillingness to be simply
> decent than some of these thrifty down easters who only a short time ago
> were wondering how in the world they could manage to pay their taxes.[203]

The *Boston Herald*'s editors *misconstrued* Roscoe's remarks: he was
holding out for the best price he could get *but* was prepared to give
away his land for the benefit of Quoddy.

Roscoe responded to the *Herald*'s slander by *misdirecting* an extraordinarily vicious letter at Boyer. It was a letter he should have sat on for
a day to cool off and never sent. Its effects would be to make him look
worse and to make enemies:

> On the day you came to my office I had many callers. They ranged from
> a down-and-outer seeking work, to the Governor of a New England state
> who drove some distance to renew with me a friendship of long standing. I
> treated them all, high and low, with equal courtesy. . . .
>
> I disliked you the moment you slouched in the door, but I swallowed my
> disgust at your buttock-face sloppiness because you claimed to be a news-
> paper man. . . . Had I recognized you as the liar and blackguard you proved

yourself to be, I would have booted you into the street, the mud of which is pure and sweet compared to the rotten thing you are.

You made a sneering reference to my weekly newspaper. It may be small, but it is clean and honest. . . .

Your description of me as resembling a rural city slicker is probably actionable, and if it is, I promise to obtain for it whatever satisfaction the libel law may allow.[204]*

Equally outraged by Boyer's defaming of Eastport, Roscoe asked the city's attorney, "Is it outside the range of legal practice for a municipality to sue for libel?"[205]

Boyer treated Dexter more kindly, saying everyone in Eastport felt "personally affronted . . . that Cooper has been forced out of the picture." Nonetheless, Dexter "retains an Olympian aloofness to the politics. . . . He still is obsessed with the idea. It's his and he is overjoyed to see it near consummation no matter who directs it." Dexter

says nothing about the fact that he once hoped the idea would make him a rich man and that he is now reduced to the stature of an outsider watching his own creation be carried out by another. He is given to long silences and occasionally he fingers a black mustache, indulging in the mannerism of grimacing with a heavy featured face. He is slow and cautious in statement and one cannot help but like him. . . .

As Mr. Cooper talks, his eyes acquire the gaze of a man who sees a vision. "This is bigger than politics," he says and in his voice there is a certain contempt. . . . "We should forget about politics and think about the people and what Passamaquoddy will do for them."[206]

Rice Farm was now the army's canvas. On it bulldozers were the broad brushes, saws for shaping forms, and drafting pens for the fine points. Architects John P. Thomas and John Stevens of Portland were hired to design the idealized town to be known as Quoddy Village.[207] Everything would be prim, proper, and appropriate to a New England village and US Army base. Every detail would be planned and executed to government spec.

Site work began early Monday morning, July 19, on the 50-acre tract of field and pasture that had been Rice Farm[208] and, half a mile away, at what had been Spring Farm.[209] The former was to be the location of Quoddy Village and the latter to be one of two "camps" of worker barracks. Each of several dozen barracks would hold 50 men, and each camp would have its own mess hall. Quoddy Village was to include 120 *temporary* housing units for staff. At a separate site atop Redoubt Hill, the army planned to build 9 *permanent* residences for the project's officers.[210] Contractors had fifteen days in which to prepare their bids.[211] By August 7, the Army Engineers had final approval to construct their $1 million outpost in Eastport.[212]

In army parlance, "permanent" buildings would have concrete foundations and full basements, whereas "temporary" structures would sit on wooden posts over crawl spaces. There were to be six types of temporary houses at Quoddy Village: 20 Type A single-family houses, 15 Type B single-family houses, 15 Type C single-family houses, 20 Type DD "Dutch-double" duplexes, 20 Type EE duplexes, and 30 Type FF four-family houses.[213] The bid invitations specified work had to be completed within 45 days. When the bids were opened, the army discovered a problem.

First, let's have the *Sentinel* set the scene at the army's temporary headquarters:

> It is three o'clock in the afternoon. The room is smoke-filled and crowded with people leaning against the walls and cramping the seats to double capacity. On every knee an open notebook lays below poised pencil. Piled on a desk in the front of the room are fat, brown envelopes most of them with Western Union messages clipped on the outside. Behind the desk stands Captain Donald J. Leehey, head of the Quoddy Administration Division. It is an opening of bids.
>
> Captain Leehey announces the article called for. Pencils and pens scratch. An office secretary fills in a space on a long yellow chart, the official record. . . .
>
> He leans over the desk, slits open the top envelope, reads the name, address, price, and time period of the offer. A cryptic wire confirms or modifies the bid; sometimes it withdraws the offer as the company's experts refigure at the last minute and find their profits on too small a margin. So on down through the list as the pile grows smaller and the blank charts of the secretary blacken with figures. With the opening over, totals are hurriedly added and a sweeping check mark on each memorandum pad indicates the low bidder. Then begins the rush for Western Union as company representatives, insurance agents, and press correspondents race with the news to the home office.[214]

The problem appeared on August 20 at the opening of bids: *all* the bids came in above the army's estimates. The bid submitted by

J. Slotnick & Co. of Boston, Massachusetts, was the lowest (by more than $100,000), offering to construct the entire lot of 120 buildings for the sum of $697,000: 41%, or $203,008.15 *more*, than the army had estimated.[215]

It was a bad omen for the *low* bid on the first big contract to be 41% more than the army had estimated. Many wondered what that meant for the army's $30 million estimate for the entire tidal power project.

Nonetheless, the army decided to build all 120 houses on its own. Captain Sturgis would be in charge. The army would now have to source all the materials and hire and manage all the employees, which those in charge estimated would add at least three weeks to the project—thus pushing the completion date into the late fall, when ground and exposed pipes would be freezing and snow falling. The builders would have to hurry.

Four hundred workers set to work the next day:

> Construction began Friday [August 23] on one of each of the six types of houses called for in the temporary housing plans. . . . All of the structures are being built in a consecutive line on the connecting curve between Kendall Road and Kennebec Road. . . .
>
> Near the edge of the field by the county road, water boys for the 400 workers wait by a small spring for the water to clear before scooping their next pailful. Directly beside them a well-driller pounds rhythmically down into the earth for a more useful and abundant supply. . . . Smoke from the great bonfires burning brush and clearage materials curls in and out through the trees and blankets the region with a clean-smelling blue haze. . . . The big mechanical dirt scoop . . . is now deep in the forest ripping out roadbeds.[216]

The army planned to build the 120 houses in assembly-line fashion, so it organized six teams, one for each type of house. For each type, there was a specialized "gang" for each task—excavations, foundations, walls, roofs, plumbing, electric, painting—working in two shifts, from 6:00 a.m. to 8:00 p.m.[217] Just like a factory, all the "parts" were delivered precut, labeled, and waiting at each house each morning for that day's work.

To accomplish this feat of orchestration, the army set up a staging area next to the railroad spur and kept it running 24 hours a day, illuminated by huge arc lights. There the workers handled everything as few

times as possible—unloading raw lumber from railroad cars directly to
sheds where electric saws cut them to the exact size and shape required
for each element of each house, and the pieces were then stacked in neat
piles for selection and delivery each night. Roving gangs of "make-up
men" would finish off at night any tasks not done during the day.[218]

The experiment was a success, and mass construction began at
Quoddy Village on September 25. Meanwhile, road construction
continued, with ditch diggers and water and sewer pipe layers everywhere. At 5:30 each evening a whistle blew, and those near the drilling teams knew to step back and look up as explosions blasted away
outcropping rock.[219]

It had been the hottest summer on record,[220] but fall was coming on
fast, and work had not even begun on any of the larger buildings—or
the central heating plant.

Bid invitations for group two buildings went out September 14:

- A two-story, 40,000-square-foot administration building
- A two-story, 85-room dormitory with a dining room for 125
- Two two-story apartment buildings, each with 40 apartments[221]

Still being drafted were the requirements for the group three buildings:[222]

- A two-story, eight-classroom schoolhouse for 350 students
- A 20-bed hospital with eight private rooms and an isolation ward
- A seven-bedroom guesthouse with a lounge and gymnasium
- A central heating and power plant with two 500 kW diesel genera-
  tors[223]
- A firehouse

As an insufficient number of qualified bidders had been a problem,
Captain Lord went to New York to roust up more qualified contractors. Nothing was heard from him for three days, and time for bids was
running out. Then came a telegram directing a fleet of cars to meet the
daily "milk train" at a specific siding 30 miles away at 4:30 a.m. No one
knew quite what to expect, but as Lord liked the dramatic, curiosity and
a fleet of 20 cars met the train as directed. At the tail end of the humble
train were four "swank private cars" provided by as many railroad

presidents, and aboard them were what proved to be very interested bidders from eight of the nation's largest construction firms, including Stone & Webster. How that august group and their deluxe transport had been assembled and brought the distance on such short notice was the captain's secret, as Lord moved in mysterious ways.[224]

MAKE HASTE

How high and how fast are Quoddy's tides? The US Coast and Geodetic Survey began answering those questions in 1807. Admiral Owen made his systematic tide and current surveys in the 1840s. Dexter Cooper spent half of the 1920s and much money to do it in even greater detail. Now in 1935, the US Army Corps of Engineers commissioned their own tidal surveys to accompany their topographic (above sea level), bathymetric (below sea level), and subterranean (below ground level) surveys. The plan called for 50 tide stations to be set up all around Cobscook Bay, Passamaquoddy Bay, and the Bay of Fundy. The gauges were to be read at 15-minute intervals, 24 hours per day, for two months,[225] including the fall equinox on September 23, when the highest tides of the year usually occur. To this they would add a margin for safety when calculating how high and how strong to make their dams, gates, and locks.

While reading tide gauges on shore was tedious, doing so in a small skiff at sea was also dangerous:

> A party of five engineers . . . were anchored off Esty's Head early in the morning during the flood tide. . . . They had been on the shift since twelve o'clock midnight, sinking weighted lines and conducting current and velocity tests against the powerful flow. As it happened, all of them were looking in the same direction for some time during the course of their work. When the first one turned again, looming almost beside them and looming over their small craft was the big three-masted schooner *Helvetia* drifting broadside with the [out]going tide. It had cast loose from the pier in Deep Cove and was taking advantage of the current to coast out into the harbor before starting the engines. All was quiet. High up on the bow of the schooner a man peered over and said almost casually, "Look out, we're going to hit you." One of the frightened engineers climbed into the dinghy fastened at the stern of their boat and shouted to

cast him loose; he was not a sailor by nature. The head of the party hur-
riedly loosened the anchor rope and the two vessels, giant and pigmy in
comparison and each out of mechanical control, edged by each other.[226]

It was on Sunday, October 13, that the army learned the full size of
the forces it was up against. On that day the tides reached a record 28.8
feet high. "Hordes of rats were driven from their waterfront habitudes,"
observed the *Sentinel*, "and debris of various kinds was swept into the
shore smokehouses." A few hours later, the low tide at Lubec was "the
lowest tide in the memories of the oldest inhabitants and permitted men
in hip boots to wade across the bay to Campobello Island, N.B."[227]

The point of the sword for construction of the tidal power project was
the hardened steel bit of the boring drill. It was to penetrate the unknown, to drill through mud, dirt, and solid rock to determine whether
the ground on which the dams, gates, locks, and powerhouses were to
rest was in fact solid enough to build on. Drillers operated on land and
at sea. Although Dexter had done extensive boring and worked for a
decade figuring out the optimal building sites and configurations, the
army didn't trust him, so they hired Paul F. Kruse, a civilian, and put
him in charge of resurveying and redesigning everything.

Kruse had spent five years working for J. P. Morgan's[228] Niagara
Falls Power Company, followed by a dozen years in New York City as a
consulting engineer.[229] As the most senior civilian on the Quoddy staff,
he was granted a superior house on Redoubt Hill—an attitude he apparently adopted and passed down to his surveyors, who sneered when
they came across benchmarks set by Dexter.[230]

Boring operations on land were noisy, but safe and straightforward. It
was not so at sea. Outside experts and large, specialized equipment had
to be brought in for the pioneering work, in deeper water, higher tides,
and faster currents than had ever been bored before. Sprague & Hen-
wood of Scranton, Pennsylvania, was the primary contractor.[231] The men
worked 24 hours a day, seven days a week, regardless of the weather.

Water jets were used to dig through mud and loosen sand.[232] Once
down to more solid stuff, hardened steel or diamond-tipped[233] drills
would cut four- or six-inch continuous cores. First, however, they had
to invent a system to keep constant pressure on the drill bit despite
the tides changing the level of the barge by an inch per minute—twice

as fast as the drills were penetrating the rock.[234] The drilling was in effect prospecting and revealed potentially valuable deposits of lead and graphite.[235] The workers also discovered some things they didn't want to find.[236]

The fishermen knew how deep the water was between Treat Island and Eastport. They didn't know how deep the mud was. That's what the boring teams discovered.

The dam the army was going to build between the two islands would have to go all the way to where sea creatures crawled on the bottom, 130 feet down, and then some. Unlike a lightweight lobster trap or even a 30-ton ship's anchor, which would hardly dent the sea bottom, a 9-million-ton dam[237] would sink deep into the goo. The goo was almost 200 feet deep.

Like stepping into a mud puddle, the dam would sink as the goo squirted out the sides. In this case, however, the dam's "footprint" was not half of a square foot; it would exceed one hundred acres.[238] Some feared it might have to be half a mile wide.[239]* And the dam would not be an inflexible monolith, a single piece of concrete like Hugh Cooper's confections, but five *million* dump truck loads of rock and clay.

The dam would sag like a bean bag, its "feet" slipping away from its center while its rump landed with a bump in the middle. This description isn't just picturesque but based on careful tests with models made of sand and stones, layered clays, and gelatin parfaits.[240]

The vast amount of rock candy needed to fill Treat-Dudley's gluttonous gut enticed the *Boston Herald* to quip, "It is entirely feasible in an engineering way, as the leveling of the Rocky Mountains is."[241] A more apt analogy was constructed by writer Kenneth Roberts, who likened the tidal dams to the Egyptian pyramids all placed in a row with the spaces between filled in.[242]

The army changed the plan. Dexter's plan had been for a one-pool system with Cobscook Bay as a low-tide pool, a pumped storage reservoir and auxiliary power plant at Haycock Harbor, and a sixteen-mile transmission line connecting the two. If sufficient industry didn't develop locally, additional transmission lines would be built so that Quoddy's excess power could be sold in other states and provinces. This "immediate project" was phase one of Dexter's long-term plan to build the two-pool International Plan with Passamaquoddy Bay as the high-tide pool.[243]

The whole plan was thrown in doubt in November 1935 when the boring teams discovered the ground where levees were to be built for the reservoir was unstable and thus unsafe without expensive remediation. Dexter's budget for the reservoir and auxiliary generators had been $9.1 million; the army's new estimate was $18 million.[244]* This huge jump in cost set off a region-wide search for an alternative site for the reservoir and other auxiliary power solutions. The most promising alternative reservoir was to enlarge Howard and Nash Lakes near Calais.[245] It would provide a higher head; flood less land, thus reducing the acquisition and clearing costs; and require fewer levees, potentially saving several million dollars.[246]

Ned Wadsworth and Donald Wilson, civil engineers on the job, thought there must be something wrong with the army's surveys. John Sweeney, another civil engineer, didn't agree with the army's analysis or plan. He said he could build the Haycock reservoir with a "sucker dredge" for less than $1 million. He was transferred to another division.[247]

Dexter also didn't agree. On October 3, he sent a long letter to Colonel Fleming arguing that the project's long-term profitability depended on creating a two-pool system so continuous power could be produced and that Canada would not tolerate having its waterfront resort town of St. Andrews being left high and dry if Passamaquoddy Bay were made the low-tide pool.[248]

Cutting the Gordian knot, Colonel Fleming said the slack-tide power problem could be solved with a diesel-fueled auxiliary power plant costing only $3 million.[249]* Many wondered: *If it's so much cheaper to build a diesel power plant, why build a tidal power plant?*

"Cut the power transmission line to Massachusetts," ordered Samuel Fergusson, president of the Hartford (Connecticut) Electric Light Company. Apparently he didn't care about disrupting service to his customers in Massachusetts or the effects on their businesses, nor did he have the touted "social instinct of cooperation"[250] of super-power grid builders. Apparently he wanted to avoid scrutiny of his interstate commerce under PUHCA, which FDR signed into law on August 26 and would become effective on October 1, 1935.[251] Cutting off the power to spite the regulators was not a successful strategy. President Roosevelt ordered the FPC to investigate and publicize Fergusson's draconian and possibly illegal actions.[252]*

At the height of Mayor Emery's political prestige, he did the unthinkable. He gave up his power and recommended his office be abolished. His reasons were simple and sound. Eastport's practice of electing a mayor *every year* was not conducive to continuity in town planning and operations. It also fostered antagonism between parties, with too much effort going into the primary elections. Last but not least, a growing town required more than the part-time attention of an amateur administrator (even a very competent one).

Roscoe recommended the city change to a five-person council with a full-time professional manager. The councilors would be at-large, on staggered terms, based on votes cast regardless of party affiliation. As a further effort to get the politics out of the process, the first manager had to be from out of town. The reorganization referendum passed by a wide margin. Three members of the previous administration were elected as councilors, including Roscoe.[253]

Roscoe had another reason. Back in July 1934, he had offered the *Sentinel* for sale, but it had sat unsold for a year. Now everything was different. Eastport was booming, and the *Sentinel* was the newspaper of record for the largest construction project in Maine. Ad sales were way up. The small ads offering the *Sentinel* for sale disappeared.[254]

Instead of selling the *Sentinel,* Roscoe reinvested in it. Page 1 of the September 21 edition announced, "A Prosperity Campaign . . . to increase circulation—We are giving away a New Chevrolet Car or $500 in Cash; $200 in Cash; $100 in Cash." There were also weekly rewards, so the total of perhaps $1,000 was a colossal amount for Roscoe to offer in the middle of the Depression, when a laborer on the Quoddy Project was grossing $44 per month.

The campaign was not a raffle; it was a sales contest in which citizen-salesmen could earn points by selling new subscriptions or renewals and even collecting debts. As Roscoe explained, "It is the belief of the publisher that this is a good time for business to undertake new enterprises as nothing can be accomplished by idly waiting for better times."[255]

The number of men employed and the amount of money spent on Quoddy were the important numbers counted every week by the army and published in the *Sentinel*. These numbers enumerated the project's immense scale and give insight into the thousands of lives that would be affected by the federal government's controversial deficit spending to sustain its citizens during an unprecedented economic disaster.

Every man directly employed at Quoddy (most were men, but not all) was one household sustained. A fair guess is that every man employed represented an average of three people fed: the man, his wife, and a child. While Quoddy would be entirely in Washington County, much of the labor force would be from other parts of Maine, and much of the money earned by them would be sent home to their families.[256]*

Every dollar spent on supplies and equipment was indirect employment for someone somewhere. Most of the money spent on such things went out of Washington County or out of state. It was mounds of money, mountains of materials. As the *Sentinel* said, "Evidently there is an intention on the part of Major Fleming and those associated with him to see that communities from one end of Maine to the other share in the profits of the huge government undertaking."[257]

Work on Quoddy officially began on July 4, 1935. Three weeks later, there were 442 people employed.[258] Three weeks after that there were 1,069 people employed, earning a total of $17,910 per week.[259] Just imagine, for one minute, an organization growing from zero to one thousand employees in a little over one month. Unlike the civilian world, where the employer just manages and pays its employees, on the Quoddy Project the federal government also had to house and feed everyone it hired.

Housing was job number 1. Housing was needed for five thousand people—housing for a project that would last for years and need to accommodate not just men in barracks but whole families, year-round. According to the calendar, winter was four months away. By the thermometer, it was two months. No matter how much Eastport spruced up, added on, and elbowed in, there was no way it could accommodate the influx. The Army Corps of Engineers would have to design and build a whole new city—*twice* the size of Eastport—before the snow fell.

Who were the workers? They were unemployed farmers, factory workers, tradesmen, and professionals of all kinds, and flocks of women, whose fast hands kept scores of typewriters chattering out letters, specifications, and reports in an endless flurry of carbon paper and onion skin copies. By the end of August, 1,415 men and women were employed at Quoddy.[260]

And more workers kept coming.

By the end of September, the number surpassed 2,000. There were so many paychecks to issue that Captain Leehey purchased a "Signograph" so he could sign five at a time.[261]*

To accommodate as many as 100 new employees arriving each day, at all times of the day, the employment office was kept open until 2:00 a.m.[262] Job seekers came in all conditions: destitute and hungry, some without shoes, many too aged or unfit to work or in need of medical attention. All were cared for.[263]

Quoddy needed workers. Whoever was on the relief rolls and capable of work was ordered to show up or their relief payments would be stopped.[264] The federal government did not discriminate on race, only on need. The government hired heads of households first. Quoddy was the only PWA relief project in Maine.[265]

There were problems. The relief rolls were not accurate; the needed skills didn't match the supply; and when the crunch came in September, less than 30% of those called up showed up. The main reason for the no-shows was workers' unwillingness to leave their families for the low "security wages" being paid. Consequently, the original federal requirement that 90% of the workers be from Maine's relief rolls proved impossible to meet, thus requiring multiple federal waivers, causing frustration and delays.[266]

Forty Passamaquoddy Indians joined the workforce and were congratulated on their work by Captain Lord.[267] Among two truckloads of the unemployed who arrived were nineteen "negroes."[268] They were fed and housed with everyone else, a ground breaking of another kind.

The *Sentinel* announced, "[T]he honor of being the FIRST QUODDY BABY goes to Shirley Ann Campbell, 8-1/2 pound girl. . . . Mr. [Gilbert] Campbell is a senior draftsman on the Quoddy Project."[269] In her infant frame, baby Shirley embodied the hopes and dreams of a whole city.

By October 19, Moose Island's population had doubled, with Quoddy employing 2,973 men and women.[270] Housing them all was the big issue, but hunger was the acute one. At the hectic start, the army had simply bought whatever groceries it needed from local merchants, but soon the demands exceeded local suppliers' capabilities, and the army began issuing bid requests for food in bulk. The first order gives a flavor of what was cooking in Quoddy's mess halls: "13 tons of beef, two and one-half tons of veal, five tons of pork products, two tons of bacon, three tons of lard, two tons of hams, one and one-half ton lots of butter, sausage, and chicken, and 1,200 pounds of cheese, and 400 pounds of mutton . . . [and] six thousand dozen eggs."[271]

For most, the mess was fine dining. The menu for October 4 read, "Breakfast—apricots, corn flakes, boiled eggs, fried potatoes, coffee, bread and butter, fresh milk. Dinner [lunch]—rice and tomato soup, baked pollock, boiled potatoes, cream peas, vegetable salad, apple pie, coffee, bread and butter. Supper—meat loaf, mashed potatoes, tomato sauce, succotash, pineapple fritters, tea, gravy, string bean salad, bread and butter." Said the *Sentinel*, "The big chunks of butter placed on each plate would shame any restaurant."

The cost to the government for this repast for which all were grateful was 39 cents per person per day.[272] The cost to a worker for room and board at the barracks was 20% of their wages. For a common laborer making the minimum of $44 per month, shelter *and* sustenance cost $8.80 per month, or 30 cents a day. You can be sure prayers of thanks were said before many of the meals were devoured, with the names Roosevelt and Cooper on their lips.

"Can the government work fast?" the *Sentinel* asked on October 2. "Absolutely," it answered, and then it described the furious pace of construction at the barracks: "Long rolls of roofing material were given a shove and unwound along the flat surface of the roofs, while the carpenters nearly kept up to the moving roll with their swinging hammers fastening the material down. Below them, the frames and springs of the cots were set in the buildings before the roof was completed or the widows in."[273]

The workers on the job had to keep ahead of the workers coming for a job.

"Snob Hill" is what the locals called it.[274] Officially it was "permanent" houses for the officials in charge. Whatever you called it, it was not part of Quoddy Village, but a mile away, atop Redoubt Hill, overlooking the site of the future powerhouse, and superior in every sense.

Although one-third of Americans still didn't have an indoor toilet,[275] Snob Hill houses had *three* thrones *and* wine cellars.[276]

While $100,000 was to be spent to house 3,000 laborers in "temporary" barracks (about $33 per person), and another $300,000 to house 1,000 professional staff and families (about $300 per person) in "temporary" dorms, apartments, and houses; $102,088 was going to be spent to build nine houses for army officers and their families[277] (more than $4,000 per person). That was the initial estimate. The true cost of "Snob Hill," in dollars, would eventually be $245,232,33.[278]* Its cost in terms of goodwill was even higher.

Also elevated were the top officers' ranks and salaries. At the same time the construction contracts for their houses were awarded, Philip Fleming was promoted to lieutenant colonel, and Lieutenants Lord and Leehey to captains, retroactive one month.[279] Quite a coup for a group that had been on the job for less than four months and for a project that was already over budget and behind schedule.

CHAPTER 10

Yellow Lights (1935–1936)

RESISTANCE

Maine's governor Brann started his Labor Day weekend catching fish from the *Sea King* and having them fried for breakfast.[1] It was a perk of the job, having government assets available for his personal pleasure. He was, however, supposed to earn the privilege. Instead, it appeared he was *sinking back contented*—on the fantail of the army's yacht—not *sweeping the ground clear of snakes*, as the letter in the *Sentinel* had advised.

On September 30, 1935, President Roosevelt proudly dedicated the Boulder Dam, the biggest in the world, reprising his pledge to build "Government power developments . . . in each of the four Quarters of the United States."[2] Anchoring the Northeast Quarter, Quoddyites were pleased with their president's promises.

But snakes were in the grass, the lobby, the cabinet, the legislature, and the newsroom. Rumors appeared in several Maine dailies about Quoddy's need for more funds.[3] Then, on October 1, a bigger problem appeared in the *Bangor Daily News*:

> According to the *New York Times*[4] the Federal Government plans to send soon to Governor Brann an ultimatum that if he does not call a special session of the Legislature to create an authority for taking over the Passamaquoddy tidal power project when it is completed, work on the $36,000,000 undertaking will be halted. The understanding here has been that no special session would be necessary. . . .

As the *New York Times* hears many things that are not so, we are not worrying much over the financial prospects of Quoddy. . . . We don't think any
administration would drop Quoddy after the start that has been made. To
do so would be neither good business nor good politics. And when it comes
to business affairs, we feel that the advent of a Republican administration
would hurry the job along, instead of wasting so much time in backing and
filling, peanut politics and dragging the work along with dribbles on the
installment plan.[5]

Rather than worrying about "peanut politics," Roscoe focused on the
positive, and two weeks later he quoted a supportive article by Mary
Simkhovitch[6]* in the *American City Magazine:* "[N]o one can foretell
exactly how the plan will turn out in these uncertain days when imagination and a certain amount of daring are necessary. The immediate
effect of this great undertaking is to reestablish hope and confidence in
this vicinity. The first evidence of this is the increased tidiness of Eastport. People, hitherto unwilling to use up last reserves or to borrow, are
now painting their houses and in general spending a bit more freely."[7]
But what if the Times *was right?*[8]

"Dribbles" were, in fact, what Maine was getting, according to a
letter from J. E. Barlow, city manager for Portland, to Mayor Emery of
Eastport, and even that was going to be cut by 68%.[9]
As Mayor Emery ran down the columns of numbers in Barlow's
complaint to the Federal Emergency Relief Administration (FERA), a
knot may have formed in his stomach. The New England states, all of
which had voted Republican in the 1932 national election, weren't getting a fair deal from the New Deal. On average, southern Democratic
states were receiving about 50% more FERA funds than Republican
states. As a result, New England states and municipalities were going
deep into debt or going bankrupt trying to make up the shortfall. For
example, Maine had been paying a monthly average of $27 per family
on relief, the third highest in the nation, after Massachusetts at $33.29
and New York at $34.21. While Mississippi didn't need to worry about
wintertime fuel costs, its $3.96 was still shamefully low.[10]
The disproportionately low relief money was not the only problem.
Favoritism in the selection of projects and contractors caused numerous complaints. Maine's FERA director warned he would immediately
"remove the administrators from their duties without notice" if they

engaged in such favoritism.[11] Likewise, he demanded full value for the work he was paying for—no "loafing" (*the landlubber should have said "slacking off"*) when the tide was too high or low.[12]

"Quoddy Funds Cut in Half!" cried the *Sentinel* when $5 million was taken from Quoddy's original allocation on October 22.[13] The reduction was ordered by the Federal Bureau of the Budget.[14] Quoddy's army and civilian leaders were caught off guard by the cut and rushed to find out what was going on and to allay everyone's fears. Colonel Fleming said the cut was part of a $16 million reduction in the work-relief funds for rivers and harbors projects and was made to release funds for use by the Works Progress Administration elsewhere. As only $2,145,666.81 of Quoddy's $5,000,000 had been obligated so far, there would be plenty of money to carry on the work through the winter, and he contradicted reports that 1,500 men would be laid off.[15] Captain Casey, head of Quoddy's Engineering Division, added his assurance: "The work is not being stopped nor is it even being slowed down."[16] Mayor Emery declared, "We are not disturbed or even much surprised over the cut in the allocation for Quoddy for the current fiscal year . . . $5,000,000 will probably be sufficient . . . for the time being and that is all anyone can ask."[17] Governor Brann said, "I feel not the slightest concern about the future of Quoddy."[18]

Portland is closer to Boston—geographically, economically, and politically—than to Eastport and 250 miles closer to Washington, DC. Perhaps that proximity gave the *Portland Evening News* a clearer view. "There should be," the *News* recommended, "definite assurances from Washington of the restoration . . . of the full amount originally earmarked for Quoddy before those interests that would like to see Quoddy halted get in with their under cover work. The friends of Quoddy need to be vigilant now quite as much as when they were battling for the start of the project. The fight for the tide harnessing development has been too hard and too long to allow it to be hampered under any pretext."[19]

Again, Dexter had to fight the army to get paid. It had been three months, and he still hadn't been paid anything for his rights in the Quoddy Tidal Power Project. On October 9, he had agreed to a revised contract that split the $60,000 sale price into two payments: $50,000 on signing, and $10,000 on delivery of the outstanding shares in his company held by Westinghouse on behalf of GE, Boston Edison, and Middle West Utilities (MWU). He had signed the contract—but the army didn't pay him. After two weeks, Dexter complained to the White House.[20]

The responsibility for getting the cooperation of Westinghouse, GE, Boston Edison, and MWU was put on Dexter. The investors' lack of cooperation, however, was caused by the government, for the government failed to reimburse them for their $500,000 invested in Quoddy.

As you will recall, when Dexter publicly revealed the identities of his investors in 1934 at the Public Works Administration (PWA) Board of Review hearing, he also announced that they had agreed to "cash out" their investments if reimbursed by the government. According to their "Five-Party Agreement" of October 31, 1927, the investors' rights in Quoddy would become null and void if the project financing or construction hadn't "commenced" by June 1, 1935—but the project was partially financed on May 16, 1935, by President Roosevelt's $10 million initial allocation. Thus, Westinghouse, GE, Boston Edison, and MWU had a legitimate claim *against the government* (the army was in charge financially) for refusing to reimburse them. Given the amount of money involved—5% to 6% of the approved budget—it was foolish for the army not to pay it, for instead of encouraging the cooperation of the most powerful companies in the country, *and Maine*, it ensured their enmity.[21]*

It is interesting, and possibly telling, that the investors never made the details of their claim public. If they had, it seems likely that they would have been paid, thus settling the issue. Instead, the public was led to believe the claim was against Dexter, when in fact *the party that created the conflict, and the only one that could resolve it*, was the federal government.[22] This situation cast Dexter in a negative light and allowed Maine's power trust to continue attacking Quoddy as unnecessary and uneconomical—whereas the trust would have had nothing to stand on *if* it had been shown that the investors had been investing in the project continuously for seven years (1924–1931); had encouraged Dexter to seek government funding in 1933; *and had been fully reimbursed, as Dexter recommended.*[23]*

"High Tension Jitters," tittered the title of the *Collier's* article about Quoddy on October 25, written by Walter Davenport. Roscoe mentioned it in the *Sentinel*[24] but didn't say that it was brilliantly written, persuasive, clever, funny, punny, sarcastic, derisive, and damaging.

Davenport called Eastport "oddly frontierish" and portrayed its residents as yokels, Maine's politicians as spineless and greedy, and the federal government as incompetent and dishonest. While he can be credited as having a good dose of journalistic skepticism and some of his criticisms were justified, he could justly be charged with being a master propogandist: asserting, overstating, preconditioning, cherry-picking, mischaracterizing, and mismatching facts and opinion in an effort to sow confusion, fear, and doubt. Nary a verb, adverb, or adjective is neutral; all are underlain with negatives. To use Davenport's own phrases, his arguments are executed with "exceeding skill" and "thickly cocooned . . . in phraseology." His pen is less a righteous rapier of investigative journalism than a stiletto of assassination. To fully appreciate what the nation was reading about Quoddy, it's worth quoting, with emphasis added to show his slanted reporting:

> The outlook for the great Passamaquoddy tide-harnessing and *moon-hitching* enterprise is not *universally conceded* to be *cloudless.* . . .
>
> Now it *may come to pass that,* some day not yet visible on our horizon, we shall rejoice that one of the fruits of this *honeydew revolution* of ours was Passamaquoddy. Similarly our children may sing the praises of the Administration which gave us such power producers as Bonneville, the Grand Coulee, Wheeler, Norris and other dams to light us our destiny. But it will be just as well to remember that the adverse reports made on Passamaquoddy (or Quoddy as they call her fondly and otherwise) were the work of exceedingly practical engineers *who probably like a bit of dreaming* as well as the next man but who were at extraordinary pains to weave none of *such gossamer stuff* into their literary burlap.
>
> . . . Being engineers and therefore almost axiomatically poor politicians, they set down what they believed to be fact—*today's fact and not tomorrow's possibility.* There is no evidence that any of them knew that such completely Maine gentlemen as Governor Louis J. Brann and Congressman Ralph O. Brewster, respectively a New Deal Democrat and an Old Guard Republican, were *wearing out the White House steps and the knees of their trousers praying* to President Roosevelt for Passamaquoddy and the $36,000,000. . . .
>
> None [of the engineers], of course, contended that the mighty tides of Passamaquoddy Bay could not be hitched to electric generators. Men like these, given a few wheels and a bucketful of water, could turn out a power plant in no time. But *almost unanimously* they agreed that as a *self-liquidating project,* to say nothing of a dividend producer, the Passamaquoddy project was destined to be a *reverberating flop.* . . .

In such circumstances, torn between today and tomorrow, Mr. Ickes' PWA report on Passamaquoddy was of great length and *exceeding skill*. And yet, *thickly cocooned as it was in phraseology*, it was clear to all readers that Mr. Ickes harbored doubts as to the advisability of an *immediate* Quoddy. . . . *Virtually all* of the engineers contributing to the several reports were insisting that Quoddy power would not be cheap power, the *tide-harnessing feature being an expensive process and the results being possibly less than dependable.* In Quoddy there are *experimental features. One may not say now whether power generated by the tides will be as dependable* as that ground out at the conventional hydro plant. And the cost of maintaining and operating *a moon-governed hydro plant might be bad news.* . . .

To facilitate the work of the government, to which Governor Brann is properly grateful, he is being urged to call a special session of the legislature to create a State Power Authority. Such a body would, the government hopes, find some means of *disposing* of Quoddy's power. *Just how this might be done,* in view of the Fernald Law and the charter granted to Mr. Cooper, is still something of *an enigma.* But a similar deal was made with the State of Washington. In the latter state the Grand Coulee power dam *is rising in the wilderness with no visible consumers of its kilowatt output.* . . . And that will *leave the government holding a highly electrified bag.*

At this moment Maine's private power interests are undecided how they will meet the Power Authority issue *if and when* it arises. *If they can control the proposed body and make such use of Quoddy's power as they see fit at the rates they choose to name, they may withdraw their opposition.* The *federal government will, in any event, find itself in a nasty situation.* It can't comfortably permit *Quoddy to lie idle, a monument to an incomplete power policy, nor may it under the laws of Maine invade the private power field.* . . .

All of which would seem to leave poor Quoddy pretty much isolated for years. And it is an immensely baffling situation.[25]

Davenport's surrendering to the private power trusts the *right* to "control" the proposed public power authority, to use Quoddy "as they see fit," and to charge "rates they choose to name" makes plain where his allegiance lies.

The *New York Times* countered Davenport's deceits with intelligent insights the following week. The article began with a worrisome comment: "New England is filled with rumors that [Quoddy] . . . will never be completed. Just how far these rumors are mere wishful thinking on the part of Republicans, private power interests, and other anti-New

Deal groups, and to what degree they are real political, economic and engineering obstacles it is difficult to determine."

Commenting on the Roosevelt-Brann relationship, the *Times* said that "little affection exists between Augusta and Washington, because of the usual jealousy over the expenditure of relief money. . . . But it had been assumed that the two Democratic Executives would at least cooperate to the extent of carrying through their party's most spectacular enterprise in New England."

"Such is the dark side of the picture," the *Times* continued, "On the other side are the facts that Quoddy is proving highly successful as a work-relief project . . . that the people of eastern Maine are generally enthusiastic about the project, and that the army engineers, to whom any large and unusual assignment is a challenge, are becoming keenly interested in Quoddy and would be disappointed if it were called off." In conclusion, "it appears certain that work will continue at Eastport for at least another year. Even if Governor Brann should decline to call a special session, or if the Legislature should refuse to enact a power-marketing law, it is politically inconceivable that the Roosevelt administration would admit failure before November 1936. Eastport seems assured of twelve more big months of crowds, excitement and prosperity."[26]

Governor Brann may have seen the *Times* article and concluded he needed to clear the air between himself and President Roosevelt. On his gold-embossed official stationery, Brann wrote a six-page letter to Roosevelt on November 5 explaining why he was angry about the administration's allowing Congressman Brewster to take credit for Quoddy and then endangering it in the fight with Corcoran—but assuring the president he fully supported Quoddy.[27]

While shortsighted satirist Davenport derided engineers' dreams, the farsighted journalist Elizabeth Craig encouraged Quoddy's completion as a "pioneer project":

It may take much longer than engineers think, to get the best results from Quoddy. It may cost a good deal more than the 36 million dollars. . . . But the trails of the United States from the Atlantic to the Pacific were white with the bones of pioneers before we won the West. It is true that the rates that may be commercially charged for Quoddy power may not cover the cost of this pioneering. . . . But if we learn from Quoddy how to get power from the eternal tides, constant and everlasting, then Quoddy will be worth anything it costs, and future generations will call us blessed, that we gave them this gift.

. . . Washington remembers how poor old professor Langley went down to the Potomac River to try out his first airplane. It fell down in the river amid the jeers of the newspaper reporters and the hardheaded business men and the unimaginative scientists and the curiosity-seekers. And Professor Langley was dragged out dripping and muddy and died bankrupt and broken-hearted. But now we are building the China Clipper, for regular passenger service across the Pacific, and we know this Italian-Ethiopian war will be decided by airplanes.

. . . Dexter P. Cooper tried for years to get private capital to do it and they wouldn't because they wanted to see surety of return and there isn't any security for pioneers.[28]*

Craig's praise for tidal pioneers was more elegant but less effective than Davenport's damning them. She was Quoddy's Cassandra.

The army changed the plan, again. On November 12, Captain Hugh J. Casey, chief of Quoddy's Engineering Division, announced the army was considering three different plans, each with a pumped storage reservoir: (1) Cobscook Bay as the low-tide pool, powerhouse for inward flow from the ocean; (2) Cobscook Bay as the high-tide pool, powerhouse for flow out to the ocean; or (3) Cobscook Bay as the high pool, with bidirectional flow.[29]

The army thought Plan 2 was best because making Cobscook the high pool would improve navigation therein and would make the bay's 140 miles of mostly undeveloped shoreline more suitable for industrial and summer resort development. More important, it estimated 7% higher power output because (1) the bay's area is 30% larger at high tide than low tide, and the high-tide plan would increase the average operating head by 0.4 feet; (2) a high pool was less turbulent; (3) electricity could be generated for fifteen minutes longer; and (4) the turbines would be subject to less wear because high water carries less sediment.[30]

Cobscook as the low-tide pool had been the plan because it was reasonably cost efficient as a stand-alone project *and* could be expanded into the International Plan by damming off St. Andrews Bay as the high-tide pool, thereby making a larger and more efficient power project. Conversely, making St. Andrews Bay the low pool and leaving its waterfront resort towns fronted with mud drew dirty looks from Canada.

While financial troubles loomed, the impending crisis in Eastport was housing, and winter was coming. With too much construction still to do, on November 13, Colonel Fleming declared a housing emergency.[31]

Fleming got approvals from bureaucratic Washington for a 45-day period during which employees could work up to 56 hours each week. This increased the labor available by 40% without increasing the number of workers. Wages were also increased by 45%. To a common laborer who had been earning $44 per month (27.5 cents per hour), the new rate was 40 cents per hour[32] (essentially matching the commercial labor rate). In combination, a laborer's gross earnings for the 45-day period could shoot up from $66 to $134.40. It was a tremendous incentive and would make a wonderful Christmas bonus.

The payroll department added a night shift.[33] Fleming also announced the project expected to reach a peak of 5,000 employees by Christmas and to stay at that level through the winter.[34]

Quoddy's countdown to Christmas was like no other. The first door of Quoddy's virtual Advent calendar[35]* was the hour and wage increases on November 13. Eastporters snuck a peek inside the second door that same day when the *Bangor Daily News* reported President Roosevelt had allocated $7 million more to Quoddy as a result of a conference with Governor Brann, General Markham, and Lieutenant Colonel Fleming—though there was not yet official confirmation.[36]

By Thanksgiving, Quoddy was in fact employing 5,042 grateful souls.[37] Quoddy was powering through. The army had hit its stride. The Village was growing by leaps and bounds. Hammers, sawdust, and materials of all kinds were flying. Before Quoddy, fewer than five railroad freight cars per month arrived in Eastport. The total for November 1935 was 385.[38]* As bathtubs and toilets were hustled out of the depot, the yellow snowplow that had been half-hidden between the enameled masses became an ever more obvious reminder of winter's approach, an incentive for speed; these were, after all, the workers' own homes they were building.

Looking inside *American Magazine*, wise men found Thomas Sugrue's article, "Santa Comes to Way down East," which likened Quoddy to a New Deal Christmas tree growing in Eastport.[39] Opening the *New York Herald Tribune* revealed a cartoon of a fisherman frying his dinner in a tide-powered toaster.[40]

Maine's foxy power barons opened their own Advent calendar door—the door to the chicken house—on November 20. It was a cause for celebration in Augusta, Bangor, and Portland, but not in Eastport: "Friends of government of and for the people in Maine deeply regret Governor Brann's retirement of Col. Albert Stearns from the Public Utilities Commission. He has been an independent, courageous, honest and very capable servant of the public in his office as chairman of that most important agency. Friends of Quoddy regard with dismay the Governor's choice of Mayor Frederick Payne"[41]—that "skunk" who wanted federal funds spent in Portland instead of on Quoddy.[42]

Opening their mailboxes on November 24, *Boston Post* readers saw a new attack on the PWA and Quoddy.[43] The *Sentinel*'s Joseph F. Preston, the expert from Niagara, fought back: "When you read these effusions written by half-baked economist, reporters, special and editorial writers, that have been fed up on propaganda put out by those comic opera power magnates, Samuel and Martin Insull, that cherubic Pickwickian chap whom they call Hobson, pass it up for what it is worth,—pure bunk."[44]

Real new doors opened on November 28 when Lieutenant Colonel Fleming and his family celebrated Thanksgiving in their new home on Redoubt Hill. Captain Lord and his family did the same. Redoubtable doors opened for Captains Leehey and Sturgis on December 5, Blair and Bruce that weekend, and Casey and Kruse the next week.[45]

On December 3, there were 5,122 employees furiously working away.[46]

Meanwhile, Quoddy's 1,000 professional staff wrote out their wish lists specifying their employer (government or private), position, wages, family size, and preferences for housing types and locations. Then the administrators worked late into the night trying to be as fair and accommodating as possible. The smallest houses, Type A, were single-family residences with one bedroom and a dining room, living room, bath, and kitchen—for $48 per month. The largest, Type C, were 1½ stories, with three bedrooms, at $65. Apartments were $35.[47] Dorm rooms ranged from $20 per person for an eight-person suite to $30 for a single.[48] "Evidence of the luxurious furnishing to be in the Quoddy Village dormitory," the *Sentinel* remarked, "could be seen in the invitations for bids." The orders included artwork and love seats.[49]

Then the pilgrimage began. Wives and children who had been away from their husbands and fathers for months packed up their belongings and came to the Promised Land by car, bus, train, and truck. There, in the far east, the families were reunited. On December 1, the first family moved into Quoddy Village.[50] Edward Tokheim, of the purchasing department, chose A-14, the first house built at the Village.[51] Every day thereafter, joyful, grateful, excited families moved into their new homes, their new lives. They unpacked their suitcases into new dressers. They cooked their meals on new stoves. They ate their dinners at new dining tables. And they lay their weary heads on new beds.

The houses at Quoddy Village weren't tenements or block houses; they were homes. The streets had graceful curves to them. They had side yards and plantings. Many had water views. There were 2,246 windows. They were gayly adorned with bright red, blue, and green shutters, and matching shingles. At Quoddy Village, the living was easy. There were no outhouses, woodsheds, or iceboxes. Quoddy residences had indoor plumbing and steam heat. They had electric lights, stoves, and refrigerators. They had it all.

Residents of Quoddy Village were lucky and happy. It was the Depression, yet they had food, a job, and a new home. They had neighbors and friends. They had money to spend and places to spend it. They had purpose and community and the cohesion of shared experiences like few are blessed to enjoy. They did everything a community does: baseball and basketball games, theater and music performances, sewing and garden clubs, church services, picnics and Halloween costume parties. "They met, fell in love, and married"[52] and were expecting.

For the Army Engineers, the peacetime mission of mercy and the huge construction project was a happy time. They also looked forward to the years ahead. In mid-December, the Associated Press gave Eastport the happy news that President Roosevelt was proposing a half-billion-dollar public works program for the next fiscal year, and he said the money would be used chiefly for large projects already underway, including Quoddy.[53] The army strung blue and green Christmas lights in the masts of the *Sea King*.[54]* Colonel Fleming decorated his front door with red and green lights that could be seen, high on Redoubt Hill, by everyone who drove into Eastport.[55] A large evergreen tree was set up outside the Administration Building.

On Christmas Eve, the Army Engineers sent a letter up the chimney to Santa.[56]

SURPRISES

Santa wasn't pleased.

The letter from the Army Engineers said they needed not just $35 million to build the Quoddy Tidal Power Project but *$27 million more!*[57]*

Increasing the budget by 77% was not acceptable. Rather than opening Santa's sack of cash, the president played the Grinch and told the army it would get nothing if it didn't figure out how to build Quoddy for less.[58]

The phone rang at the Army Engineers' offices in Eastport at 9:30 Monday night, January 6, 1936. It was Lieutenant Colonel Fleming calling from Washington, D.C., for Captain Casey. It was urgent. A few minutes later, Captain Casey called a meeting of the senior staff. They stayed up all night. In the morning they put out the word to the foremen: reorganize your work plans, lay off 1,800 workers.[59]

When workers learned there would be no more work for them, at least for a while, fear grew that the whole project would be cut.

It wasn't so, Captain Casey assured them; it was only a temporary layoff, one driven as much by the usual winter slowdown and completing the housing construction and surveying as by the temporary lack of funds. More funding and a return to full employment should come in six to eight weeks when the president's public works bill was expected to be approved by Congress. The *Sentinel* said, "No fear of halting or suspension of the project is held by the Engineers."[60] Not publicly.

How 1,800 workers, most of whom were common laborers who had been earning "subsistence wages," were supposed to subsist *without* wages for two months was not reported. Not publicly.

Casey reminded those who could still listen that winter layoffs had been planned. Trying to reassure the thousands who were or might be affected by layoffs, he said, "Quoddy is far better off financially than many others, especially the Florida ship canal project."[61] That was no comfort. Not in Maine. Not in January.

The army had already used up $4,156,094.51 of Quoddy's $5 million initial allocation.[62] That left only $843,905.49, or about seven weeks of funds at the current payroll. Assuming even modest expenses for materials and equipment, the funds would run out in half that time.

Quoddy's leaders didn't have any choice but to cut staff. But cutting staff, whether from the relief rolls or from the trades, would be terribly hard on the people who had come in good faith, or were ordered to, and expected a secure job. It just wouldn't be fair to kick them out.

There was also that worrisome issue of Maine's lack of action on creating the Quoddy Power Authority. Without it, Roosevelt risked another Supreme Court case, and he was already in hot enough water there because of the power trusts' lawsuits over the Tennessee Valley Authority (TVA) and FDR's very unpopular attempt to "pack the court" by adding new members in a bald attempt to outnumber and overrule the conservative-leaning judges already on the bench. Governor Brann had just visited the White House and tried to assure the president he had a "gentlemen's agreement" with the Maine Legislature to use the "best endeavor" to enact the Quoddy Authority bill at the next regular session,[63] sending the president an elaborate "memorial" and 148 telegrams to that effect from legislators.[64] This, the governor hoped, would settle the issue. Roosevelt, however, realized that legislators' promises were no more reliable than Quoddy cost estimates.[65]*

There were other political angles as well. Most critical was the national election coming up in the fall. Republicans would be attacking the New Deal, and traditionally Republican Maine was going to be a hard fight.[66]

The layoffs triggered frantic calls to the White House. Lubecers wired the president on Friday bewailing the "unprecedented burden" that had been thrown on their community by the wholesale dismissal at a time when relief was most needed. "Distress is most acute," it said, "and many are desperate."[67] Senator Hale carried a copy of Lubec's message to the White House on Saturday morning, where he met Governor Brann, also pleading for Quoddy.[68]

On January 11, President Roosevelt relented and reinstated $2 million to Quoddy's budget.[69] He had heard the cries from his old friends and neighbors and wanted to ease their pain.

Just as important, he heard from General Markham, who outlined several cost-saving options. One option—with a half-sized main power plant, and without the pumped-storage plant—could be built for $36.5 million.[70]* That option, however, was incompatible with the International Plan, because it made Cobscook Bay the high-tide pool.

The $2 million allowed 4,000 relief workers to be paid through April 1, at which time $5 million more would be available. Then for fiscal year 1936–1937, which would begin on July 1, the War Department was recommending $20 million.[71]

When forced to reconsider their plans, the Army Engineers realized something: Why build an expensive new pumped-storage reservoir and auxiliary power plant when there were dozens of existing reservoirs with power plants on rivers all over Maine? Why not "export" all of Quoddy's periodic tidal power to the emerging New England electric grid—and "import" power from the grid when Quoddy wasn't producing? When Quoddy was exporting, the existing river reservoirs could be filling (and their attached generating stations idle), so they would have lots of water with which to generate electricity when Quoddy wasn't exporting.[72]

It was brilliantly simple. It was Samuel Insull's idea: balance supply and demand—through the grid.

It would mean cooperating with the enemy—the power trusts— because they owned *everything*: the reservoirs, the generating stations, the transmission system, and the distribution system.

Having been a victim of dirty dealings by the power trusts, when Dexter learned of the army's suggestion, he was furious. He saw it as evidence of the army's betrayal, its collusion with the power trusts.[73]

Cooperating with the power trusts—Westinghouse, General Electric, Boston Edison, and Middle West Utilities (Central Maine Power, CMP)—was made even more difficult than usual because the army had stiffed them, just as it had stiffed Dexter.[74]

On the same day that President Roosevelt granted $2 million to Quoddy, he also met with the enemy, the power trusts. Though the scope of his discussion was national, every item on the agenda related to Quoddy, especially item 5, "On what basis power can be exchanged between government owned power plants and private power plants."[75]

Four days later, the president quietly asked the US Reclamation Service, the builders of the Boulder Dam, to confidentially estimate the potential cost of building Quoddy.[76]*

The very next day, January 16, 1936, the president got impressively good news about Quoddy. He kept it secret, for the time being.

Snow began falling on Moose Island on Saturday, January 18, and continued on Sunday. Winds up to 53 miles per hour blew the snow into large drifts.[77] Luckily, Quoddy Village had its bright yellow snowplow ready to go. Luckily, Eastport was able to replace its old clunker with a new snowplow—a 40-hp Caterpillar diesel weighing 7½ tons—for $6,650 with only $2,500 down.[78] Even with the new equipment, it took until Tuesday to clear the streets.[79]

There was an unusual amount of traffic on those snowy days; they were moving days for the Quoddy offices. While motion picture cameras whirred and shutters clicked, the key to the new 40,000-square-foot Administration Building at Quoddy Village was presented to Lieutenant Colonel Fleming on Saturday morning by J. Slotnick Co., the builders of the impressive white structure that covered half a football field and had cost $132,814.[80]

With the doors flung open, moving began. The weekend and the following week "saw a steady stream of furniture-laden trucks churning through the snow toward the Village, while churches, halls, and wharves about Eastport were gradually divested of office equipment."[81] The Operations Division moved from McNichol wharf; the Army Engineers evicted themselves from the V.F.W. hall. Out of the Armory and the Orangemen's hall marched the Administration Division. More operators opted out of the Unitarian Church. The Congregational church was depopulated of the Personnel Office. The Baptist church was abandoned. The Eastern steamship wharf disbursed its purchasing agents. The Camp and Land division decamped from the Knights of Pythias hall. The finance group may have shivered as they left the Burr building.[82]*

That was just the beginning, just the offices. From hotels, rooming houses, and private homes all over town, renters packed up their things and moved out of Eastport and into Quoddy Village's houses, apartments, and dormitory. Downtown Eastport was no longer the center of commercial activity, cultural activity, or population. And a great deal of the revenue—the room and board that had been collected by Eastport landlords and restaurants—left too.

The army's Administration Building was impressive. The two-story, 234' × 125' building had a pedimented entrance with Doric columns, high-relief sculpted tympanums, and a clock tower with lantern. The layout was in the form of an E, with the entrance in the center of the flat side and the three wings to the rear. More than 200 windows provided natural light for 480 desks and 1,000 people who would work within its walls.[83]

As visitors entered the two-story lobby, they would find the information booth and post office. The emblem of the Army Engineers, the rook, was carved into a wall ornament at the upper level. Opposite it, a carved eagle, wings spread wide, perched over the double staircase.[84] Telechron electric clocks clicked from several locations.

Colonel Fleming's office, in the southwest corner on the first floor, was "strikingly-finished." The walls were covered with "a beautiful mahogany veneer. Built in bookcases and closets, a fireplace of black and gold marble, and a lighting system of the latest design add the touch of order and ease."[85] Captains Casey, Lord, Leehey, and Sturgis also had corner offices.

On opening day, scores of letters were mailed by collectors eager for the precious first-day-of-issue postmark: "Quoddy, Maine, Jan. 20, 1935." It was technically a sub-branch of Eastport's post office, but mail was to be delivered to it twice daily.[86]* While the building did not have "gold-plated bath-tubs," as the *Philadelphia Inquirer* suggested would be appropriate for such "boondoggling de luxe,"[87] it did have "188 gold-steel mail boxes."[88]

Gertrude thought the building was "so ostentatious that it resembled a state capitol in some western city."[89] To others, it seemed like another "White House."[90] Parked outside was a fleet of 60 new cars—Fords, Chevrolets, and Oldsmobiles—each assigned to an individual staffer, many with chauffeurs.[91]

The building was not "permanent," like the army officers' private residences on Redoubt Hill, but "temporary." Its Doric columns and whitewashed walls were not made of stone or brick, as Sidney Stevens had imagined, but of wood. The 40,000 square foot edifice was not even built with a concrete foundation; rather, it sat on wooden piers.

Hadn't the army heard of "The Three Little Pigs"?

The fiscal restraints imposed on Eastport were in stark contrast to the army's lavish spending. Eastport's new City Council had much to take care of in January 1936. Roscoe was unanimously elected chair.

He was pleased to report the state had agreed to reimburse the city $3,000 for some school expenses resulting from the influx of new students, but nothing yet had been received to offset extra expenses for fire and police services.

There were the usual petitions for building and business permits, but the discussion held everyone's interest because the city was going to start charging for them. Doing so, the council hoped, would generate some income from a source other than the already heavily taxed property owners. From now on, if you wanted to open a business in Eastport, you would have to pay a fee based roughly on how much hassle your business would create. Since everyone had to eat, a victualer's license was only $10. Recreational activities such as bowling or billiards were $15, whereas shooting ranges and dance halls were $25, and your ticket to open a movie theater would be a blockbuster $50. To serve beer, you would be tapped for $50 and get soaked $100 for wine or liquor permit. All in, Eastport hoped to reap $2,000 a year in cash up front.[92]

While the army had spent freely feathering its nest, it, too, now had to pinch pennies. It couldn't afford to build all of Quoddy's Group III buildings except the hospital. The school, firehouse, and guest house would have to wait.[93]

The army also strung along Dexter Cooper. So far, it had paid only $50,000 of the $60,000 due to Dexter. The payment of the remaining $10,000 "awaits performance of certain conditions."[94] For the first time the public got some idea of what those conditions were when the *New York Times* reported, "Boston Edison, General Electric, and Westinghouse companies still contend that they have claims against the Dexter Cooper Company,"[95] although Dexter did not cause the problem or have the funds to solve it—the US Army did.

The army was also dickering with the owners of the landing sites it needed for the dam abutments. The owner with the most sites wanted $435 per acre. They settled for $237, about twice what the army had paid per acre for Rice Farm, on which Quoddy Village was built. Roscoe got the higher price for his very small but strategically located pieces of property on Carlow Island because he had both landing sites for the dam and the quarry needed to build it.[96]

"Roaring out of seven side-tilted railroad dump cars," tons of rock poured into Passamaquoddy Bay from the Carlow Island trestle on

February 7, 1936. It was the first load of a routine that would continue for 150 days or until the dams connecting Pleasant Point, Carlow Island, and Moose Island were completed. The dramatic event was witnessed by many dignitaries, including the Army Corps's top brass, Major General Markham and Brigadier General Spalding.[97] Like many of its peers in the press, the *Sentinel* was there:

> Soon after high noon, Friday, locomotive Engine 246 steamed into the new quarry at Carrying Place Cove with five loaded dump cars, a flat car, observation car, caboose, and two coaches. While officials who had been gathering at Eastport for 24 hours watched, the big Bay City gasoline shovel swung steady loads of 1 ½ cubic yards each from the hillside to fill two additional cars. . . . Despite bitter cold wind as the train whistled by Quoddy Village, government photographers remained on the flat car and ground out motion and snapshot action pictures. As the cars slowed to a halt on the trestle, the spectators clambered from the seat-less observation coach to points of vantage on the shoreline. The engineers' cameras recorded the first load of rock-material as it slithered into the swift-moving, deep, tide-water below to inaugurate a steady chain of similar loaded trains which will fill the water gaps.[98]

The army had spent seven months building housing before starting to build the actual tidal power project, but now drilling and blasting for the dams would go on 24 hours per day. The initial order was for 80 tons of dynamite. It rained and it poured rock. Even the train had to hide in a shed from the hail of real stones.[99]

The chessmen were moving as rapidly in Washington as the blasters in Eastport. Quoddy had a certain amount of funding and a certain momentum but not a certain future. Those opposed to it and the New Deal—the power trusts and the Republican Party—wanted to stop it before it gained too much mass, before it became too big to fail.

Maine's congressional delegation had to figure out how to pilot the big project through the dangerous waters of the legislative process to the promised land of federal funding.

Roosevelt had asked for $9 million[100] to be included for Quoddy in the fiscal year 1936–1937 appropriation, as part of a $29 million appropriation for five river and harbor projects that were estimated to ultimately cost a total of $219 million.[101]* But on February 11, the House Appropriations Committee announced that it was in thorough accord with what seems to be the future policy to completely finance such projects out of specific regular annual appropriations, but only after such projects have been authorized by law. . . .

If for no other reason, the committee feels that the total ultimate cost involved is too great for it to recommend an appropriation for further prosecution of work upon such projects until they have run the usual gauntlet of scrutiny by the corps of engineers, the legislative committees having jurisdiction of such matters, and the legislative bodies themselves.

The committee's action does not mean in any sense that these five unauthorized projects are to be abandoned. Two of them, as has been previously stated, already have been favorably recommended by the Chief of the Engineers. It is entirely possible that all will be.[102]

That move gave the administration two choices: either subject the projects to the slings and arrows of outrageous congress members or use the president's executive authority to fund them out of previously approved discretionary funds—thereby taking sole responsibility for the controversial projects. The *Sentinel* rationalized the conundrum this way: "[M]any friends of the project feel that its future continuation is better assured in the hands of the President, where it apparently has been placed by the authors."

The authors of the plan included Maine's Republican congressmen, who didn't want to publicly advocate for the quintessential New Deal pet project of the Democratic president. *They preferred to have Roosevelt pay for their cake.*

The *Boston Herald* hurrahed the news: "Funds for Quoddy Refused." The *Sentinel*'s page 1 headline was inscrutable: "Optimism over House Report Quoddy Funds: Future Now with President; Deficiency Appropriation Money Possible"; its statement on page 8 was more understandable: "Quoddy on the Mat."[103]

And the reality? Lieutenant Colonel Fleming said his instructions still read "go ahead." As if to reassure everyone, he added that he would invite bids the next week for the navigation locks, with construction to begin on April 1.[104]

The army's extravagant expenditures (and the *Sentinel*'s frequent characterization of the furnishings at Quoddy Village as "luxurious")[105] provided ample ammunition to the opposition and their mouthpieces. As it was considered inappropriate for the army to respond directly to the newspapers harping on the issue, General Markham arranged to reply to a request for clarification from Congressman Brewster.[106]

Among the clarifications were the following. The 142 residents in the dormitory paid a total of $3,000 per month in rent. The income over the four-year project[107] would be only $25,000 below the total cost; however, "the salvage value of the building would be considerably more than that." The same was true of the officers' houses on Redoubt Hill: the $100 average rents and resale values would be self-liquidating.[108] The "old masters" artwork on the walls were prints costing $1.04 each, including the frame. Half of the personnel in the dorm were women, which is why half of the blankets were pastels, and they didn't cost any more than the men's red and black blankets. The two grandfather clocks cost $77.80 each; the two radios cost $57.75 each. The "'love seats,' which the War Department wishes they had never been called," were just two-person couches that fit in the apartments. Furniture for the permanent residences was bought on the open market, and "the War Department invites housewives acquainted with prices of furniture to look over these figures."[109]

The new acting chief of the Army Engineers, General G. B. Pillsbury, addressed another point of contention with the press, politicians, and power trust. It seemed ridiculous for the army to be building an auxiliary diesel power plant when they were building a huge tidal power plant—and when *it was said* there was already an excess of electric power in Maine.

Not so, said Pillsbury; "the present electric service is subject to frequent interruptions," and "over the life of the project it [the diesels] will provide electricity at a cost per kilowatt hour somewhat less than the best quotation received for power delivered to the site."[110] Accordingly, the first of five 500kW Winton diesel generators was installed in February in a wooden building just north of the central heating plant in Quoddy Village.

But, said Pillsbury, the army was "deferring the purchase of the remaining units so that the contractors may elect their own sources of power." And although it was referred to as a "stand-by," it would be "required for almost constant operation."

Thus the army contradicted itself on the use and value of the diesel power plant.[111]*

Where the *New York Times* came up with the claim that the president "now knows" that the whole job would not cost the estimated $37,732,000, but up to $100,000,000,[112] is anybody's guess. The paper's political assessment, however, was again on point:

> Once Congress, in its present mood, becomes imbued with the idea that the cost is too high the whole project may go. Politics complicated the situation. If Governor Louis J. Brann sticks to his avowed intention not to run for a third term the Republicans will carry Maine in September as well as in the general elections in November. An adverse Supreme Court ruling on TVA might outlaw Quoddy. With nothing to gain politically from the investment of further funds at Eastport the administration might abandon the enterprise. All Maine outside that one county would not likely protest vigorously. Eastport would be put on sackcloth and ashes, and one may sympathize with the people there without at all endorsing Quoddy as an economic proposition.[113]

Roosevelt had invited the lawsuit, literally, seeking to settle the constitutionality of the TVA and public power one way or another.[114] The power trusts were happy to oblige. Both parties thought they would win.

On February 17, 1936, the US Supreme Court sided with the president: the federal government could operate electric power production and distribution businesses, and Alabama Power Company had "no constitutional right to insist that it shall be the sole purchaser of the energy generated."[115]* Roosevelt's great socioeconomic experiment in regional planning, the TVA, was legitimate in the eyes of the law.

The Supreme Court decision cleared Quoddy of potential legal challenges by Maine's power trust. No longer could the trust challenge Quoddy as a form of unfair competition or as unlawful taking of land by eminent domain.[116]* The decisions also obviated the need for the Quoddy Authority, a helpful thing since the Republican-controlled Maine Legislature was still balking. If the legislature didn't approve the Quoddy Authority, the administration could operate Quoddy as a federal project.

The "game,"[117] though, wasn't over, as the *New York Times* explained:

TVA DECISION LEAVES POWER WAR STILL ON:
. . . What the aftermath of the TVA decision will be no one can be sure. One outcome is almost certain to be the appearance of many projects for

regional planning based on Federal sponsorship of hydroelectric systems.
. . . States as far apart as California and South Carolina have ideas for an
imitation TVA. . . .

What the next move of the private companies will be remains in the
dark. Hugh S. Magill, president of the American Federation of Investors,
has suggested that, so far as he is concerned, the legal fight is only getting under way. On the other hand, Mr. [Wendell] Willkie said things that
may mean that he is prepared to turn to Congress for relief, rather than
to the courts.[118]*

On May 29, 1936, Wendell Willkie, as head of the C&S holding company, the owner of the Tennessee Electric Power Company (TEPCo),
filed another suit challenging the constitutionality of the TVA.[119]

It was war.

The long-awaited fisheries report from the International Joint Commission was finally hauled to the surface for everyone to see at the end
of February 1936. The *Sentinel* noted, "In general it contains little not
already known by every resident of Quoddy who is familiar with the
sardine fisheries. The cost of the investigation was $90,000 and the
time required was about two years. The net result is about the same
report the *Sentinel* volunteered to write for nothing, at the time the
Commission was appointed."[120]

The exhaustive report put a lot of science behind Maine's assertions
that the environmental impacts of the proposed International Project—
which would have dammed off *all* of Passamaquoddy Bay—would
probably not be as severe as New Brunswick's "mixing bowl scientist,"
Dr. Huntsman, had predicted.

If the tidal dams were to enclose *all* 150 square miles of Passamaquoddy Bay, "The herring fishery inside of Passamaquoddy Bay would
almost certainly be reduced to negligible proportions."[121] This, you will
recall, the Canadian fishermen were willing to concede in 1928 if they
were financially compensated for their losses (though they changed
their minds in 1929, possibly having been encouraged to do so by the
power trusts). Meanwhile, the six-times-larger American sardine industry was willing to accept possible losses on the theory that the 4% of
the fish caught in Passamaquoddy Bay would go somewhere else, and
the project would have other economic benefits.

The effect of the dams on the Bay of Fundy was more complicated. The commission agreed with Dr. Huntsman that the "mixing mechanism" was important but noted:

> Any influence of the proposed dams upon this supply [of zooplankton from outside the bay] would probably be insignificant. . . .
> Many changes in the set of the tidal streams may be expected, and probably every little change would have an effect on the fishery of nearby weirs. Some weirs would be made richer, some poorer. It cannot be foretold whether the total effect of disturbance of tidal streams would be deleterious or not. There appears little probability of the proposed dams affecting the sardine industry along the coast of Maine or even seriously at Grand Manan.[122]

The All-American Plan of enclosing just the 37-square-mile Cobscook Bay would presumably have proportionately less severe effects than the International Plan. The fisheries report itself did not appear to have any effect.

ALMOST

As the *Bay City* shovel clawed its way into what had been Roscoe's property on Rice's Hill,[123] and as Engine 246 dragged the remains away and dumped them, the tidal currents that had raced between Pleasant Point and Carlow Island since time immemorial—ceased. The first dam of the Passamaquoddy Bay Tidal Power Project was above the high-tide line. It was March 8, 1936. "The trestle was swarming with visitors Sunday who were curious to see the progress of the work" and witness the end of an era. Many of them no longer lived on an island.[124]

The dam was not fully complete. The army still had to build it up another dozen feet or more with quarry stone, coat the sides with impervious clay, and armor the top 50 feet with enormous boulders to break the waves. While the dams were no bigger than thousands of similar "cut and fill" projects on roads and railroads everywhere, the stopping of the tides was a major milestone in the history of the Quoddy project, Eastport, and the Pleasant Point Reservation, whose futures were now even more intimately connected.

Eastport's future was being debated in the Bahamas. There, President Roosevelt was on a fishing vacation with his uncle, Frederic Delano. Franklin went to get away from the noise of Washington and plan his own future.

The 1936 presidential election was coming in the fall, and Roosevelt's New Deal programs were both hugely popular and vehemently opposed. To the beneficiaries, the programs were salvation and Roosevelt their savior. To the opponents, the dole and government competition with private enterprise was socialism, and Roosevelt a despot and "traitor to his class."[125]

The opposition view had been bolstered when the Supreme Court struck down some aspects of the New Deal as unconstitutional.[126] Since then, the Republicans had regrouped from Herbert Hoover's embarrassing loss and were becoming hopeful their new candidate, Alf Landon, governor of Kansas, could capture the White House.[127]

Sitting by the rail of the USS *Potomac*, his new presidential yacht, Franklin fished, smoked, drank, soaked up the Bahamian sun, and talked with Delano. Their discussions were interrupted when Franklin caught a 75-pound fish he didn't recognize. He sent it to the Smithsonian for identification. It was an amberjack tuna.[128]*

Delano was head of the National Power Policy Board and an advocate for Dexter Cooper and the Quoddy Tidal Power Project. He believed it would work and would attract new industries and that he, Delano, a builder of railroads, knew how to get the dams built economically. As the *Sentinel* said, "Maine was very fortunate" in having Delano aboard.[129]

The opposing view, however, was more widely held, owing to the anti-Quoddy propaganda. The confusion and doubt sown by Quoddy's enemies, along with legitimate concerns, gave the Republican Party a ripe target for attack. Quoddy was vulnerable. The president's pet project—the *goat* to Republicans—could be made to look like a Democratic *ass*. Thus, despite his sincere belief in the project and his sincere desire to help the desperate people of Downeast, President Roosevelt was having second thoughts.

Harold Ickes, head of the PWA, the organization paying for Quoddy as a relief program, was also having second thoughts, as he confided in his *Secret Diaries* on March 3, 1936:

The President is trying to let go of the tail of the Passamaquoddy bear too. I told him that there was a good deal of local dissatisfaction with the way the project is being run. The Army engineers have bought a $17,000 yacht,[130]* and according to what Dexter Cooper says, they have built seven houses costing $30,000 each. . . . The natives of Eastport and that part of Maine seem to think that the Engineers are spending money with a lavish hand and, of course, that goes against the Yankee grain. I told the President that the special board that I appointed some time ago to investigate the cost of this project had not reported yet.[131]* We commented again on the fact that the Army engineers had accepted Dexter Cooper's figures on cost without ever checking them until after the Government was committed; and work was actually started. I remarked that this was like accepting a promoter's estimate of cost, and that Dexter Cooper, with all due respect, was a promoter so far as this project is concerned.[132]

"Honest Harold" Ickes was either forgetful or not being honest with his diary or his president. The funds for the project came from PWA's relief funds and thus were Ickes's responsibility. He also knew of Dexter's $43 million PWA loan request for the project, including $4 million for metallurgical plants. Neither the army's original nor Dexter's estimate included building an army base, which is essentially what the army was doing at Quoddy Village, complete with poop decks on Snob Hill. The parsimonious Dexter, you will recall, preferred his offices Spartan, and had proposed using old hulks for temporary housing. Dexter therefore could not be held accountable for the army's plans, budget, or execution. He was just the consulting engineer and not consulted.

The private power trusts attacked Quoddy through Congress. There, Wendell Willkie, the head of C&S holding company, found the corporate relief he sought and a friend in Arthur H. Vandenberg, a Republican senator from Michigan with presidential aspirations. Vandenberg took up the megaphone and led the attack.

In February 1936, Vandenberg started a drive against further appropriations for Quoddy, the Florida Ship Canal,[133] and three other relief projects, on the grounds that they had not been properly vetted or authorized by Congress. Vandenberg used that reasonable (though not fully accurate) premise to precondition his assertions. On March 9, in a letter to the Senate Commerce Committee, Vandenberg included a summary of an *unidentified and unpublished* report, which purported to show that Quoddy power was too expensive, there was no market for it, and the energy could be developed more cheaply from Maine's rivers. "If the conclusions are sound," he wrote, "the project isn't. I respectfully submit that these presumptions must be successfully rebutted before your committee is justified in lending its approval to the enterprise"[134]—although, of course, any *unidentified and unpublished* report should be vetted *first*.

Let's be fair. The Roosevelt administration *was* hiding things: the still-secret PWA-Hunt Report and the not-yet-completed report Howard Ickes mentioned in his diary (the Reclamation-Durand Report).[135] But those reports probably weren't what Vandenberg was referring to. Given its purported content, he was probably referring to one of the reports palmed by the power trust—likely the one forged by Murray & Flood. But it really didn't matter who created the report or how reliable it was; once the senator suggested a cover-up, it was as good as true. And a fully legitimate complaint was that by funding these projects incrementally, the total cost was never apparent, each increment of funding tended to ensure the next, and so on to an unknown and undoubtedly expensive result.

"Cooper Announced Sale of Entire Quoddy Power Output," proclaimed the *Sentinel* in a full-page banner headline the day after Vandenberg's attack, and he "could sell more if available." Continuing, the article noted, "Mr. Cooper did not name the firms he has interested in Quoddy, this being obviously not a matter for publicity at this time, but they are understood to be firms of national standing. They are to locate factories at or near Eastport that will create employment for two thousand people."[136]

As Representative Brewster had explained in June 1935, Dexter's job as consulting engineer was *not* to help the army with the engineering but, in the next three years, to find buyers for Quoddy's power. Now, just ten months later, *and without any of Quoddy's functional elements completed*, Dexter had succeeded. But as Roscoe explained, "The signature of final contracts, of course, must await the establishment of the Quoddy Authority or Commission competent to sign them, and determination of the date at which power will be available."[137] Since it would take only six to twelve months to build industrial facilities, and Quoddy's completion was still three years away, the companies in question wouldn't want to divulge their plans to *their* competitors, nor would Downeasters want to disclose their opportunities to *Quoddy's* competitors.[138]

According to Dexter, *all* of Quoddy's power would be purchased by just three companies, strongly suggesting that an even larger demand would develop as more companies followed the leaders into eastern Maine to service all those employees and their families.[139]

So much for Vandenberg's claims.

So much for the power trusts' claims.

Of course, what everyone wanted to know was: *Who were the companies? When would they "sign on the bottom line"? When would they start hiring?*

The answers didn't matter. The Senate was interested in exercising political power, not developing electric power. Rather than consider the new evidence of Quoddy's economic feasibility, the very next day, March 11, the Senate Subcommittee on Appropriations voted 6 to 5 against providing the $29 million President Roosevelt had requested for river and harbor projects, including $5 million for Quoddy, based on the Republican contention that Congress had never sanctioned the projects.[140]

"Once bitten, twice shy," was Dexter Cooper's reason for not divulging the identities of the companies interested in Quoddy power.[141] He might have been referring to the Senate's resolution of the week before, or to the power trusts, or to what had happened when he divulged his plans to build an aluminum manufacturing plant.

Maybe Colonel Fleming knew the identities of the companies; maybe he didn't. In the same March 18 article in the *Sentinel* that quoted Dexter, Fleming was quoted as saying, "In my brief stay here, I have come to like Maine, to like its people. I would be content to spend the rest of my life here. . . . The eyes of the entire nation are turned toward Quoddy. There has been much unjust criticism of the work here. The time has come for some sort of defense. We have faith and pride in this project, and we are not ashamed of it."[142]

"Don't open them," was the order phoned in from Washington. Again, it must have been a shock to the Army Engineers in Eastport. This time the order was not to fire people but to not hire them. Don't open the bids for construction of the navigation lock—as was scheduled for the next day, March 19, with work beginning two weeks later—but send the bids back.[143]

For the army, with its right foot on the gas 24 hours a day, the order was like stomping on the brakes with its left foot, sending everyone into a tailspin.

Colonel Fleming and Mayor Emery tried to reassure everyone it was just another bureaucratic bump: "There is no cause for alarm about Quoddy's future, however. It is reported that Acting Director of the Budget Bell has ordered that ERA [Emergency Recovery Act] funds cannot be used on contracts which will not be completed by June 1st. Quoddy's present funds come under this classification."[144]

Captain Leehey, Quoddy's head of administration, was more candid. He said the $1.5 million the army had on hand was enough to continue construction only until July 1, and delays now would cause problems at the end of the building season and another race against winter. He also acknowledged the political reality: "The war department considers it necessary to permit no further commitments until the uncertainty as to future funds is removed. But to us here on the job it is tragic."[145]

Vandenberg was trying to script Quoddy's tragic end. He had his reasons beyond pious concern for fiscal conservativism. As the *New York Times* proclaimed, or should we say *anointed*, with oil and ink, "Vandenberg Becomes the Real Senate Minority Leader: . . . It is the Michigan Senator who, after long urging by friends interested in his career and in Republican prospects, has lately made the real opposition to the President. It is he whose initiative his party colleagues await and follow. . . . The new, dominant and effective activity will undoubtedly put Mr. Vandenberg in the forefront of the Cleveland convention."[146]

In the meantime, the presidential wannabe sat in the second row on the Republican side of the US Senate chamber next to Maine's senator Frederick Hale.[147]

Officially, spring arrived on March 23. But it also seemed to have been delayed. Roscoe could find "no tangible proof" of the supposed change in the season until the following day, when crocuses finally bloomed in Joanna Davis's garden on Ray Street. The next day, Roscoe noted that "the home of Raymond E. Gest, Quoddy Village, was host to a robin at seven-thirty o'clock this morning. Arise and shine Old Man Sol!"[148]

Old Man Sol had his usual effect on Maine in March—he made mud—while Jack Frost heaved and buckled the pavement. "The highway grew worse each day, with cars constantly getting mired." A crew of government men went to work, happy to have something to do.[149]

Elsewhere, the rain caused problems. It melted the snow, filling the streams, then filling the rivers, then overflowing their banks, then flooding fields, roads, and towns. As recounted in CMP's official history, "CMP's Brunswick Station flooded twice to the top of its windows. The generators at Shawmut hydro were under water, and a six-ton generator at Union Falls was pushed off its foundation. Substations flooded out: Bath was out of power for days. Fred Clement of the West Buxton hydro station cheerfully worked from the roof, using a crowbar on a rope to break windows and ease the flood waters' pressure against the building. He said he always wanted to get paid for breaking windows."[150]*

The Kennebec River crested at 30.7 feet at Augusta, about 20 feet above a typical spring freshet, with 200 times the rate of flow of the summer low.[151] The unpredictability of the weather, and hence the dangers of relying too heavily on rivers for hydroelectric power, could not have been more dramatically demonstrated.

There was severe flooding all along the East Coast of the United States. In Washington, truckloads of Civilian Conservation Corps men were rushed in to save the city: "With ten-foot dikes of sandbags erected across the street within one hundred yards of the White House to keep out the flood waters of the Potomac, Washington took on the appearance of a European capital in a state of siege."[152] The flooding was so widespread and the damage so great that Captain Casey and fourteen other Quoddy engineers were temporarily transferred to emergency duty in Massachusetts, Rhode Island, and New York to help with the cleanup and repairs.[153] The *Sentinel* noted with pride on April 8, "RECENT FLOODS OF NO MENACE TO QUODDY PROJECT."[154]

Roscoe was wrong: the flow of flood waters wasn't a threat to Quoddy, but the redirected flow of federal funds was.

Forked tongues were a menace to the Quoddy Project. There were essentially two electric power monopolies in Maine: CMP in the southwest, and BHE in the northeast. Since Quoddy's state charter precluded it from selling power to consumers in their service districts[155] or to any of their existing commercial accounts, Quoddy was no threat to them (other than as a "yardstick" for price comparison). Indeed, an influx of people attracted by new industry to Maine would swell CMP's and BHE's businesses. Thus, the companies *officially* stated they were in favor of Quoddy.[156] But that wasn't true, as previously reported by many newspapers. And when Dexter was in Boston, he asked the president of one of the largest banks why the bank wasn't supportive of Quoddy when it would lead to the spending of $36 million of federal funds and many millions more of private investment, all of which were growth opportunities for the bank. The banker replied that Quoddy was "stepping on the toes of some of his clients, specifically two Maine power companies."[157]

The Republican Party also wasn't giving up its campaign against Quoddy, and another flood of forked words came from Senator Vandenberg. Roscoe began his coverage of the story (on April Fools' Day) with a bit of editorial counterspin: "Vandenberg, the Gentleman from Michigan who is winning a little cheap publicity by attacking projects as far removed as possible from his home State, rushed into print last week with an assertion that Quoddy had been repudiated along with the Florida canal, and added that it had been killed by its own friends . . . [Maine's] Senators Hale and White."[158]

Vandenberg was impugning Hale and White for their efforts to remove Quoddy from the War Department appropriations bill and hence remove it from attacks in Congress. They must have been under tremendous pressure to follow the Republican Party line and vote against New Deal spending like Quoddy and yet act according to the wishes of their constituents, who wanted Quoddy. In a craftily worded statement, Senator Hale said he had always been for the Quoddy Tidal Power Project as a relief project—though he voted against the Relief Bill—and that he had never argued for or against the Florida Ship Canal but had voted against its appropriations.[159]

Quoddy's supporters were given an unusual opportunity to respond to an unfavorable commentary on Boston's WNAC radio station. On April 10, William Beale of Eastport tried to make the case for Quoddy. As well reasoned and fact filled as it was, it was a plaintive cry. The most memorable line was "What favorable publicity on Quoddy has leaked through into the open—has been wrung out of the situation by a group of local individuals—men who were compelled to pass the hat for gas, stamps and printer's ink."[160]

The army also had a public relations problem. Two pastors in Eastport wrote to the president telling him of drunkenness, waste, and nepotism on the part of the Army Engineers. They encouraged him to take action to prevent the few bad actors from spoiling the whole project.[161] The PWA sent undercover agents to investigate.[162] Their report was kept secret.[163]

Nonetheless, Gertrude got wind of the investigation, and in April she wrote a series of intriguing letters to "Dear Franklin," one of which said:

> [T]he man I wrote about—It is Roscoe Emery, for many years Mayor of Eastport, and a very clever man who must be able, he has so many enemies. I feel that it will be a real public service to see him at your very earliest convenience. He really has something to tell. It would seem to me that the less publicity the better, so he will go alone, whenever you send for him. You can notify him direct, or let me know and I will pass on the message.[164]

The president responded that he would be happy to meet with Roscoe, adding, "I am terribly disappointed over the present situation but it is not an irrevocable defeat."[165] Nine days later, Gertrude wrote to Franklin again: "If I can see you, as I did before, without anyone knowing it, it would be best, I am sure. Perhaps Eleanor would ask me for lunch or tea and I would see you for a few minutes afterwards."[166]

Roscoe Emery or Gertrude Cooper may have met with President Roosevelt, but if so, it was done clandestinely, for there are no records of such a meeting in FDR's official calendar for 1936.[167]

April was the cruelest month. The spring rains had come, but the money didn't. The first bit of bad news was a statement by Harry L. Hopkins, the new head of the newly renamed and reorganized PWA, now known as the Works Progress Administration (WPA), who gave "positive assurances" to the House Appropriations Committee that no WPA funds would be used for "tide-harnessing."[168]

In the April 21 edition of the *Sentinel*, Roscoe admitted (but only on page 6) that "[t]he big and distressing news . . . was the statement of the President regarding Passamaquoddy";[169] on April 15, President Roosevelt had said Congress would have to approve the Quoddy Tidal Power Project before he would authorize any more money for it.

In other words, President Roosevelt had the executive authority to continue to fund Quoddy, but he wouldn't.

Vandenberg's fusillade had hit its mark.

There was $300 million of the $4.8 billion public works appropriation remaining in the president's hands. Use of these funds was entirely at his discretion. But Quoddy wasn't going to get any of it, not unless Congress agreed.

Congress was not likely to agree. Presumably, that was why Maine's congressional delegation had taken Quoddy out of range of congressional attack and put it in the president's lap. The president, however, punted it back. Evidently he did not want to be solely responsible for the project, which most of the country regarded with suspicion, if not derision. Quoddy, the president's "baby," was becoming his "tar baby."[170]*

According to the *Sentinel*, "The Army Engineers have prepared a modified plan which provides for the completion of the project within the original estimate as to cost and is now ready for consideration by the President."[171] Presumably the budget revision should have taken care of that politically embarrassing problem, if that was the issue. But as the *Portland Sunday Telegram* explained, "People suspect some obscure political strategy but they do not know what it is. Maybe he wanted to make Maine and Republicans sit up and beg for Quoddy and then he will give it to Brann as a Democratic card to use in the Maine campaign. Best description of the situation is the remark of Rep. Hamlin . . . as to what was to be tried, 'I dunno,' says Mr. Hamlin."[172]

So Quoddy's worn-out supporters returned to Maine to "gather advice on what should be done next to save the big tidal project from the guillotine."[173]

The *New York Times* seemed pleased to report that even Quoddy's existing funding—about $1 million—might be taken away and used for other WPA projects in Maine, a number of them having already been approved with such a shift in mind.[174] A few days later, the rumor became more specific and threatening when the *Times* reported, "[T]he flood-control program, which is being rushed through, may force taking away what is left of Quoddy's 'mite.' The State WPA Administrator, with $2,000,000 ready for flood-control use, has asked for a third million. Whether Quoddy will suffer is the important question now."[175] Alluding to Quoddy Village, the *Times* concluded, "[T]he well-furnished stage apparently may never be occupied."[176]

THE FATAL ATTACK

Secret government reports held the answer to why President Roosevelt seemed to lose faith in Quoddy—and the Army Engineers: the cost.

The cost to construct Quoddy would ultimately determine the price of the power produced. Back in the summer of 1935, the army had estimated $30.1 million. In December 1935, it had confidentially upped the ante to $61.5 million—nearly causing the cancellation of the project.[177]*

The lack of clarity on cost caused the *Sentinel* to complain on April 29:

> [T]he so-called Hunt report on which the President finally decided to go forward with the project . . . has never been made public or communicated to Congress or the public in any way. Meanwhile all the other adverse reports have been fully publicized. The result has been that the unfavorable aspects of Quoddy have been fully exploited while the grounds upon which the President felt justified have never been made public. Even now publication of this report is refused. Passamaquoddy has apparently been made the football of the relief situation and has received critical attention in the press throughout the county out of all proportion.[178]

President Roosevelt wasn't the only person with a copy of the PWA-Hunt Report. Governor Brann had a copy. The Sills Commission members had copies. And its authors—Dr. Hunt, Dexter P. Cooper, Moses B. Pike, and H. T. Cory—had copies.

Why wasn't one of them releasing it?

There was also the Reclamation-Durand report, which was delivered to the president on May 8. Its estimates, and even its existence, were still secret.[179]*

Friendly papers in Maine defended Quoddy. The *Portland Press Herald* declared, "QUODDY WORK SHOULD GO ON":

> Work relief for Maine's unemployed is still required and will probably be true for a long time to come. Why stop the Quoddy job while this condition

prevails? . . . The State of Maine has so far obtained comparatively little of its share of this original work relief appropriation. Why shouldn't it be given its fair proportion of this money? . . .

All of Maine is interested in having this work at Eastport carried on to completion. It will be the greatest reflection on the Administration at Washington if it is now dropped, the blackest mark on its record. It is unthinkable that such a thing as abandoning the entire Quoddy project should even be contemplated.[180]

The Guy Gannett–owned *Waterville Sentinel* printed what many in Maine must have been thinking:

It's time to end the ridiculing of the Quoddy project. . . . That time has passed and the issue now is whether the Government will keep faith with the people who have staked their all on the promises made by Washington that this enterprise would be completed. . . . Yet apparently the people who relied on the promises of their Government are to be left flat, money gone . . . and nothing left but black despair.

How the Boston papers love the Quoddy project. They lose no opportunity to point out that it is all a foolish waste of money and that anyway the money should not be spent in Maine, we do not deserve it here, Massachusetts is where the money should be placed. . . . What would Maine people do with electric power anyway, they wouldn't know how to use it? Then to people who are in the habit of receiving their illumination at night by means of a tallow candle, an electric light would be far too bright.[181]

The *Sentinel* quoted its archrival, the *Lubec Herald*, as saying:

Quoddy . . . has been made the football not only of politics but of the relief situation, and has been held up as a mark at which all the slings and arrows of outrageous fortune have been directed. . . . Quoddy had as its chief opponents the power companies who still think they can sell electrical energy at a too-big price and find people putting in the hundred and one appliances which they cannot afford to operate at the present rates. . . . [I]t is still too early to let the poison spray of political venom kill something that will be of the utmost value until the moon decides to go out of business, which is still a long way off.[182]

Newspapers and magazines loved Quoddy. It provided unequaled opportunities for irony, mockery, and grandstanding. It was a comedy and a tragedy. It was full of politics and money, skullduggery and backstabbing, fact and fiction. There were local angles aplenty, national significance, and a cast of famous and witty demigods hurling salvo after salvo. It was a circulation-boosting, life-and-death serial thriller.

One example of the repartee appeared on April 26 when the *Portland Sunday Telegram* ran a large article with three aerial photos of Quoddy Village under construction and the headline:

> **Quoddy Supporters Now Making Greatest Fight Ever Waged for Project:** . . . Rank and File of Citizens Feel President Roosevelt Forced to Act Contrary to Personal Desires Because of Politics.

Several paragraphs in the article addressed the charge that Maine just wanted money: "[S]ome, to be sure, may have seen the possibility of increasing their own wealth as a result of the building of the project, but if so, it was after all a human and natural enough reaction, particularly since anything approximating wealth in a larger sense had for years been a stranger in Eastern Maine."[183] Describing Quoddy's advocates, the *Telegram* commented:

> Calumny, slander and distortion of fact was almost entirely ignored as being immaterial and irrelevant. . . . And so it went until the President announced a bit more than a week ago that no more relief funds would be made available for Quoddy. Then the buttons were removed from the foils and the duel continued more grimly. . . .
>
> Some ten days ago, Bainbridge Colby[184]* . . . made this declaration regarding Quoddy, that "New Deal spending could not buy the state of Maine." The statement did not pass unremarked. Out came the snickersnee in Eastport, and Roscoe C. Emery.[185]

To which Roscoe, the "snickersnee," sniped back:

> Mr. Colby has apparently never recovered from the fact that he was for a few months Secretary of State under Woodrow Wilson. . . . It follows that he may be one of those alleged Americans who believed in extreme liberality to foreign nations and are exceedingly parsimonious with their own. If Quoddy were being built in Czechoslovakia with American money he probably would be for it. . . . Maine neither wants or appreciates advice from the cold storage vaults of the Wilson era, nor from political hermaphrodites who are neither Republican nor Democrats, and are running around the county trying to attract attention to themselves by attacking a President whose shoes most of them are unworthy to unlace.[186]

Calling Colby a "political hermaphrodite" was a queer charge for Roscoe to make, given that Roscoe was a Republican and a vocal supporter of the Democratic president's New Deal.

The red-blooded Republican Alice Roosevelt Longworth, daughter of former president Theodore Roosevelt, disparaged Quoddy in a nationally syndicated article. She was rebutted in a letter signed "Johnny Tidewater." "While lacking the rapier-like penetration of Mr. Emery's commentary," the *Telegram* noted, "it nonetheless has its points." Johnny Tidewater asked, "[P]lease tell me what the popular Republican word 'boondoggling' means—I have heard the word before and have looked all over the dictionary, but I can't find it. Has it anything to do with keeping 5,000 poor respectable American citizens and their families from all over Maine supplied with the necessities of life?"[187]

When all else fails, plead on bended knee. That was the best idea Maine's political leadership could come up with. So, on April 29, Governor Brann flew to Washington to beg the president to reconsider. He took with him 700 *real* telegrams from *real* people who wanted the relief program to continue and Quoddy to be completed.[188]

The model village was taking on a "doll-like appearance" by the first week of May 1936. "Underbrush and refuse among the trees is being removed; in its place go clothes lines, white-laden each week with the washings of the housewives." Grass, flowers, and vegetable gardens were being planted. General Markham toured the works and declared himself "well satisfied." He was escorted by Governor Brann, who was "optimistic."[189]

All around Eastport were signs of progress, prosperity, and normalcy. A high-tech, "Tom Thumb"[190]-sized hospital at Quoddy Village opened in mid-May with an operating suite, a lead-lined X-ray room, and an emergency battery backup system that would switch on instantly if needed.[191] Farther up-island, the old drawbridge that had connected Eastport to Pleasant Point was removed as no longer necessary, now that road and rail traffic could travel over the nearly done dam-causeway.[192] To the south, "the ages old silence of revered Treat Island is no more," the *Sentinel* recorded. "Today, twirling, clanking drills, motivated by five air compressors vibrating side by side in a syncopating roar, are piercing the bowels of cliffs which never before felt the steel of man"[193]

There were 2,096 people on Quoddy's payroll.[194]

Governor Brann came up with a better idea: take Quoddy out of the political fray, as he had privately tried to do once before.[195] This time he proposed that two "presidential commissions" be established, each with three independent engineers, to review the two largest political foot-balls, the $40 million (or so) Quoddy Tidal Power Project and the $150 million (or so) Florida Ship Canal. These commissions were to publish their reports no later than June 20. If their reviews were positive, then the president would be authorized to allocate $9 million and $10 million respectively for the fiscal year beginning July 1, 1936. "Friends of Quoddy were disposed to view the resolution and its special provisions with optimism, basing this feeling on confidence in the ability of Governor Brann to 'bring home the bacon.'"[196]

Quoddy's presidential commission was to address four issues: (1) engineering feasibility; (2) economic justification, including industrial development; (3) scientific value of the pioneering tidal experiment; and (4) costs. It was as straightforward and apolitical as possible. Senator Robinson (D-AR) introduced the resolution in the Commerce Committee.

Vandenberg objected. "I've heard about the three-shell game," he said, "but this is the first time I've heard of the three-engineers' game." Vandenberg called it a "back door scheme"[197] to circumvent congressional review (*politicians in his mind being better qualified than professional engineers*). Brann, Brewster, Hale, and White defended Quoddy as a desperately needed work-relief project. Vandenberg ridiculed them.[198]

To prevent the engineers from doing their review, Vandenberg introduced sixteen amendments to the Robinson resolution to delay the report until the next session of Congress and to force Cooper to publicly identify the "alleged" purchasers. *Vandenberg didn't want a confidential financial review of Quoddy; he wanted a show trial—starring Arthur Vandenberg.*

On May 19, the *New York Daily News*, then the largest newspaper in the world, with over one million subscribers, published a much-appreciated editorial: "Yes, Go On with Quoddy: . . . The humorous cracks which did so much to put Quoddy to sleep are accordingly being revived, chiefly by Senator Vandenberg, who invented most of them. But this isn't a question that should be solved by wisecracks."[199]

That same day, President Roosevelt announced he would spend part of his summer vacation at Passamaquoddy Bay[200] It seemed to be an endorsement.

None of Vandenberg's amendments were accepted by the committee. Sensing his defeat, the *Sentinel* snickered, "VANDENBERG FEELING BLUE." On May 26, the Commerce Committee voted 12 to 5 in favor of the Robinson resolution.[201] Governor Brann's strategy was working. Quoddy's fortunes were improving.

The next big question was: *What would happen in the full Senate?*

In Eastport, the army began digging the excavations for the navigational lock. Out at the ballfield, "The Wives" outhit the typewriter-tapping "Secretaries" 22 to 10 in "seven picturesque innings" of softball, while a "large group of spectators grinned on the sidelines." BHE invited the ladies of Eastport to classes on cooking with electricity. And Mr. and Mrs. Nebiker welcomed an eight-pound baby boy into their model home in Quoddy Village.[202] He was, of course, conceived nine months earlier, during Quoddy's halcyon summer of 1935 when dreams were becoming real.

Vandenberg's venom continued to spray. The unrepentant viper kept up his wisecracking:

"It sounds to me as though it were out of the Wizard of Oz,"[203] *said the senator from Michigan about the project in Maine he had never seen.*

"A totally unexplored, unilluminated undertaking,"[204] *lied the unenlightened one, ignoring sixteen years and tens of thousands of pages of technical reports.*

"It remains for a generous, gullible old man named Uncle Sam to rush in where others feared to tread,"[205] *said he whose state was getting much more federal money.*

"I'm perfectly willing to let Congress vote and the people decide whether to approve this sort of indefensible exploitation, extravagance and recklessness,"[206] *though he was doing his best to prevent a straightforward vote.*

"President Roosevelt was trying to take the treasury moondoggling on Quoddy Bay,"[207] *howled the Republican attack dog.*

The "Robinson resolution" would face a vote in the full Senate on May 30. It was Quoddy's last hope for congressional approval and funding.

Vandenberg led the Republican attack. He was armed with a "huge brown envelope . . . filled with what he called 'poison gas and dum

dum bullets.'"[208] He talked nonstop for two hours. "He [the President] started all these projects," roared Vandenberg, "but when he gets into hot water, he wants us to share the bath!"[209] He criticized the independent commission as "convenient chloroform for the Congressional conscience." He quipped that "we are back boondoggling on Quoddy Bay and pipe dreaming on the phantom Florida canal." He charged, "There never was any justification for these projects and there isn't any now." They were, he exclaimed, "an economic extravaganza."[210] "It's perfectly absurd to authorize another year's work," he scolded, "unless we are prepared to carry them through. No amount of camouflage or weasel-worded argument can make it appear otherwise."[211]

While Senators Hale and White of Maine had consistently opposed the Florida Ship Canal like most Republicans, they had gone along with the Robinson resolution, in which Maine partnered with Florida to get the two work-relief projects objectively evaluated and *maybe* funded.

But then during the debate Hale did something unexpected. Instead of pushing for a package deal, he split the baby. Instead of having one vote for the resolution, Hale insisted there be two votes, one on each project. He succeeded. Then he tried to have Quoddy acted on first, but he was overruled. Thus, the Florida Ship Canal vote came first.[212]

After 48 votes were cast, the resolution was dead even: 24 in favor, 24 opposed. "Then came a succession of "ayes" from staunch Administration supporters and the canal's Commission was approved 35 to 30."[213]

But Maine's senators Hale and White had both voted "no." They had double-crossed Florida. Not only did they split the baby, but they also tried to kill Florida's half.

There was a furor in the Senate chamber as members considered the sudden turn of events. Moments later, it was time to vote on Quoddy.

Senator Bailey of North Carolina had been a supporter of the Robinson resolution, but since the pact had been broken, he changed his vote to "no," and as the *New York Times* recounted:

When he was followed by Senator Byrnes of South Carolina [D] it was evident that Southern blood was up. They were followed by Senator Russell of Georgia [D] and Senator Thomas of Oklahoma [D]. These changes would have been one vote short of balancing the change of Senators Hale and White from "no" to "aye," but Senators Murray of Montana [D], Norris of Nebraska [R], Schwellenbach of Washington [D], Wagner of

New York [D], and Wheeler of Montana [D] also changed over. Senator Bullow of South Dakota [D] and Senator Minton of Indiana [D], who had not been present for the first roll-call, also voted against Quoddy.[214]

The southerners weren't going to let the northerners get away with their traitorous behavior. Even Senator Norris, the renegade Republican and "Father of the TVA," voted against the traitorous Republicans from Maine.

The final vote was 28 in favor of Quoddy, 39 opposed. Quoddy was killed. Quoddy's last hope for congressional approval and funding was over. Finished. Done. Sliced from its life-sustaining southern sibling, Quoddy was dead again.

Just as Senator Vandenberg had said back in March, Quoddy *was* killed by its friends, Maine's very own senators, Frederick Hale and Wallace White.

The Republican Party rejoiced. The power trust triumphed.

Governor Brann denounced Senators Hale and White:

I cannot let Senator Hale's defense statements go unchallenged. His defense before charges were presented is an indication of a guilty conscience. . . . Senator Hale's insistence upon the separation of the Maine and Florida projects was fatal and was the betraying kiss of death to the Maine project. Senator White's speech in opposition to the spirit of the Robinson resolution and against the Florida ship canal was the fatal stab of annihilation. Senator Hale inserted the first screw in the lid of the coffin and Senator White drove it home, that the corpse of Quoddy might not escape.[215]

"Eastern Maine, with new businesses and an enormous influx of population depending on the fate of the project," said the *New York Times*, "is bitter and bewildered at Quoddy's Senate defeat. . . . The little city of Eastport, now overgrown and clustered with shiny new storefronts, is ready to collapse."[216]

The consequences of the double-crossing congressmen's actions, Mayor Emery said, would be "a disaster, from which this section of the State could not recover during the present generation. The fortunes of many thousands of people are tied up with this project and its abandonment would bring ruin to most of them."[217]

Full Throttle: The just-launched coal-fired 293', 10" USS *Flusser* (DD-20) during sea trials in 1909, making twenty-six knots. (*Source:* https://www.history.navy.mil/our-collections/photography /numerical-list-of-images/nhhc-series/nh-series/NH-63000/NH-63257.html)

Strait and Narrow: Lubec and Lubec Narrows from Campobello (c. 1910). (*Source:* https://i.ebay img.com/images/g/kqAAAOSw575gDfiO/s-l1600.jpg)

The Brass: Assistant Secretary of the Navy Franklin D. Roosevelt, unidentified commander, Rear Admiral Charles J. Badger, Secretary of the Navy Josephus Daniels, Secretary of Commerce William C. Redfield, and Secretary of War Lindley M. Garrison aboard the USS *Wyoming* in 1913. (*Source:* National Archives)

High Point of the Sardine Industry: A sardine (herring) cannery in Eastport, Maine, in 1887, at low tide (averaging eighteen feet). (*Source:* George Brown Goode, *The Fisheries & Fishery Industries of the United States* [1887], plate #134; author's collection)

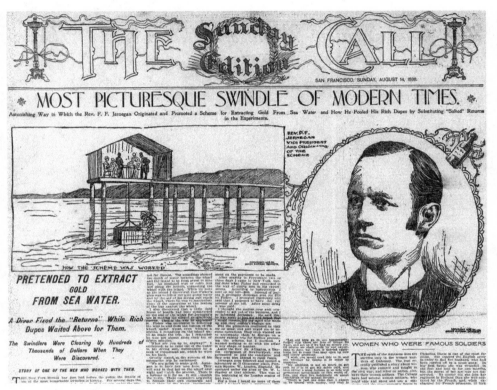

Under Water Industry: Reverend Prescott Jernegan, who deceived his investors in the Lubec Gold Swindle of 1898. (*Source:* https://www.newscientist.com/article/dn15012-eleven-of-the -greatest-scientific-hoaxes/, accessed September 16, 2019, crediting California Digital Newspaper Collection/Centre for Bibliographic Studies and Research/University of California)

Small Fry: Original caption: "Three cutters in Factory #7, Seacoast Canning Co., Eastport, Me . . . Grayson Forsythe, 7 years old. Middle is George Goodeill, 9 years old, finger badly cut and wrapped up. Said, 'the salt gets unto the cut.' . . . Clarence Goodeill, 6 years . . . 1911." (Spelling according to Wayne Wilcox, Eastport, Maine, who knew George Goodeill.) Note Clarence's bare feet. (*Source:* https://morningsonmaplestreet.com/files/2014/12/GraysonForsyth.jpg, accessed October 25, 2018)

Funding a New Industry: Full-page (15" x 21") advertisement for Central Maine Power Company stock sales, from the *Eastport Sentinel*, October 13, 1920. Note the image of Yosemite's Bridal Falls partly obscured by electric transmission lines. (*Source: ES* October 13, 1920)

"Sit:" Dexter P. Cooper and Push, his Newfoundland retriever, on October 31, 1925. (*Source:* Photo by World Wide; author's collection)

Full-color and Full-page: A 16" × 24" illustrated map of the army's Plan D for the Quoddy Tidal Power Project, published as an editorial feature in the *Boston Sunday Post* on July 28, 1935. (The red ink headline "Quoddy" has since faded.) (*Source:* BHS/Tides Institute)

Superior: The "permanent" housing for Quoddy executives on "Snob Hill," 1936. (*Source:* BHS, Barracks Museum collection)

High, Wide and Handsome: The administration building and central section of Quoddy Village, June 8, 1936. (*Source:* https://catalog.archives.gov/OpaAPI/media/23940773/content/stillpix/018 -aa/Box_68/Folder_551/18-AA-68-035.jpg)

The Big Scoop: Members of the press watch the loading of the side-dump railroad cars at the quarry on Rice Hill for construction of the Pleasant Point to Carlow Island dam, February 1936. (*Source:* Steam Shovel at Dam Site, Creator: War Department. Office of the Chief of Engineers. Boston District. 1/18/1883-10/1/1946 [Most Recent], Date Created: 1936-02-07. External Link: National Archives Catalog, https://catalog.archives.gov/id/7350917)

Going Deep: The drill ship *Eastern Chief* in one hundred feet of water at mid-tide on October 28, 1935, just east of downtown Eastport. (*Source:* BHS, Barracks Museum Collection)

Cut-and-Fill: This quarry on Carlow Island supplied some of the stone for the dams, like the Pleasant Point–Carlow Island dam seen in the background, on May 26, 1936. (*Source:* National Archives at Boston)

Making Tidal Power Concrete: The "action model" of the Quoddy Tidal Power Project being built, on July 13, 1936, by the Army Corps of Engineers. The twelve- by sixteen-foot, wood and cement model was primarily the work of Fred J. Nebiker (center, holding yardstick). (*Source:* https://catalog .archives.gov/id/7350883)

CHAPTER 11

Red Lights (1936–1938)

CONSEQUENCES

Recriminations followed Quoddy's death. Predictably, the *Boston Herald* did *not* print Governor Brann's denunciation of Senators Hale and White, but it did quote White calling Brann "ill-mannered, ill-tempered and . . . ill-informed." White also asserted, "No member of the Maine delegation in the Congress, Republican or Democrat, was consulted as to the wisdom of the course adopted or as to the form of the resolution of the amendment. The entire delegation had heretofore sought to prevent a legislative battle over Quoddy and had appealed to the President to go forward as was his present clear right." Summing up, White said, "This amendment was designed to shift responsibility from the President and those with whom he advised and to furnish the President with an alibi for whatever happened."[1]

More accurately, *White wanted Quoddy—without responsibility*.

Governor Brann, who was running for White's Senate seat, challenged him to publicly debate, "Who betrayed Quoddy?" White evaded him.[2]

Senator Hale also avoided cross-examination, so let us now examine his written statement. One's first impression is of a man acting logically and in good faith. But under scrutiny one realizes Hale blamed Quoddy's demise on the Democratic senators who had voted against *him* on the basis that *they* should have voted for Quoddy, and the president should have used *his* executive authority to fund Quoddy.[3]

Hale wanted Quoddy—without responsibility.

307

The cynical view was that the whole series of political maneuverings—from the surprise removal in February 1936 of the $29 million the president had requested for river and harbor projects to the separation of Quoddy from the Florida Ship Canal in the Robinson resolution—had been a Republican Party setup.[4] The Republican's motives were to embarrass Roosevelt and Brann, to strike a blow at the New Deal and the president's public power policy, and to boost Republican chances in the upcoming elections. Senators Hale and White were allegedly co-conspirators in the plot: the well-placed assassins. The cold-blooded calculus was that if Quoddy lived, the Democrats would sweep the elections—but if Quoddy were killed, the Republicans could regain control of Maine and the nation.[5]

When Hale's statement is put in context, a convincing case can be made that the suspects, the Republican Party and Senators Hale and White, were in fact guilty of killing Quoddy. It wasn't that Quoddy lacked merit. It wasn't an accident. It wasn't manslaughter. It was premeditated murder. Murder for personal profit and political advantage.

There was, however, no trial, no competent court to hear the case. Only the court of public opinion, and public opinion at the time was dominated by the power trusts and their co-conspirators in the newspapers they controlled. So let us now take a critical look at Hale's statement, point by point:

> As a matter of fact the dividing of the amendment had in my belief little or nothing to do with the result of the vote on the project. The whole issue depending on whether or not I was willing to give my support to the Florida ship canal with its cost to the government of at least $150,000,000 in exchange for the support of the friends of the ship canal for the Passamaquoddy project.[6]

Note first Hale's preconditioning for his argument, "As a matter of fact . . . in my belief," thereby trying to turn opinion into fact. Second, he falsely and egotistically frames the issue as his personal support for spending $150 million on the Florida Ship Canal—when the issue was the creation of a nonpartisan commission of independent engineers to review Quoddy and the canal. Hale's contention that by voting for the Robinson resolution he would be "selling" his vote is specious: if an independent review of Quoddy was valid, then it was also valid for Florida.

Third, both projects were work-relief projects. Work-relief funds had already been authorized by Congress, so funding for these two projects would not be incrementally more money; it would simply be for a specific use instead of an unspecified use. Regardless of what the money was spent on, the need for relief existed, especially in Maine.

> I did not feel that I could consistently or honorably become a party to any combination requiring my support of the two projects, in one of which I did not believe. To have done so would have amounted to a sale of my vote.[7]

Hale's statement is disingenuous for two reasons. He had *already* made himself a party to the combination by going along with it for a month. Proposing the separation at the last minute was not an act of honor but an ambush.

> It remained for me to select my course of procedure. Three courses were open to me. I could, without defining my position, wait until the Robinson amendment in the form in which it was introduced came to a vote in the Senate and then vote against it. Whether or not any trade had been made beforehand, and I have already stated that it had not, I should, and I think properly, have been open to the accusation, through my silence, of acting in bad faith and the amendment in my opinion never could have carried under such circumstances with the Passamaquoddy project included.
>
> The second course that I might have pursued was to declare my position and allow the amendment to come to a vote with the Senate having the knowledge that I should vote against it. There is little doubt that this would have led either to defeat of the amendment or to its being divided at the request of some Senator hostile to the amendment or friendly to the ship canal.
>
> The third course was the one followed and to my mind was the honorable course to pursue. I gave notice very definitely of my position on the floor of the Senate. I urged the merits of the Passamaquoddy project in which I believed and I asked for a separate vote on the two projects, at the same time announcing openly that I could not support the ship canal proposition.[8]

Hale implies there were *only* three courses of action. This is patently not true. There were many possible actions during the four weeks leading up to the vote, the clearest being to vote for the Robinson resolution because it would authorize only the nonpartisan reviews of the projects—not the approval of the Florida Ship Canal, as he falsely implies.

Hale's contention that splitting the Robinson resolution was *inevitable* was obviously not true, but it *might* have happened. Likewise, his assertion that a *stand-alone* Quoddy would thereafter be voted down was not a foregone conclusion. According to Governor Brann, "Had Senators Hale and White supported the Robinson resolution as it existed or in its spirit, it would have passed the Senate with a majority of at least nine votes. Quoddy was doomed in the Senate when the Maine senators made it a sectional fight. The success of Senator Fletcher's measure upon two occasions in the Senate indicated that, had Quoddy been joined, both projects would have passed successfully."[9]* Or, as an anonymous analyst in the *Sentinel* said, "Any Senator could compel a division of Quoddy from Florida. But only a Maine Senator could hurt Quoddy by making that motion. So Hale made it, instead of leaving it to Vandenberg or Hastings."[10]

As for Hale's concern that he not be charged with "acting in bad faith," you be the judge.

Many people hoped or believed President Roosevelt would come to the rescue and save Quoddy. They tried to go on with their lives as normal, despite the turmoil. Dam construction continued. The playground and the tennis courts were being finished. Softball was in full swing, the fencing team on point, and the reading group engrossed. Quoddy Village's Little Theater was staging its second production: Oscar Wilde's comedy, *The Importance of Being Earnest*.[11]

For several months, informational material about the tidal power project had been on display in the second-floor lobby of the Administration Building. It was popular with residents and visitors alike. Given the display's popularity, the army decided to erect an Exhibition Building for the growing collection and crowds.[12] The main attraction would be a twelve-by-sixteen-foot "action model" of the project. It would be filled with water, which would rise and fall like the tides; miniature sea gates would open and close; and a miniature generating plant would light up. The building would also serve as the bus station waiting room, with "comfort stations" and tourist information.[13]

One visitor was Representative Oliver H. Cross (D) from Waco, Texas. After the criticism of Quoddy in Congress, he decided to see it for himself. He and Mrs. Cross made the long trip by car and were taken to all the relevant sites by Captains Sturgis and Lord. Cross was impressed: "From a commonsense viewpoint, it is one of the most worthwhile projects that could be put through. . . . Quoddy will be a great asset to this part of the country. It is very logical that furniture factories and pulp mills will locate here when the power is ready for wholesale production." Mrs. Cross, who was especially interested in Quoddy Village and the dormitory, where they were overnight guests, said she was very much surprised at the furnishings in the dorm, which were in "such good taste yet so evidently economical."[14]

Senator Frederick J. Hale, from Portland, Maine, never visited Quoddy.[15]*

Normal lasted two weeks. In mid-June came the dreaded phone call: cut another 1,800 employees by July 1. Keep only 300 to finish up.[16]

It was a steep decline from a high of 5,484 employees in December 1935, of which 3,095 were from Washington County.[17]

There would be no Independence Day celebration in Eastport this year. Instead of a joyous parade of floats and bands, there were lines of cars and trucks grumbling away, a funeral procession for a dying town. Instead of explosions of joy kicking off construction, there would be the hollow sound of a community imploding.

Quoddy Village was emptying fast. To keep community services up as long as possible—the hospital, fire department, police, and dining halls—the army asked those staying to move out of the apartments and model homes, into the dormitory so they could close sections of the village, save some money on maintenance, and keep more people employed for a little longer. Cash from the sale of now unneeded equipment and supplies could also keep a few more people employed and housed for a little longer, maybe through September, maybe until Christmas.[18]

Besides mothballing, the work was to finish, as far as funds would allow, the small dams already begun. But there wasn't enough money. The dams wouldn't get their clay blankets or riprap armoring. They would be left unprotected for the waves and tides to worry.[19]

Worry was what everyone did. They had lots to worry about.

Under the gloom there was still a spark of hope. Hope that their champion, President Franklin Roosevelt, the man who had the power of the purse, would pull them through. It was not a false hope, but real hope, one that *could* come true.

The rumor from Washington on June 29 was that the president would make the final decision on Quoddy's fate—to fund it fully, to force the issue on Congress in the next session, or to shut it down completely—after his sailing vacation to Campobello in July.[20]

With so much uncertainty, Roscoe the booster used the *Sentinel*'s front page on July 1 for an editorial encouraging everyone:

HAVE FAITH IN THE PRESIDENT!
News concerning the continuance of the Quoddy project is becoming more and more disheartening with each day's developments. . . .

Very considerable private investment in property improvements and extensions of mercantile stocks and equipment throughout all this region will be very largely lost if the project is stopped, and bankruptcies, failures and worse can be surely expected to follow such a disastrous surrender to an interested opposition. . . .

[The president] still retains entire freedom to allocate any sum he may determine, from the abundant funds still available under the Relief Act of 1935; and he may also allocate under the current Deficiency Bill subject only to the provision that the sum so allocated be sufficient to complete the project. . . . It is to be hoped that he will choose the latter alternative . . . and take it definitely out of the range and charge of political maneuvering.[21]

"As for the cost," the *Lubec Herald* pointed out, "one first-class battle-ship, good for only about eight years, costs about $35,000,000. After eight years it is blown up. Fair minded men however opposed to Quoddy must admit that it would be good for a longer period than that."[22]

The president was expected to sail to Campobello on July 24. Quoddy supporters had to stay calm and carry on for three more weeks.

"Shut it down." The dreaded order arrived in Eastport on July 7, three weeks before the president himself.[23]

With unusual sympathy, the *New York Times* reported, "President Roosevelt, apparently with good intentions toward the $36,000,000 work relief engineering job at Eastport, but harassed by Congressional disapproval, has had to say 'no' when asked for more money. . . . Since the open announcement of Quoddy's death, Washington, and the construction scene itself, have buzzed with ideas regarding the use or disposal of the million-dollar village."[24]

"In this town of disillusion," the *Boston Herald* said, *oh so sympathetically*, "none feel the end of Quoddy more bitterly, or more silently, than the army engineers who were hampered and hamstrung in their efforts on a job that almost immediately became the football of politics."

Nonetheless, the Army Engineers immediately said who was to blame for what the *Herald* dubbed "the most colossal of the New Deal's failures." According to the *Herald*, the army blamed Dexter Cooper: "They found Cooper's plans virtually useless" and "too expensive." According to the *Herald*, the army also blamed the president: "[T]he project is a failure because irregular procedure was followed in assigning money for its construction." The army also wanted the *Herald*'s readers to know that "it was not the army that lost $7,000,000 of the public's money."[25]*

A few of the Army Engineers continued working, on double shifts, on the one thing that could be finished in time for the president's visit that might change his mind: the "action model."[26] It was hoped the action model would help experts and laypeople alike understand how tides could be used to generate electrical power. It, the blueprints of the powerhouse, tide prediction graphs for years into the future, and photographs of construction could explain Quoddy's practicality. Together with a tour of the model village, Quoddy could demonstrate what prosperity in Maine could look like. That, it was hoped, would have an intellectual impact on the president as much as the emotional appeals on each and every face of the thousands who would greet him.

Colonel Fleming said Quoddy could generate 50% more power than previously reported. Fleming's statement was "surprising and of intense interest to the general public," according to the *Sentinel*.[27]* That the army's Plan D could produce up to 450,000 hp—at 2 mills per kW— was *huge* news, so *why wasn't it reported until July 15?*

Perhaps because a lower output would make the unit cost higher— and higher output would make the unit cost lower—and some liked high-priced power.

The most recent report on Quoddy's output had been published by the *Portland Press Herald* on April 25: "Quoddy power will cost 8 mills per kw. . . . Cost from the international project would be 4 mills," while a kilowatt of Niagara power cost 3 mills and Bangor Hydro-Electric (BHE) was charging 15 mills.[28] Even with the army's huge cost overruns, Quoddy power would still be about half the price of BHE's power. Such an accomplishment would completely disrupt the electric power market in New England and establish a yardstick for fair prices.

Taken together, Dexter's March 10 statement that he had commitments to purchase Quoddy's *entire* power output, *and more* if it were available, and Fleming's statements about Quoddy's huge power output and low cost should have been enough for any *independent* commission to find in favor of Quoddy's economic feasibility, even without considering its justification as a work-relief project.

Information is valuable if it is timely. For Quoddy, it was *deliberately* released too late.

Paul F. Kruse, the top civilian engineer working on Quoddy, jumped ship from his habitude on Snob Hill. He landed at BHE.[29]

The president's cruise in July 1936 had the same purposes as his cruise in 1932: both were pre-campaign vacations and publicity stunts to show Franklin Roosevelt was fit to be president. Navigating Maine's notoriously rocky coast and the open ocean would demonstrate his skill and stamina—and the press would eagerly cover every moment. This trip would include a visit to Campobello. From there he could see the half-done Treat-Dudley Dam, one and a half miles away, and Eastport's old factories, a mile farther. Now and forever after, when sitting on his porch he would face his New Deal baby, his problem child, fondly or otherwise called "Quoddy."

On this trip he would be sailing with his sons James, Franklin Jr., and John, aboard James's 56-foot schooner, *Sewanna*. The flotilla to follow included the 165-foot presidential yacht USS *Potomac*, which would carry his executive staff; the 314-foot destroyer USS *Hopkins* as military escort; two 50-foot navy speedboats for picket duty; two seaplanes to ferry messages; two whaleboats for close-quarter guards; and the press, aboard the schooner *Liberty*.[30]

The press knew where the *Sewanna* was being fitted out but not what ports it would visit. That kept the crowds down and satisfied some of the security concerns, while keeping much of Maine and the Maritimes hopeful for sightings.

The trip got off to a bad start. As the boys headed *Sewanna* out of Portsmouth, New Hampshire, it was hit by a fishing boat. The damage was not too great, so they kept sailing. After they arrived at Rockland, Maine, a repair team worked through the night, and *Sewanna* was ready to sail again the next morning when "Dad" arrived.[31]

The longest leg of the cruise was a 125-mile, 30-hour crossing from Maine to Cape Sable Light, Nova Scotia. The last leg was a four-and-a-half-hour sail, tacking all the way, from L'Etang Harbor, New Brunswick, ten miles southwest to Welshpool. As the flotilla came into view at the north end of Campobello, most reporters believed the president was aboard the *Potomac*, and they wrote their stories *just so*. But the playful commander in chief had continued sailing the *Sewanna*, with the Secret Service following like ducklings.[32]* He was welcomed home on July 27 by an honor guard of Royal Canadian Mounties, resplendent in their red dress uniforms.[33]

The president of the United States had serious business to talk about with the premier of New Brunswick, and there was no better place to do so than home turf for both: Campobello. To make the mood even more convivial, on July 29 Franklin hosted a picnic on the beach and invited several hundred[34] friends, neighbors, local dignitaries (including Councilman Emery), and the press. There, the elite broke bread with the commoner and ate roasted frankfurters and cake off paper plates.[35] "As the President talked, while seated on the sand, with his back resting against a boulder, a strong breeze blew his hair over his forehead, which was deeply tanned from the fortnight's cruise," observed the *New York Times*.

At this "home folks" reception, the president said, "Quoddy will be completed. I believe in Quoddy and I believe you do, too." He added it would be necessary to "educate the people of this country and of Canada" on its benefits.[36] In a loquacious mood, the president then revealed information about Quoddy that was news to everyone but Dexter Cooper. It was first published in the *Times* on July 30:

He pointed out that the Passamaquoddy project was based on a conception that industry would continue to increase its demand [for power] by 10 to 15 percent each year, thereby requiring constant enlargement of all sources. Use of the thirty-foot tides in the Bay of Fundy was considered as long ago as 1921, when small experiments in tidal power development were begun on the Severn River in England and at St. Malo on the coast of France. Mr. Roosevelt said that at that time, well before he was either Governor of New York or President, he talked with Owen D. Young, chairman of the board of the General Electric Company, with the result that the company made a rough survey and reached a conclusion that the topic would be well worth studying later, although the lack of demand for power thus created would not justify the work then. In 1926–7 Dexter Cooper, an engineer who was largely responsible for the current operations, made a survey financed by the General Electric Company, the Westinghouse Company, and the Aluminum Company of North America and reached substantially the same conclusion. Finally, in 1933, government engineers made a third survey and pronounced the project feasible, the President said.[37]*

This was the first time it had been *reported* that General Electric had considered financing the Quoddy project in 1921 but turned it down. However, we know it was in early 1924 that Dexter moved his family to Campobello, having received substantial investment from unnamed sources. At the Public Works Administration's Technical Board of Review hearing in 1934, it was revealed that GE, Westinghouse, Boston Edison, and Middle West Utilities (MWU) had invested in Dexter's company to study Quoddy's potential, but not the dates or amounts of those investments. It is also interesting that the president said Alcoa had contributed funds for the Quoddy survey in 1926–1927, as Alcoa was not mentioned as an investor in reports of the PWA's review. Conversely, the president did *not* mention the investments that were made by the Insulls through MWU and Central Maine Power.

In 1933, when President Franklin Roosevelt began considering Quoddy as a public works and public power project, Quoddy shifted from being a huge *competitive* threat to Maine's private power monopolies to an *existential* threat to the entire private power industry. If publicly owned power projects like Quoddy and the Tennessee Valley Authority (TVA) were allowed to live, and replicate, they would reduce or eliminate the monopoly pricing power of the privately owned utilities. Quite possibly, this situation could also lead to the end of the private power monopolies, through financial pressure, nonrenewal of their charters or dam site leases, "un-bundling" of generation business from the transmission and distribution businesses, expropriation, or even nationalization. Much changed for Quoddy—for the worse—when it became President Roosevelt's pet public power project, for Quoddy became a key battle in the national life-or-death fight for electrically generated *economic power.*

Let us now consider the last sentence of the *Times*'s article: "Finally, in 1933, government engineers made a third survey and pronounced the project feasible, the President said."[38] This is also confusing, as the only report on Quoddy produced by the federal government in 1933–1934 was the negative Federal Power Commission (FPC) report (by McWhorter) that resulted in Dexter's loan request being denied by the PWA in March 1934. It wasn't until January 1935 that the positive reports—Sills, PWA-Hunt, and Maine State Planning Board—were delivered to President Roosevelt.

When President Roosevelt made these statements, he knew purchasers were waiting for *all* of Quoddy's power—*but he didn't say who they were or even that they truly existed.* He also would have known of Quoddy's larger-than-expected generating capacity and the relatively low cost of it. Thus, there on the shores of Quoddy Bay on July 30, 1936, FDR had all the facts he needed to support an executive order to go ahead with Quoddy, and he had the funds available to do it. . . .

The picnic over, the president was asked whether he would visit Eastport to inspect Quoddy in person. "No," he said; he would probably just look at the dams from aboard the *Potomac* as he passed by on his way to a meeting with Canada's governor-general in Ottawa. Thus, Franklin Roosevelt would never see the model village with "doll-like" houses, which was becoming a tourist attraction; or the "ostentatious" Administration Building; or the dormitory with "love seats"; or the thrones on "Snob Hill"; or the "action model" the engineers were working overtime to have ready for him—and he would never face the people of Eastport.

It looked as though he, the president of the United States, would be like Senator Hale of Maine—and never bother to look on his "baby."

While Franklin was sailing, his advance team was scoping out Eastport and Lubec. Rexford Tugwell was presumably considering whether Quoddy Village could be repurposed to house displaced persons from the dust bowl, but he wouldn't say. Harry Hopkins, the new head of the Works Progress Administration, also took the tour, after which he said,

"I feel that Quoddy has generally been misunderstood, by myself and the general public. I am certain that something can be done with the village and that it will not be left to rot to pieces just because Congress has refused the funds needed to finish the power project."[39] The following day he went further: "I think it would be tragic if this magnificent undertaking should be halted in mid-air. I was very much impressed with what I saw and I have every confidence in its engineering possibilities. If more people, particularly the critics, could see the place as I did, I am sure that much of the opposition would be minimalized." When pressed, he admitted, "It is true that I talked about Quoddy to the President today at Campobello, but it is impossible to talk about anything else in Maine. Governor Brann eats, sleeps, and lives Quoddy."[40]

Eleanor Roosevelt arrived at Campobello the week before Franklin. In her hugely popular, nationally syndicated column, "My Day," on July 25, she remarked, "All the way up through Maine people kept telling me things about Quoddy Dam, what strange things had been done there, why did the houses have no chimneys, how could anyone expect to keep warm without fires in this climate, how could anyone expect ordinary people to pay for houses which cost $15,000 apiece, and what was the point in getting old masters to hang on the walls, and the most expensive furniture and blankets!"[41] Since confusion about Quoddy was still rampant, the First Lady went to investigate for herself.

Colonel Fleming escorted the First Lady and her friends to his residence on Redoubt Hill for a luncheon and then on a tour of Quoddy Village. Of the furnishings there, she said, "Mrs. Blodgett who was in charge deserves great credit for she has done a charming job economically."[42] She then went to see the "action model," accompanied by Roscoe:

> She made herself thoroughly familiar with the working and processes of harnessing the tides for generating electric power. Pleasant and conversational to reporters who approached her at the Village and in Eastport, Mrs. Roosevelt repeated that she had had a very different conception of Quoddy thru the widespread misrepresentations in the press. Seeing the project as it really is, she was keenly interested, asked many questions, and apparently left with a distinctly favorable impression of the development. To a *Sentinel* reporter while she was shopping in Eastport that afternoon, Mrs. Roosevelt made the interesting remark: "Why is it that neither of Maine's Senators have ever visited the project? One would think that this being in their own state they would have found time to look it over."[43]

The week before, another famous person had stepped ashore at Wadsworth's dock in Eastport. Unannounced, the prime minister of Canada, William Lyon McKenzie King, arrived on a Canadian police boat, bought a copy of the US National Ocean Service Chart no. 801 of Passamaquoddy Bay at Wadsworth's store, and then headed out into the bay.[44]

The First Lady of the United States came to see Quoddy. The prime minister of Canada came. A senator from Texas came. Numerous congressmen came. Untold thousands of tourists trekked all the way across Maine and much of the nation to see Quoddy. But not Senator Hale of Maine.

And the president of the United States?

"You should see it," the First Lady of the United States must have said to her husband. *"Take the tour. You will regret it if you don't."* The next morning Franklin told the Secret Service to take him to see Quoddy Village and the dams.

As Philip Fleming recalled in an interview years later, there had been some "apprehension" about the president visiting Eastport since "the people were in an ugly mood" at the sudden cancellation of the project and the financial hardships they now faced. But that was all the more reason for the man responsible to go in person rather than shy away.

Colonel Fleming anticipated his commander in chief's interest. When the president's secretary, Marvin "Mac" McIntyre called Fleming to requisition a seven-passenger touring car for the president, Fleming told his old friend Mac, "Go to Hell—the nearest car of that description is in Bar Harbor, 150 miles away."

"Well," said Mac, "is there any car at all available?"

"There is one sitting out in front of the President's home at this very moment: my Oldsmobile."

Sometime before, James Mullens, who had never been outside of Maine before becoming Fleming's driver, had asked his boss, "[D]o you think the President will ever drive in our car?" Understanding the hope in the question, Fleming had replied, "Jimmy, you go over to the President's home and park outside the door for any driving call he might give you."

Following Mac's call for a car, Fleming went to Campobello to join Mullens as the president's tour guide. "I arrived in time to witness the President being carried out to his Oldsmobile." Fleming also discovered

Mullens had a three-day growth of beard on his face, having slept in the car, waiting for his chance to meet the president.

"Mr. President," said Fleming, "this is James Mullens. He's been waiting for an opportunity to drive you around."

"Hello, Jimmy," said the president and shook his hand.

Fleming sat in the back and they drove down the road to pick up the president's mother. While they waited for her, Fleming happened to look at the gas gauge. It was near empty.

"My stomach sank," Fleming recalled, "and I'm sure Mullens' did too. But he thought we would have enough for the fifty-mile trip, and in the absence of any other alternatives, we had to take the risk of suddenly being stalled in the middle of the road for want of gas."

They left Campobello followed by a Secret Service car and crossed Lubec Narrows "aboard a small and very old ferryboat" that could carry just two automobiles. Little did they know that ten days before a truck loaded with fish had crashed through the floorboards of that very same "very old ferryboat."[45]

Safely across the Narrows, the president found Lubec decked out with flags and buntings, and "it seemed that everyone in the area knew we were coming," recalled Fleming. "The streets were all lined with cheering crowds, waving greetings to the President. Roosevelt was sitting on the right side of the car and this brought a frequent reminder from Mrs. James Roosevelt, 'Franklin, there are people on the left side, too. Wave at them.' And he did."

On the long road around Cobscook Bay to Eastport, the president asked Fleming numerous questions. "I am sure I responded with the most stupid answers," admitted Fleming. "My eyes were glued on the gas gauge."

Leaving Perry, they traveled over the Pleasant Point and Carlow Island dams, entered Quoddy Village, and drove past the neatly trimmed model homes, the big white apartment buildings, the dormitory, and the stately Administration Building, finally coming to a stop at the Exhibition Hall. There the president got out of the car, walked into the building, and spent twenty-five minutes studying the displays. The "action model" particularly interested him. As he watched the artificial tide fill the "high pool" in the miniature Cobscook Bay, the filling gates close, the emptying gates open, and the powerhouse light up, he asked many questions about the All-American Plan and what would change if the project were to be expanded into the International Plan. Obviously enthused, the president waved to the crowd gathered around his car and called out to them, "When I get back next year I hope that it will be in operation."

The president's car, now followed by many cars,[46] drove up Redoubt Hill and around the army officers' houses. "My mother-in-law was visiting us at the time," recalled Fleming, "When we drove past our home, she left the front porch where she had been sitting and went into the house rather than look at the President. She was the staunchest of Republicans."

Heading into Eastport proper, the president was greeted by most of the town's population, who had waited for hours, eager to see the most important person in the land. He waved but did not stop until he got to the Customs pier, where the *Potomac* stood ready. There he greeted a small delegation of dignitaries before boarding and then waved from the bridge as the big gray yacht pulled away.

"At last, the trip was over," recalled Fleming. "I left the dock, got back in the car and directed Mullens to take me home. Two blocks away from the dock, we ran out of gas."[47]

And Quoddy?

DEMOBILIZATION

In the politically charged summer of 1936, Quoddy's supporters were shocked by hot news flashes, alternately enlightening, then infuriating, then vindicating, then humiliating.

The New England Power Association (a holding company)[48] recorded on July 15 its all-time highest demand for electricity. "This is remarkable," the company admitted, "because such gains are not commonly to be looked for in mid-summer. The only adequate explanation would seem to be that there is a heavy demand for power for industrial uses."[49] This, during the Depression.

Apparently the president was right about America's ever-growing demand for power.

The TVA announced on July 23 that Alcoa was the first company to sign on the dotted line for the purchase of electric power, 20,000 kW, from the TVA's year-old Norris Dam.[50]

Apparently Dexter Cooper was right about the aluminum industry's demand for power.

The Light Metals Bureau, an American trade association, on July 28 said Soviet Russia would produce 80,000 tons of aluminum in 1937, or three times as much as the United States. "Russia's amazing achievement," they explained, "is based mainly on ingenuity in making use of low-grade ores and in utilizing huge resources of cheap hydroelectric power. . . . [T]he Russians plan to speed up production to manufacture 300,000 tons of aluminum annually" at two plants, particularly the Dnieper Dam—built by Hugh and Dexter Cooper. The Russians had pioneered the new smelting process seven years before—when the Coopers were in Russia.[51]

Apparently Hugh Cooper was right about Russia's ability to catch up with the West, and Dexter Cooper was right about Russia having an efficient way to make aluminum.

The *New York Times*, which had mocked Quoddy as akin to "bathroom fixtures of platinum,"[52] ran an extraordinarily different article (also without a byline) on August 2, 1936:

THE PRESIDENT ON POWER

Though engineers have damned it for economic reasons, the President has lost neither faith nor interest in Passamaquoddy. It would be unjust to attribute his declared intention of harnessing the tides solely to improvement of his political chances in Maine. If he had his way the whole Bay of Fundy's rise and fall would turn the wheels of industry, drive trolley cars, milk cows[53]* and wash dishes. . . . Power engrosses the President. There can be no doubt about it.

In his insistence on publicly owned hydroelectric plants—"never shall the Federal Government part with its sovereignty and control over its power resources while I am President." . . . he recognizes . . . the generation, distribution, and widest possible application of electric energy transcend any private interest.

"Electricity is no longer a luxury—it is a definite necessity," was the burden of the Portland [Oregon] address of Sept. 21, 1932. . . . The President has gone far beyond the public utility companies in his conception of what the consumption of electricity can be. The Tennessee Valley, Passamaquoddy, the Northwestern projects have all been condemned because they seemed economically unjustified. The President believes in them and others to come because of a conviction that there cannot be enough electric energy, provided it is cheap. He relies not on statistics but on his profound faith in the need of power.

About the best piece of long-time planning for which the President should receive credit is that which gives expression to his power policy. The development of hydroelectric energy is seen as part of a larger problem that involves flood control, soil erosion, afforestation, elimination from agricultural use of marginal lands, the decentralization and diversification of industry. . . . To the President the abundant life is evidently the electric life.[54]

Apparently President Franklin Roosevelt's vision was transcendent, not lunatic. He was leading the American people out of the Depression and into the Abundant Life—the Electric Life.

Consider now what Bostonians were reading in July 1936. They would benefit the most from the low price of Quoddy power, and yet they were being told:

- Quoddy is "a symbol of federal competition with private industry. . . . It is a reminder of the President's willfulness and obstinacy, his insistence on going ahead with a pet project regardless of solid advice to the contrary. . . . This chimerical teat at Eastport was designed primarily to 'sweeten' the Maine voters for the September elections."—*Boston Herald*[55]*
- "In this national administration so assiduously devoted to the commission of outstanding blunders, the Passamaquoddy Tidal Power Development sticks out like a sore thumb. The thumb is attached to the 'Gimmee' hand of the tax collector stretched palm up across the map of the United States."—*Boston American*[56]

Those who benefited from high prices, no competition, and scarce supply were successfully controlling the narrative in most of the nation. The public was caught in the echo chamber of lopsided reporting and reverberation.

More "porky pies"—Cockney for "lies"—were published about Quoddy. The stinker first raised its snout in the *Bangor Daily News* on August 3, without a byline. It was then bellowed, echoed, and fattened for market by papers across the country, including the front pages of the *New York Times* and the *New York Herald Tribune*. The tale wagged that when Quoddy was fully staffed, the army ordered 500 pies and 800 loaves of bread to be delivered daily. But when the army's staff and appetite waned, their order for pies didn't. With pies piling high, they fed them not to hungry people but to Farmer Pottle's pigs in Perry. High on the hog, the swine dined in style on eight kinds of pie—apple, peach, rhubarb, quince, squash, plum, cherry, and cream. The press gobbled it up without checking whether there was any meat on the pigs' tale and couldn't resist larding it further.

Even the Los Angeles–based sportswriter George E. Phair took a swing at far-off Quoddy with a witty ditty that began, "In dear old Maine a ruin stands / That once was Quoddy Dam . . . " and ended with, "The butcher, as he stops to weigh / A porkchop with his thumb, / Will wear a honeyed smile and say: 'What flavor—peach or plum?'"[57] Likewise, the Republican National Committee hee-hawed it into their daily press release, dressing it up with "This little piggy went to market / At a price too fancy and high."

The army denied the fib with facts and figs; 'twas naught but "malicious slander." Pigless Pottle of Perry protested his innocence, and the actual garbage gatherer solemnly swore he had never seen anything so sweet in the army's swill. But as usual, "the corrections never . . . caught up with the canard."[58] *Time* magazine finally got the dirt on who had started the journalistic swilliness:

> The story of pigs' pie was originally broadcast to various newspapers by the arch-Republican *Boston Herald*. The *Boston Herald* received it from Lawrence Thomas Smyth, of the *Bangor (Maine) Daily News*, who got it from John McFaul, an oldtime *News* correspondent in Calais, Me., who got it from a "farmer over in Perry." Said Newshawk Smyth in Bangor: "The way the yarn came in from McFaul it was written in a kind of wooden style. So I polished it up a bit before sending it along to the *Boston Herald*. Of course, if I was writing for the *New York Sun*, I'd have polished it up a bit more. . . . As a matter of fact the way the story came through it said six kinds of pie, but I jacked it up to eight. I figured the story must be pretty near right. There isn't anybody in Washington County with enough brains to think up a thing like that."[59]

The most sentient words to be heard, but only in Eastport, were in the *Sentinel*: "In reference to not feeding the hungry, Quoddy authorities ironically stated that they have been feeding 5,000 hungry workers during the winter and more would be able to eat now, except for such news articles."[60]

The army admitted defeat in the fight *for* Quoddy on August 12, 1936. In the press release, it didn't use the word "defeat," but it was abandoning its position and recalling the troops. Call it a strategic retreat. The Army Engineers who did believe in the project had been outgunned and outmaneuvered by the power trusts. They had been sabotaged by their own countrymen in the opposing political party, and maybe in their own ranks. More money may have been spent fighting Quoddy than building it. Quoddy's still-fresh corpse now lay dead on the battlefield, waiting for scavengers to pick it apart.

Understandably, the army focused on what it had accomplished. It had fulfilled its primary mission: It had provided 5.5 million man-hours of employment to thousands of men and women who desperately needed jobs. It had built an entire town, in no time flat, for 2,400 laborers in bunkhouses, and another 2,000 staff and family members in Quoddy Village. It had kept them all warm and fed over the bitterly cold winter. It had amassed a mountain of information about Passamaquoddy Bay and how to build the world's first tidal-hydroelectric power plant. The army had built all the infrastructure it would need for the heavy construction phase of the project. And it had built several smaller dams to provide valuable experience for the big dams yet to be built. The army may have started the project reluctantly, but it was now ready, willing, able, and apparently eager to finish the job. *But*, said the army,

> [i]t is apparent that a vast amount of misinformation has reached the public, including a complete misstatement as to the aims of the federal government. . . . It is confidently expected that the future will bring a more complete understanding of the public benefits involved, and will see a congressional authorization to carry Quoddy on to completion. . . .
> [O]riginal expectations . . . have been greatly exceeded. These studies have indicated . . . [Quoddy's] entire feasibility and practicability. . . . Though the Engineer Department's designs vary in many major respects from those developed by the early promoters of the project, their investigations fully confirm the public benefits to be derived from the development.[61]

Point blank: the army *could* build Quoddy; it was economically justified.

Few noticed the army's press release. Many more saw Kenneth Robert's article in the *Saturday Evening Post*, the nation's largest magazine, with three million readers. Disclaiming bias, Roberts wrote skillfully and evocatively, but his libertarian bent warped the piece, while he ignored root causes and exploited the victims to condemn the New Deal.[62]

Quoddy was to make its movie debut in the award-winning newsreel *The March of Time*. "The general tenor of the film," said the *Sentinel*, "is expected to be friendly . . . although some unfavorable material must probably be included since an effort was made to present the undertaking and general situation here with the utmost fairness. . . . Its publicity value . . . can hardly be overestimated."[63]

The exciting opportunity to tell their side of Quoddy was heightened when Eastporters were recruited to reenact receiving the news of FDR's $10 million allocation, and Roscoe got to voice his frustration on "another fool story about Quoddy." Eastport and its momentary movie stars could hardly wait to see themselves on the Acme Theater's silver screen.

The eight-minute episode, "Passamaquoddy," ran with set pieces on "The 'Lunatic Fringe'" (about Gerald Smith and similar demagogues) and the "U.S. Milky Way" (the milk industry's triumph over typhoid). It debuted on September 2 in 6,200 theaters throughout the English-speaking world. Tens of millions saw it.

Here are a few excerpts from the film, with emphasis and commentary added to highlight *Time*'s editorial choices:

[Narrator:] Eastern Maine's dismay at the collapse of the Passamaquoddy project is voiced by the State's Republican Senator, Wallace H. White [shown with posters for Republican candidates behind him]: ". . . the people of Eastern Maine looked to the construction of Quoddy with high hopes. They saw in the project employment for the distressed people of this area, they saw the rehabilitation of Eastern Maine, and Quoddy has become *a monument to disappointments*" [caused by the saboteur, Senator White!]. . . .

[Music: "Auld Lang Syne"]

Genesis of the Passamaquoddy power project was the friendship of three long-time members of Eastport's famed summer colony. . . . A life-long friend was his [President Roosevelt's] next-door neighbor, the former Gertrude Sturgis who married the *visionary promoter* [not "hydroelectric engineer"] named Dexter P. Cooper. An object of wonder to *promoter* Cooper as to every Eastport visitor was the tide. . . . *Puttering* [to go about casually or haphazardly] about the Bay, Cooper worked out a way to do it and *broached* [to raise a difficult issue; or when sailing, a sudden, uncontrolled, and dangerous turn broadside to the wind] his plan to private power interests. But after a $400,000 investigation, they rejected it. . . .

[Music: "Happy Days Are Here Again"]

When federal funds began flowing *freely* out of Washington for public works, Dexter Cooper [actually Senator Hale] took his Quoddy project to Administrator Ickes of PWA [shown tearing papers into pieces], who *promptly* rejected it. But one year later, Mrs. Cooper made a pilgrimage to the White House to try to persuade her good friend, the President. [Positive recommendations by the Sills, MSPB, and Hunt commissions are not mentioned.] . . .

[Music: upbeat]
 . . . $5 million as the President's *first ante* in this $37 million *game* [pejoratives]. *Discarding* Cooper's plans, with typical Army dispatch, they press into action. . . . Quoddy must have a *costly* auxiliary plant to operate. [shows image of the huge main powerhouse] . . . as . . . millions of *New Deal dollars go tumbling into the swirling tide.* . . . As expenditures *soar* to $7 million [only $6 million had been spent]. . . .

[Music: ominous]
 . . . [T]he *brief exciting dream* of Eastern Maine's Yankee Republicans fades.
[Shot of Roosevelt waving from yacht, followed by a staged shot of angry looking people on shore.]

[Music: "Auld Lange Syne"]
 . . . [H]owever *fine a gesture* [again alluding to vote baiting] to Downeast Maine, their Quoddy was squashed as a New Deal boondoggle ["boondoggle" is used five times]. Today finds the disappointed Eastporters returning to their old dependable ways of getting a living from the sea. [Movie ends with images of townspeople lobstering (in dresses and ties), people fishing from a small boat, and a man scratching in the mud for clams.]

Was this film made with "utmost fairness"? *Or is it condescending and deceptively edited propaganda? Is it balanced reporting of strengths and weaknesses, causes and consequences? Or is there a subliminal message: eastern Maine was led astray by promoters, Democrats, and New Deal dreams? Was the pairing of "Passamaquoddy" with the "Lunatic Fringe" and "U.S. Milky Way" just coincidence, or was it implying Roosevelt was a demagogue and Quoddy lunatic— whereas industry provided trusted solutions?*
What's your vote?[64]

Maine's elections for statewide offices were the earliest in the country and ballyhooed as the national bellwether, so the Republicans were determined to do whatever it took to win Maine. As the ads in the *Sentinel* demonstrate, the Republicans' rhetoric was clearly against the New Deal, but with little indication of what they would do instead, and their tone was strident and derogatory, bordering on desperate. In an unusually large one-third-page ad, Maine's Republican Party declared with words as lacking in substance as sentence structure, "That choice lies between the NEW DEAL with all its foolish experimentation. And that tried and proven form of Constitutional Government under which we of America in the past have enjoyed more liberty and attained a higher standard of living than any other people in the world." It then criticized "the staggering record of NEW DEAL waste and extravagance. Or its unfairness and discrimination."[65] The following week, there was a one-half-page ad imploring voters to not split their ballots but vote for the full Republican ticket, and the *Sentinel* printed its first-ever full-page political ad.[66]

The money spent on advertising did not change the minds of Washington County voters. They voted in record numbers for the Democratic candidates. But it didn't matter; they were outvoted by the rest of the state, and, consequently, all Republicans were elected: Brewster to the House, White to the Senate (by only 1,200 votes over Brann), and Lewis O. Barrows as governor.[67]

The *Portland Sunday Telegram* trumpeted the Republican victory in told-you-so fashion, saying the president's promise to continue Quoddy if reelected "fooled Eastport but no one else." Roscoe's reply was to criticize his own party's lack of a coherent plan, its ethics, and its methods, and to disagree about who was being fooled and bought. As for Eastport, "Confidence here in President Roosevelt's sincerity is and will continue unshaken by any of the events of the past five months." The people were grateful for what he had done and was trying to do, even if it wasn't all successful.[68]

Maine got an inkling of who was being bought and for how much when the required post-campaign reports were published. Republicans spent $116,338.21 ($2.1 million in 2020 funds), while Democrats spent $53,231.22. The Republican funds were mostly from out of state, with thirteen wealthy donors (at the legal limit of $5,000 each) contributing two-thirds of the total. White outspent Brann by four to one.[69]

As Maine went Republican, so Republicans hoped would go the nation.

Senator Vandenberg caught the scent of another secret report about Quoddy and an even more expensive estimate. He wrote to the president on May 18 asking for a copy.[70] The president evaded him.[71]

On September 17 Vandenberg wrote to the War Department asking for a copy of the as-yet-unidentified report.[72] The army asked the president what to do.[73] He had the army send another evasive letter stating that the report "accepted in general the plans proposed by this Department, and estimated the cost of construction to be slightly higher." In reality, the Durand report estimated the army's Plan D could cost as much as $72 million[74]—an $11 million increase over the army's estimate, and $40 million more than Dexter's—and coyly pointed to an alternative plan costing only $37 million.[75]

One should consider the circumstances of the Durand report. The Durand committee had been asked in January 1936 to critically review a never-before-attempted project purported to be impractical, unnecessary, and wildly expensive, and to estimate what it might cost. Thus, the committee assumed the worst and estimated high. Among other items, it included contingent allowances of up to 70% for construction of the dams based on concerns the dams would settle 60% into the clay seabed; plus they figured 10% of the fill would be lost to current "drift." The committee also decided the size of the navigation lock should be increased by 140%.[76]*

No wonder the president didn't want to release the Durand report. He knew only the worst-case estimate would be reported or remembered.

The 1936 national elections were in six weeks. The Republican candidate for president was Alfred "Alf" Landon, an oil company executive and governor of Kansas. Although politically moderate, his position relative to Quoddy was negative, and his policy vague: "[W]hatever I do in the matter of utility and power projects, when elected President, I will not use the money of the people to attempt to harness the tides!"[77] Contrast Landon's ambiguous policy with this analysis by the *New York Daily News*:

> One thing almost everyone agrees on—that President Roosevelt is the greatest orator to be heard today; and that in the last couple of weeks he has made speech after speech to which no Republican has given adequate answer. Of course, the President's foes go on to say, after agreeing to that much, that (1) he doesn't mean what he says, or (2) when he is sincere, the ideas he expresses are unsound, un-American, socialistic, communistic, Red, Asiatic-horde, etc.
>
> Anyway, the latest of these Roosevelt speeches state the electric power issue so clearly that anybody could understand it.
>
> This speech was delivered in Washington last Friday, before the Third World Power Conference. . . . The President, incidentally, spoke more like

an expert than did any of the experts whose remarks got into the newspapers. He also pressed the golden key which started Boulder Dam's turbines rolling out power.

The nub of the President's speech was that in electric power production the consumers' interests should be considered first, not the producers'.

The private producers of power take the opposite position. They insist that, having staked out claims on what were once the people's rivers and waterfalls, they are entitled to an absolute monopoly on those power sources. . . .

The President's insistence on breaking up the power monopoly for the benefit of power consumers is what makes the private power producers hate Roosevelt so bitterly. And their hatred has increased ten-fold or more since they have learned that he means business in this matter: that he can neither be bought nor intimidated—and that he can state his side of the argument more clearly and convincingly than any speaker or writer the private power interest have yet been able to dig up.

Reminiscent of the fight for Muscle Shoals, the *Lubec Herald* summed up the political-economic contest this way: "The Democrats say that only Roosevelt can and will finish Quoddy. The Republicans say if Landon is elected, private capital will finish it."[78]*

Would the power trust dig up a convincing speaker to challenge President Roosevelt? Would Quoddy be revived—and then stolen?

Franklin Roosevelt was reelected in a landslide, garnering 523 electoral college votes to Landon's 8, and sweeping every state in the nation but two: Vermont and Maine. As Roscoe the disillusioned Republican, conflicted conservative, and closet Progressive said, it was:

A REMARKABLE ELECTION

[T]he endorsement accorded Franklin Roosevelt's Administration is so overwhelming as to leave observers in both parties amazed, and perhaps a little troubled with all its implications.

The first and most impressive of these is that this country has made a definite and calculated choice between two widely divergent theories of government. For the time being at least, the people are demanding and will continue to receive leadership which emphasized the responsibility of the government to the individual, rather than that of the individual to the government. . . . Conservativism is out at this moment. The county is liberal and progressive, but not yet radical.

The second is the tremendous personal popularity of the President. To a degree not equaled even by that other magnetic and liberal Roosevelt who was President thirty years ago, he commands the confidence and affection of the people.

Only rabid partisans approve all he has done; but satisfaction with most that he has done, with the way in which he has done it and especially with the spirit in which he has attempted it, is the dominant note in the attitude of the electorate toward the President.

A third is that it is distinctly a personal, not a party triumph. Without Roosevelt, there would have been no landslide, probably no victory. He has developed a following that has broken clear of established political alignment, and become distinctly his own. The Democratic party, in a sense, has ceased to exist as such. It is now the Roosevelt party. . . . It was he that swung the party into office, and with this prestige goes, as it probably should, power and distinguished responsibility.

The American people believe Franklin Roosevelt will use that power with wisdom, sanity and moderation, otherwise they would not have conferred it on him, and he will continue to carry on as President with the good wishes and support of a nation whose course has been determined and may well lead to renewed strength and a more substantial, more equally distributed prosperity.[79]

Roscoe followed this editorial with a second, one-liner: "Yeah. Yeah. As Maine goes, so goes Vermont."[80]*

Rock-ribbed Republican Maine, still mostly controlled by the Old Guard, may have voted for Landon, but oddly frontierish Washington County voted for the frontiersman, Franklin. Maine picked the loser; Washington County, the winner. Thus, Roscoe could have said, "*As Washington County goes, so goes the Nation.*"

Perhaps Washington County's leadership was unappreciated because the people knew Franklin Roosevelt. They knew what *he* had gone through and that he appreciated what *they* were going through. They knew he had used every opportunity to help the people of Quoddy Bay. They knew his interest in Quoddy was sincere, not cynical. They saw what fierce determination had accomplished for a cripple. He was imperfect, but he had pulled through. He could pull them through. He could pull the nation through.

Perhaps, too, the people of Quoddy Bay understood President Roosevelt's public power policy better than any other community because they were small and frontierish and therefore felt the need more acutely and could see the solution more clearly. They had the contrast of private water and power monopolies in Eastport and public water and power companies in Lubec to help them understand the realities and possibilities. Their choices were clear, their decisions based on facts and experience, not propaganda.

Perhaps, too, Quoddy Bay benefited from a newspaper that still strove to live by its founding principles: *"Here shall the Press, the People's rights maintain—Unmov'd by influence, and unbridg'd by gain."*[81] While the *Sentinel* was subject to selection bias,[82] it was real news, not fake news.

Perhaps Washington County voters objected to money in politics. They hadn't been bought. It was Maine's politicians who had been bought.

Perhaps, too, FDR saw Quoddy as the quintessential New Deal project: It was Relief for the unemployed, Recovery through economic development, and Reform of corrupt industries. It was the exemplar of his effort to *democratize* power.

That's how the power trusts saw it. And that's why they would never stop fighting it.

"Demobilization" was the US Army's term for stopping what it had been doing. There were three main elements: transferring staff and equipment, closing up Quoddy Village, and wrapping up the dams. Captain Leehey assured the press, "There would be no loose ends."[83] Or loose change. When the Army Engineers officially left town on November 1, there remained $973,000 in Quoddy's account. They took it with them.[84]*

By most accounts, the professional and clerical staff had enjoyed working on the tidal power project and living at Quoddy Village.[85] While they were "assets" who could, and hopefully would, be reassigned rather than simply laid off, they were also friends whose partings would be sweet sorrow. Rather than drifting away, the engineers decided to go out in style. Just as they had officially taken possession of Quoddy Village at a midnight ceremony on New Year's Eve, they would kiss each other goodbye at another ball, on September 30, and hand the keys and their baby, just nine months old, to Quoddy's foster parents, the National Youth Administration (NYA).[86]

The Roosevelt administration had offered Quoddy Village to the University of Maine as a satellite campus, but it wasn't interested. Neither was the Massachusetts Institute of Technology or Philadelphia's Franklin Institute. But George W. Paulsen, principal of the Dwight Morrow

High School in Englewood, New Jersey, was interested. On August 6, he wrote to the president with an idea. After praising FDR's Civilian Conservation Corps as "one of the salvations of this period," Paulsen made his pitch:

> My proposal is that Quoddy be worked over into a great National Vocational Training School on par with the aforesaid [army and navy] defense schools. Quoddy is ideally suited for such a project—invigorating climate, rugged environment, broad physical adaptabilities such as power sources, natural agricultural facilities, industrial and commercial history. A National Vocational Training School might draw suitable personnel from all quarters of the United States in much the same manner as does West Point. . . . In the future a similar institution for women might be in order.[87]

President Roosevelt evidently liked the idea, as he marked each of Paulsen's six paragraphs with blue pencil. No doubt the president would have talked with the First Lady about it, and no doubt she liked it, as she was running a miniature vocational training program at the Roosevelt's estate in Hyde Park, New York.

For the Army Engineers, the most important task was making history. Having been promoted in August to acting district engineer on Lieutenant Colonel Fleming's reassignment,[88] on October 23, 1936, Captain Sturgis issued the army's final report on Quoddy, a 40-page narrative with many thousands of pages of appendices. The narrative is a summary of the project's history designed to fix for posterity the world's perception of Quoddy, and to justify the army's actions and expenditures.

The army's report began by noting it was unusually detailed because of the extraordinary public interest in the project, the "unusual manner in which the War Department was charged with undertaking the Project, and from the exceptional conditions that affected its prosecution."[89]

First the report fixed Dexter Cooper as the one to blame for what the army contended were inaccurate surveys, faulty plans, and unbelievably low cost estimates (despite acknowledging the army had done a preliminary review and accepted them).[90] The narrative didn't mention the army's shamefully low offer to Dexter but said the purchase negotiations caused delays and increased costs.[91] It asserted that "$25,000 *was to be* paid annually to Mr. Cooper *and his staff* to investigate the market possibilities for the power to be produced."[92]* But that wasn't what happened.

Repeatedly the army made the point that it had been ordered to build Quoddy *without* congressional approval (but by legal executive order), and this had short-circuited normal procedure, which would have included more detailed field research and cost estimates before approval.[93]

The army misleadingly summarized "the reports prepared during 1934 and the early part of 1935 by two groups appointed by the Governor of Maine" as saying they "recognized the project was costly and was not economic from the point of view of cheap power"[94]*—when in fact the Maine State Planning Board and the Sills Commission had concluded that Quoddy *might* produce competitively priced power, and markets *could* be expected to emerge, and the project was needed as a *relief* measure.

The report also stated some facts that were, at the time, nearly impossible to refute:

> Although Mr. Cooper has frequently stated on various occasions that . . .
> he has been able to sell all the power that could be generated in the initial
> development, there is no documentary evidence to substantiate this state-
> ment, nor has he made known to the War Department any information
> in regard to the identity of the proposed customers for power or the type
> of market to be supplied. His final report to the National Power Policy
> Committee [NPPC] dated July 1, 1936, contains no information whatever
> as to specific power customers for the Passamaquoddy Project market.[95]

Dexter's report to the NPPC *does* contain specific information about the companies that would purchase the power: two chemical companies and a "cellulose" (i.e., paper pulp) company. And he explains the competitive reasons for not naming the companies—to the army or its agents within the NPPC.[96] He did, however, reveal the names to someone more trustworthy.

The report detailed the "exceptional conditions" the army had to deal with: the fast schedule, Eastport's remoteness, the work-relief requirements, the civilian workforce, and the curious public. While not the largest or most important, the latter expenses were first in the project's cost summary and provided in the greatest detail.

The seven-page cost summary is the most revealing section of the report, though the categorizations are at times opaque. For instance, the officers' houses on Snob Hill are not in the cost of housing but listed as "operator's residences" in the costs of the tidal power plant; overhead is a big cost, but electricity is not itemized. And only with separate knowledge of the army's original estimates is it clear the army was going over its own budget on almost everything.[97]*

The third subject covered in detail was the army's decision to switch from Dexter's plan to the army's Plan D: switching Cobscook Bay from a low pool to high, abandoning the pumped storage reservoir, and building a diesel-fired auxiliary power plant. Again, the army skewered Dexter, despite its own flip-flops and illogical implementation.

Although "considerable propaganda had been aimed at the project," the army admitted it had "made few efforts to combat or refute same." Conversely, a "surprising acceptance and approval of the basic ideas involved in the project was evoked in the majority of visitors" by the "action model."[98] In the summer of 1936 alone, those visitors, who drove 300 miles through Maine to reach Eastport, patronizing gas stations, gift shops, restaurants, and motels all along the way, amounted to an astounding 180,000 individuals. Quoddy was in fact a huge benefit to southwestern Maine, even if southwestern Maine refused to admit it.

Hard for anyone to get their hands on, and never until now published, appendix A-3 revealed a lot. It is the audit conducted by the PWA in May 1935 of Dexter's finances from 1924 to 1935. Therein are the amounts invested in Dexter P. Cooper, Inc., by Westinghouse, GE, Boston Edison, and Middle West Utilities: $15,000 in 1925, $80,000 in 1926, $79,000 in 1927, $110,000 in 1928, $55,000 in 1929, $31,250 in 1930, and $5,000 in 1931, for a total of $375,250.[99]*

But the most important thing in the army's report is what *isn't* there. Unlike the press release issued by Colonel Fleming on August 12, the final report issued by Captain Sturgis says nothing about the relative cost of the power that could have been produced or Quoddy's economic justification. Quite the reverse. While the report is presented as black-and-white facts, *it invokes a tidal wave of red ink.*

Christmas Present, 1936, was like Christmas Past. Scrooge was hunched over in the cold counting pennies and the unemployed. The need for relief was great—and the resources tiny.[100]

In addition to a thousand charity and relief cases in Eastport, there were 500 people certified for work relief, but due to the federal quota system intended to wean the population off the dole, only 189 had been offered jobs. When they were on work relief, they had received "subsistence wages" and subsidized room and board. But the wages were not enough to accumulate any savings, and now there was "nothing to tide people over winter."[101] What's more, residents of other towns were still in Eastport, unwilling or unable to return home for lack of transportation, jobs, or hope. Quoddy, they figured, could still be revived, and if it was, they wanted to be the first in line. In the meantime, Eastport had to figure out how to feed and house these hapless hundreds.

It was just as cold and bleak on Campobello. There Dexter waited. It had taken him six months, but he had somehow convinced his corporate investors to turn over their expired shares in his company (now that the project was dead), and with that accomplished, he had expected prompt payment from the army for the $10,000 contingent on his clearing up the matter. After ten days of his request for payment being in the hands of the same army clerk who had dragged his feet for months on the previous payment, Dexter again wrote to James Roosevelt to complain.[102]*

Quoddy's closing ceremony echoed its opening, but with fewer people and more dynamite. This time the explosion would be more purposeful, but less dramatic. It would create 20,000 cubic yards of rock to repair the unfinished and already eroding dams between Eastport and Perry. The smaller material would just be dumped, and the larger material—boulders up to eight tons—would be fitted together like a giant jigsaw puzzle to form the dams' outer armor.[103]*

Rather than using successive rounds, the army decided to do it in one shot. Doing so would require round-the-clock drilling for more than a month, boring 50 to 60 holes 1,200 to 1,500 feet deep. For the opening ceremony, 600 pounds of dynamite was placed in shallow holes in the dirt to create a big show of flying debris. This final blast would make the first look "Lilliputian." This time, William Neptune, the former Passamaquoddy chief and now chief of the blasting team, would use 12,000 pounds of dynamite, placed deep in the drill holes. In theory, the explosive force would be directed outward, not upward. In theory, there would be little visible explosion and shock, but the interesting spectacle of a hill collapsing like an avalanche into a pile of rubble.[104]

Eastporters weren't convinced. Even in town, two miles away, they were expecting to get their "back teeth jarred."[105] Five miles away,

Lubecers tucked their precious bric-a-brac into their feather beds, and wired their mirrors to the walls.[106] Then they went to see what would happen. The explosion had been planned for New Year's Day. There were delays; it was now Sunday, January 3, 1937:

> A slight tremor of the ground, the sudden flash of a billow of white smoke . . . a dull double roar, and the much-heralded giant blasting operation at the Carrying Place quarry was over. It wasn't much to watch, especially as the driving rain of the forenoon had changed to a thickening fog, that obscured everything to the hundreds of watchers in cars parked a safe distance away. . . . A flying particle of stone struck the helmet of William Neptune . . . [and] a five foot long piece of six-inch pipe hurtling in a straight line like a bullet landed with terrific force at the side of the shack from which work on the quarry has been directed. No one was there, but had the projectile varied its course by about ninety degrees toward the south, it might have caused some few casualties. As it was, it didn't land far from the ambulance stationed there to take care of such contingencies.[107]

The show over, and the project over for all but a few, Downeasters drove home in the cold rain and wondered what the new year would bring.

Would their reelected and nationally popular president revive Quoddy?

Was the NYA program a temporary use of Quoddy Village or Eastport's booby prize?

LIMBO

"Watchful waiting" is the medical profession's prescription when they don't know what to do. That's what Quoddy's supporters in Maine were forced to do in February 1937 while the political machinations in Washington played out.[108] With the election over, it was hoped the partisan distortions would be tamped down, and realistic plans could be developed in the near term for Quoddy Village and in the long term for restarting the tidal power project.

A "blue-ribbon" committee recommended $100,000 to study Dexter's International Plan, *contingent* on Canada's cooperation. The committee included five members of the president's cabinet and two business executives. It was a clear indication of the administration's interest in Quoddy—but it wasn't $30 million for construction.[109]*

President Roosevelt was "going ahead full steam" on other public power projects, as Elizabeth May Craig noted in the *Portland Press Herald*, "regardless of opposition of private companies, and of the advice of some who think that cooperation might be the best method."[110] The president, she continued,

> is expecting a report in a week or so from the new Ickes committee on [the] sale of Government power at the Bonneville dam, and another report on general policy in regard to the sale of Federal power.
>
> That immediately brings to the Maine mind the question of Quoddy.
>
> Proposal for a Quoddy Authority gets a fishy eye, here in Washington, whether because the Administration doesn't want a Quoddy Authority at all, or whether it wants to wait until the Ickes committee formulates its general power-sale policy. . . .
>
> Dexter P. Cooper's arrival in Washington roused up conjecture as to whether or not he was called in to talk about . . . Quoddy power; . . . utilization of Quoddy Village; . . . or about the resumption of construction; or in relation to the general Federal power picture.
>
> But Cooper is like Ol' Man River—he don't say nothin'.[111]

"Among the endless pieces of equipment . . . shipped out from Quoddy," the *Sentinel* noted wryly,

> the dismantling of the Diesel generating plant at Quoddy Village is perhaps one of the most interesting since it affords plenty of food for thought for those whose indoor sport consists of weighing the motives and possibilities of each new development. . . .
>
> The big [53-ton] gray power unit, all nicely installed in a neat and even artistic station of its own, was designed to be used eventually for the production of power after the dams were completed, during those hours when production was nil. This meant that the unit was purchased and set up approximately four years before it was to be needed in its real capacity. In the meantime, it was put in order as an auxiliary plant in case of emergency,— i.e. sudden failure of the regular power supply line. In that role only has it functioned to date, and but a negligible number of hours are believed to have been chartered to its credit.[112]

So instead of generating its own power—from a prototype tidal plant as Dexter had proposed, or from the diesel plant the army had actually built—the army bought *all the power used at Quoddy* from BHE![113]* No wonder the army didn't specify what it had spent for electricity in it final report!

The indoor sportsmen wondered: *If the diesel plant wasn't needed, was it because the two-pool International Plan was being considered again? Had President Roosevelt found interest in Ottawa?*

The NYA had gotten off to an impressive start, constructing a new 1,000-person mess hall in September 1936,[114] and was about to break ground on a new gymnasium in November when the phone rang, and it was told to stop.[115]

Indoor sportsmen didn't know what to make of yet another sudden reversal. Was it bad news—there would be no NYA school and no jobs? Or was it good news—the administration was focusing on the long-term objective and deciding not to create near-term complications? On February 17, 1937, a headline read, "N.Y.A. Is Definitely Out."[116] One month later NYA was definitely in: Quoddy Village would be used for vocational training of up to 1,000 young men, age 18 to 25, starting June 1.[117]

The indecision in Washington was not just about Quoddy. While Roosevelt, the progressive Democrat, had overwhelmed the conservative Republican Landon, there had been five other parties on the national ballot: Prohibition, Union, Socialist-Labor, Socialist, and Communist.[118] Their presence indicated a deep dissatisfaction with the status quo, manifesting most dramatically in violent strikes at automobile manufacturing plants. Representative Brewster lamented, "Congress moves in an atmosphere of gloom and suspense. More and more of the menacing murmur of the social and racial discords that rend Europe are being heard in the United States. More and more citizens begin to sense that 'it can happen here' and may happen here as passions become more tense." Looking for the bright side, he encouraged everyone to "renew their efforts to have Quoddy . . . understood. This understanding must . . . be spread to all our visitors who shall come to see the project at first hand and be convinced by what they see. Thousands have come to Quoddy in the last year or two to scoff and have gone away with a perhaps unspoken prayer that the splendid vision of Dexter P. Cooper shall be converted into a reality that shall forever serve."[119]

Haste makes waste. "Quoddy Village has gone to the dogs," barked out-of-town papers. "But in reality it is on its way to the cleaners," whistled the *Sentinel* to the cathartic work:

> Yesterday, twenty [houses] were thrown wide open to the warm spring air, the first step in the reclaiming process. And it's no dream—those houses have got to be reclaimed before they can again be occupied by human beings. The extent of the damage wrought by leaky roofs, condensation, poor construction and the remarkable shortsightedness of somebody in not installing shut offs for steam and water for each house . . . is unbelievable.
>
> A casual stroll thru those open abodes revealed shocking conditions. Smooth hardwood floors harbor pools of blackish water that has in many instances, stained the wood beyond restoring it. . . . Walls are green with mold; wall-board is wet and mushy—it collapses at a touch; window and door casing have turned black as tho [*sic*] originally painted that color. Ceilings drip water; electric fixtures are rusty and window fittings are so imbedded in rust, complete replacement will doubtless be the only remedy. Those compact and model kitchens with their enameled electric ranges and refrigerators are a sorry mess. . . . Some of those formerly jolly little bathrooms are pitiful sights now. . . .
>
> Some of the roofs frankly leak—have ever since construction—some Engineering! More than one village resident grew clever at estimating just about where the rain would start to drip over the baby's crib—even in the previous palmy days of Village life. That was due to lack of cement on the seams of the roofing. But the greatest damage of all has been done by condensation,—steamy air in the tightly closed rooms. With no means of ventilation provided, the houses just shut and locked as occupants left them, air became stale to the point of foulness. One can scarcely breathe in one of those newly opened houses. . . .
>
> When Quoddy Village was built, water pipes were laid into the houses with no means provided for shutting off the supply. Consequently, when the houses were closed and winter shut in, there was nothing to do but keep those houses warm to avoid burst pipes all over the place. So the steam plant has been operating steadily all winter keeping an empty village warm and cozy. That plan uses no less than 90,000 gallons of fuel oil monthly . . . to warm the ghosts of departed souls. . . . [F]or want of a small shut-off at each house . . . the greater part of Quoddy Village is a wreck. . . .
>
> [B]y Wednesday night, the consumption of fuel oil in the heating plant will have been lowered . . . to about 40,000 gallons per month. . . . Economy is a lusty youngster out at the village, these days. Eastport has been so accustomed to the "high, wide and handsome type" of administration of project and village affairs, it can scarcely assimilate the quiet but effective program of economy that one sees on every hand there today. . . .

The writer remembered during the building of those houses, of seeing trees cut down to make way for post holes—and also remembered that several of those stumps, being right in the proper places for corner post, were merely leveled off and utilized in all their green and growing state for the purpose. The house was calmly set on those cut off trees. . . . It is rather entertaining to visualize the possibilities. Some . . . of those houses may take a shoot skyward when the sap begins to flow in those tree stumps. And what a pity it would be if one corner of a cottage were the only section to be "posted." It would get rather an upsetting. But in case all four corners were properly located on growing tree stumps, in a mere eight or ten years, the 50,000 visitors recorded at the village last summer would be lost in the maze of the hordes that would trek a path to the door of this wonder village. Provided the houses hang together as long as that. Whichever way you scan the situation, Quoddy Village still has claims to being the most unique collection of dwellings known to politicians and engineers. It retains the capacity for earning publicity if nothing more.[120]

The *Sentinel* was criticized for publishing such a negative story about Quoddy, but Roscoe rebuffed the criticism: "The story was written by a reporter whose ability to observe accurately and of whose loyalty to Quoddy's best interest there can be no question. . . . The time has come, if it isn't in fact long past, when a sharp and unmistakable distinction should be drawn between Quoddy as a project and the manner in which it has been handled."[121] Roscoe wrote privately to Dexter about the incident, saying:

W.P.A. rather desired, apparently, that the truth be told, and I used it with the feeling that this might be a gentle hint as to the attitude on the highest authorities,—that some justification may now be welcomed for definitely washing the Army out of the picture. [Captain] Sturgis got red-headed over it, and intimated that such reports hurt the project. My viewpoint is that his concern is for his own reputation and that of his gang, and we should not worry about that.

We want to play the game always with the larger objective of what is best for Quoddy in mind, and I would appreciate your advice, strictly in confidence, as to the proper course to take. We have plenty of material showing up the Army and will start shooting if that is the right thing to do. If it isn't, we're entirely willing to "suffer in silence" a little longer, but not otherwise.

I have discussed this matter guardedly with the local correspondents of the daily papers and can give assurance that they will cooperate.

Anything you write or wire will be handled with the utmost care. In fact I am not entrusting the typing of this letter to my office girl even, but am doing it myself, which will explain some of the extra commas and periods.[122]*

The WPA purchased 13,500 bundles of shingles to give almost every building a new "lid," a job that took 75 men three weeks to complete.[123] The most unpleasant spring-cleaning chore was at Barrett Farm, where one building was turned into a "de-lousing plant" and where "thousands of blankets and mattresses, more or less thickly populated" were "subjected to sulphuric [sic] fumes over periods of forty-eight hours, then aired, wrapped in protective covering and stored in such manner that no further damage of any sort can result."[124] Summing up the cleaning and inventorying at the dormitory, the *Sentinel* used a simile: "The building with its widespread wings is like a satisfied biddy with her brood gathered in and counted."[125]

A real brood of 250 NYA "boys" began arriving in June, a scene nicely described by the *Sentinel*'s Grace B. Carter:

> The first contingent arrived Friday night by train, piling off the lone railroad coach, famished and excited over the peculiar stroke of fate that brought them to this far-away spot called Quoddy—a spot familiar to them in the headlines of the newspapers some of them have been selling on Boston and Portland streets. Some of them appeared to be inclined to pinch themselves during those first few hours. . . .
>
> The boys were led at once to the Leadurney Apartment house, where quarters had been arranged for them, barracks style. A dash across the street to the recreation hall that is now functioning as a dining room was next in order and in less time than it takes to tell, thirty-three boys were making heavy inroads on cold meats, platters of salad, bread, butter and coffee.[126]

Quoddy Village was again full of life and excitement.

Married life was to begin for Nancy Cooper (age 23) as Mrs. Brian Hamilton on May 2, 1937, at Saint Anne's Episcopal Church on Campobello. The *New York Times* announced the nuptials on April 28. The reception would be at Flagstones, her parents' house a few hundred yards from the church.[127] The young lovers had met in England while Dexter was working in Russia.[128]

Brian didn't show. It was another heartbreaking disappointment for the Coopers.

Life ended for Hugh Lincoln Cooper on June 24, 1937, at age 72. The *Sentinel* noted, "Col. Cooper had more than once visited Quoddy and it was known that he strongly approved its construction although he preferred private ownership to government control, to which he was opposed on principle."[129]

"We can't land there," said the pilot to Congressman Brewster, *"it's not an officially approved seaplane harbor."*[130] Instead of landing at the public wharf in downtown Eastport—where small boats for ferrying passengers, Coast Guard and police escorts, cars, drivers, and dignitaries were all waiting—they landed at Broad Cove, three miles distant.

Congressman Brewster led this bipartisan junket, which included two seaplanes filled with one senator, seven representatives (all members of the Naval Affairs Committee), and a Marine Corps colonel.[131]* Brewster borrowed a punt and rowed to shore to telephone those waiting downtown. It wasn't an auspicious start to Brewster's showing off the Quoddy Tidal Power Project, the NYA's Work Experience Program at Quoddy Village, the city of Eastport, and the harbor that could hold the entire US Navy.

Although time was lost in the shuffle, with Roscoe, Dexter, and NYA staff as tour guides, they managed to see almost everything. "The distinguished visitors . . . were very favorably impressed . . . and several of them were emphatic in asserting that having been started and brought to this stage of advancement it should now be completed." Then they hurried back to the planes for the return flight:

> The Coast Guard amphibian was soon in the air, but the Navy plane could not get her port engine in action, and finally after drifting toward Treat's Island, taxied to the Thorofare at North Lubec, where she tied up for the night . . . [so] the party . . . left for Bar Harbor by automobile at 8:45. The plane was able to take off early Monday morning . . . This is the plane used to deliver mail to the President when he is in on vacation trips at sea.[132]

According to the *Sentinel*, the 1,300-mile jaunt "demonstrated the possibility of air travel and that eastern Maine has now come within commuting distance of New York and Washington." But since it wasn't realistic to fly every US senator and representative to Eastport—not to mention their millions of constituents—why not bring the mountains and bays of Maine to the masses and to the Muhammads on Capitol Hill? *Why not*, wondered the congressmen, *set up a model of Quoddy in Washington?*[133]

Moving the "action model" *wasn't* realistic. The Army Engineers had built it with the material they knew best: steel-reinforced concrete. Besides the excessive weight, it was too wide to fit on a truck. A lighter-weight, portable model would have to be built.

Misinformation was still a big issue. In an editorial on October 3, the *Portland Sunday Telegram* questioned the wisdom of the president's power policy and asserted electricity could be generated from steam more cheaply than from hydroelectric dams. But as the *Sentinel* pointed out, Maine's private power companies were spending $3 million to build a new hydroelectric power plant on the Dead River and proposed to spend another $2 million for a hydroelectric power plant at Mattaseunk on the Penobscot River.[134] The people in Maine who knew the truth were investing in waterpower, *not* steam power.

President Roosevelt also knew the long-term cost efficiency of waterpower. "People needn't have been so surprised," remarked Elizabeth May Craig, "when the President . . . said he thought giving cheap electricity to small communities and farmers was more important than booming huge industries in huge cities. He is more interested in giving light and comforts to people than in building 'more Pittsburghs' and he always has been. It's part of his original plan for 'the forgotten man and woman.'" And, she reported, "he was going ahead with the plan for eight regional authorities. . . . Indications are now that Quoddy would be an international project with the All-American plan abandoned."[135]

While the power trusts were painting the president's power policy Communist Red and the government's competing with private enterprises (i.e., monopolies) as "totalitarianism," New Dealers held the opposite view. The *Sentinel* reckoned, "Those who controlled the supply and price of electric power have by such control a power very close to that of taxation."[136] Indeed, in Eastport and elsewhere, the private monopolies that controlled the necessities of modern life—water and power—were taxing the public more effectively than their democratically elected governments. Unlike other public services—such as national defense, roads, and primary education—which are deployed universally irrespective of remuneration, public utilities are deployed individually and shut off if not paid for by the individual. And unlike most government tax policy, which aims in part to *redistribute* wealth, monopoly taxing *concentrates* wealth.

Where the Hoover administration had favored private enterprise, particularly large corporations and centralized power production, the Roosevelt administration favored democratizing power. The TVA and Bonneville were, as the *New York Times* said, "a sort of social laboratory. Here not only a large block of humanity but cheap energy itself is to play a part, and the theory that decentralization will follow the generation and distribution of energy will be put to the test."[137]

The assertion that steam power was cheaper than waterpower was propaganda fed to the press by the Committee of Utility Executives in a pamphlet titled *Administration Experts Declare Water Power Is More Costly Than Steam*. Therein, P. R. Gadsden,[138]* vice president of the United Gas Improvement Corporation, declared, "It would follow from the testimony of the [Roosevelt] Administration's own experts, that the federal power program is economically unsound." The pamphlet's publication was timed for when there would be maximum press coverage: President Roosevelt's dedication of the $83 million[139] Bonneville Dam on September 28, 1937.

How could the private power trust make such claims? The New York State Power Authority, whose efforts to develop public power on the St. Lawrence River were also being attacked, decided to find out.

Gadsden and his confederates based their false and misleading assertions on a report by A. A. Potter, dean of the Purdue University School of Engineering and a member of the Edison Electric Institute's (EEI) Prime Movers Committee. Potter's falsehoods were based on deliberately distorted calculations by F. F. Fowle, the former consulting engineer for EEI's predecessor, the defunct National Electric Light Association. Mr. Fowle's negative numbers were based on more foul play by the US Chamber of Commerce. Disassembling the deceptions, New York's detectives found:

> The figures . . . represents gross exaggeration when applied to cost of the government's hydroelectric projects and gross understatement of the cost of modern steam generation. Correction of the fallacious economic and technical assumptions upon which their assertions are based completely reverses the conclusions as to the comparative economy of government hydro as opposed to steam generated power. . . . [Hydroelectric power] can be delivered to markets at approximately half the cost of alternative power generated in the most modern private steam stations.[140]

Among the power trusts' widely skewed assumptions, according to New York, was a 50% utilization, when 80% would be more realistic (while they assumed 100% utilization for their steam plants); extremely low fuel costs; high transmission costs (since public power plants were located far from the population centers where private steam plans were usually built); unrealistically low construction costs for steam plants; and including the costs for the flood control, navigation, and irrigation portions of the public projects in the hydroelectric costs.[141]* That was like including the cattle ranch in the price of a hamburger. Snapping back, Wendell Willkie accused the Roosevelt administration of using a "yardstick" made of "rubber."[142]

New York's analysis gives us an idea why Dexter's cost estimates were so much lower than those of the FPC. While professionals may honestly disagree, the estimates produced by both sides—private power and public power—were heavily influenced by their political philosophies and financial interests. The differing estimates are also the difference between optimists and pessimists, and between men who want to succeed—and those who don't.

Thanksgiving was thin in Downeast Maine. Although residents were grateful for one year of the tidal power project and a semester of the NYA program, their continuation was getting less likely with each passing month.

In early November, the *Sentinel* noted, "The President . . . ordered . . . the Reconstruction Finance Corporation and the Public Works Administration—to make no further commitments. . . . By sending the two agencies into the limbo of history, the President declares, in effect, that the depression is over and the emergency period has passed."[143]

The Depression may have been "over" for the nation, but it wasn't over for those on the fringes of society, either racially or geographically. In his annual Thanksgiving message extolling unity and equality, the president failed to mention any gratitude due to the guests of honor at the original Thanksgiving feast. Rather, segregation was ascendant, as illustrated in the *Sentinel* by the nine-foot-tall marble statue of Jefferson Davis being raised in the rotunda of Kentucky's Capitol.[144]

The city on the edge of the country had taken on debt to pay for expanded services and facilities for the influx of Quoddy workers and families. Despite assurances that federal loans and grants would be available to help offset Eastport's extraordinary expenses, nothing—*not a penny*—was provided.[145] With the NYA employing only dozens of the needy, not thousands, Eastport fell into an economic crisis.[146] Its annual budget for 1936 was $96,000, but it managed to collect less than half of the taxes due and was already $170,000 in debt.[147]* What little the city had, it couldn't put in the bank for fear creditors would attach its funds, so it kept what little cash it had in a strong box at city hall.[148]

It must have been painful for Roscoe—the unanimously elected chair of the City Council and Quoddy booster extraordinaire—to recommend the city of Eastport be placed under the control of the state's Emergency Municipal Finance Board. It was that or bankruptcy.[149] Eastport's City Council again voted unanimously, and on December 30, 1937, at 10:00 a.m., Eastport became a ward of the state.[150]

After Eastport residents read the grim news about their town on the front page of the *Sentinel* in the last issue of 1937, they would have seen another article from which they might have sensed the end of an era. Dexter Cooper was surveying the St. Croix River for a traditional hydroelectric dam for the St. Croix Paper Company. But only Roscoe Emery and Oscar Brown knew the story's full significance.[151]

Six pages later, *Sentinel* readers would have come to an end-of-year co-op ad like they had seen many times before, with a score of companies thanking them for their patronage and wishing them a happy new year. This year, along with the pleasantries was a more pointed message: "NEVER ADMIT DEFEAT."[152]

THE LAST FERRY RIDE

Dexter P. Cooper died on February 2, 1938, age 57. He was defeated.

The shocking news came to Quoddy in a brief telegram: "Please tell Eastport Dad passed away this morning—Elizabeth."[153] Gertrude explained what had happened in her memoir:

> One day in the autumn of 1937 . . . while walking in Boston, he suffered a heart seizure. He proceeded to his doctor's office, received an immediate electrocardiogram, and was sent home to bed. After some months he was nearly well, up and around and seemingly out of danger, but with the knowledge that he would be forced from then on to lead the life of an invalid. . . .

His physician was a practical, down-to-earth person. Doctors never say a patient died of a broken heart, he said later, but if there was ever a man who did, it was Dexter Cooper. His magnificent dream had been crushed and it was more than he could bear.[154]

Dexter's remains were carried across Lubec Narrows on the rickety old ferry and laid to rest on Campobello. The sad scene was beautifully described by [Maud[155]] Maxwell Vesey of the *St. Croix Courier* and reprinted on page 1 of the *Eastport Sentinel.*

A TRIBUTE TO DEXTER P. COOPER
"And the Lord said, write the vision and make it plain upon tables,
that he may run that readeth it."
–Habakkuk II, 2

The rain dropped heavily through the pine needles and upon the roof of the little Anglican Church of St. Anne's at Welshpool, Campobello. Inside, in the dimness of that February afternoon, the pall of sorrow lay so oppressively that the atmosphere itself seemed as tangible a thing as the benches or altar.

Before the chancel was a draped table behind which stood two large flags drooping on their standards, the Union Jack and the Stars and Stripes. Upon the table was a little copper receptacle.

So small that bit of shining metal, so trivial a thing to hold all that was left in an earthly sense of him whose brain had held a Vision, and made it plain upon the tables of maps and charts and countless blueprints for all the world to read. . . .

The tolling of the bell fell through the sad darkness of a day which resembled April rather than that of a Canadian February. There came a stir about the door and the congregation stood with bowed heads as Mrs. Cooper and her three children entered and took seats. The simple impressive service of the Anglican communion was read by the rector, and all joined the choir in the heart touching words of "Abide With Me." Then the rector with deep feeling read the lines so appropriate to the one in whose memory they gathered. No eulogy was given, or needed. This sufficed:

> Now the labourer's task is o'er;
> Now the battle day is past;
> Now upon the farther shore
> Land the Voyager at last.
> Father, in Thy gracious keeping
> Leave we now Thy servant sleeping.

. . .

Not a breath of wind stirred the air that was so strangely warm. The rain fell like drops of lead, blotting out everything but the immediate surroundings. Up to the hilltop behind the church the cortege moved, all following on foot to the place prepared to receive the dust of one who was a true friend to Campobello and whose loss to it and the surrounding communities is literally irreparable. . . .

Directly facing the tall Celtic Cross and the white fence which marks the grave of William Fitzwilliam Owen is that of Dexter P. Cooper. Once before and eighty years ago, on another dark winter afternoon when rain was falling had Campobello laid that other beloved and striking figure to rest beneath these pines. Campobello men had carried the remains of the Admiral from a boat, up by the same winding road to the same little church which he had built, and laid him, also a maker of charts and plans, on the same hilltop of this enchanting island. There they lie, the sturdy old English Admiral, one of the foremost of Naval surveyors, whose work still endures, and the younger man, an American, great Engineer of a science not known in the days of the Admiral, that of modern hydro-electrics.

But the Admiral died at the end of a long earthly journey, Dexter Cooper before his work was done. So much there was for him to do, so great the need of his knowledge, his enthusiasm, his creative urge. For those who hold him in remembrance, the surge of the tides through Lubec Narrows to meet the onrush around Campobello and up Quoddy River will recall his Vision of the harnessing of these waters for the benefit of mankind. The wind off Fundy, the evergreens dark on the hilltop, the winding road he traversed so often, all will speak of him. . . .

There were hopes, and there was bitter disappointment, but always there was the joy of the Vision, the thrill of the task. And yet, it seemed sometimes as if the whole continent was arrayed against this lone Engineer and his idea. The gigantic power of vested interests, opposing, always opposing, and the deadly thrust of adverse propaganda. . . .

Is it not for money that such Engineers as Dexter P. Cooper work? Yes, and No. Yes, in the sense that they have need of bread and shelter. No, in that there is something in their souls which is far beyond. "AND THE LORD SAID, WRITE THE VISION AND MAKE IT PLAIN UPON THE TABLES." That command has been obeyed since the days of the Chief Engineer of the Great Pyramid down through the ages to those who carried the line of the C.P.R. through the mountains, swung the spans of the Quebec Bridge, dug through at Suez and Panama, pierced the Alps with the Simplon Tunnel, buried underground for subways, built the power plants at Niagara and Muscle Shoals, dammed the Mississippi and the Nile at Assouan.

The compulsion to create, to write upon the tables, drives such men. To what result? Railroads, canals, harbors, fortresses, tunnels, bridges, dams, and the waterworks and sewerage constructions without which the enormous modern cities could not exist. Whatever we enjoy of comfort, cleanliness and speed has come from the brain of an Engineer. Built first from out the Vision and then written upon tables, that Vision has become transmuted into terms of steel and concrete that other workers who read may run to the furtherance of the task. . . .

Perhaps God who ordained that the Vision be written has so ordered that the rewards may await them in another and far different world.[156]

The *Sentinel* published a "SPECIAL COOPER MEMORIAL" on March 16, 1938. Among the eight pages were a 212-item chronology of the project from 1924 to July 1937,[157] reprints of the *Sentinel*'s first reporting on the pioneering project, and many testimonials and memorial ads. Therein, Roscoe the reporter, mayor, and advocate said, "We, who worked with Dexter Cooper in the promotion of the Quoddy project, are grieved by his sudden death, but are going forward, as he would want us to, until his great plan for producing power from Quoddy's tides is consummated. That is the only adequate or appropriate memorial that can be established for him."[158] The *Sentinel* went on to say, "To Mr. Cooper belongs the unique distinction of such sheer loyalty to an idea that, though submission to it cost him his life, it at the same time gave him rank as a pathfinder."[159]

Lieutenant Colonel Fleming echoed those sentiments, recalling, "In the early days, P.W.A. had placed before it many bizarre projects which on their face appeared wholly impractical. To harness the tides seemed fantastic. But a project backed by an engineer of Mr. Cooper's standing could not be ignored. . . . He was equipped with a technical education perhaps superior to that possessed by any other engineer in the country."[160]

Congressman Ralph O. Brewster, another tireless (if erratic) advocate for Quoddy, eulogized Dexter: "Cooper will have [a] unique place in [the] history of Maine. . . . The Quoddy tides will ultimately be harnessed as surely as the moon shall rise and wane, and Quoddy will stand as a monument to the preserving and patient spirit of one who labored during a lifetime in the service of his fellow men. All sense of loss is dimmed in the inspiration of his spirit as we re-consecrate ourselves to carry on."[161]

Conspicuous by its absence was any comment from President Roosevelt. Roscoe had asked for one but was told it was contrary to the president's established policy.[162]

Ironically, the report of Dexter's death published in the *New York Times* on February 3 was accompanied by a photograph of—Philip B. Fleming.[163]*

Gertrude, Elizabeth, Nancy, and Dexter Jr. were bankrupt.

Dexter had invested in his own company, and when he sold his interests in the Quoddy Tidal Power Project to the federal government in 1935, he was owed $24,723.79 for unpaid expenses and $45,400 of unpaid salary for the year 1929–1935[164]* (about $450,000 and $850,000 respectively, in 2020 funds).

Of the $50,000 paid by the US government in 1935 to Dexter for his rights in the Quoddy Tidal Power Project, all but $4,676.22 went to pay debts.[165] I could find no record that the government ever made the final payment of $10,000 due to Dexter Cooper. The records I could find indicated the army was trying to avoid making the payment.[166]*

While it was reported by the *Lewiston Daily Sun* on March 6, 1936 (and subsequently repeated by others), that Dexter was receiving an annual salary of $25,000 while working on the Quoddy project, that wasn't true.[167]* Dexter earned consulting fees of $25 to $50 per day working for the National Power Policy Committee while trying to find future customers for Quoddy's power. But like so many who worked on the project, his wages were too low and too short-lived to pay off old debts or accumulate a nest egg sufficient to support a widow and three children. White House records show that from the federal government Dexter earned in 1934 a mere $1,000; in 1935 a total of $11,900; and in 1936, only $7,600.[168]*

Since Gertrude had no income of her own, her brother took her in but proposed that she work as a governess and that her children be farmed out to her wealthier sisters. Gertrude refused to give up. She would not allow someone else to raise her children. As a stopgap, she went to work at Jay's lingerie store in Boston.[169] Nancy took a clerical job with the WPA. Elizabeth and Decks were still in school.

Like Eastport, Gertrude Cooper needed a rescue.

Dexter Cooper took his secrets to the grave. *Ol' man river* didn't *say nothin'* about who would purchase all of the power Quoddy could produce. He never betrayed the confidences invested in him—so he was branded a liar.

President John F. Kennedy with NASA's chief scientist Wernher von Braun, ▶
Cape Canaveral, Florida, on November 16, 1963

PART IV

THE DREAM THAT WOULDN'T DIE

CHAPTER 12

The War Years (1939–1945)

THE TURN OF EVENTS

Dexter and Quoddy were dead. Gertrude was alive and angry. She went on the offensive. She didn't think she had anything left to lose. Gertrude called the president of the United States and leveled three sets of explosive charges:

1. Malfeasance by White House staff and elected officials
2. Malfeasance and sabotage by the US Army Corps of Engineers
3. Extortion and sabotage by Quoddy's competitors

The charges could bring down the president.

The president needed to know, *Were they true?* If they were, *What should he do? What would he do?*

President Franklin Roosevelt had the WPA hire a private investigator, Frank Reel, a Boston-based attorney, someone believed to be effective and discreet.[1] Reel spent a week in Eastport interviewing witnesses and looking for evidence; then he wrote a 29-page preliminary report labeled "Personal" and "Confidential." To keep it beyond the reach of inquisitive senators, he did not deliver it to the WPA or to the president, but rather to the president's personal secretary, his son James. "Jimmy"—as Gertrude called her daughter Nancy's former beau[2]—kept it in his private files.

Gertrude charged that Louis Howe and Governor Brann had planned to use Quoddy as political bait during the 1934 midterm elections. Howe, who died in April 1936, was Franklin's closest friend and chief political adviser. Their plan, said Gertrude, was for the federal government to buy a six-month option on Quoddy for $30,000 to make it at least *appear* that the Roosevelt administration was going to spend $30-plus million in Maine. Dexter had agreed to the plan because he thought they honestly intended to investigate Quoddy. Dexter had documented what he thought were the terms of the deal in a letter addressed to President Roosevelt dated June 29, 1934. But as Reel reported:

> According to Mrs. Cooper, her husband did call Mr. Howe who said that he could send no money. Howe challenged Cooper to expose the scheme saying that if it were exposed, Brann would be defeated for Governor in the coming Fall elections of 1934 and then Quoddy would never be built. Cooper evidently did nothing more about this. . . .
>
> According to her, the President said he knew nothing about the option letter that Cooper had written to Louis Howe. Mrs. Cooper says that she showed him a carbon copy of it and he seemed very much surprised and said that such a method would have been illegal.[3]

Reel was unable to find a copy of the letter in Dexter's files on Campobello. Had Reel looked in the White House files, he would have found two copies of the letter.[4*] Had Reel examined the files of the Sills Commission, he would have found the same letter,[5] thus seriously undermining FDR's statement that he knew nothing of it.

Also in the White House files is a copy of Dexter's letter to Howe dated July 5, 1934, which concludes:

> My proposal was made one week ago, which should have given you ample time to decide whether you wish to exercise the option or not. Therefore, if I do not hear from you by Saturday morning, July 7th, I shall conclude that something has happened to change your mind and that you are no longer interested. This will mean that I shall have to resume my work abroad and that before leaving I must give to my people of Maine an account of my endeavors to get favorable action on Quoddy in their behalf.[6]

Regardless of who proposed the option or whether it was agreed to, there was enough substance to the vote-baiting charge for any political opponent to make a lot of trouble for President Roosevelt.

Gertrude charged the US Army Corps of Engineers with malfeasance.[7]* While investigating the charge, Reel heard complaints of waste by the army: the houses on Snob Hill; the ostentatious Administration Building; fancy yachts but worthless workboats; sixty-plus staff cars, many with chauffeurs; overstaffing; and favoritism.[8]* He heard complaints of incompetence and shoddy workmanship: the washed-out Carlow Island Dam and the plumbing and roofing of Quoddy Village. He heard Colonel Fleming was a "rum hound" and often under the influence. He heard that the army's public comments—such as "if we finish" and that they were "just buying votes"—had undermined the project. And he found bad blood between Eastporters and the US Army, which he described as the unfortunate outcome of different mores, soured land deals, mutual distrust, poor communication, and the relentless anti-Quoddy propaganda.[9]* None of these complaints appeared serious enough to warrant further action.

Gertrude's charge of sabotage by the army was much more serious. It began with Gertrude recounting the three warnings given to her in 1935—by General Hof's wife, Major Fleming, and the anonymous tipster in Boston.[10]* Reel looked for corroboration—and *may* have found it.

He heard repeatedly that the test borings for the Haycock Harbor pumped storage reservoir *may* have been falsified to make construction appear unsafe or more expensive. These suspicions were reinforced by two subsequent events: (1) the army had transferred John Sweeney, the engineer who publicly challenged its analysis and plan, *just as Hugh Cooper had been transferred from Muscle Shoals for challenging his superiors in the army*; and (2) immediately after Quoddy was canceled, two of the three suspected falsifiers—Paul Kruse and Leonard Hunt—were given jobs by Bangor Hydro-Electric (BHE), at salaries several times their earnings while working on Quoddy.[11] Kruse was the most senior civilian employed at Quoddy and one of the Snob Hill nine. *If* he was a mole, he was perfectly positioned to undermine the project.

If John Sweeney corroborated the reports of sabotage, or *if* there was hard evidence the soil samples were tampered with—especially *if* there was collusion by the army—then it would cause a national scandal for the army and its commander in chief.

Why would the Army Engineers want to sabotage Quoddy? Roscoe told Reel he believed the army had been "captured" by the power trust, pointing to a statement by Captain Eliot: "The Army is not in the power business and we don't intend that it shall be."[12]* Roscoe, however, understood this was smoke, not fire, and not a smoking gun.

Gertrude charged Quoddy's commercial competitors with extortion and sabotage:

> Mr. Cooper obtained an agreement to finance investigating operations from four utility corporations, namely,—Westinghouse, General Electric, Boston Edison, and Samuel and Martin Insull's Mid-West Utility Company. According to his widow, Mr. Cooper did not want the Insulls to be connected with this because they had some control over the utilities in Maine and he was afraid that their interest was merely one of being able to know what was going on rather than the desire to realize the project. After two years of investigations, the utilities hired a firm of New York Engineers, known as Murray & Flood, to make a report on the Quoddy Project. According to Mrs. Cooper, Murray & Flood told her husband that if he promised them the electrical engineering, they would make a favorable report, which promise he refused to make. At any rate, Murray & Flood made a report on the water power resources in Maine and recommended certain waterfall developments ahead of the Quoddy Project. Mr. Cooper evidently felt that their report was not worthy of credence because during the months that they had supposedly investigated the rivers in Maine, they were entirely frozen over. . . . Murray & Flood's report was evidently not followed by the Utilities because they gave Dexter Cooper another $100,000 to continue his investigations.[13]*

Murray & Flood (M&F), you will recall, was exposed by the Federal Trade Commission in 1930 as having been paid by the National Electric Light Association to write the misleading report that disparaged the public power projects at Niagara Falls. Now we have attempted extortion by M&F, along with a third deliberately misleading report written by M&F—intended to benefit Maine's power monopolies, CMP and BHE.

The first half of M&F's extortion attempt—the illicit gain—was unsuccessful: Dexter refused. The second half of the extortion—the threat—was not *immediately* successful, as his investors provided some additional funding, despite M&F's estimate that Quoddy would cost $125 million to build, as opposed to Dexter's $75 to $100 million estimate.[14] *Ultimately* the threat was deadly: M&F's expensive estimate for Quoddy and report on Maine's river power would be used against Quoddy forever after.[15]

M&F said Maine's rivers offered both more power and more economical power than Quoddy's tidal power. Not coincidently, most of the potential dam sites in Maine were already owned by CMP. This intentionally misleading report was delivered to Gerald Swope, president of General Electric, on April 15, 1927,[16]* apparently to negatively impact Dexter's chances of obtaining private funding. And it was also apparently delivered to the Federal Power Commission (FPC) to sabotage Dexter's chances for public funding from the PWA in 1934.[17]*

Though believable, given Murray & Flood's history of subterfuge, there is no evidence to support Gertrude's charge of attempted extortion.

Reel concluded his preliminary investigation by advising, "The one big possibility that should be further explored is the question of the soundings [i.e., falsified borings] at Haycock Harbor Reservoir." In addition to his findings, Reel briefly described what he saw at Quoddy Village and Eastport: the promise—unfulfilled:

> It is needless to describe the village. I was impressed with its beauty, utility, completeness of its hospital, fire and police services . . . [but] . . .
>
> Eastport is the most bleak and "down at the heel" town I have ever seen. . . . There seems to be no reason for the existence of a town at that point now.[18]

Would President Roosevelt agree there was no reason for Eastport's existence? Would he look at the mess in Reel's report and want to wash his hands of it? Or would he be moved by Gertrude's appeal and fulfill the promises made?

Shortly after Frank Reel delivered his report, Gertrude, Nancy, Boo, and Decks returned to their big house on Campobello. They were shocked to discover that "when we looked in his [Dexter's] filing cabinets and map cases, which had always been full of his charts and drawings, we found them completely empty. The only remains were a few torn scraps of maps scattered on the floor."[19] What Dexter had *made plain upon the tables* so that others *may run that readeth it* was gone. All his business records were gone. All his plans were gone. All his letters were gone.

Nothing else was taken from the house.

The thieves were not interested in things: they wanted memories. They wanted to prevent an idea from being revived, to hobble those who would *run* with what they could *readeth*, to destroy the evidence of what had been conceived, said, and done.

They left none of Dexter's technical papers or financial documents. No correspondence remained to or from the Insulls or Walter Wyman that would indicate if or when they had turned against Dexter or tried to help him. None of Dexter's private notes on what the army did and what he would have done differently. No notes of Dexter's conversations with Louis Howe, Governor Brann, or President Roosevelt.

Gone. All gone.[20]*

Was **"Something Stirring In Quoddy Project?"** wondered the *Sentinel* on June 22, 1938, due to a flurry of visits by army officials, including Colonel Fleming and Captains Sturgis and Leehey. "It is not reasonable to assume that all these visits are inspired by a sentimental attachment to the scene of their former labors, or by merely morbid desire to 'view the remains.'"[21]*

The stirrings were the result of President Roosevelt's feeling confident about his national power policy after a string of court victories culminating at the end of March when the Supreme Court upheld the constitutionality of the Public Utility Holding Company Act (PUHCA), thus allowing the Securities and Exchange Commission to regulate all utility companies and to break up holding companies, if need be, to enforce transparency and fair competition.[22]*

Quoddy *was* stirring, Congressman Brewster was pleased to report in August: "The President is sincere in his attitude toward Quoddy. . . . It is still close to his heart, and proof of this lies in the fact that within the past two months he discussed with me the possibility of working out a new way of financing it. Since that time I have been making contacts and getting additional information from a great many sources toward that end."[23]

That end was taking a long time to come. In the meantime, Maine's midterm elections took place in September. All the Republican candidates were reelected,[24] and they were now all vocally *for* Quoddy, the quintessential Democratic New Deal project. Presumably such a bloc would give the president an unusual opportunity for bipartisan action. He took no action.

Or did he?

The FPC shed some light on the public versus private power contro-versy when it published a report for each state comparing the cost of electricity in each town and whether it was served by public or private power companies. For Maine towns with fewer than 2,500 residents, the municipally owned power plants in Lubec and Madison provided the lowest priced electricity, saving their residents as much as 76% compared to towns that got their power from private companies.[25]* Eastport, which bought electricity from privately owned BHE, was among the most expensive.[26] Eastporters hoped this objective presenta-tion of facts was part of the president's plan to gain public support for Quoddy's revival.

President Roosevelt may have been thinking of Quoddy, but in 1938 his thoughts turned more frequently to the ominous news from Europe. Germany had annexed Austria on March 13. On Septem-ber 30, in an attempt to appease Hitler, the Munich Agreement was signed by Germany, France, the United Kingdom, and Italy, allowing Germany to annex the Sudetenland part of Czechoslovakia. Neville Chamberlain, Britain's prime minister, hailed the agreement as ensur-ing "peace for our time."

"PRES. ROOSEVELT REVIVES PASSAMAQUODDY PROJ-ECT." The headline stretched across the *Sentinel* on January 18, 1939, and was itself a stretch. The article was more circumspect: "Redeeming his oft-repeated assurance that the Quoddy project will be built," the president recommended Congress make an appropriation to investi-gate building a small experimental tidal-hydroelectric power plant in American waters[27]* and restart negotiations with Canada for building the International Project:

> [T]he news was the only subject of conversation everywhere and every-where the same sentiments were expressed,—that the unshaken confidence held here in the President's intention to go forward with the project has been justified; quiet gratification that the long, dreary period of uncertainty may presently come to a satisfactory ending; and that if it does, renewed prosperity is in sight, for all of Eastern Maine, and with it vindication for Dexter P. Cooper, who gave his life to the project. There was, however, no tendency to be unduly optimistic, since the tremendous strength of the political and power opposition has been thoroly [*sic*] demonstrated and

there was a general, sobering realization that a hard fight is ahead before construction is finally authorized.

It was pointed out by various leaders that the President now cannot be accused of political motives.[28]

The "tremendous strength of the opposition" was immediately demonstrated by yet another derogatory editorial in the *New York Times*, the headline of which howled "ECONOMIC MOONSHINE." Hacking away with its old saws, the *Times* derided "the late Dexter Cooper [who] spent fifteen years of his life trying to convince capitalists that the moon could thus be made to drive trolley cars, milk cows, suck dirt out of carpets,"[29]* and ridiculed the foolish president for rushing in "where bankers feared to tread." The anonymous author then confused the facts, saying "as much as $100,000,000 may be spent to make the Moon generate 200,000 horsepower," citing the full cost of the vast International Plan but only the power generated by a small experimental plant. Even that small power would be "for industries which do not exist, and which, if they did exist, could be served with electric energy at lower cost by steam plants erected in more favorable localities than Eastport, Me."[30]

Unfortunately, the identities of the companies interested in purchasing all of Quoddy's power had still not been made public, as confided by the *Portland Press Herald*: "There was a Cooper report which was never divulged and rumors of a still more secret report, with names and places and amounts of the market but this is one of those Washington mysteries."[31]

Whether the experimental tidal power plant was to be built depended on how the politics played out in Washington. The *Sentinel* asked journalist Elizabeth May Craig to analyze Quoddy's chances. Her analysis was discouragingly complex:

Some think the President wants Quoddy power for a power reserve for national defense. Some think it is merely a gesture. . . . He must know that Congress is hostile to Quoddy. To lay it on that unwelcoming doorstep means either that he knows it won't be taken in and nurtured and doesn't care, or is willing to force it through. He must have a strong reason, if he means to force it through.

Some think he just threw it into the boiling cauldron of New England power and flood control. Just as a sort of hint to the power interests which are back[ing] of the "states rights" stand of the New Englanders, that if they do not take Vermont flood control power, they might have to take Quoddy power. . . .[32]*

It is to be hoped that Maine does not fix its hopes too firmly on this action. Congress knows little about the Quoddy project and regards it with a mixture of amusement and hostility.[33]*

Trying to redeem himself, Maine's Senator White introduced a bill in January 1939 for a $100,000 appropriation to investigate the "small experiment" proposed by the president. It was approved by the Commerce Committee and then went to the Rivers and Harbors Committee, where its chances appeared good. Senator Vandenberg, Quoddy's wise-cracking critic, then introduced Resolution 62, nominally asking the FPC to update its (negative) 1934 economic analysis of tidal versus riparian and steam power. But his preamble was so pejorative that it was clear what conclusion he wanted. With the preamble removed, both bills passed.[34]*

In the House, Congressman Brewster pushed the small experiment, but progress was delayed.[35] Winter turned to spring and then summer. With the congressional session coming to a close, Brewster tried everything he could, but "the Democratic leadership in the House has absolutely refused to permit the Quoddy Resolution to be brought up for consideration except under a procedure that would require unanimous consent. In a House of four hundred and thirty-five members this presented an almost impossible situation."[36]

Most Democrats and Republicans in the House were opposed to Quoddy, and given their objection to any federal project—flood control, navigation, or scientific experiment—that produced power, it was clear the power trusts were backing the opposition.

Maine's Republican governor, Lewis O. Barrows, called the legislature to enact the Quoddy Authority so the Roosevelt administration wouldn't face legal challenges if the small experiment were actually constructed.

Joe Gallagher, who had been Colonel Fleming's legal adviser in 1935 and helped draft the original Quoddy Authority Bill, and who now worked for CMP,[37] explained the amended draft to the legislature. Moses Pike covered the engineering issues and presented a petition signed by 500 Lubec residents supporting the bill and saying their municipally owned utility was not threatened by it. CMP's lobbyist,

Edward F. Merrill, and BHE's lobbyist, Harold Murchie (both Dexter's former lawyers),[38] assured the committee their employers "did not and *never had*, opposed the Quoddy Project." The amendments, explained Gallagher, were "only to protect the Fernald Law and the stockholders of his company from *unwarranted competition*."[39]*

To *protect* CMP and BHE from *unwarranted competition*, the Authority's geographic scope was limited to a 50-mile radius of Eastport but excluded all existing electrical customers.[40] Quoddy was allowed to sell locally to new commercial and industrial customers, export power, and operate as a water utility and as a port authority, and it also had the right of eminent domain, but not to any property of the Bangor and Aroostook Railroad. "Just why this outfit should have any prior rights as to shipping facilities at Quoddy is difficult to explain," commented the *Sentinel*, "unless it is accepted as fundamental that the B&A must be protected at all costs against any possible threat to its fading monopoly on potato shipments."[41]* Most important, the bill set up Quoddy under the control of Maine's Department of Public Utilities *and hence of Maine's power barons*. Under this plan Quoddy would never be a "yardstick" with which to measure the true cost of power in Maine.[42] When the chair called for those in favor to stand, 75 stood. Those opposed: none.[43]

Measured on a butcher's scale, the Quoddy Authority Act of 1939 was a prime cut from the political sausage factory and either an artful compromise or a sellout.

Gertrude Cooper started revealing Quoddy's secrets one year after Dexter's death. To Elizabeth May Craig of the *Portland Press Herald*, Gertrude said, "It was General Gerry [Guy E.] Tripp, chairman of the Board of Westinghouse at that time, who persuaded Dexter to give up his other work and put all his time on Quoddy. He offered to find the money for a thorough engineering report, which he did. General Electric, Boston Edison and later the Middle West Utilities, went into it with Westinghouse and put up half a million."[44]*

Responding to an erroneous article by one of his colleagues, Fulton J. Redman of the *Portland Sunday Telegram* revealed more about Quoddy's financial backers. It is impossible to tell, but it seems likely Redman's source was also Gertrude, for there were few people who would have known the details of that trip to Maine, made just before the stock market crashed:

In the summer of 1929 a group of notable persons in the electrical field made a trip to Eastport. These included Gerald Swope, president of General Electric; E.M. Herr, president of Westinghouse Electric; Charles L. Edgar, president of Boston Edison Co.; Martin Insull, of Middle-west Utilities, and others. The evidence is convincing that these men were enthusiastic about Quoddy as a power development project, as well as a future industrial center. . . .

If the project were not feasible as an engineering proposition, it is difficult to believe that these men would have shown such an interest in it. They would have been making themselves look almost laughable to make the journey from New York on a private car attached to the *Bar Harbor Express* and left at Ellsworth, where they were met by automobiles for a triumphant trip through Washington County, if Quoddy were just a visionary scheme to harness the moon and the tides—"economic moonshine" as the conservative *New York Times* now suggests. These men do not make trips of this nature unless they have something quite convincing to go on before they start. They never had the reputation for being "economic moonshine" chasers. . . .

The fact that private interest not long thereafter dropped Quoddy "like a hot potato" as is now claimed by its opponents, is understandable in view of the fact that since 1929 there have been no large developments by private interests either in power generation or in general industry anywhere except in the Tennessee Valley. In the latter area private interests have already located . . . new plants to take advantage of the power development.[45]*

How would the history of the Quoddy Tidal Power Project have changed if these facts had been revealed in 1935–1936, when the project was in operation? Would it have quieted the critics? Would these gentlemen have been willing to cooperate with the federal government in a public-private partnership if they had been asked by President Roosevelt? Would they now?

And what about the companies Dexter said would contract for all of Quoddy's power if there had been someone authorized to sign contracts? Who were they? Why didn't Gertrude say who they were? Would they still be interested? Or had they gone to Tennessee?

Death causes redistribution. When the occupant dies, the house is sold. It is sad for the deceased's family but joyous for the new owners. Real estate agents like Roscoe facilitate the transaction and earn a fee.

This was different. This was not just the end of a life but also the end of a shared dream. Gertrude couldn't afford to keep the family home on Campobello, so she asked Roscoe to sell it. They must have been fighting back tears when they shook hands.

Never before had Roscoe published a display ad for a single property. The *Sentinel*'s ad for the Cooper residence highlighted its proximity to President Roosevelt's summer home. It described the house, floor by floor, and then the outbuildings, including a large hay barn with a fireproof garage for five cars; a workshop; two horse stalls; tie-ups for four cows; and, ironically, an electric milk machine. Dexter had invested over $30,000 in the property. Roscoe offered it for $12,500.[46] There were no buyers. A year later, the National Bank of Calais wrote to President Roosevelt:

> We made a loan to Mr. Cooper shortly before he died, taking a mortgage on his home as security. Mrs. Cooper was unable to pay the loan and deeded the property to us. . . .
>
> As the "Cooper" home is comparatively close to your own, the Directors thought that you, or some member of your family, or acquaintance congenial to you, might wish to acquire the property. . . .
>
> A 30 day option is hereby extended to you to purchase it for $7,000 [in] United States money.[47]

President Roosevelt declined.[48]

Cutting off Washington County, towing it away, and sinking it was the way the S*entinel* described how little Down-Westers cared about far-off Downeast Maine.[49] Conceptually, though, there was some merit in the idea of making a chunk of Maine mobile. That idea was first proposed by the congressmen who visited Quoddy in June 1937 and proposed a portable model be made.[50] Congressman Brewster acted on the idea immediately, but it took a long time for it to percolate through bureaucratic Washington.

Two years later, and with President Roosevelt's blessing, it was announced that the National Youth Administration (NYA)'s Quoddy boys would build a portable model under the direction of Hector Poirier, one of the country's foremost model makers. Unlike the army's model of the All-American Plan, the portable model would demonstrate the full International Plan. It would also show places of interest, such as the president's home on Campobello, with color photographs and locator lights.[51] Plans were made to display the model in the most prestigious and popular places in the country: under the rotunda of the US Capitol and at the World's Fair in New York.[52]

Alas, the Quoddy model was not displayed in the Capitol—presumably because some thought it would be *too* visible—so it was exhibited on the National Mall.[53]

President Roosevelt himself opened the World's Fair on April 30, 1939. Its theme, "Building the World of Tomorrow," showcased real and imaginary products of the future, including sparkling new cities, cars, airplanes, televisions, and robots. For seventy-five cents admission, you could forget your troubles; be entertained by shows, rides, and gizmos; and dream of a better future. All the razzle-dazzle depended on power. In the Hall of Industry, the Quoddy model showed four million visitors how that bright future could be generated from clean, everlasting tidal power.[54]

Europe was a greater concern than Downeast Maine during the summer of 1939. Nazi Germany had annexed the German-speaking Sudetenland region of Czechoslovakia in September 1938 and then invaded the Czech portion in March 1939. Meanwhile, Austria, Hungary, and Poland each seized parts of the country, leaving only Slovakia as a tiny and teetering holdout. Hitler said Sudetenland was "the last territorial demand I have to make in Europe"[55] in his drive to reunite the German Empire that had been purposely broken up—"Balkanized"— by the victors after World War I. It seemed obvious to many that he wanted more, including the German-speaking portions of Poland and the natural resources in Soviet Russia.

So far Nazi propaganda and intimidation had been enormously effective, and there had been little use of their massive military force. To deal with the threat of war in eastern Europe, American policy makers promoted two strategies: (1) isolation, staying out of any war as long as possible; and (2) preparedness, getting ready for the worst. While many favored the "ignorance is bliss" form of isolation, President Roosevelt was quietly pushing preparedness.[56] This could be seen all over the country, and very clearly in the part of the United States closest to Europe: Downeast Maine.

The NYA's programs at Quoddy Village had started out focused on the civilian trades: carpentry, plumbing, electrical, auto mechanics, and printing. As the Nazi threat increased, NYA's programs shifted to the defense trades. NYA boys began building a seaplane landing base.[57] They took classes in aerial mapmaking and airplane mechanics.[58] Two houses in the Village were converted into a radio broadcasting station, the first link in a chain that would eventually connect NYA stations along the entire Atlantic and Pacific coasts.[59] Quoddy's signal was strong enough to reach into the Antarctic, where Admiral Richard Byrd was exploring its uncharted expanse.[60]

Early in 1938, the National Association of Railroad and Utility Commissions had issued a warning: "it would be disastrous to await the actual coming of a war" to begin building more electrical generating capacity. On July 1, the FPC and the War Department sent a confidential report to the president saying the same thing and that costs would quadruple if the power plants had to be built during a war.[61]

Albert Einstein and fellow physicist Leo Szilárd sent a letter to President Roosevelt on August 2 warning that Germany might develop atomic bombs.[62] In response, Roosevelt began the Manhattan Project and the American effort to harness the power of the atom. The project took place at the newly created city of Oak Ridge, Tennessee, at industrial facilities powered by the Tennessee Valley Authority (TVA) and built by Stone & Webster.[63]

Canada was closer to Europe. In August 1939, President Roosevelt went there for vacation and to get to know his North Atlantic neighborhood. As he had in 1933 and 1936, he traveled by navy cruiser, this time aboard the USS *Tuscaloosa*. The *Tuscaloosa* rounded East Quoddy Light on August 12 at 12:30 p.m. and anchored directly opposite the Roosevelt cottages. By 5:00 p.m. the president was back aboard the *Tuscaloosa*. "So far as is known," the *Sentinel* was disappointed to report, "the President saw no Maine people during his visit."[64] The commander in chief was headed farther east, to Halifax, Labrador, and Newfoundland, and thinking further ahead.

To everyone's surprise, Nazi Germany and Soviet Russia signed a nonaggression pact on August 23, despite inherent doctrinal antagonisms between fascism and communism. Nonetheless, the pact freed Hitler to direct his military resources to intermediate targets in the east without fear of the Big Bear. Many viewed the pact as a threat to the

United States. The *Sentinel* reprinted an editorial by Mark Sullivan of the *New York Herald-Tribune* reacting to the news:

> Some—and observe carefully that I say "some," and not "all"—some of the New Deal is identical with, or is parallel to, or has precedent in, or takes inspiration from, the Nazi-Fascist-Communist conception of society and government. . . .
>
> If we let part of the Nazi-Communism take root here and grow, we shall presently get the whole of it. The section of American thought that most endangers us is that fatuous section which thinks we can take some economic parts of the Nazi-Communist system, but escape the part that is deprivation of spiritual rights, the suppression of free thought, the persecution of individuals and minorities.
>
> What needs to be done, needs desperately to be done, is to go over the New Deal with the most intelligent care; to separate and hold that part of it which is consistent with the American system—which is indeed, as some of the New Deal is, a needed reform and improvement of the American system—to keep that part of the New Deal and make it workable by improving the administration of it.
>
> And to take the other part of the New Deal—the part which if kept will bring death to the American system—to seize that part of the New Deal, and pull it up by the roots and abolish it forever.[65]

World War II began eight days later, on September 1, 1939, when Nazi Germany invaded Poland. Pledged to Poland's defense, England and France, and their dominions, declared war against Germany. The Dominion of Canada, America's next-door neighbor, was now in the war.

"The general feeling is strong for neutrality," admitted Roscoe in an editorial, "as long as it may honorably or safely be maintained, but it is also true that many, many patriotic people believe America must eventually take its stand definitely with all the strength it can muster, not so much for France and England as nations, as for what they represent, for human liberty."[66]

GRAY DAYS

Europe was on fire. Clouds of smoke were encircling the globe. America was in a quandary: How could it help England and France without being drawn into the war? The isolationists argued for neutrality in absolutist, black-and-white terms, while the internationalists argued for supporting the Anglo-French Allies with military materiel and eventually fighting men. The *Sentinel* framed the debate like this:

> [T]he average citizen is at a loss to understand why we should not sell arms and munitions. . . . If it is argued that selling to France and England is helping them to Germany's disadvantage, why, that is about what everybody wants; and it is equally fair to say that a refusal to sell to France and England is hindering them to Germany's advantage. When reasons for and against a specific course of action balance, it is just about as wise, and certainly much more satisfactory, to decide in favor of what one really would like to do.[67]

In other words, the *Sentinel* was arguing for a US foreign policy in shades of gray—the colors of aluminum and steel.

The need for American metal became painfully obvious between May 26 and June 4, 1940, when 338,000 Allied troops were forced to evacuate from Dunkirk, France, abandoning their armaments to the Germans as they fled across the English Channel to England in mostly small boats manned by civilians. It was a miraculous rescue and, in Winston Churchill's words, "a colossal military disaster" that left England "naked before her foes."[68]

As the last Allied troops reached the safety of England, Churchill took to the airwaves to declare his defiance to the Nazis, and to rally his troops—and the rest of the world—against fascism. Churchill promised,

> We shall go on to the end . . . we shall defend our island, whatever the costs may be, we shall fight on the beaches, we shall fight on the landing grounds, we shall fight in the fields, we shall fight in the hills; we shall never surrender, and even if, which I do not for a moment believe, this island or a large part of it were subjugated and starving, then our Empire beyond the seas, armed and guarded by the British Fleet, would carry on the struggle until, in God's good time, the new world, with all its power and might, steps forth to the rescue and liberation of the old. . . . The Battle of France is over. The Battle of Britain is about to begin.[69]

"In all the world, only the United States had the ability to resupply the British military,"[70] explained historian Doris Kearns Goodwin. And yet after a decade of Depression, we, too, were nearly naked and hardly knew how to clothe ourselves, much less care for another continent.

The Anglo-French Allies were now in a "tonnage war" against the German-led Axis in which the objective was to starve the enemy of food, energy, and materials. England had advanced manufacturing capabilities, but as an island nation, it depended heavily on imported raw materials. Consequently, the Battle of the Atlantic, the longest, most complex, most destructive naval conflict in history, became a deadly duel of German submarines against Allied merchant ships and their defenders.

German Wolf Packs prowled the Atlantic with horrifying success. During *Die Glückliche Zeit*, the Nazis' "happy time," from June until October 1940, German subs sank 270 Allied ships.[71] The losses were devastating. Somehow, Americans had to muster their strength and replace everything—hundreds of new ships had to be built, thousands of sailors trained, millions of tons of supplies manufactured—or the Allies would lose the war, and the Nazis would rule both Europe and the Atlantic. From there, North America would be next.

Many Americans were still struggling. Ten million were unemployed; two and a half million more were dependent on government support.[72] War preparedness was reducing the ranks of unemployed men,[73] but the opportunities for women were far leaner. Gertrude Cooper, now a widow with one child still in school, was one of the women who needed a job at which she could earn enough to support a family. Her $15 per week[74] salary at Jay's lingerie[75]* store wasn't sufficient. Like millions of women, she had no formal education beyond high school, no work experience outside the home, and thin prospects for a position that would satisfy the family's needs. So Gertrude wrote to her longtime friend, Frederic Delano, for help finding a job as a companion or housekeeper. On receiving her plea, Delano walked to the White House and handed it to the president's secretary, saying something should be done for Dexter's widow.[76]

Gertrude got a better placement than she could have imagined.

The president was at the time working on an unusual project, very near to his heart and to his home on the Hudson River. He was trying to convince Mrs. James Laurens Van Alen to donate some of her excess property—the Vanderbilt Mansion at Hyde Park—as a cultural

heritage museum and trying to convince a recalcitrant Congress to provide $20,000 per year for the maintenance of the $2 million property
they were going to be given. Once he succeeded, the park would need
a manager, and Franklin was happy to have Gertrude as his neighbor
once more. Gertrude was "sailing around on air" at the news and
thought it very wise for the federal government to "preserve what we
have of the Old World, especially now."[77]

The bureaucrats in the National Park Service were not pleased to be
passed over and complained to the president that Gertrude had no apparent qualifications for the job. To the critics, and to Mrs. Van Allen,
President Roosevelt said Gertrude had good taste and knew how to
manage a large country house.[78] Since the Park Service wouldn't hire
her, President Roosevelt appointed Gertrude S. Cooper by executive
order, and "without Regard to the Civil Service Rules," as the superintendent of the Vanderbilt National Historic Site, the *first woman* in
charge of a national park.[79]* She was sworn in on July 17, 1940.

Job one for Gertrude was to open the site to the public in just two
weeks; she did. To restore, adapt, and run the estate for the public, the
Park Service estimated it would need eight employees. At its peak,
Gertrude managed 94. For the first year the Park Service expected 300
visitors per month; Gertrude welcomed an average of 5,612.[80]

As an example of Gertrude's creative problem-solving, she imported
a flock of sheep to mow the grass and a troop of Boy Scouts to tend the
sheep in return for the wool and meat. She, of course, got the approval
of the president, who was quite comfortable with her idea, having
played golf among the free-range flocks of Campobello.[81]

In her mind, Gertrude worked for and reported to Franklin, not the
Park Service. According to Harold Ickes, secretary of the interior, of
which the National Park Service is a part, Gertrude's personal relationship with the president upset the entire chain of command.[82] When
Gertrude needed to get something done, she wrote to "Dear Franklin,"
made her request, and signed off, "Affectionately, Gertrude." Sometimes she reminded him of their youthful flirtations. If she couldn't
catch him as he drove himself around the neighborhood[83] in his specially modified car with motorcycle-style hand controls, she might
threaten to "camp out on your door."[84] Decades later, the Park Service's
official history of the mansion conceded that Gertrude was an unusually

effective manager, possibly more so than anyone else could have been, *given the circumstances.*[85]

Concluding her memoir of her years at Hyde Park, Gertrude wrote, "I have always said that I would like to live in a Dictatorship, provided that I knew the Dictator personally."[86]

Exceeding expectations happened frequently as the prewar economy heated up. In Maine there was so much demand for electric power that in June 1940 CMP announced plans to build a new 20,000 kW steam-turbine plant for Cumberland County.[87] In October CMP said it would expand its hydropower plants on the Kennebec River by 7,000 kW,[88] as well as build a new 16,000 kW steam plant at Bucksport;[89] then in December CMP announced it was going to build another 27,000 kW steam-power plant "somewhere on the middle seacoast of Maine." Meanwhile, the Great Northern Power Company completed a new 18,000-hp hydropower development on the Penobscot River at Matta-seunk. And since 1937, the Cumberland County Power and Light, partially and soon to be fully owned by CMP, had added 9,000 hp at its plant on the Saco River in Portland.

In total, 97,000 hp was being added to Maine's electric power re-sources "since the power companies of this State helped to kill the Quoddy project 'because there was no market for power,'" the *Sentinel* noted dryly. "Time, and their own activities have amply demon-strated the insincerity of these arguments, of the power barons, but in the meantime, the damage is done. . . . [W]e have only what we had before,—a county denied development and kept in the status almost of a colony, of concern to the remainder of the State only insofar as it contributes revenue, market, or jobs for favored political hangers-on."[90]

The political power of the power trusts was a problem everywhere,[91]* as Senator Homer Bone (D) of Washington made clear: "For 25 years out in the Northwest most of the corruption in public life could be at-tributed directly to the private ownership of government exemplified by the control of the private power companies. . . . There is not a well-informed politician in my neck of the woods who did not know that the State of Washington, like other Northwestern States, was owned in political sense by the private power companies."[92]

Since the power trusts "owned" many of the states, it wasn't unreasonable for them to think they could "own" the nation. All they needed was the presidency.

Who would be the next president of the United States? The leading Republican contenders in the summer of 1940 were Ohio senator Robert A. Taft (son of the former president), Michigan senator Arthur Vandenberg, and New York district attorney Thomas E. Dewey. Taft, an isolationist, was losing ground as the Nazis overran Europe. Dewey, still politically wet behind the ears, didn't look like he could withstand Nazi aggression.[93]* That left Vandenberg in the lead, but in a field where two-thirds of the delegates were not pledged to any candidate.

Enter Wendell Willkie, the dark horse, with a lot of horsepower behind him.

Wendell Willkie was a big midwesterner with a law degree and Wall Street smarts. He had never run for public office, though he had been a pro-Roosevelt delegate at the 1932 Democratic National Convention. Willkie turned against President Roosevelt when the scope of his trust-busting public power policy became evident. Now Willkie was the president of Commonwealth & Southern (C&S), one of the largest utility holding companies. C&S had failed in its attempt to have the Supreme Court declare PUHCA unconstitutional and thus became the first victim of its "death sentence." C&S was forced to divest some of its assets, including TEPCo, the Edison Electric Institute (EEI)'s award-winning company in TVA territory,[94] which was sold to the TVA at a significant profit, earning Willkie a reputation as a brilliant and tough negotiator.[95]

Willkie was highly motivated and articulate. He fiercely opposed government involvement in business but supported many of the New Deal social welfare programs.

While Democratic and Republican political campaigns were going on in the United States, by mid-June 1940, Germany's military campaigns had succeeded in conquering Norway, Denmark, Belgium, Holland, Luxembourg, and France. Only the English were resisting Hitler's advance, but they were suffering the Blitz (September 7, 1940–May 11, 1941), the aerial bombing of the civilian population. For the English, it was a question of survival. They desperately needed food, fuel, arms, ammunition, ships, and planes.

For Americans, the issues of the 1940 presidential election included the European war; the role of government raised by Roosevelt's New Deal and public power policies; and Roosevelt's precedent-busting, king-crowning, dictator-like decision to seek a third term as president.

On August 30, by executive order and *without* congressional knowledge or consent, President Roosevelt announced the Destroyers-for-Bases deal, in which he had traded 50 older US Navy destroyers to cash-strapped Britain for 99-year leases on military bases in Newfoundland, Bermuda, the Caribbean, and South America.[96]* Isolationists rightly complained the deal was in violation of the US Neutrality Act and risked Germany's considering it an act of war—while giving away America's means of defense.[97]* Willkie denounced the action as "the most dictatorial and arbitrary decision of any President in the history of the United States."[98] He then changed his stance to isolationist and made the war (and fear of it) the central issue of his campaign,[99] saying "you may expect we will be at war by April 1941 if he [Roosevelt] is re-elected."[100]

In a dramatic turn of events at the Republican National Convention, Willkie, who had skipped the primaries, went from nowhere to winning the nomination on the sixth ballot. Thus, Willkie became the official champion of *both* the private power companies and the Republican Party.[101] He went on to wage a vigorous campaign, which was enthusiastically supported by hundreds of "Willkie Clubs" that sprang up all over the country. His campaign featured tremendous media coverage, with hagiographic propaganda appearing in the Republican press, including the *Sentinel*, despite Roscoe's presumed personal preference for Roosevelt.

Nonetheless, Roosevelt won the presidential election with popular majorities in 36 of the 48 states. As usual, Maine was not among them, but Washington County was.[102]*

Roscoe attributed Willkie's defeat to intrigue at the Republican convention: "His nomination was accomplished by methods not wholly clear, and without adequate preliminary presentation to the voters. While much capital was made of the apparent spontaneity with which he was named, there remained always some doubt as to whether his victory was one of the people over the politicians, as reported, or one of power politicians over all others."[103] Roscoe's suspicions were correct: the Willkie Clubs had been covertly organized and supported by the EEI and "public" utilities—at ratepayers' expense.[104]*

Thrice crowned as their president, Franklin Roosevelt spoke to the American people by radio on the evening of December 29, 1940. "My friends," he said, "this is not a fireside chat on war. It is a talk on national security." He told them they had met the financial crisis of 1933 "with courage and realism" and must do the same with this new crisis. "Never," he said, "has our American civilization been in such danger as now":

> The Nazi masters of Germany have made it clear that they intend not only to dominate all life and thought in their own country, but also to enslave the whole of Europe, and then to use the resources of Europe to dominate the rest of the world. . . .
>
> If Great Britain goes down, the Axis powers will control the Continents of Europe, Asia, Africa, Austral-Asia, and the high seas.
>
> And they will be in a position to bring enormous military and naval resources against this hemisphere. It is no exaggeration to say that all of us in all the Americas would be living at the point of a gun—a gun loaded with explosive bullets, economic as well as military. . . .
>
> Some of us like to believe that even if Britain falls, we are still safe, because of the broad expanse of the Atlantic and of the Pacific. But the width of those oceans is not what it was in the days of clipper ships. . . . Why, even today we have planes that could fly from the British Isles to New England and back again without refueling. And remember that the range of the modern bomber is ever being increased. . . .
>
> I make the direct statement to the American people that there is far less chance of the United States getting into war if we do all we can now to support the nations defending themselves against attack by the Axis than if we acquiesce in their defeat, submit tamely to an Axis victory, and wait our turn to be the object of attack. . . .
>
> The people of Europe who are defending themselves do not ask us to do their fighting. They ask us for the implements of war, the planes, the tanks, the guns, the freighters which will enable them to fight for their liberty and for our security. Emphatically, we must get these weapons to them, get them to them in sufficient volume and quickly enough so that we and our children will be saved the agony and suffering of war which others have had to endure. . . .
>
> Our national policy is not directed toward war. Its sole purpose is to keep war away from our country and away from our people.
>
> American industrial genius, unmatched throughout all the world in the solution of production problems, has been called upon to bring its resources and its talents into action. . . .

But all of our present efforts are not enough. We must have more ships, more guns, more planes—more of everything. . . .
We must be the great arsenal of democracy.[105]

The **"arsenal of democracy" needed more power.** Vast power. Overwhelming power. Five million kW more power.[106] The Roosevelt administration's public power projects—Hoover, Bonneville, Grand Coulee, and the TVA—were how America would generate that enormous increase in electric power.

The Bonneville Dam on the Columbia River began transmitting electric power in 1937. On September 1, 1940, Alcoa began making aluminum at its plant in Ampere, Washington. When fully up to speed, it would consume 65,000 kW (*twice* the output of CMP's Wyman Dam)[107] and produce 60 million tons of metal per year, increasing America's output by more than 50%.[108]*

We would need all of it. And more.

In 1940, *all the strength we could muster* was only a few hundred combat airplanes. To defeat the Nazis, President Roosevelt estimated that America would need to produce 50,000 aluminum airplanes *per year*.[109] Edward Stettinius, chairman of U.S. Steel, speaking for the National Defense Advisory Commission, assured the nation that Alcoa—which had enjoyed a patent-protected total monopoly on aluminum production in the United States for 48 years—could produce more than enough of the essential metal for every foreseeable need.[110]

Quoddy could still be a huge asset in the war *for* American power—if the power trusts would fight for America rather than for themselves. But Big Business didn't trust the trust-busting president. They viewed his call for cooperation and coordination in the name of defense as a government takeover in disguise.[111]

UNITED THEY STOOD

The **"democratic way of life [is] being directly assailed** in every part of the world," President Roosevelt warned in his State of the Union address on January 6, 1941. World security, he said, was founded on four freedoms: freedom of speech, freedom of religion, freedom from want, and freedom from fear. It was "no vision of a distant millennium . . .

[but a] world attainable in our own time. . . . To that high concept there can be no end save victory."[112]

But in our efforts to achieve victory, we were "behind schedule in turning out finished airplanes" and must work "day and night . . . to catch up."[113] We were behind because Alcoa had lied about its capacity.

Since 1937 the Justice Department had been investigating Alcoa for antitrust violations. Interior Secretary Harold Ickes called Alcoa "one of the worst monopolies that has ever been able to fasten itself upon American life."[114] His assistant explained how: "First, they had patent control, then they moved to control power, then mining. Everything they did was calculated to keep the supply down and the prices up."[115] As historian Doris Kearns Goodwin relates in *No Ordinary Time:*

> Limited supply was damaging enough in peacetime, in wartime it was catastrophic, for, even as Stettinius[116] was heaping praise on Alcoa's unlimited capacity, Alcoa was unable to fulfill the defense orders already on its books. . . .
>
> As one airplane company after another testified at hearings to delays in manufacturing as a result of shortages of aluminum, the Alcoa men finally had to admit that they simply could not keep up with the demand. The government responded to the admission by bringing the Reynolds Metals Company into the business with a generous Reconstruction Finance Corporation loan to finance construction of a large new plant. But invaluable time had been lost. "When the story of the war comes to be written," Harold Ickes testified at the hearings, "if it has to be written that it was lost, it may be because of the recalcitrance of Alcoa."[117]*

Nazi Germany was getting close to victory over Europe. Only England was still actively fighting it. *But how long could England defend itself? How could the Brits ever defeat Germany?*

Wendell Willkie went to England in January 1941 to see the struggle firsthand. He went as the personal envoy of his former adversary, Franklin Delano Roosevelt, who gave him a handwritten letter of introduction to British prime minister Winston Churchill:

> Dear Churchill,
>   Wendell Willkie will give you this—
>   He is truly helping to keep politics out over here.
>   I think this verse applies to you people as it does to us:

"Sail on, Oh Ship of State!
Sail on, Oh Union strong and great.
Humanity with all its fears
With all the hope of future years
Is hanging breathless on thy fate."

As ever yours
Franklin D. Roosevelt[118]

The quote was from "The Building of the Ship," an allegorical poem about the United States on the verge of the Civil War by Maine's poet laureate, Henry Wadsworth Longfellow.[119]* Roosevelt's letter was "an inspiration" to Churchill. He framed it and hung it in his home.[120]

Even with inspiration, the British couldn't build or buy enough of anything. To survive the Nazi onslaught, they needed more of everything.

While at sea aboard the USS *Tuscaloosa*, on a hasty and (his advisors thought) ill-timed vacation to the Caribbean in mid-December 1940, the commander in chief thought deeply about *ships* and *the hope of future years*. Once again, Roosevelt drew strength from the sea—and had an epiphany. It was, as Secretary of Labor Frances Perkins later described, a "flash of almost clairvoyant knowledge and understanding."[121]

Returning to the White House "refueled" on December 17, President Roosevelt proposed the Lend-Lease program, a dramatic expansion of the Destroyers-for-Bases deal, through which the United States would lend civilian and military materiel to the British, who would repay it in kind when they could. Part of the value we would receive in return was the increase in our own security: by our keeping the British fighting, they kept us out of the war.[122]

Cleaving the waves of criticism and the confused sea of uncertainty like *Tuscaloosa*'s well-formed bow, the president used a parable to explain the desperate urgency that required our immediate moral response: Our neighbors' house was on fire, and we had the fire hose. We must not hesitate to provide it. We must not dicker about price. We must give the aid now, when needed. And when our neighbors are out of danger, they will return or replace the fire hose.[123]

President Roosevelt was calling for action the American people would understand and support. But the *Sentinel* recognized that the president's plan was "almost an informal declaration of war,"[124] regardless of his assertions to the contrary,[125] and vehemently opposed by isolationists and the Republican Party.

But not by Willkie.

Life magazine chronicled Willkie's trip to England, which made him "an international celebrity whose prominence in news columns of two hemispheres exceeded at times that of the man who beat him. . . . In the last ten days . . . he scored a personal triumph of fabulous proportions. He went everywhere, saw everything, and met everybody at a breathtaking tempo that completely diverted the embattled British from bombs and U-boats."[126] Said *Life*, "Wendell Willkie left no doubt that he was 100% for Britain and the cause for which it fought."[127]

Returning on February 11 from his eighteen-day, 26,149-mile, four-continent tour of Allied countries, Willkie went straight from the airport to the US Senate to give his report.[128] To hear him speak, 1,200 people wedged themselves into a room designed to hold 500. He dismissed his former fearmongering as "a bit of campaign oratory,"[129] urged passage of the Lend-Lease bill, and advised America to "send Britain five or ten destroyers a month and all its bombing planes except those needed for training." Sitting on and leaning over the table to forcefully make his point, he warned the United States would be at war in 60 days if Britain fell to the Nazis.

The shining star of the Republican Party was now a pariah to its isolationist leaders.[130]* Nonetheless, Lend-Lease was signed into law on March 11. In the words of Walter Lippmann,[131]* Willkie's advocacy for aid "made it possible to rally the free world when [Britain was] almost conquered. Under any other leadership but his, the Republican party in 1940 would have turned its back on Great Britain."[132] And Great Britain would have fallen.

Arthur Vandenberg got what he asked for in Senate Resolution 62: for the second time the FPC killed Quoddy.[133]* The FPC delivered its report on April 7, 1941. The report's first sentence did the trick: the cost to build a tidal-hydroelectric power plant would be $89 million to $91 million.[134] Sentence three delivered the punch line: a coal-fired steam-electric plant of the same output would cost only $15 million. Quoddy was dead again. No need to read further.

Unless you cared about *the truth, the whole truth, and nothing but the truth.*

Let's pick apart what the FPC said and didn't say.

Sentence one compared the *capital cost* of a tidal-electric plant versus steam-electric—*not* the *cost of the power generated*. All hydroelectric plants are more expensive to build than steam plants but cheaper to operate because they don't need fuel. Thus, the FPC was comparing apples to banana peels. *Very slippery.*

Sentence three did address the cost of the power: 6.65 to 10.47 mills per kWh for tidal power, versus 4.49 to 5.57 mills for steam power. Note that the 6:1 differential in the capital costs of tidal power versus steam power has dropped to about 2:1 for the power generated. This would seem to have been a fair comparison, *but was it?* Just as the army had *literally* done when building Quoddy Village, the FPC gold-plated its tidal power straw man. Thus, its estimate was high.[135]*

Second, the FPC assumed the cost of coal would stay the same, indefinitely, despite knowing fuel costs would go up due to the war in Europe.[136]*

Senate Resolution 62 didn't ask for it, so the FPC didn't report in any detail on the small experiment tidal power project proposed by President Roosevelt or on the most efficient version, Dexter's International Plan,[137]* and dismissed the idea of operating Quoddy in tandem with existing hydroelectric reservoirs, as Colonel Fleming had suggested in 1936.[138]

Senate Resolution 62 didn't ask for it, but FPC did use a large part of the report to promote building seventeen hydroelectric dams on Maine's rivers, at a total cost of $105 million (*about the same as Quoddy's International Plan*). According to the FPC, the seventeen river dams would generate five times as many kWh as the FPC's tidal straw man. The cost of the river power would be only 1.87 to 3.26 mills per kWh, or about one-third of the estimate for tidal power.[139]*

The FPC recognized the demand for electricity was growing, but not so rapidly that "orderly development" of river power wouldn't take care of it.[140] It dismissed using tidal power for smelting aluminum because it would be more expensive than the 2 to 3 mills per kWh usually paid for the purpose.[141] Therefore, the FPC concluded that there was "no local or export market at this time" for the power, regardless of how it was generated.[142] While the commission acknowledged a potential increase in defense-related demand for electricity, it dismissed Quoddy as a solution because of the projected costs.[143]

The report did throw bones to Quoddy's supporters, acknowledging that "no informed person" would challenge the engineering feasibility and there were no "insuperable obstacles to the construction."[144] The FPC also said tidal power was the "most permanent source of power known to Man" and not affected by droughts or floods—an important point since 93.8% of Maine's electricity in 1940 was river powered.[145] Then, in the most frustrating way, the FPC concluded:

> The natural and doubtless justifiable presumption that the construction of a large international tidal power project at Passamaquoddy will not be economically feasible or desirable in the near future should be no bar to a thorough exploration of the possibility of such a project jointly by the Governments of the United States and Canada. It may confidently be assumed that the power potentially available in the Passamaquoddy tides will ultimately be developed. The event seems certain; the only uncertainty is in point of time. . . .
>
> The idea of generating electricity from the enormous latent power of the tides along the shores washed by the Bay of Fundy is fascinating, interesting, entrancing—a challenge, as it were, to the imagination and ingenuity of Man. As high-grade fuels become scarcer and dearer concurrently with greatly increased demand for electric power in the northeastern States and in contiguous Canadian territory, attention will surely be directed to the problem of harnessing the tides.[146]

The FPC's 1941 report was not the whole truth. It was, perhaps, some hard truths—and certainly some convenient assumptions, deliberate omissions, and dexterous deferrals.[147]*

Regardless of the FPC, America's power position was about to change.

Spring rains didn't come, and by June 1, 1941, much of the United States was in a drought. Reservoirs of water—and hydroelectric power—were dropping. Seeing another chance to redeem himself, Senator White introduced legislation for the construction of the International Plan, figuring the power shortages and war concerns would convince his colleagues in the Senate of Quoddy's merit. Two days later, power trust–influenced members of Congress passed a joint resolution prohibiting the Reconstruction Finance Corporation (RFC) from funding the Florida Ship Canal, the St. Lawrence Seaway, or Quoddy.[148]

Trying again, Senator White called the FPC, which confirmed the power shortage in Maine. The next day, Walter S. Wyman, president of CMP, asserted that Maine had a power surplus (despite the FPC's report promoting the construction of seventeen new power plants in Maine) and—*Believe It or Not!* —declared his company had foreseen the drought![149]

Even the TVA was experiencing water and power shortages and was seeking $50 to $75 million—*about the cost of Quoddy*—to construct more *hydro*electric power plants.[150]* That other liquid fuel—oil—was now in such short supply and so expensive that it was being rationed,[151] thus combusting one of the core arguments the FPC had used against Quoddy just two months before.

Meanwhile, Charles W. Kellogg, former president of the EEI, who for years had said, "We have ample power and there is absolutely nothing to worry about," was now showing his benevolence as a "dollar-a-year man" working for the government "trying to find enough power to fill the 800,000-kilowatt shortage in order to keep defense industries going"[152]—*all of which power Quoddy could produce*. Quipped Ickes, "He's worth all of the $1 a year he's being paid."[153]*

The missing power meant America couldn't make enough aluminum. Recycling one of Samuel Insull's World War I ideas,[154] a nationwide aluminum donation drive was started by the Kiwanis service club and three Congresswomen, including the newly elected representative from Maine, Margaret Chase Smith.[155] To this "teakettles for airships" campaign, Eastport housewives contributed 500 pounds of used pots and pans,[156]* a veritable drop in the ocean of molten metal Dexter Cooper could have *already* smelted using Quoddy power and Russian smelting technology.

To deal with the shortage, the RFC financed the formation of another new aluminum company, Kaiser, as well as an additional Alcoa plant; both got their electricity from the public hydropower projects at the Bonneville and the TVA.[157]*

Even the formerly antagonistic *Portland Sunday Telegram* was beginning to wonder, "Could it be that the Nation might be a little better off if they [the 'wild dreamers'] had their way and Quoddy was turning out badly needed power? Would defense industries now be moving Maineward away from areas of power shortage to a region where power was plentiful?"[158]

Germany invaded Russia on June 22, 1941. Many had expected it to happen eventually, but it came with the terrifying speed of the

Blitzkrieg. The Nazis' objectives were to seize territory and resources from their eastern neighbors in Soviet Ukraine. In a conquered Ukraine, Hitler imagined more *Lebensraum* or "living space" for his superior Aryan race—a place where his army could feed off Ukraine's fertile fields and where he could harvest its coal, oil, and hydropower for his ravenous war machine. Operation Barbarossa was the largest invasion force in history. It was led by 2,700 planes followed by 3,400 tanks and 3.5 million troops. Germany destroyed most of Russia's air force before the planes got off the ground. Within a week the Germans were 500 miles into Soviet territory. In two weeks, they reached the Dnieper River. Hundreds of thousands of Russians were killed; 600,000 were taken prisoner. The Nazis seemed unstoppable.[159]

Fifteen years earlier, Hugh and Dexter Cooper had built a secret gallery deep in the body of the Dnieper River Dam. It was designed for use only once, only in an extreme emergency.[160] That day came on August 20, 1941. "Acting under personal orders of Premier Josef Stalin," Soviet army marshal Semyon Budenny did the unthinkable to prevent the worse: he lit the fuse to tons of dynamite stuffed in the secret gallery and destroyed the dam to prevent it from falling into the hands of the Germans. The resulting flood killed thousands of civilians.[161] Blowing the dam also blew the bridge over it,[162] covering the Russian army's retreat.[163] To further hamper the Germans, the Russians took with them as much of the power plant equipment as they could cart away. American journalist H. R. Knickerbocker, who had won the Pulitzer Prize in 1931 for his coverage of Russia's Five-Year Plan,[164] explained the significance:

> The Russians have proved now by their destruction of the great dam at Dniepropetrovsk that they mean truly to scorch the earth before Hitler even if it means the destruction of their most precious possessions. . . . Dnieprostroy was an object almost of worship to the Soviet people. Its destruction demonstrates a will to resist which surpasses anything we had imagined. I know what that dam meant to the Bolsheviks . . . when it was built it was the biggest on earth and so it occupied a place in the imagination and affection of the Soviet people difficult for us to realize. . . . Stalin's order to destroy it meant more to the Russians emotionally than it would mean to us for [Theodore] Roosevelt to order the destruction of the Panama Canal.[165]

Germany's double-crossing invasion forced Russia to join the Allied side[166]* and, in self-defense, to suffer immense losses. Wendell Willkie argued for extending Lend-Lease to Russia, and Margaret Chase Smith and a few other rogue Republicans voted with the Democrats in approving the extension.[167] When Harry Hopkins met with Josef Stalin in Moscow in July and asked him what he needed most, Stalin replied (1) anti-aircraft guns and (2) aluminum.[168]

Summer is for vacations, and for Franklin Roosevelt that meant sailing, though the war cramped his style. He could no longer simply borrow a sailboat, race unknown skippers he happened upon, or play cat-and-mouse in the fog with his Secret Service escorts. Now he had to stick to the big (and armed) presidential yacht, the USS *Potomac*, closely followed by a pair of navy destroyers, with the press, public, and all other potential threats kept a safe distance away. Still he went, though his itinerary was secret.

Eleanor was already at Campobello, with a gaggle of friends, including the reporter Elizabeth May Craig. FDR thought highly of Craig, saying she "stands in the first row of my press conferences."[169] Craig flew up to Eastport in a private plane with her boss, Guy Gannett, the newspaper magnate. It was her first visit to the place she had written about so often.[170] Eleanor was busy as ever, inspecting NYA's Quoddy programs; hosting dignitaries, social activists, and student leaders; and writing about it all in her column, "My Day."[171]* Odds were Franklin's island home would be among his destinations, and the *Sentinel* speculated that the US president might use the opportunity to meet with the Canadian premier: "Campobello would be an ideal place for it, for reasons too obvious to be recounted."[172]

On August 4, President Roosevelt traveled by train to the submarine base at New London, Connecticut, where he boarded the *Potomac*, spending the first night at Point Judith, Rhode Island. The next day the *Potomac* and its escorts cruised into Buzzard's Bay, where Franklin was seen tooling around in a Chris-Craft speedboat with Princess Martha of Norway[173]* before retiring to the *Potomac*, which then anchored at Martha's Vineyard, Massachusetts. When the sun came up the next day, the *Potomac* weighed anchor as its famous passenger waved to the crowds on shore. Through the Cape Cod Canal the president's flotilla went, escorted on shore by Secret Service and State Police cars and hundreds of well-wishers. With his jaunty cigarette holder, cocked hat, and big waves, the president seemed to be enjoying the journey.

Only he wasn't there. It wasn't him. It was a double. A look-alike. A decoy.

The real president of the United States and commander in chief was not on a vacation cruise to Maine; he was on a mission, far out at sea, steaming northeast with a squadron of fighting ships.[174] They were going to Newfoundland, to the new naval base at Placentia Bay that the United States had acquired in the Destroyers-for-Bases deal. There, a secret rendezvous would take place—under the protection of an equally large squadron of British ships—with Winston Churchill.

No one, not the secretary of state, not Eleanor or Sara Roosevelt, not even the Secret Service, knew in advance about the charade the president had planned and staged.[175]

On August 10, 1941, the British battleship HMS *Prince of Wales* came alongside USS *Augusta*, where Roosevelt awaited his distinguished guest. Churchill and Roosevelt, who had not met before, regarded each other in silence for a moment. Then Churchill said, "At long last, Mr. President," and Roosevelt replied, "Glad to have you aboard, Mr. Churchill."[176]

Over the next five days Roosevelt and Churchill worked out the Atlantic Charter. The charter made clear the United States would support Great Britain in its war against Nazi tyranny but focused primarily on the peaceful world the two nations wanted after the war. The Charter fit on one page. It was simple. It was powerful. It was never signed. It was honored.

President Roosevelt returned to Maine and a hero's welcome. While in Portland he met with Guy Gannett, the newspaper baron who was once opposed to Quoddy but who called the president's attention to the need for tidal power, "now sharply emphasized by the present unhappy situation of Maine with reference to power and fuel."[177]

The "unhappy situation" was caused by the simultaneous occurrence of several things—*assumptions*—that were each previously thought (or said) to be remote possibilities but now conspired to create conditions far worse than expected. In this case, the first problem was the rapidly growing consumer demand for electricity due to the proliferation of electric devices and the improving economy. Second was the skyrocketing demand for electric power by industry, particularly for defense. Third was the increasing expense and scarcity of coal and oil for steam plants because of defense needs for *portable* fuels for ships, planes, and tanks and the concurrent interruption of supply lines due to the Battle of the Atlantic. Fourth was the deepening drought and the resulting reduction in hydroelectric power production.

The "unhappy situation" was about to become much worse.

On September 4, 1941, a German submarine fired on the destroyer USS *Greer*. On September 7, Sara Roosevelt died. On September 11, with a black band of mourning on his left arm, President Roosevelt ordered the US Navy to "shoot on sight" anything that *could* threaten any US naval or merchant ship.

By October, New Brunswick was forced to extend daylight savings time to save power. At CMP, Walter S. Wyman admitted things were "getting very bad," with the demand for electric power up 20% but water reservoirs one-third below normal.[178] Even if enough fuel could be found to run the steam-power plants at full capacity, they could not make up for the shortage of waterpower. Some factories in the state shut down for lack of power, others were forced onto shortened schedules, and plans were made for rationing.[179] Maine and the Maritimes now woke each morning to the harsh dawn of wartime economy.

In November, the United States signed an agreement with the Dutch government in exile[180]* for the US military to occupy its colony of Surinam in South America to secure for Allied use its bauxite mines. The bauxite was shipped to Portland, Maine, and then sent by rail to Alcoa's huge hydroelectric-powered aluminum smelter in Ontario, Canada, thus avoiding Nazi U-boats in the North Atlantic.[181] Remarking on the irony, the *Portland Press Herald* said, "Tidal power, like its companion, time, in the greatest duet of the infinite, goes on inexhaustibly. . . . Dexter Cooper had something. It is not yet too late to make his dream come true."[182]

It was too late.

On December 7, 1941, Japan attacked the United States.

CASUALTIES

Edward Haven, a "Quoddy Boy" turned navy man, was killed by the Japanese when they attacked Pearl Harbor.[183] He was the first Quoddy Boy to give his life for his country in World War II.

Louis Vitelli was the second Quoddy casualty of war. He was stationed on the USS *Pillsbury* in the Java Sea, which was lost without a trace in March 1942. Effectively an orphan, when Louis wrote letters "home," he wrote to Mrs. Fountain in Eastport, his "adopted Mother."[184]*

On the front lines of national defense, on the point of land closest to the enemy, exposed to attack by air and sea, and with lives already lost, Quoddy Village became a crucible, forging men from boys, soldiers and sailors from ballplayers. Quoddy Boys built a forge and cast bronze parts for submarines. They organized air-spotters and staged air raid drills. The air raid wardens carried billy clubs and were authorized to smash any light bulbs or headlights not turned off. The fire and ambulance crews rushed to put out bonfires and care for the injured in mock bombings.[185]

Food was part of the fight, and Maine provided two ingredients in the Allies' recipe for success. Highly nutritious, easily transported and stored, and then eaten right out o' the can or cooked in it, sardines were a perfect food for an army. In February 1942, the US government announced it would purchase at least two-thirds of Maine's production—at least 2 million cases—all to be shipped to England through Lend-Lease.[186]

On the other side of the world, as US forces fought their way across the Pacific, the soldiers developed a hunger for home. They wanted pie. Good old American pie. Blueberry pie. Rich in flavor and highly nutritious blueberry pie. Spread 'em on toast or bake 'em in a crust o' crackers—it didn't matter; they wanted blueberries. The army asked, and Downeast delivered.[187]

You can be sure Mr. Pottle's pigs didn't get a lick.

The Quoddy Tidal Power Project was a casualty of war (or at least among the wounded). President Roosevelt made it official in April 1942: Quoddy would not be on active duty until after the war since it could not contribute immediately, and its exposed position put it at risk.[188]

Roscoe put as good a face forward as he could: "There will be no disposition to find fault with his decision . . . and there is quiet satisfaction with, and confidence in, the implied promise."[189]

NYA's Quoddy Village Work Experience also succumbed. It had been created as work relief, but now that the nation was fully employed in the war effort, it was no longer needed, and it shut down in August 1943.

Due largely to President Roosevelt, who thought it would be "a crime to abandon Quoddy Village,"[190] the navy took it over on August 31, 1943—not as a naval base but for training construction battalions, or Seabees.[191] They rapidly expanded and renamed the facilities Camp Lee-Stephenson, bringing thirty-five hundred transient residents to Eastport.

Disaster struck on March 3, 1944. The Administration Building, the most visible evidence of the Quoddy Tidal Power Project, caught fire. It began in the library sometime after midnight. Thirty-mile-per-hour winds fanned the flames. In minutes, Eastport's wooden "White House"[192] was ablaze. The inferno lit the night sky and George Rice's hay barn. The nearest water hydrant didn't work. Soon it was gone. All gone. The cause of the fire was suspicious.[193]

With the navy's training needs winding down in late 1944, the numbers of Seabees stationed at Quoddy Village began to drop.[194] Soon they, too, would be gone. All gone.

Wendell Willkie died on October 8, 1944. He was 52. His book, *One World*, written after a 49-day, 30,000-mile journey as President Roosevelt's unofficial envoy, sold a million copies within a month. He had transcended politics.[195]

By January 1945, the Allies were beginning to win. In two years, American industry and labor had produced a manufacturing miracle, doubling, then doubling, then doubling again their output. Innumerable ships, tanks, planes, and material of every description roared out of factories and into the fight. The prewar work-relief programs had helped solve the Depression's unemployment and showed the nation—previously segregated by gender, race, class, politics, and geography—*how* to work together.

The Royal Air Force had repulsed the German Luftwaffe and gained air supremacy over the English Channel. That allowed American, Canadian, and British troops to launch the greatest amphibious invasion in history. They broke through Hitler's Atlantic Wall on D-Day, June 6, 1944, in Normandy, France. While the Allies pushed eastward toward Berlin, the Russians pushed westward.

Acknowledging the critical role Soviet Russia was playing in defeating Germany, the Allies decided the second conference of the Big Three would be held on Soviet territory. That meant President Roosevelt, 63 years old and in poor health, would have to travel through more than 5,000 miles of active war zones.

Under cover of darkness, President Roosevelt slipped out of the White House, took a train to Newport News, Virginia, and boarded the heavy cruiser USS *Quincy* on January 23, 1945. With the anti-submarine nets lowered to let it exit, the *Quincy* headed for the North Atlantic, escorted by a light cruiser and three destroyers. Roosevelt was invigorated by the trip, greatly enjoying the time at sea and the rough play of the sailors. After steaming into the Mediterranean, they stopped at Malta to meet Winston Churchill on February 2.

Roosevelt and Churchill then boarded the "Sacred Cow," the first presidential plane, a four-engine Douglas C-54 transport. Specially modified for President Roosevelt, the plane included an elevator to help him into the plane and another to lift him into the cockpit so he could sit with the pilots. They landed on a snow-covered runway at Saki, about 80 miles from their final destination, the czar's palace at Yalta in the Crimea. The final leg took eight hours traveling by car through bombed-out villages, passing destroyed tanks, and over roads lined by Russian soldiers. They were but 150 miles from the remains of the Dnieper Dam. The American assistance building Russia's industrial infrastructure was not forgotten.

The eight-day conference was hard on everyone, especially the ailing American president who had traveled so far. Nonetheless, Roosevelt chaired the meeting, and together, Roosevelt, Churchill, and Stalin planned the postwar world.[196]

They reached an agreement. They made promises to each other.

Downeasters began acting on the promise of postwar prosperity. In March 1945, Maine's legislature approved a new, further watered-down

Quoddy Authority Bill, to replace the lapsed 1939 version. The Authority's board now required the approval of the power trust–controlled Executive Council. And its scope of operations was further limited to eastern Washington County, "except," as the *Sentinel* noted, "Machias and Dennysville which for some obscure but presumably interesting reason were omitted."[197]

All was in readiness for when peace and President Roosevelt returned to Passamaquoddy Bay.

Franklin Delano Roosevelt died on April 12, 1945. He had been president for twelve years, one month, and nine days. The United States had been fighting World War II for three years, four months, and five days. Throughout the war, Franklin Roosevelt had been the nation's commander in chief and its spiritual leader, never faltering. At the time of his death, both Germany and Japan were collapsing. He was within months, maybe weeks or even days, of total victory. The *Sentinel* put the sad news in context:

> Fate denied Franklin Roosevelt the chance to enjoy the fruits of victory over the Axis. Yet history seemed destined to enshrine him as one of the immortal American Presidents. And every citizen who mourned the passing of the Commander-in-Chief felt that he was a casualty of the war just as every G.I., Marine and Sailor who had fallen in battle.[198]

Testimonials poured in from around the world, but as Roscoe related in his editorial:

> [I]t is as a friend that he is mourned here. Notwithstanding the fact that he had reached the highest pinnacle of eminence and power that the world now affords, the people of Quoddy never at any time felt that there had been any change, to the slightest degree, in the pleasant, unassuming, sincerely friendly and thoroly [*sic*] democratic personality he displayed as a summer resident at Campobello. They were sure that he could walk with kings and never lose the common touch which so endeared him to them, and they counted him as one of their own, as undoubted he was and desired to be.[199]

Memorial services were held all over Downeast Maine and attended by the *Sentinel*'s staff:

While the notice was short for the schools, Friday afternoon was devoted to suitable memorial programs. On Saturday afternoon the stores and all other places of business were closed from four to six . . . and in Eastport a five-minute period of silence and suspension of traffic was observed starting at ten o'clock Sunday morning, during the committal service at Hyde Park. . . . The chapel [at Welshpool] was filled to capacity, no less than a third of the population of the island filling every available foot of room there or standing in silence outside.[200]

Gertrude went to Franklin's funeral at Hyde Park:

I went to the funeral as one of his oldest friends. . . . It was bitterly cold but a beautiful clear day and the garden looked lovely and peaceful. We waited for two hours and then heard the horses' hoofs and the wheels of the caisson as they drove slowly along the outside of the hedge. The West Point Cadets played Chopin's Funeral March and the casket was carried through the narrow entrance to the garden. There were very few dry eyes and I saw many internationally known figures, statesmen, diplomats, and high-ranking officers of the Army and Navy blowing their noses and surreptitiously wiping their eyes at the same time. I personally was able to keep a pretty good hold on my self control until the three volleys were fired across the grave and "Fala" gave three short, perplexed barks. That seemed too much and I didn't try to hide the tears running down my cheeks.[201]

Nazi Germany surrendered unconditionally on May 8, 1945, just 24 days after Franklin Roosevelt's death. Despite Germany's capitulation, Japan refused to surrender. Concluding the war with Japan would stretch on and cause too many more casualties, US president Harry S. Truman decided to put an end to the war with a demonstration of America's supreme power. To that end, he ordered the US military to detonate atomic bombs over Hiroshima and Nagasaki. Confronted with its potential annihilation, Japan surrendered on August 15, 1945.

The war was over.

We won.

He won.

CHAPTER 13

The Atomic Age (1946–1969)

DISMEMBERMENT

The Passamaquoddy Bay Tidal Power Project had been conceived in 1919 by Dexter Parshall Cooper, one of the most experienced hydroelectric engineers in the world. It had been championed by Franklin Delano Roosevelt, the most popular and longest-serving president in US history. Both were now dead.

Who would pick up Quoddy's torch: the electric light of clean power for the people? What man, or woman, would have the vision, the knowledge, the skill, and the stamina to make tidal-hydroelectric power— "moonpower"—more than a "moondoggle"?

Were Dexter and Franklin ahead of their times? Or was Quoddy quixotic, a romantic but foolish dream?

Would atomic energy power a world of peace and prosperity?

What did the future hold for the people of Passamaquoddy Bay?

The immediate future was bleak.

Gertrude Cooper was fired by the National Park Service immediately after President Roosevelt's death.[1]* Flagstones, the house on Campobello Island she and Dexter had built for $30,000, and which had sat empty for six years, was finally sold in May 1945.[2]

Roscoe Emery sold the *Eastport Sentinel* in April 1946.

Quoddy Village was abandoned by the Seabees on May 31, 1946.[3]

Sara Roosevelt's dilapidated cottage was dismantled in August 1946.[4]

The fire started on May 21, 1947, in downtown Eastport. Oliver W. Holmes Sr. sounded the alarm at 7:00 p.m. His house was the first to be consumed by the flames. Luther Gardener's carpentry shop, Carroll Gardener's grocery, and Harry Jollotta's plumbing shop were total losses. The fire wiped out the Holmes Packing Company owned by Moses Pike. When the Mearle sardine plant exploded, it blew out the windows of three stores and BHE's transformer building. "[I]t looked as if almost all of the North End was doomed as sparks and embers dropped like snowflakes on the rooftops." To prevent more fires, BHE cut the electric power to all of Eastport. Then, *thank God*, a heavy rain fell. Contrary to early reports, no one was killed. Property losses were estimated at $1 million; 500 lost their jobs.[5]

All of Maine seemed to be burning in October 1947. Months of drought had dried the tinder, so when sparks flew—from campfires, cigarette butts, or whatever—they found fuel. Once a fire started, there was little water to douse it. Forty-five-mile-per-hour winds fanned the flames, sparking new fires that outran the thousands of volunteers who raced into Maine to help.

There were a dozen forest fires of more than 100 acres each. A quarter of Mount Desert Island—17,000 acres—burned. The largest fire burned 109,110 acres. Fires destroyed 851 homes and 397 seasonal cottages, leaving nothing but empty foundation holes and smoldering ash. All told, more than 200,000 acres burned; fifteen people were killed, and 2,500 were left homeless.[6]

The Reconstruction Finance Corporation's regional chief, in Boston, recommended Quoddy Village houses be ripped off their foundations and trucked to other parts of Maine as emergency shelters. Major General Philip Fleming, now head of the Federal Works Administration, looked for better solutions.[7]

As the dry weather continued, Maine's reservoirs dropped, and waterpower with it. Electric power shortages became acute. Nonetheless, in July the Edison Electric Institute declared, "[R]umors on an impending shortage of electric power in the United States are unwarranted and untrue . . . all customer demands will be met in 1947 and in the years following."[8] In December, BHE took out a quarter-page ad in the *Sentinel* to counter concerns about the company's inability to meet demand. While first excepting "surplus power customers" in Lincoln, Howland, and Old Town, they then said, "No," there had not been any curtailment of power. But "Yes," they admitted, they did think the problem was "likely to be more critical." In the meantime, their customers should "AVOID WASTING ELECTRICITY"—*using all those appliances BHE had sold them*—so rationing wouldn't be required.[9]

BHE got the "splendid customer cooperation" it sought. Eastport merchants reduced their business hours,[10] and residents turned off their Christmas lights.[11] The *Sentinel* did its part in the conservation effort, setting its pages "solid" rather than "leaded," thus squeezing in three more lines of type per column inch and reducing the paper from eight pages to six.[12]

With their hydroelectric output cut in half,[13] CMP and BHE couldn't make up the loss, even when running CMP's steam-electric plants at full tilt. To increase capacity, two US Navy destroyers were chartered, tied to the pier at South Portland; with their engines running at the equivalent of 21 knots, but with the propellers disengaged, their diesel engines pumped out electric power while sucking up 30,000 gallons of fuel a day. Wired to shore, the power was transmitted more than 200 miles to Aroostook County. Operation Kilowatt lasted from November 1947 to March 1948[14] and got lots of positive press.[15]* Some of the press were more circumspect, pointing out that the lack of power was a self-imposed hardship courtesy of the monopolies.[16]*

Meanwhile, the newly built and independently owned hydroelectric plant on Meddybemps Lake near Eastport offered to sell its power to BHE for less than it was paying for the destroyers.[17] No deal. The reason is easy to guess: After the crisis the navy would leave. If paid for its power, Meddybemps would stay and destroy BHE's monopoly.

Operation Kill-a-Watt succeeded.

Unable to find a use for Quoddy Village, the federal government offered it for sale. Quoddy Village—which had been built by the army for $2.4 million,[18] expanded by the NYA, and further expanded by the Seabees for a total cost of about $3 million ($33 million in 2020 funds)—was appraised at $396,055.[19] Bids for Eastport's "famous white elephant" would be accepted starting in July 1947.[20]

The imminent sale of Quoddy Village, along with the drought and resultant power shortages, motivated Quoddy loyalists to try to revive the tidal power project. After all, *all* the arguments against the project had been disproven: public power wouldn't lead to communism; there was a power shortage in Maine; river hydroelectric power couldn't be relied on for all of Maine's needs; supplies of fossil fuels were unreliable and subject to huge price volatility; public power could be developed cheaply; and industrial markets had emerged for cheap power everywhere it was created, even in the desert.

If Quoddy Village were taken over for some other use, broken up, or razed, it would make later efforts to revive the tidal power project more expensive and difficult.[21] Regardless of what was done with the property, the power trusts could point to the cheap liquidation price as evidence of the foolish waste of money the lunatic project had been. It would be another screw in the lid of the coffin, and the power trust didn't want Quoddy's corpse to escape.[22]

Having lost so much, so many times, Eastport was desperate to do *something.* Dozens of ideas were proposed, including an international military base[23] and an aeronautical laboratory.[24] When hearing of the cold shoulder given by wealthy New York City suburbs to hosting the newly formed United Nations, the *Sentinel* proposed Quoddy Village as the home for the world's peacemakers.[25] There were plenty of land, warm welcomes (if not weather), and an airport and harbor at their doorstep. True, it was the closest spot to Europe, but too far from the halls of power in New York and Washington to be of interest to the world's diplomatic corps. Another proposal was to move Baltic University—all 2,000 students and faculty—from Germany to Eastport.[26] But *if you're going to pour millions of dollars into the Baltic*, you had better *think about it pretty carefully beforehand*, as BU was already dependent on US subsidies.[27]

Resettling "displaced persons" was an idea many tried to exploit. Frank Cohen, a Daddy Warbucks industrialist, proposed to take Orphan Annies and their parents from Germany to Eastport, train them for six months in a tractor factory he would set up, and then ship the trainees and tractors to South America. Cohen's plan was said to be "primarily philanthropic," but when it was learned he planned to keep the profits, it started to sound like "slave labor,"[28] and the International Association of Machinists complained it would be unfair competition.[29] While Cohen said he would invest $200,000 or more[30] to rehab Quoddy Village, he was willing to pay only $52,500 for it.[31] One Eastporter said that was like "wishing for the moon in a registered envelope."[32]

The most serious contenders for Quoddy Village were the liquidators: the scavengers. The high bid was a measly $105,555 from Grossman & Sons of Quincy, Massachusetts, a liquidator.

The lowest bid, $1, was made by some sophomoric Harvard students who said they would set up a displaced persons training center to extract minerals from seawater.[33]

In January 1948, the War Assets Administration (WAA) rejected all the bids. Quoddy Village would not metamorphize over the winter and emerge a beautiful new butterfly in the spring but would rot for at least another season.[34]

"I am the successful bidder" for Quoddy Village, declared Samuel Greco of Lebanon, Pennsylvania, in a telegram to the *Sentinel* on January 22, 1948. "I have fully discharged my financial obligations to the United States," he asserted, "and will take possession of the property." He wasn't clear, but his plans had something to do with offering veterans business opportunities, and he attempted to marshal his "comrades" to write to their senators on his behalf.[35]

Greco showed up in Eastport on February 10 to stake his claim in person. First, he presented the town manager with papers he said demonstrated his $1,500 bid for Quoddy Village had been accepted by the WAA, and as such, he was the rightful owner of it. He then marched over to the *Sentinel* to publish notices to that effect. When questioned by the *Sentinel* about a telegram from the WAA's Boston office denying his claim, Greco said he was dealing with WAA headquarters in Washington, and Boston was evidently uninformed.

Greco then went to the vacant sentry shack at the foot of Redoubt Hill and formally declared his ownership, like a fifteenth-century European explorer setting foot on "undiscovered" lands. Thus ennobled, Proprietor Greco presented a writ (of his own drafting) to the Village managers asserting his rights, and said, by the way, he would no longer pay their salaries. The Little Seizer was politely advised to take the next train home, but he insisted he was going to remain to "look after my property."[36] He stuck it out for a week.[37]

The WAA stuck it out for seven months before acknowledging its failure and suspending efforts to sell Quoddy Village, pending, it said, the outcome of discussions with Canada about the tidal power project.[38]

For the third time, Quoddy Village was offered for sale, in September 1949.

No bids were received from either the sophomoric Harvard students or Greco.

Unit 1, the nine "Executive Homes" on Snob Hill, went for $52,100 to a group of nine locals.[39]

Unit 2, 143 acres with 34 commercial and 180 residential buildings, was sold to Grossman & Sons for $76,155.55 (about $817,000 in 2020 funds).[40]

Unit 3, consisting of a pier, 44 acres on the waterfront, a house, a barn, and a few small buildings, was sold to Roscoe Emery for $3,500 (about $38,000 in 2020 funds).[41]

All told, the government received $131,755.55, or about 4% of what it had spent building Quoddy Village, and only one-third of the anticipated liquidation value.

The good news was the properties' change of legal status, as indicated by Eastport's tax assessors, "who were seen gleefully rubbing their hands in anticipation of the revenue."[42]

The bad news was that there wasn't an industrial company moving in with hundreds of jobs. Of Washington County's 14,800-member workforce, 3,800 were unemployed at the beginning of the year. By year's end, 5,800 were unemployed.[43]

DECLINE

"President Truman favors revival and completion of the Quoddy power project," the *Portland Press Herald* announced on August 30, 1947.[44]* By February 1948, George L. Brady of the once vehemently anti-Quoddy *Boston Herald* was advocating for Quoddy:

> Too many people in years past have tried vainly to conduct the funeral for Passamaquoddy. You and I are presiding tonight at the rebirth of this vitally essential project. . . . We cannot afford to sacrifice great natural resources like Passamaquoddy on the dark altars of provincialism, special

privilege or subservient politics. If we persist in such policies we shall write our own economic obituary.[45]

The next month, Margaret Chase Smith introduced a bill to appropriate $100 million for Quoddy's construction. Her colleagues in the US House ignored it.[46] That same month, Edward Graham, president of BHE, said his company was "not opposed" to the Quoddy Tidal Power Project,[47] *having gotten control over it by capturing Maine's Executive Council and Public Utilities Commission.*

Getting nowhere in their efforts to directly fund construction, Quoddy's proponents met with President Truman in April 1948 and asked him to initiate a new two-phase investigation of the International Project. Phase one would be a preliminary review and an estimate of the cost to prepare a definitive "full-dress" survey of Quoddy's feasibility. The estimate would cost $60,000, to be shared equally by the United States and Canada. The full survey would probably cost several million dollars. The estimate was to be prepared by the International Joint Commission (IJC), of which the Federal Power Commission (FPC)'s Roger McWhorter was now a member,[48] and the survey by the US Army Engineers. President Truman liked the idea,[49] as did Canada.[50]

In June, former Maine state senator Roy L. Fernald[51]* wrote an editorial in the *Sentinel* warning that saboteurs were again trying to kill Quoddy, and "opposition to Quoddy has come in the past from the private power interests of Maine, who for more than a generation exercised almost complete and unbroken control over our State government. That there is such a super-government[52]* in Maine may seem to be a startling statement, but everyone who knows anything at all about Maine politics knows that it is true," and "so true" that only the lowest-level state appointments could be "made or confirmed without the approval of the select few."[53]

By September power shortages were so acute Ontario had to curtail use of electricity,[54] prompting the Canadian government to fund its half of the Quoddy estimate. Meanwhile in Washington, D.C., the latest quarterly reports required of registered lobbyists revealed the majority of spending was by members of the National Association of Electric Companies, and most of that was spent on efforts to amend (i.e., weaken) the Federal Power Act.[55] Consequently, Congress wouldn't even schedule a vote on its $30,000 share of the estimate.

Fed up with the shadow boxing in Congress, the Maine Legislature did something extraordinary. In a voice vote on May 9, 1949, it approved $30,000 to cover the US federal government's share of the Quoddy estimate.[56] Sufficiently shamed, and probably piqued by its power of the purse being usurped, Congress ponied up in June.[57*]

That "monumental" appropriation was, to the Edison Electric Institute, another existential threat. At its annual meeting in June, it announced a new "grassroots" campaign (i.e., propaganda) to stem the "trend towards nationalization" of the electric industry. By "nationalization," EEI meant the federal government taking over private power companies, at presumably unfairly low prices.[58*]

The IJC began fact finding for a new International Tidal Power Project in August 1949 with a public hearing at the Algonquin Hotel in St. Andrews, New Brunswick. Another hearing was scheduled at Eastport's high school auditorium the following week. They were both packed. Interested citizens came from all over Maine and the Maritimes. Interested parties, expert witnesses, and the press came even farther.

The US Army sent Philip Fleming, now a two-star major general,[59] from Washington.[60] He was reunited with his former assistant, Hugh Casey, also a major general, who flew in from Japan.[61] Paul Kruse, Casey's former assistant, came from BHE.[62]

The testimony was overwhelmingly in favor of Quoddy. Some who had opposed it in the past now favored it, notably the former sectionalist "skunk"[63] and mayor of Augusta, Frederick Payne. He was now the governor of "rock-ribbed Republican Maine" and "pleaded" for Quoddy's completion.[64] Even Maine's electric power companies said they had seen the light: Paul Kruse reiterated that BHE had "no opposition to the project,"[65] and a telegram from the Maine Development Commission signed by its chairman, Harold F. Schnurle, vice president of CMP, urged favorable action on the survey.[66] The only opposition expressed in St. Andrews was by the Canadian fishing interests,[67] but it was muted by assurances of compensation.[68]

While skeptical of Quoddy's economics, Lincoln Smith, a Maine native and professor at the University of Southern California, supported the survey, as it would provide the facts needed for sound policy decisions. He also addressed the issue of power shortages, quoting the New England Council (NEC) report on "Dependable

Reserve Capacity" in Maine. It was at a high of 10.3% in 1943 and had fallen to 0.0% in 1946. NEC said the reserve should be 15%. NEC estimated the need for 1.1 million kW more power within a decade—but noted that the planned expansions amounted to only 79,000 kW spread over fifteen new plants, 63% of which were to be steampowered.[69] In other words, all the expansion planned by the local monopolies would amount to only 7% of the need—which Quoddy could fulfill completely.

An "oft-quoted fact" from the hearing was Quoddy's ability to deliver power to Eastport for 2.3 mills per kWh and to Boston for 5.1 mills. That compared rather favorably with the average of 8 to 12 mills paid by most commercial customers in New England. The consequences of high electric rates were also presented. The governor of Massachusetts sent an analysis that New England spent $176 million more *each year* for electric power than if the same power were purchased in the TVA area and said the high cost was driving energy-consuming companies out of business or out of New England.[70] Stated in the abstract: a chartered geographic monopoly is the ultimate "competitive barrier to *entry*." Consequently, for consumers located in that geography, their only alternative is to *exit* the territory. And that is exactly what the textile industry did to escape the high cost of power in New England. As a case in point, Textron moved from New Hampshire to South Carolina because it could get power at 2.1 mills.[71]

The most surprising facts were the enormous disparities in federal funding for rivers, harbors, and flood control for different regions of the country. For 1950, New England's appropriations were $3.7 million; the mid-Atlantic $40 million; the Mississippi River Basin $350 million; the Pacific Coast $550 million; and the South $660 million. The contrast was even more striking when density of population was compared.[72]

Evidently New Englanders were paying too much for power, not getting their fair share of the money they paid in federal taxes, and falling behind economically. Or, more bluntly, Maine was being sucked dry by southern and western states. Instead of becoming a "power colony of Massachusetts," Maine had become a tax colony for the rest of the nation.[73]

More pointedly, Frederick Hale, Maine's US senator from 1917 to 1941, *had not brought home the bacon* to Maine (since he didn't live there),[74] and, consequently, Maine's economy was starving. The accumulated effect of decades of lost tax revenue and expensive energy was as yet unknown—but staggering.

In March 1950, the IJC's Engineering Board submitted its report, based on an analysis by the Army Engineers. The report concluded that the International Tidal Power Project was physically feasible[75] and that the "critical feature" determining the economic feasibility depended on whether the "dams can in fact be built at the particular locations required and at a cost which is economically practicable," just as Roscoe had written in 1925.[76] Building such dams would entail construction challenges "unprecedented anywhere in the world,"[77]* and "there [was] no short-cut" to the "costly" drilling surveys. Having thus puffed up the challenges, the IJC engineers estimated the "full-dress" survey would cost $3.6 million and take eighteen months, while additional fisheries investigations would require $300,000 and three years.[78]

The fight for power now moved from Passamaquoddy Bay's rocky shores to Chesapeake Bay and the swamp of lobbyists in Washington, D.C.

America's World War II allies, the Soviets, became our Cold War enemies. They blockaded the Allied-controlled city of West Berlin on June 24, 1948, necessitating the Allies' airlifting in all the supplies needed by the city's inhabitants. Fourteen months and 200,000 sorties later, and just after the Soviets secretly conducted their first successful atomic weapons test on August 29, 1949,[79] they ended the blockade.[80]

The fight for power **continued to rage in Washington.** In April 1950, the Third National Electric Power Survey documented shortages in Maine, New Hampshire, and Connecticut.[81] In May, President Truman criticized Congress for not appropriating funds for the St. Lawrence Seaway and Quoddy public power projects as he had requested.[82]* Meanwhile, the assistant secretary of the interior said opposition from New England's power companies to public power projects had penalized the region's industrial growth: New England's economy (from 1929 to 1947) had dragged along with a power increase of just 48%, while the country as a whole enjoyed a 77% increase, and in regions of federal projects the increases were over 100%. He then put this in pocketbook terms: while total income for New England had risen by

91%, the national average was 129%; for the TVA states 198%; and for the Pacific Northwest 202%.[83]

It wasn't FDR who was in a bath of hot water *because* of Quoddy, as Vandenberg had fumed,[84] but New Englanders, for the *lack* of Quoddy's power. New England's economy was like a frog placed in a pot of water and slowly heated to boiling: it didn't recognize it was being killed until it was too late. *All the while the private power companies kept saying to the frog, "Trust us."*

Realizing its states were being boiled, the New England Council surprised many when it contradicted the FPC's assertion that 3.1 billion kW of hydroelectric power could be produced from New England's rivers. Instead, the council said, only about 0.5 billion kW—one-sixth of the total—was economically justifiable if one counted "practical cost" such as the farms and towns that would be flooded.[85]

Robert McWhorter of the FPC and IJC picked a fight when he proposed the IJC's investigation of the International Project be sped up and scaled down from a "full-dress" survey to just a "quickie."[86] (*He already knew the conclusion he wanted.*) Unsuccessful, McWhorter then tried to force Canada to pay for half of the survey, even though it was expected to use only about one-third of the power. Meanwhile, someone managed to get what should have been well-informed and independent professional organizations—including the national associations of civil, electrical, mechanical, mining, and chemical engineers—to issue a joint statement in opposition to doing an updated survey of Quoddy's potential. They based their objection on McWhorter's biased and obsolete 1941 FPC report, which called for periodic updates.[87]*

The fight for power took on greater urgency on June 25, 1950, when North Korea, backed by Russian and Chinese Communists, invaded South Korea. Taking a stand against Communist territorial expansion, the United States launched a "police action," an undeclared but fullfledged war. Again, the United States needed more electrical power for manufacturing military materiel, but patriotism wasn't the primary consideration of the private power industry: profit was predominant. *Private power seemed to think public power was a greater threat than real Communists with guns, tanks, planes, and atomic bombs.*

Roscoe and his colleagues met President Truman at the White House on January 31, 1951.[88] After the 20-minute meeting, it was reported that

"They didn't have to sell [Truman] on Quoddy because he was already sold."[89] Recognizing most of Congress wasn't "sold," Quoddyites published a 24-page brochure and sent it to every member of the US Congress and Canadian Parliament.[90]

Sentiment for Quoddy warmed in June when BHE applied for a 12.8% rate increase.[91] Then in July the city of Bangor got a hot flash of remorse for its own lost opportunity when it was reported that an aluminum[92] company would have built a new plant in the area "if assured an adequate supply of low cost power," but the firm's requirements were "well beyond the capacity of the existing facilities and . . . if new facilities were constructed, the cost of the power would be higher than the company is willing to pay." Thus, the *Bangor Sunday Commercial* concluded, "[W]e have another instance of power presenting an obstacle to industrial development of Eastern Maine. . . . Once again, it seems clear that the people of Maine in pure self-interest should throw every ounce of support behind Quoddy—and especially behind the effort to obtain the economic survey at the earliest possible moment. Delay may be costing us untold millions of dollars."[93]

Fake news and obsolete information were still a problem. A special Committee on the New England Economy appointed by President Truman's Council of Economic Advisors released a report in July 1951 that included two sentences on Quoddy, saying it "is apparently not economically feasible under present conditions." It seemed to have based its assertion on McWhorter's 1941 FPC report despite its having been proven wrong by the droughts and power shortages in the Northeast in 1947 and 1948. While (perhaps deliberately) ill informed about Quoddy, the report's authors revealed the average cost of industrial power in New England exceeded the national average by a whopping 59%.[94]

Meanwhile, the Columbia River basin was experiencing the worst drought in 25 years, causing hydropower production to peter out. To keep the lights on, the federal government restricted the use of electricity by nine aluminum and chemical companies despite the power being needed by the defense industry.[95] On October 1, *Newsweek* reported the government was advising the "Big Three" aluminum manufacturers—Alcoa, Reynolds, and Kaiser—to consider moving their plants to somewhere with more reliable power.[96]

Also in October, the *Christian Science Monitor* ran an article about a test by the US Geological Survey and the US Navy of new electronic depth-sounding equipment that might eliminate the need for more expensive submarine drilling in the Quoddy survey.[97]* In November, the *New York Times* reported the test was successful and the estimated cost of the survey was reduced from $3.9 to $3 million.[98]

Come December, *New England Town and City* magazine reported Kaiser and other "key processors of vitally short basic materials are turning a hopeful eye toward New England and the vast, unchanging power potential of the Passamaquoddy Bay tidal development." Based on 500 interviews, the magazine declared, "From its grass roots, from its congressional and business leaders, from its newspaper editors, and from its 'men on the street,' New England has recorded itself as virtually unanimous in the desire to get to the bottom of the big, challenging enterprise."[99]

Later in December, the Canadian government said it was willing to let American engineers survey its side of Passamaquoddy Bay, but it wasn't willing to pay for its share of the survey.[100]* The reasons for the refusal were the government's beliefs that (1) the tidal power project was uneconomical,[101]* (2) abundant and cheap electric power could be generated from rivers in New Brunswick and Maine,[102]* and (3) there was no foreseeable market for power in eastern Canada—despite the Canadian Aluminum Corporation planning to spend $600 million to get more electric power.[103]

The last word on Quoddy for 1951 came from Professor Lincoln Smith. In his new book, *The Power Policy of Maine*, he argued that Maine's sacrosanct Fernald Law had miserably failed to make Maine an industrial state and failed to keep the multistate monopolies from gobbling up Maine's electric companies.[104]

On January 20, 1953, General Dwight D. Eisenhower, formerly the Supreme Commander of the Allies' D-Day invasion of Nazi-controlled Europe, took office as the 34th president of the United States.

Here was another of Franklin Roosevelt's war-winning team, the man who had launched the largest armada the world had ever seen. Operation Overlord involved 5,300 ships and 11,000[105] planes that landed 200,000 fighting men on the beaches of France in a single day. The entire operation was timed to take advantage of the light of the full moon and the extra-high and extra-low tides that would accompany it. During that critical 24-hour and 50-minute tidal cycle, specially designed landing craft and amphibious tanks fought their way through the slippery intertidal zone and the Nazis' deadly booby traps and guns. Within a few hours, the Allies towed in and began assembling immense artificial harbors with quays and bridges that floated up and down with the 20-foot tides so men and materiel could be unloaded continuously. Eisenhower was a man who would appreciate the power of the tides. So Quoddy's supporters hoped.

But Eisenhower called the TVA "creeping Socialism."[106]

Coca-Cola's sales went flat, and the company left Washington County in April 1953.[107]

Central Maine Railroad blew the whistle on service to Calais in July 1953.[108]*

The *Eastport Sentinel* folded on July 23, 1953. After 135 years, Maine's longest-living newspaper died, a victim of declining population and business activity. There was no longer a "publication of record" for the Quoddy Tidal Power Project. Although he had sold the newspaper seven years before, it still must have been a heartbreaking day for Roscoe Emery.

TRY, TRY AGAIN

"One of the most controversial engineering projects ever conceived by man," *Newsweek* announced in April 1954, "is once again under serious discussion."[109]* While Quoddy was "considered by opponents of the New Deal to be the biggest boondoggle of all (save, possibly, the Florida ship canal), some basic facts have been overlooked." *Newsweek* then listed ten points, starting with "Quoddy can be built. Despite its grandiose conception, it presents no insurmountable engineering problems," and ending with the question "Has the whole concept of 'cheap' tidal power been by-passed by the advent of 'cheap' atomic power?" *Newsweek* continued:

These questions, and many more, add up to the single $3 million question: Should Quoddy be built?

"Let's find out," says [Roscoe] Emery. "I don't want to chase a cotton rabbit any longer. . . . I've been on this since 1924."

That seems to be the mood of most of Maine. "Our purpose . . . is to have a thorough survey," Senator [Margaret Chase] Smith has said. "We don't want anything unless it is complete enough to give us the final answer on whether or not we should go forward." . . .

Summer T. Pike, former member of the Atomic Energy Commission[110] and now Chairman of the Maine Public Utilities Commission, who comes from Lubec, said: "I'd like to have the baby either shot or brought up. I don't like seeing it squalling around all the time. . . . It's certainly worth spending 1 per cent (of his construction cost estimate of $300 million) to find out if it is worth going ahead."[111]

"Honest Harold's" ghost came back to haunt Quoddy when his diaries were published posthumously in 1954. Therein, Ickes contradicted his public statements about the purpose of his trip to Quoddy in August 1934: "The plan to go up [to Maine] at this time was set out by Colonel Louis Howe who is anxious of the results of the election in Maine on September 12. The Public Works Administration does not have the money necessary to finance the project, even if we should approve it, but Louis thought that at least a graceful gesture could be made with beneficial results on election day. So he asked me to go up and look at the thing, and the President was also in favor of my going."[112] Ickes's *Secret Diaries* partially confirmed Gertrude's charge that Louis Howe had proposed using Quoddy as political bait. It also contradicted President Roosevelt's disclaiming any knowledge of the ploy.

The way to "harness" atomic power—instead of "unleashing" it in an explosion—is to keep enriched uranium from "going critical," to keep it glowing "hot" but not so hot as to cause an uncontrollable "meltdown." The hot-but-not-too-hot fuel then boils water, expanding it by 1600:1 as it turns into steam, which then forces its way through turbines to generate electricity. In effect, an atomic power plant is an extremely complicated teapot.

Back on January 28, 1946, David Lilienthal, a founding director of the TVA, wrote in his journal, "No fairy tale that I read in utter rapture and enchantment as a child, no spy mystery, no 'horror' story, can remotely compare with the scientific recital I listened to for six or seven hours today. . . . I feel that I have been admitted, through the strangest accident of fate, behind the scenes in the most awful and inspiring drama since some primitive man looked for the very first time upon fire."[113]* President Truman appointed Lilienthal as the founding chair (1948–1951) of the Atomic Energy Commission and charged the latter with developing atomic technology for civilian and military uses. When building the first atomic power plant in 1954, AEC's new chairman, Lewis Strauss, said atomic-electric energy would soon be "too cheap to meter."[114]

In 1955, New Brunswick's premier acknowledged, "The ever-increasing demand for electrical power remains one of the principal problems faced by the government." To solve the Northeast's power shortage problems; to end Maine's forty-six-year, self-imposed industrial isolation; and to improve the chances for Quoddy's revival, Maine *finally* repealed the Fernald Law power export ban.[115]*

In August 1956, the United States *finally* **put up $3 million**, and the Canadian government *finally* put up $300,000, for the IJC to *finally* do the "full-dress" survey of the International Tidal Power Project. The IJC established an Engineering Board with representatives from the United States and Canada. The American section was headed by the recently retired chief of the Army Corps of Engineers, none other than three-star Lieutenant General Samuel D. Sturgis Jr.—the man who had headed Quoddy's Camp and Land Division in 1935–1936, and the army man least trusted by Eastporters (or at least Roscoe).[116]

On October 4, 1957, there was a *new* **moon,** the Soviet-made satellite *Sputnik*. The space race was on.

In July 1958, the St. Lawrence–Franklin D. Roosevelt Power Project at Massena, New York, *finally* began generating power, 27 years after Governor Franklin D. Roosevelt began the project and 40 years after Governor Al Smith proposed it. The joint US-Canada dam was finally built because power-starved Canada threatened to do it on its own.[117]*

In October 1958, the IJC's Engineering Board finished its "full-dress" *survey* of the International Tidal Power Project. Its conclusion: spending some $500 million on Quoddy would be "economically justified."

Sixty arrangements of dams and turbines were considered. The best arrangement was found to be a two-pool plan like Dexter Cooper had proposed in 1924. Thirty turbines would generate 95,000 to 345,000 kW, for a total of 1,843 million kWh annually, or *four to seven times* the output of the army's Plan D from 1936.

To make Quoddy's varying power output more consistent and valuable, several auxiliary power sources were considered, including river-power, pumped-storage, and steam-electric plants. Building a new hydroelectric plant at Rankin Rapids on the upper St. John River in Maine, 175 miles away, was selected. The 333-foot-high Rankin Dam would generate 1,220 million kWh annually. Quoddy-Rankin would cost $630 million ($5.5 billion in 2020 dollars). All 3 billion kWh of power could be readily absorbed by Maine and New Brunswick.

Differences in interest rates and fuel costs in the United States and Canada made it necessary to compute separate benefit-cost ratios for the two countries. Consequently, IJC's Engineering Board concluded Quoddy-Rankin was economically feasible for the United States, but not for Canada, thus leaving Quoddy in limbo yet again.[118]

There was some potentially good news for Quoddy: the changing of the old guard. Roger McWhorter, the chief engineer for the FPC and a commissioner on the IJC, retired.[119]

A RISING TIDE

In 1961, another sailor took the helm of the United States. JFK, as he liked to be called in homage to FDR, followed Franklin's lead and was investigating Quoddy as a public power project even before his inauguration.[120] What he learned was not encouraging.

In 1961, the IJC disagreed with its Engineering Board's economic analysis that Quoddy-Rankin was economically justified because *the tidal project by itself* was not economically justified[121] as a power project (without considering employment or economic development benefits),[122] whereas the Rankin Dam by itself was. *Once again Solomon cut the baby in half—to be fair (and to kill it).*

The IJC did concede that tidal power might be worth looking at in a decade or two when less expensive sources of power had been exhausted.

The IJC's damning Quoddy with faint praise left a trickle of hope. Besides the maneuver to excise the auxiliary plant and leave the tidal plant's power too variable to be valuable, the IJC had based its conclusion on assumed interest rates and fuel costs. So President Kennedy asked Interior Secretary Stewart Udall to change the assumptions and determine whether a different combination would make the tidal power project viable.[123] Udall's study would take two years.

Meanwhile, President Kennedy called on James M. Landis—who had helped Thomas Corcoran draft PUHCA, was the second chairman of the SEC (after JFK's father), and would soon become dean of the Harvard Law School—to report on the "capture" of federal regulatory agencies. His study was a "merciless dissection of the [Federal Power] commissions' failures, informed not only by the author's years of study and experience but also by a bitter sense of . . . betrayal of the regulatory ideal." Landis wrote that the FPC was "the outstanding example . . . of the breakdown of the regulatory process," accused it of acting "as a virtual Chamber of Commerce for the oil and gas companies," and noted that staff positions "appear to have been sought frequently as stepping stones to further preference, or to positions of importance within the industries subject to regulation."[124]

The United States got another shock from Soviet Russia on April 12, 1961, when Soviet cosmonaut Yuri Gagarin became the first person to orbit the earth in space. Only a dozen years before, the United States had been the supreme power on the earth. But now Russia had caught up in the atomic arms race and was winning the space race.

After consulting with America's leading rocket scientist (formerly Germany's leading rocket scientist), Wernher von Braun, about how the United States could take the lead in space, on May 25, Kennedy announced the American plan:

> I believe that this nation should commit itself to achieving the goal, before this decade is out, of landing a man on the Moon and returning him safely to the Earth. No single space project in this period will be more

impressive to mankind, or more important for the long-range exploration of space; and none will be so difficult or expensive to accomplish.[125]

The critics were quick to condemn the visionary plan, for all the usual reasons.

In the Rose Garden of the White House, on July 16, 1963, President Kennedy announced the results of Udall's study. "Recent developments in electrical technology," JFK said, had made the International Project economically feasible, despite the $1 billion cost. Quoddy was expected to deliver 1 million kW of power (*3 to 10 times what the IJC proposed in 1959*)[126] at 4 mills per kWh, or 25% *less* than the *average* cost of all hydroelectric and steam-generating stations in New England.[127]* In New England, where electric rates were still among the highest in the nation, demand was now expected to double during the 1960s, and almost double again in the 1970s,[128] *just as President Roosevelt had predicted in 1936.*[129]

It was reasonable to expect, the report said, economic development around Quoddy, just like other government power projects. Addressing Canada's previous concern, the report stated, "[T]here will be no adverse effects on fish and wildlife . . . [and it] creates an opportunity to enhance the fisheries of New England by removal of existing small, inefficient hydro projects which now block the migration of anadromous fish."[130]

President Kennedy noted a tidal power plant was under construction in France and one was being planned in Russia.[131]* He concluded, "Each day over a million kilowatts of power surge in and out of Passamaquoddy Bay. Man needs only to exercise his engineering ingenuity to turn it into a great national asset. . . . The combined efforts of science, private industry, and government, will surely keep this nation in the forefront of technological energy and electrical power. I think this can be one of the most astonishing and beneficial joint enterprises the people of the United States have ever undertaken."[132]

The tidal power race was on, and President Kennedy wanted to win. He ordered the secretary of state to begin negotiations with Canada and the Army Engineers to begin detailed planning for construction.[133] Quoddy was coming out of the grave.

At long last, Quoddy had a new national champion, one full of "vig-ah."

Republicans attacked the plan, accusing the Democrats of using a "billion dollar boondoggle" to help Senator Muskie's reelection chances.[134]

While President Kennedy was full of "vig-ah," Maine was losing its political "pow-ah." Because of Maine's slow population growth relative to the rest of the country, in 1963 Maine's apportionment in the US House of Representatives dropped from three to two.[135]

After three decades of depression in Eastport, the once tightly packed city was now gap-toothed, with vacant lots where houses and stores once stood.[136]

"A rising tide lifts all boats," said President Kennedy. He was speaking on October 3, 1963, at Greer's Ferry, Arizona, where Army Engineers were building another hydroelectric dam. The president used the aphorism to explain how the construction of a dam in Arkansas created jobs in that state, which in turn led to purchases of products from other states, thus raising the quality of life for people everywhere.[137] It was such a simple and powerful metaphor that it has been picked up and reused innumerable times.

The metaphor is even more interesting when we consider its genesis. Ted Sorensen, Kennedy's speechwriter, claimed to have lifted it from Quoddy's nemesis, the New England Council, the formerly power trust–controlled industrial oligopoly in Boston that had done everything in its power to kill the idea of economic benefit from rising tides.[138]*

President Kennedy flew to Maine to see Quoddy Bay. With him aboard Air Force One on October 19, 1963, were Maine's senators Margaret Chase Smith and Edmund Muskie, as well as Colonel Robert C. Marshall of the Army Engineers. Expectation was in the air as they looked through the plane's windows to the bay below and planned a brighter future for Maine.

Four weeks later, the light was snuffed out.

President Kennedy was assassinated on November 22, 1963, in Dallas, Texas. With its champion killed, Quoddy was dead again.

Five presidents of the United States—Herbert Hoover, Franklin Roosevelt, Harry Truman, Dwight Eisenhower, and John Kennedy—had actively promoted the Quoddy Tidal Power Project. To paraphrase Will

Rogers, *when five US presidents can't get Quoddy built*, "it gives you an idea of the pull in the legislation that the Power Trust exerts."[139]

IMPOSSIBLE DREAMS

In the United States in the 1960s, dreams were dreamt, realized, and killed. The assassination of President Kennedy ended the dream of an "American Camelot." Martin Luther King Jr. awoke the consciousness of the world when he said, "I have a dream," and spoke of personal freedom and racial equality. The centuries-old novel *Don Quixote* debuted as a Broadway musical, and its theme song, "The Impossible Dream," captured the spirit of striving for good. Martin Luther King Jr. was killed, dashing our hopes of seeing the "promised land." The Clean Air Act committed Americans to making a world in which we could breathe free. Through all this, Quoddy's fortunes waxed and waned, in syzygy[140]* with society's advances and retreats.

President Lyndon B. Johnson received Secretary Udall's second report (*Udall-65*) in July 1965.[141] It recommended Quoddy be built as Dexter had envisioned: a two-pool International Plan, but with a twist. Instead of a stand-alone project producing continuous power, it would be integrated into the regional power grid and managed for maximum power output of 1 million kW during the two-hour peak demand each evening. Meanwhile, dams on the Upper St. John River at Dickey-Lincoln[142] would provide another 760,000 kW for the ever-growing "base load" demand. The total cost of Quoddy-Dickey-Lincoln would be $845 million to $1 billion.[143]

But the week before *Udall-65* was published, Congress raised the interest rate charged by the government for such projects by one-eighth of 1%, just enough to make Quoddy uneconomical.[144] Dickey-Lincoln was still deemed economical and was approved by the House, Senate, and President Johnson as the first public power project in New England—where electric rates were still 28% higher than the national average.[145]

In response, America's Independent Light & Power Companies cranked up the anti–public power propaganda machine and pumped out the poison. Included in the offensive was an ad showing Communist soldiers at the Berlin Wall with guns and barbed wire in which AIL&PC claimed public power projects were "dangerous" steps toward totalitarianism. In ironic doublespeak, another ad said, "[W]hen government owns business, it has in its hands both economic and political power . . . with power thus concentrated it can become difficult indeed for individuals to preserve their basic freedoms." And like all the propaganda campaigns of the previous fifty years, the cost of this offensive was charged to operating expenses and thus was borne by the rate-paying public.[146]

By November 1967, detailed construction planning for Dickey-Lincoln was 90% complete—when it was killed by the power trusts. For the next six years in a row, funding for Dickey-Lincoln was approved by the US Senate but blocked by the power trust's henchmen in the House.[147]

In 1966, the French proved tidal-hydroelectric power wasn't an impossible dream. At the La Rance estuary on the Atlantic coast of Brittany, their half-mile-long dam, 28-foot tidal range, and 24 "dual-effect" turbines generated 240,000 kW of electricity during both flood and ebb tides. Fifty-plus years later, the publicly owned power plant is still going strong. A financial success as well as an engineering success, the $100 million[148] project paid for itself in 20 years[149] and now provides the cheapest electricity in France.[150] The nine-square-mile estuary enclosed by the dam has had some significant negative environmental impacts, especially sedimentation and reduced species diversity. The dam attracts up to 40,000 tourists per year.[151]*

The French success prompted reconsideration of Quoddy, but at a time when the atomic dream was ascendant. In 1967, Gertrude attended a lecture about the Millstone Nuclear Power Station (Central Maine Power was a minority owner)[152]* then under construction in Waterford, Connecticut. Gertrude described the scene in her memoirs:

> In the middle of the speech, for no apparent reason, he mentioned derogatorily Passamaquoddy, referring to it as a "crazy dream of Franklin Roosevelt" and making other misstatements. After he had concluded, I approached him and asked if he had ever heard of Dexter Cooper. Of course, he had. I then identified myself and stated that I knew of at least four men

who, if present today, would contradict every word he said about Passama-quoddy. The speaker asked if I meant Secretary of the Interior Udall.

Indeed not, I told him, and named the four: Gerard Swope, president of General Electric; Guy Tripp, president of Westinghouse; Charles Edgar, president of Boston Edison; Samuel Insull, president of Middle West Utili-ties. Furthermore, I pointed out, the plans were first approved during the Hoover administration—Mr. Hoover was not only a Republican but an engineer before he was elected President.

Just then we were joined by another gentleman who corroborated my statements, adding that he had been working as an engineer in Washington and saw Dexter's plans when they were originally submitted. "It was the power companies who paid to have Passamaquoddy stopped," I said.

"Yes, and they're still paying," the speaker admitted; immediately it was obvious he regretted making the remark.[153]

Millstone no. 1 began commercial operation on December 28, 1970. In 1995, George Galatis, a senior nuclear engineer at the plant, filed a whistleblower report about safety concerns. In July 1998, Mill-stone no. 1 was permanently closed by the Federal Energy Regulatory Commission.[154] It had lasted only twenty-eight years. The radioac-tive waste will last thousands of years. A similar story played out with Maine Yankee Atomic Power Plant in Wiscasset, which closed after just twenty-five years.[155]* Meanwhile, most of the hydroelectric plants built by Hugh and Dexter Cooper have been generating power continuously, with no deadly nuclear waste or carbon emissions, for close to one hundred years.

On a glorious midsummer day in 1969, everyone was inside watch-ing television—to see whether the dream dreamt by millions, for thou-sands of years, would come true. On that day, July 20, 1969, if all went well, men would walk on the moon.

All around the world, families crowded around televisions and radios. Grandparents reminisced about being dazzled by the first electric lights. Parents remembered the thrill of their first airplane ride. Children were still perplexed by the little people in their television cabinets. All were enthralled by the events they were witnessing, that very instant, from out of this world.

Among the millions were Gertrude Cooper, age 80, and her ado-lescent grandsons, Dexter and Peter. They were at a friend's house on nearly uninhabited Indian Island, in Passamaquoddy Bay, between Campobello and Eastport. They had brought a tiny black and white TV for the occasion, the first time such a device had been allowed to interrupt their rustic isolation. They plugged it into an electric outlet connected to the submarine cable that briefly surfaced on its way from the oil-fired generating station at St. John to Campobello.

In 1961, when President Kennedy had announced the technological tour de force as America's national objective, the idea of men walking on the moon was lunacy. Now, just eight years later, the impossible was on the verge of becoming possible. And still some people couldn't believe it.

They refused to believe the Apollo program was real, and not a grand deception.

They didn't accept that lights traveling across the night sky were satellites.

They were at a loss to understand the value of satellites circling the globe.

They couldn't conceive how computers could change their lives.

They laughed at investing billions in dreams.

The naysayers sought a precedent: another gigantic government project, a ridiculous overreach, a monument to folly, an economically unsound and unethical threat to society. They wanted an epithet to damn Apollo.

They called it a "moondoggle."[156]

CHAPTER 14

Who Killed Quoddy?

First, let's dispense with the question: *Did Quoddy die from lack of merit?*

The La Rance tidal power station in France proved that a tidal hydroelectric power plant could be built, generate power cost efficiently, operate safely, and outlast most (if not all) fossil fuel and atomic power plants. Quoddy did not die for lack of merit. It was killed *because* it had merit. It is true Quoddy's merits were never proven—because it was never built. But the five-decade fight is evidence enough that Quoddy was a threat its enemies wanted to kill.

There were *eight* plans for the Quoddy Tidal Power Project. Each was killed. The circumstances were different in each case, but one of the suspects discussed in this chapter had a hand in every case.

Dates	Project Version	Notes
1919–1929	Dexter's original two-pool International Plan	private enterprises
1929–1930	Dexter's two-pool All-American Plan	private enterprises
1933–1934	Dexter's Power & Metals Plan	private enterprises
1934–1935	The Hunt Plan	public power projects
1935–1936	US Army Corps of Engineers' one-pool Plan D	public power projects
1937–1939	Franklin Roosevelt's Small Experiment	public power projects
1948–1961	Quoddy-Rankin Rapids	public power projects
1963–1965	Quoddy-Dickey-Lincoln	public power projects

Three groups wanted to kill Quoddy:

1. Professional rivals who wanted credit for Quoddy or were jealous of others' success
2. Political rivals, both Democrats and Republicans, for political gain
3. Commercial rivals who might make less money with Quoddy as a competitor.

Quoddy offered fame and fortune to whoever controlled it. It was an enormous, expensive, first-in-the-world project that would ensure decades of prestige and income. Ambition, jealousy, selfishness, and revenge were motivations for killing Quoddy.

PROFESSIONAL RIVALS

Dexter contributed to Quoddy's death three times—the International, All-American, and Power & Metals Plans—because of his strategic errors: He kept too much to himself. He believed too much in his own abilities. He was reticent. He wanted to prove that he was his brother's equal (or better). *He* was selfish.

Dexter had GE, Westinghouse, Boston Edison, and Middle West Utilities as early investors, but the investments were small for the companies that made them, and Dexter's lead investor, Guy Tripp, chair of Westinghouse, died in 1927. As far as can be determined, Dexter's investors were not involved in developing the plans or managing the company. Rather, the Board of Directors of Dexter P. Cooper, Inc. was made up of him; his brother-in-law, Russell Sturgis; his secretary, Alice Hume Sleight; and Irene Hume Wadsworth of Eastport. While these "friendly" directors would not challenge his decisions or control, they were, professionally speaking, nobodies.[1]* In contrast, true power brokers, like Samuel Insull or Dexter's brother Hugh, would have filled a board with business and political leaders and given them strong incentives to ensure success. Dexter was a lone engineer; he wasn't a financier, politician, or promoter, but he needed them.

Dexter is also a suspect for killing the US Army's Plan D—because he wanted to control it. According to Gertrude, it was Dexter who asked President Roosevelt to stop Quoddy because of the army's featherbedding and intention to turn Cobscook Bay into the high-tide pool, thus killing all hope for his International Plan:

> Finally, however, he arrived at the bitter realization of what was really happening; that they were succeeding bit by bit in sabotaging the entire undertaking. . . . Thus, although there was still money available to continue for a while, Dexter went to Washington to have the project stopped. The President complied, cutting off the funds, but with an abruptness that left the workmen stranded. So another trip to Washington was necessary; this time Dexter asked that the project be eased off so the workmen could be paid.[2]

Gertrude's account could explain FDR's gut-wrenching $5 million funding cut on October 22, 1935—followed by his reinstatement of $2 million on January 11, 1936, for a "graceful exit."[3] If so, then Dexter was the proximate cause of Quoddy's demise, while the army's $27 million (77%) cost increase was the root cause.[4]

We know Dexter was in Washington on January 16, as the president directed Secretary Harold Ickes to meet with him the next day to hear "his story" and defense of his cost estimates.[5] While there is no official record of Dexter meeting with Franklin in 1936, we know the records are incomplete, and there may have been a conversation by phone or through an intermediary.[6]*

Was Dexter lying about being able to sell "all the power Quoddy could produce"? No. The proof is in the White House files. Therein is a letter from Dexter to President Roosevelt dated January 16, 1936, in which Dexter names the companies that planned to purchase the power *when available* and to build factories in Eastport to employ 2,000 workers.[7]

Dexter did not publicly say he had these commitments until March 9, 1936, and even then he did not name the prospective purchasers. The names were not publicly revealed until *after* FDR's death and the declassifying of his private papers in conjunction with the opening of the FDR Library at Hyde Park, New York.[8]* The power purchasers were

1. St. Croix Paper Company—80,000 kW

    The Maine State Planning Board report of 1934 identified St. Croix Paper as a likely customer, and *for lack of* Quoddy, St. Croix

Paper hired Dexter in 1937 to survey its namesake river for alternative hydroelectric dam sites.[2]

2. Merrimac Chemical Company—160,000 kW
3. Dow Chemical Company—80,000 kW
4. For resale through distributors in eastern Maine—100,000 kW (probable)[10]*

Compare this list to the big news from the Tennessee Valley Authority in August 1936: Alcoa agreed to purchase 20,000 kW from the *already completed* Norris Dam.[11]

Given Dexter's success in proving Quoddy's economic feasibility, Dexter would have wanted to *stop* what he viewed as the army's dead-end[12]* Plan D *and switch* to the more powerful and cost-efficient International Plan. President Roosevelt had this information but didn't reveal it—not to the public, or to the army, or to the Federal Power Commission (FPC). *He must have had powerful reasons not to trust them.*

To *some* of the Army Corps of Engineers, Dexter Cooper was the enemy. Another declassified document in the White House files corroborates Gertrude's accusation of malfeasance by the Army Engineers. It was written by Frederic A. Delano, Franklin Roosevelt's uncle, former railroad executive, founding vice chair of the Federal Reserve, and a colonel in the US Army Corps of Engineers. On December 27, 1938, Delano replied to FDR's request for his thoughts on the possibility of reviving Quoddy and the dithering responses to the same question by the heads of the army and the FPC, who wanted to do yet another study before committing to anything. Delano was "disappointed" by the army and FPC and asked, "For what purpose did the Engineer Corps . . . spend upwards of six million dollars if it was not at least to get the basic data for carrying out the project?" Delano continued:

The successful application of tidal power to the economic generation of electricity has never been solved on a large scale so far as I know. Because I had known Hugh Cooper for upwards of thirty years and his brother almost as long, I believed they were in a fair way to find a solution to this problem. When I first saw Hugh Cooper he was building the dam at Keokuk across the Mississippi River. It was the first attempt in this country to put a dam across a river of such size. It had never been attempted because it was recognized as a very hazardous undertaking, and the books did not tell you <u>how to do it</u>. Hugh Cooper, who was not a graduate of an engineering school, had native ingenuity, enormous courage and persistence; he found out <u>how to do it</u>, and I watched him during the work.[13]* Dexter Cooper was one of the young men on the staff, and Hugh Cooper sent Dexter Cooper to learn the theory of engineering at M.I.T. and supposed that by combining theory with the rugged practice which he had given him, he would turn out a great engineer. He did some very good work during his life and I think he was in a fair way to succeed with Quoddy. He was willing to give his life to it and he spent a great deal of time and a great deal of money trying to put it over, but the odds against him were too great.

I do not think it is likely that the work will be undertaken soon, but I am by no means convinced that the project is not feasible if undertaken by men who have some of the required natural gifts and who are determined to make it succeed. The history of Hugh Cooper's relations with the Army Engineers, the bitter animosities that were developed by him, the fact that his brother, Dexter Cooper, was punished by the Army Engineers who hated his brother, are facts well known to me. . . . What I was impressed with was that the building of the dams connecting the islands ought to have been an exceedingly simple task. The Army Engineers made it a monumental one, enhanced the difficulties enormously, showed no ingenuity whatever, and last but not least, did not explore the various possibilities.

When I was a boy, I read most all of the Dickens novels, and I have always been impressed that Dickens was the first author to dramatize the problems which are now dignified by the title of "social science research." It was his books that led to the early reforms of prisons, legal processes, almshouses, poorhouses, boys' schools, and the like. One of the expressions that he used to use quite frequently was that the "bureaus of government existed not to do things expeditiously but how <u>not</u> to do them." I have lived around Washington long enough to appreciate that there are lots of men in the Government who are past masters at how <u>not</u> to do things.

. . . When the problem is again tackled, it must be done by a man of superior and unusual gifts. As far as I can see, there is no one in sight. . . .

Respectfully, Frederic A. Delano,
Chairman, Advisory Committee.[14]

While Frederic Delano makes it clear the Army Engineers were at least complicit in Quoddy's death, he does not name names.

Can we figure out, "Who done it?" Probably.

None of the Army Engineers assigned to work on Quoddy in 1935–1936 had worked with Hugh Cooper at Muscle Shoals; thus the animosity Delano describes would have resided in the older and higher ranks of the Engineers' leadership. Such foes would have needed someone in Eastport to carry out their revenge.

Immediately the finger points to Philip Fleming, the engineer in charge at Quoddy, but his friendship with Hugh Cooper, Dexter Cooper, and Frederic Delano dating from Keokuk in 1910, and with Theodore Roosevelt while at West Point in 1911, as well as his early warning to Gertrude in 1935, indicate his innocence. Likewise, his statements in defense of Quoddy after the shutdown in 1936,[15] and President Roosevelt's keeping him on as an adviser for seven years thereafter, argue against Fleming's impeachment. If army bosses wanted him to kill Quoddy—and he didn't—he would have had to walk a tightrope.

From the scraps of information available, it seems the most likely agent of revenge among the army's old guard was Captain Samuel D. Sturgis Jr. As a fourth-generation West Pointer, he was probably imbued with the biases of his predecessors and was well placed to undermine Quoddy: (1) as head of the Camp and Land Division, he was responsible for much of the lavish spending at Quoddy Village; (2) he oversaw the drilling teams and could have been involved in the alleged falsification of test results,[16] as well as the transfer of the engineer who questioned the army's analysis and plan;[17] (3) according to Roscoe, the "porky pies" story was tracked back to an employee in Sturgis's office;[18]* (4) the army's *Final Report*, issued in 1936 by Captain Sturgis, took pains to pinion Dexter as incompetent and unbelievable; and (5) despite his *implying* that he approved the final payment of $10,000 due to Dexter, and that it was paid, it wasn't.[19]

Roger B. McWhorter is another suspect. He was the chief engineer for the FPC and a member of the International Joint Commission (1939–1958). McWhorter was born and died in Alabama, not far from Muscle Shoals. In 1924, he was a major in the army working under Colonel Hugh Cooper at Muscle Shoals and would have at least crossed paths, if not swords, with Dexter Cooper.[20] *If* McWhorter carried a grudge, it could help explain the negative FPC reports that thrice killed Quoddy: in 1933–1934, in 1939–1941, and in 1958.

Colonel Henry M. Waite,[21] the assistant administrator of public works in 1934–1935, is another suspect. He was the "dumb as an oyster"[22] engineer who complained that to fund Quoddy would be to "pour millions of dollars into the Baltic Sea."[23] According to Gertrude, his negative report of January 1934 and denial of Dexter's right to rebut errors therein was the impetus for General Hof's warning that "the Army would ruin the project," thus forcing Dexter to go over Waite's head and appeal to President Roosevelt for a public hearing. Waite resigned soon thereafter.[24]*

Quoddy's killer could have been any one of these or several other army officers. The US Army appears to have been against public power generally, not just against the Coopers. Its bias in favor of private power was stated by one engineer in regard to flood control in New England[25] and by another when testifying before Congress about Bonneville Dam.[26] "Honest Harold" Ickes thought the bias was widely held within the army.[27]

However, the positive press release issued by Colonel Fleming in August 1936 about Quoddy's feasibility suggests that he and others in the army were or became Quoddy believers. And to be fair to those who did their job without ever being convinced of Quoddy's merits, we must remember that Quoddy was *not* funded because it would be a profitable power project, but rather as a work-relief project. Thus, the army's primary mission was not to build a tidal power plant as efficiently as possible but to employ, house, and feed as many people as fast as possible. In that humanitarian objective the army succeeded spectacularly.

Even if Sturgis, McWhorter, and Waite are not guilty of intentionally killing Quoddy, the army bears responsibility for their "high, wide, and handsome"[28] spending that fed anti-Quoddy propaganda and sentiment. The army helped to kill Quoddy. It did not, however, act alone.

POLITICAL RIVALS

The Republican Party wanted to kill Quoddy in 1936 because they saw it as the epitome of what they were against—the social restructuring of the New Deal—and as a weapon to use against the Democrats and President Roosevelt.

The Brann and Roosevelt plan was upended in May 1936 by Franklin Roosevelt's political rival and Republican presidential candidate, Senator Arthur Vandenberg of Michigan. His rhetorical assaults spread confusion and doubt, and his constant rehashing of the FPC's misleading reports turned Quoddy into a quagmire.

The sabotage by Senators Frederick Hale and Wallace White of Brann's carefully constructed apolitical compromise (the Robinson resolution) to have independent engineers review the economic merits of *both* Quoddy and the Florida Ship Canal was, as Brann said, the "betraying kiss of death" for Quoddy.[29]

President Roosevelt had many reasons to want to kill Quoddy by cutting its funding in late 1935 and *not* using his executive authority to fund it in 1936 after Hale and White's sabotage. While there is no question that President Roosevelt saw Quoddy as a worthwhile project, it was causing him disproportionately large political problems when many other issues demanded his attention. Putting an end to it might have been a relief.

Conversely, if FDR wanted Quoddy to live, a pause might have been the best strategy, especially since he believed Dexter's assertion that the true value of the project was in its efficiency—which could only be fully realized as the two-pool International Plan.[30]

It would have been political suicide for President Roosevelt to publicly accuse the Army Corps of Engineers of sabotage or incompetence, or to override the Corps's technical decisions and order them to follow Dexter's plan, *or* to repeat the hatred-inspiring arrangement of placing Hugh Cooper's baby brother in command of the Army Engineers. Doing so would have set the army against him and given FDR's political enemies loads of political ammunition and seemingly crystal-clear examples of political patronage and FDR's "dictatorial" tendencies.

Similarly, it would have been a problem for President Roosevelt if Quoddy's costs had risen significantly above $40 million.[31]* When President Roosevelt shut down Quoddy on July 7, 1936, projected costs were ranging up to $71 million. Given the egregious lack of relevant experience in the army's team and the many credible reports of incompetence, waste, and sabotage, it is easy to see how the army's version of Quoddy could have cost the government twice as much to build as having it be built by the most experienced and *positively motivated* hydroelectric engineer in the world, Dexter Cooper.

While President Roosevelt is often credited with foresight bordering on clairvoyance,[32] he could not anticipate everything. He probably did believe he could shelve the army's Plan D for a while, get Canada back on board, and then restart the project as the International Plan, thus "washing out"[33] the army's plans and the cost overrun issues without having to override the army. FDR appeared to be following this course of action in August 1936 when he went to Canada to meet with the prime minister after his tour of Quoddy Village.[34]

Events then overtook Roosevelt as the war in Europe, and its potential to engulf the world, became his primary focus. Quoddy had to be shelved.

COMMERCIAL RIVALS

Dexter's International Plan was killed in 1929 by the commercial fishing industry of New Brunswick, Canada. The killing was so swift that Roscoe Emery thought it was a secretly orchestrated "party affair" arranged with the help of pliable politicians. The *Bangor Commercial* suspected a "red herring," and reporter Ruby Black believed it was fomented and financed by Maine's private power monopolies.[35]

Dexter's All-American Plan was killed by the Great Depression in 1930. At the time there was no private financing available for a pioneering project like Quoddy. What little chance Dexter had was undermined by Murray & Flood's extortion attempt, their inflated cost estimates, and their duplicitous promotion of river power.[36] The Great Depression itself, of course, was caused in part by the vast stock sales and asset valuation frauds perpetrated by the power trusts and their facilitators in the financial industry.

Dexter's Power & Metals Plan of 1933–1934 was killed by the aluminum trust to protect its monopoly, in the opinion of several well-informed individuals, including Frederic Delano, vice chair of the Federal Reserve. Seven years *too late*, Interior Secretary Ickes came to the same conclusion, calling Alcoa "one of the worst monopolies"[37] and adding "if it has to be written that it [World War II] was lost, it may be because of . . . Alcoa."[38]

If it bleeds, it leads **is the media's maxim.** Most of the press reports about Quoddy were either inaccurate or biased. Whether they were simply sloppy, guilty of paid propaganda or circulation-boosting sensationalism, or suborned by advertisers, the media were willing accomplices in killing Quoddy and have blood on their hands.[39*]

The power trusts are the prime suspects because they had the most to gain or lose. They feared Quoddy because it was a serious competitive threat. To combat competition, they "captured" and corrupted the FPC and many state utility commissions. Money, political pressure, and the flood of propaganda biased these public agencies against public power. The bias manifested itself most critically in the FPC's negative reports on Quoddy in 1934 and 1941.

There is no doubt the power trusts were gunning for Quoddy. The Federal Trade Commission reports (1927, 1929–1936), Senator Hugo Black's hearings (1936), the New York State Power Authority report (1937), and dozens of newspaper reports provide thousands of pages of evidence of the power trusts' decades-long campaign to kill *all* public power projects, especially Quoddy.

Evidence of criminal behavior regarding Quoddy hasn't emerged, possibly because Dexter's files were stolen, and *they aren't the only collection of evidence that is missing.*

Buried in the Central Maine Power Company (CMP) archives at the Maine State Historical Society in Portland, inclusive of the years 1853 to 2001, filling 377.6 linear feet of shelf space, are just two diagrams of the Quoddy Tidal Power Project.

In the CMP collection at the Fogler Library at the University of Maine in Orono there are another 79.5 linear feet of files plus 7.25 cubic feet of other materials, containing CMP's records on the 40 electric companies it gobbled up between 1883 and 1965.[40] Despite the focus, scope, and extent of these CMP archives, there are no records of CMP's $100,000-plus investment (about $1.5 million in 2020 funds) in Dexter P. Cooper, Inc. (1927–1931), or of the four decades thereafter that CMP would have been closely following every step in Quoddy's consideration.[41*]

Evidently everything that might reveal CMP's attitude and actions relative to its investment in tidal power and Dexter P. Cooper, Inc., was expunged.

Evidently everything that might reveal the personal interactions of CMP's president, Walter S. Wyman, with his bosses, the infamous Insull brothers, was expunged.

Evidently everything that might reveal CMP's interactions with the most famous hydroelectric engineers in the world, the Cooper brothers, was expunged.

Evidently everything about the biggest electric power development in New England for four decades—CMP's greatest opportunity and its greatest threat as either a private investment or public power development—everything before 1953, was expunged.[42]

The evidence is gone. All gone.

As Sherlock Holmes might have said, "The dog that didn't bark speaks loudly."

Like the murder of Caesar, many had a hand in killing Quoddy. The reasons were eloquently summarized by the *Boston Daily Record*[43]* in May 1947:

TRAGIC EASTPORT

Eastport, Me., is again in the news; this time as the result of a fire which destroyed more than a million dollars' worth of property and left five hundred factory workers jobless.

The recent history of this coastal community, with its unique high tides and fortunate position on the map, is as sorrowful as any story written.

Eastport might have been the wonderland of the Northeast, rivalling the T.V.A. and Niagara. It could have surpassed all other New England cities and towns in growth and progress.

Instead it became a political casualty when the champions of scarcity and the forces of exploitation combined to strangle Passamaquoddy.

Strangely, many of the politicians responsible for the scrapping of this mammoth hydroelectric development have retained the support and confidence of Maine people, and they are regularly returned to public office.

And, significantly, only a few months ago the Legislature at Augusta, bowing to the demands of the power trust, overwhelmingly rejected a bill which would have revived Passamaquoddy and given the homes, farms, and factories of the Pine Tree State the cheapest and most abundant electricity in the nation.

Perhaps Maine does not want progress and prosperity.

However, it is more likely that the State has had the bad luck to elect the wrong kind of officials.[44]

Professional, political, and commercial rivals killed Quoddy. The most guilty are the power trusts. They are the "champions of scarcity and the forces of exploitation." They had a hand in every effort to kill Quoddy. They wrote Eastport's "economic obituary."[45]

They corrupted the executive branch through regulatory capture of the Army Corps of Engineers, the Federal Power Commission, and state public utility commissions.

They corrupted the legislative branch through campaign contributions, cronyism, bribes, and voter fraud.

They corrupted the judiciary, with the "Attleboro gap" as an example.

They corrupted the press with fake news and biased reporting.

They corrupted the educational system with slanted curricula and professors paid to write misleading reports.

They corrupted the banking and financial systems through chicanery, inflated values, and fraudulent transactions.

They swindled hundreds of thousands of innocent investors with "watered" private utility stocks.

They undercut and obstructed legitimate business through self-dealing and preferential pricing, restraint of trade, racketeering, theft, and sabotage.

And they cheated hundreds of millions of people of the benefits of low-cost public power.

Quoddy was killed five times during Franklin Roosevelt's lifetime. While Presidents Harry Truman, Dwight D. Eisenhower, and John F. Kennedy tried to revive Quoddy, the power trusts made sure it never got out of the grave.

But Dexter's dream didn't die. It won't die. Men and women will continue to try—until they succeed—to harness the power of the tides. And to those who are trying, as Roscoe, Gertrude, and Franklin would say:

"More power to you."[46]*

Finis

Author's Notes and Acknowledgments

I did not know of any personal connections to Quoddy when I began my research for this book. I came to the subject through an interest in tides, which originated early in my career (1986–1996) when I created the *Embassy Boating Guide* series, covering the Atlantic Coast from Eastport, Maine, to Key West, Florida. In September 2018, I toured the Bay of Fundy to see its many tidal phenomena, tide mills, and Quoddy.

I was surprised by what I could and couldn't learn about Quoddy while in Eastport and Campobello. Returning home, I spent a week researching Quoddy and then wrote a preliminary book proposal. I sent it "over the transom" to Michael Steere, editorial director at Down East Books in Rockland, Maine. After all, I figured, you can't get any more Downeast than a story about Eastport. Four hours and six minutes later, he said, "Yes."

Almost immediately I discovered startling information and surprising connections. The story became increasingly relevant because I was creating the book during the constitutional turmoil of the Trump administration, the Texas energy freeze, the California drought, the COVID-19 pandemic and resulting economic depression, and Russia's invasion of Ukraine. The parallels are unavoidable, the lessons applicable.

Thank you, Michael, for launching me on a most interesting project, and for sticking with me as it got bigger, more complex, and more important than we had imagined.

In conducting my research on Quoddy, I was aided by the work of previous authors who had investigated Quoddy.

Kenneth Roberts rehashed his article from the *Saturday Evening Post* as the last fourteen pages of his book *Trending into Maine* (1938). He is engaging and evocative, but his libertarian bent warps the story with told-you-so certainty, making Quoddy into his exemplar of the New Deal's evils and failures without addressing the true causes of its collapse.

Lincoln Smith wrote as his doctoral thesis "The Power Policy of Maine" (1950), which includes a chapter on Quoddy. Smith's investigation of the legal and political history of riparian rights and the development of the private power utilities is excellent. Smith admits that he "does not like the Fernald law and public power,"[1] accepts the Federal Power Commission (FPC)'s reports at face value, and is therefore skeptical of Quoddy's economic justification. Smith dismisses as irrelevant the propaganda and financial fraud perpetrated by the power trust. He gives an enlightening treatment to the issues of home rule and states' rights.

Theodore Holmes wrote for his master's thesis at the University of Maine "A History of the Passamaquoddy Tidal Power Project" (1955). An Eastport native and teenager when Quoddy was under construction in 1935–1936, Holmes had the good fortune to interview Roscoe Emery (his uncle), Moses Pike, and Gertrude Sturgis Cooper, as well as the FPC's Roger McWhorter.

George Marvin wrote for his bachelor's thesis at Bowdoin College "The Passamaquoddy Bay Tidal Power Project and the New Deal" (1972). Marvin found some of the important and recently declassified White House documents. Written during the debate about the Pittston oil refinery proposal in Eastport and published as a supplement to the *Ellsworth American* newspaper in 1974,[2] he blames Senators Hale and White and the Army Engineers for Quoddy's death and concludes that Quoddy should have been built.[3]

Quoddy is included as chapter 9 in *Army Engineers in New England, 1775–1975* by Aubrey Parkman (1978). Mostly an abstract of the *Final-36* report, it also brings the story into the 1970s. It provides little context and fails to mention that a market for the power did exist in 1936, as documented by Marvin. Nonetheless, one senses the army's frustration at being whipsawed by Congress.

Two books on the private power trusts are particularly relevant. *The Public Pays* (1931) by Ernest Gruening, at the time the editor in chief of the *Portland (Maine) Evening News*, summarizes the 1927–1932 Federal Trade Commission investigation into power trust propaganda and corruption. Gruening was one of the trust's victims (though he doesn't say so)[4] and played an important role in the Brewster-Corcoran debacle.[5] Gruening made no mention of Quoddy (which was not being actively pursued in 1931). Updated in 1964, Gruening's book summarized the FPC's 1940 investigation of the power trusts and the industry's anti-government propaganda offensives in the 1960s.

Insull: The Rise and Fall of a Billionaire Utility Tycoon (1962) by Forrest McDonald is an exhaustively researched biography that focuses on Insull's business innovations and civic contributions before 1929, including the World War I origins of the power trust propaganda machine. McDonald summarizes the creation of Insull's holding companies, their subsequent collapse, and his trial. McDonald blames much of the collapse on speculators who bid up Insull stock to absurd levels and on Insull's rival, the Morgan financial conglomerate, which shorted the stock and made a killing from the collapse. The friendly and forgiving[6] McDonald concludes Insull was made the scapegoat for the abuses of the finance and power industries.[7]

McDonald blames the Insulls' troubles in 1929–1930 on "three major blunders," the first of which "came about primarily through the actions of Walter S. Wyman," president of Central Maine Power Company, who was responsible for the "spectacular and fruitless battle to repeal the state's archaic Fernald law." The second was Samuel Insull's alleged attempt to "fix" the US ambassador in Germany. The third and "most costly blunder" was that Martin Insull attacked a politician, "an extremely important one"—the governor of New York, Franklin Roosevelt. McDonald characterizes Roosevelt's position on holding companies and power trusts as inconsistent and illogical, and possibly motivated by a personal grudge against "his old friend," Howard Hobson. According to a legend circulated by power trust executives, Hobson laughed Roosevelt out of his office in 1928 when he was looking for a job before deciding to run for governor. McDonald quotes a Roosevelt adviser saying the tale did not "ring true," but McDonald thought FDR "acted as if it were the case," and the private power trusts (and apparently McDonald) believed it.[8]* McDonald makes no mention of Quoddy.

I took a personal approach to relating the Quoddy saga. I focused on three main characters and their intertwining life histories—Franklin

Roosevelt, Dexter Cooper, and Roscoe Emery—and the three people who connected them all—Gertrude Sturgis, Frederic Delano, and Philip Fleming. The primary sources I relied on were the official and private papers of Franklin Roosevelt and related materials at the FDR Library at Hyde Park, New York; letters, plans, reports, photographs, and files in the collections of the Dexter Cooper and Moses Pike families; and the Roscoe Emery collection of letters, clippings, and microfilms of the *Eastport Sentinel* at the University of Maine's Fogler Library in Orono.

I believe the confusing history of the Quoddy Tidal Power Project can only be understood if put in context. Accordingly, I framed the story at the local, regional, national, and international levels. For the local view, I relied on the pro-Quoddy *Eastport Sentinel*, from which I clipped some 2,000 articles spanning the period from its founding in 1818 until it ceased publication in 1953. For the regional view, I chose the usually anti-Quoddy[9]* *Boston Herald*, which discussed the subject more than 500 times, averaging three times per week during 1935 and 1936. For the national view, I read the *New York Times*, which, in more than 200 articles covering Quoddy, ranged from strongly supportive to viciously negative. Supplementing these were hundreds of clippings from other publications in Roscoe Emery's collection and in the National Archives.

Broadening my perspective, I found that Doris Kearns Goodwin's book *Bully Pulpit* provided the backstory on Theodore Roosevelt, the Progressive movement, and the power of the popular press. Likewise, Goodwin's *No Ordinary Time* captured the dynamism of Franklin and Eleanor Roosevelt and the social upheavals of the Depression and World War II. Jonas Klein's *Beloved Island* provided the history of the Roosevelts at Campobello. John Riggs's *High Tension*, Gordon Weil's *Blackout*, and Philip Funigiello's *Toward a National Power Policy* were very helpful in understanding the development of FDR's national power policy and its legacy. The international framework was developed from academic journals and the popular press, with additional material about Hugh Cooper and Soviet Russia from the US State Department archives.

When quoting material from the *Eastport Sentinel*, I attributed it to the *Sentinel* or Roscoe Emery interchangeably for the period of his ownership because he was normally the only staff person identified. On the rare occasions when individuals were named in bylines, I noted them.

I have many individuals and organizations to thank.

The descendants of Dexter P. Cooper were of immeasurable assistance, including Dexter P. Cooper III and his sister Carla Cooper, and Peter Homans, son of Nancy Cooper Homans. I was thrilled to have located DPC III and his collection of his grandfather's files from Boston. I am particularly thankful for the professional editing advice from and friendship I developed with Peter Homans over the course of a year of weekly emails as I began hammering the confusing elements into a coherent story.

Likewise, Moses Pike's family were a great source of documents, stories, and leads. Thank you, Davis Pike, for being my guide in Lubec, and thank you, Anne Pike Rugh, for a pile of documents and explaining the extended family relationships in Quoddy Bay. My dear and newfound cousin Kate Borton Jans was a great find and located for me all sorts of interesting tidbits, provided good feedback, and gave me much encouragement.

I have many people in Eastport to thank for their assistance over several years and many queries. Wayne Wilcox and I enjoyed many long discussions about the tidal power project, and he was a great help in identifying who and what I was looking at in stacks of photographs. Many of those wonderful images, and the tales attached, came from Ruth McInnis of the Todd House B&B. Hugh French, of Tides Institute and Museum of Art, deserves great credit for preserving Eastport's history and architecture, as does Ed French of *Quoddy Tides* for continuing Eastport's proud literary tradition. Thank you both for allowing me to be part of the legacy.

Dana Chevalier and Patricia Gardner-Theriault at Eastport's Peavey Memorial Library helped orient me and tracked down important materials, especially quotes from rare editions of the *Eastport Sentinel*, in their collection. Cory Critchley and Eleanor Norton of the Border Historical Society introduced me to Quoddy's "action model" and loaded me up with Eastport's other history. Pam Koenig opened Roscoe Emery's house for me and introduced me to John Melby and his collection. Ross Furman of Rossport by the Sea was a great tour guide for Quoddy Village of yesterday and tomorrow and the curator of another fascinating collection. Kris Johnson and Daryl Wurster of Puffin

Haus B&B made my wife and me feel at home. Susan MacNichol and Christopher Brown provided insights into Eastport's social history and lineages. John Smith, bookseller, and Joe Clabby, author, were also encouraging.

I was inspired to create this book in part by the welcome I received from the staff at the Roosevelt-Campobello International Park, who responded to my original query with an invitation to visit and a great tour when I arrived and, when they ran out of answers to my questions, by suggesting, half-jokingly, that I write a book. Little did we know. . . . So thank you, Will Kernohan, Sherry Mitchell, Jocelyn Calder, Laura Storm, and Stephen Smart—it's been quite an experience.

Others on Campobello who welcomed, informed, and guided me include the pleasant, patient, and helpful Stephanie Anthony and Stephanie Gough at Campobello Public Library. Joseph Gough provided some great stories and explanations. Dale Mallock at the Porch Restaurant kept me well fed with hearty food and stories such that I kept coming back. Mike Calder showed me around and filled in some of the Cooper family history. Sally and Glen Stephenson graciously gave me a tour of their house and barn, formerly the Coopers' Flagstones property. Vivian and Sue helped me find Dexter's grave at St. Anne's Church cemetery. Joyce Morrell of the Owen House Hotel helped with the history of Campobello's other visionary leader, Admiral Owen.

From farther afield in Canada, I received much appreciated help from Brian Usher, Minister's Island; Kurt Peacock, St. Croix International Park; Derek Cooke, Charlotte County Archives; Sandra Currie, Fundy Ocean Research Center on Energy (FORCE); Les Smith, Annapolis Tidal Generating Station; Jacqueline Foster, Nova Scotia Power; and Jim Hill, Niagara Parks.

In Lubec, I would like to thank Barbara Sellitto of Lubec Historical Society; Nancy Briggs and Jennifer Multhopp at the Lubec Public Library; James O'Neil of the James O'Neil Gallery; and especially Victor Trafford, at the Inn at the Wharf, for the fabulous view and trip to Treat Island.

Elsewhere in Maine, thanks go to Robert, Jane, and Aaron Bell of Tide Mill Farm; Charlie McAlpin, Eastern Maine Electric Cooperative; Joyce Garland, Calais Free Library; Meg Scheid, Saint Croix Island International Historic Site; Al Churchill, St. Croix Historical Society; Collin Beal, St. Croix Paper Company; Caitlin Lampman, Bates College; Roberta Schwartz, Bowdoin College; Desiree Butterfield-Nagy,

Special Collections Department, Fogler Library, University of Maine, Orono; Adam Fisher and Michelle Brann, Maine State Library, Augusta; and Valdine C. Atwood and Carla Manchester, Washington County Courthouse Archives.

To the members of the Tide Mill Institute, I owe particular thanks for their welcoming, guiding, reading, expertise, and good cheer: Bud Warren, Bob Gordon, Patrick Malone, Ron Klodenski, Earl Taylor, Lincoln Paine, Tim Richards, Todd Griset, and Jonathan White.

For information and images relating to the Keokuk Dam, I gratefully acknowledge the assistance of author John Hallwas, Angela Gates of the Keokuk Public Library, and Kathy Nichols and George Usher of Western Illinois University.

My support team in Connecticut starts with the Chester Public Library and Stephanie Romano, Pam Larson, Patty Petrus, Stephanie Rush, Linda Fox, Leigh Basilone, and Trish Guccione, who chased down dozens of arcane books. Likewise, the Russell Library in Middletown hosted me for many weeks while I reviewed 50 years of the *Eastport Sentinel* on microfilm sent to me by the University of Maine. Thank you to Denise Mackey-Russo, Kimberly Spachman, Valerie Funderburg, and Catherine Ahern. Patient readers of my early drafts include Jeff Singer, Anne Bracker-Singer, Alex Stein, Terri Wilson, and Susan Strecker. Thank you also to Shannon McKenzie at Mystic Seaport Museum, who let me into the warehouse to commune with FDR's *Vireo* and hundreds of other classic boats.

And at the Franklin Delano Roosevelt Library, Hyde Park, my thanks to Virginia Lewick and Kristen Carter for helping me understand the richness of the resources they carefully steward.

Finally, I need to thank my family for . . . well, everything. In particular, I have to thank my wife France for letting me take years to work on this project; my father Terry, my first reader and editor; and my mother Debbie, my research assistant. Likewise, I learned a lot about writing from my daughter Rebecca and my aunt Lady.

One Last Note

While I labored diligently to be complete, accurate, and fair, I have probably made mistakes of fact and interpretation and will gladly correct any such errors in future editions. Please send me your comments via Down East Books.

APPENDIX 2

The Astronomy That Creates the Ocean Tides

Tides on the earth are created by the gravitational attraction or pull of the moon and, to a lesser degree, the gravitational pull of the sun. The height of the *ocean tide* depends on which one is winning this planetary tug of war.

Gravity is the attraction of everything to everything else in proportion to its mass and inversely to the *square* of its distance ($G = M/D^2$). Gravity's tide-generating forces vary by the *cube* of the distance ($T = M/D^3$). Thus, distance matters more than mass. While the sun is 27 million times more massive than the moon, the moon is 390 times closer to the earth, and thus the moon exerts twice the gravitational pull on the earth.[1]* Our liquid oceans are free to move in response to the gravitational pull of both the moon and the sun; in doing so, they create the ocean tides. (The other planets are so far away that they have insignificant gravitational effects on the earth.)

Since the moon is always closer to one side of the earth than the other, the gravitational pull on the near side is 3% stronger than the far side.[2] The ocean responds to this pull and creates a *gravitational tidal bulge* of water on the side of earth *toward* the moon in which the water is lifted by about eighteen inches. Meanwhile, on the *far* side of the earth where the gravitational pull of the moon is weaker, there is an equally large *centrifugal tidal bulge* that swings *away* from the moon and the earth in response to centrifugal forces caused by the spinning of the earth—somewhat like an ice skater's ponytail sticking out behind her when spinning.[3]* The sun also creates a gravitational tidal bulge and a centrifugal tidal bulge, but only about half the size because of its distance from the earth.

As the earth rotates on its axis, our position on it rotates into the lunar tidal bulge (high tide) and out of the bulge (low tide). As *we* rotate in and out of the tidal bulges, it *appears* to us as the rising and falling of the tides. This rhythmic rising and falling creates the "tide wave" or *ocean tide*. Most regions of the world have *semidiurnal* tides (Latin for twice daily), meaning that there are two high tides and two low tides each day.[4]*

The *tides*—the vertical change in the level of the sea—are linked to *tidal currents*, the horizontal movement of the sea. The higher the tides, the faster the currents. Currents in the sea, also known as *streams*, are of two types. The great *ocean gyres*, like the Gulf Stream, are the result of the *Coreolis effect* and the spinning of the earth (and not technically tidal). *Currents* that flow *in and out* of harbors and archipelagos are caused by the tides. (If the flow is in only one direction, then it is probably caused by a river.)

The earth takes 24 hours, or 1 day, to rotate around its axis. While the moon orbits the earth in the same direction, it takes 27.3 days (a *sidereal month*) to do so.[5] The earth-moon rotation-orbit combination causes the moon to appear above the same point on the earth every 24 hours and 50 minutes. This is why moonrise and high tide occur 50 minutes later each day.

At least twice each month, the sun, earth, and moon are roughly in a line, during the dark "new" moon phase (sun-moon-earth) and during the bright "full" moon phase (sun-earth-moon). When these *syzygies* (alignments) happen, the *lunar* tidal bulges pile up on top of the *solar* tidal bulges, creating large *spring tides*—they "spring up" higher than usual (and drop lower than usual). Despite their name, spring tides can occur at any time of the year.

On the rare occasions when the sun-moon-earth are perfectly aligned at new moon, there will be a *solar eclipse*. When the sun-earth-moon are perfectly aligned at full moon, there will be a *lunar eclipse*. Eclipses could increase the size of the spring tide by a tiny bit, but the distances of the sun and moon have a greater effect.

During the two half-moon phases each month, the moon's gravity is pulling perpendicular to the sun's gravity, partially canceling each other out, thus creating smaller *neap tides* and less variation or *range* between high tide and low tide.

The moon's orbit is not a perfect circle but an ellipse. At *apogee*, the moon is at its maximum of 252,000 miles away. At *perigee*, the minimum, the moon is only 221,000 miles from the earth, thereby increasing the tides by about 11% from average. Similarly, the earth's annual orbit around the sun is elliptical. At *aphelion* (July 2), the sun is 94.5 million miles away; at *perihelion* (January 2), it is only 91.5 million miles away.

The earth is *declined* (tilted) on its *axis* about 23.5 degrees, creating the seasons. In either hemisphere (Northern or Southern), it is summer when the axis leans toward the sun and winter when it is leaning away. Tides are strongest during the *equinoxes* in the spring (usually March 20) and fall (September 22 or 23), when the sun is directly over the *equator.* Likewise, the moon's orbit is declined 5.2° to the *ecliptic* (plane of the earth's orbit).[6]

Every 18.6 years, the sun, earth, and moon return to the same relative distances apart; thus the global tidal cycle repeats itself every 18.6 years.[7]

The height of the tide on any given day is a combination of these and other astronomical variables, as well as the weather (winds and barometric pressure) and local geography.

Surprisingly, the time at which high tide occurs is not when the moon is directly overhead but lags due to the tide's friction with the ocean floor. The highest tides of the month can occur up to a week *before or after* the full or new moon passes overhead. This is called the "age of the tide." In the deep ocean, the tide's *age* is about an hour. Along America's Atlantic coast, the tide's age averages 1½ days because the tide has to travel over the relatively shallow continental shelf. In Santa Rosalina, California, the highest tide arrives a day before the full moon.[8]

While the vertical pull of the moon on water is minuscule, the horizontal pull is bigger, because the fluid nature of water makes it easy. (Think how hard it is to *lift* a floating boat but how easy it is to *pull* it.) These horizontal *tractive forces* draw water to the point closest to the moon, thereby increasing the tidal bulge. Thus, the ocean tides are partly the result of the horizontal tractive forces.

In summary, the height of the ocean tide is driven by complex astronomical cycles. But as discussed in chapter 2, "Passamaquoddy Bay," geography dramatically changes the size and timing of the tides.

Exceptional Dams Built by Hugh and Dexter Cooper

	St. Lawrence/ Niagara River: Niagara Falls, Ontario, Toronto Generating Station Canada	Susquehanna River: McCall Ferry, Pennsylvania, Holtwood Dam USA	Mississippi River: Keokuk, Iowa USA	Tennessee River: Muscle Shoals, Tennessee, Wilson Dam USA	Passamaquoddy Bay: Maine and New Brunswick, International Plan USA & Canada	Dnieper River: Ukraine Soviet Russia
Public vs. private	Private enterprise	Private enterprise	Private enterprise	Public power	Private 1920–1934; public 1935-on	Public
Construction Dates of operation:	**1903–1906** 1906–1974[1]	**1906–1908**[2] 1910[3]–continuing	**1910–1913** 1913–continuing	**1918–1924**[4] 1925[5]–continuing	**Announced in 1920 Construction 1935–1936:** *started but not completed*	**1927–1932** 1932[6]–continuing
Engineered by Constructed by	**E:** Hugh L. Cooper **C:** Dexter P. Cooper	**E:** Hugh L. Cooper **C:** Dexter P. Cooper, *assumed*[7]	**E:** Hugh L. Cooper **C:** Dexter P. Cooper	**E:** Hugh L. Cooper/ US Army **C:** Dexter P. Cooper	**E:** Dexter P. Cooper (original) **E & C:** Army Engineers (1935–1936)	**E:** Hugh L. Cooper **C:** Dexter P. Cooper
Water flow	200,000 cfs[8] to 210,000 cfs[9]; 20 mph at wing dam[10]	50,000 cfs[11]; 2,200 to 750,000 cfs[12]	370,000 cfs[13]	950,000 cfs[14]; 51,900 cfs average[15]	2,400,000 cfs: 24 feet per second[16]	
Horsepower (1 hp = 0.7457 kW)	200,00 hp[17]	158,000 hp[18]	150,000 hp;[20] 160,000 hp from (4) 30,000 hp units and (14) 35,000 hp units[19]	260,000 hp[21]	300,000 to 800,000 hp;[22] 464,000 hp to 1,087,000 hp[23]	750,000 hp[24]
Continuous or primary power (1 MW = 1,000kW)	91.8 MW[25]	252 MW[26]	142 MW[27] to 150 MW[28]	50 to 184 MW[29], 0.76 to 2.1 billion kWh/ yr.[30]	3 billion[31] to 3.6 billion[32] kWh per year	156 MW[33] in 1930; 157 MW[34] in 2019

Power pools	n/a	Lake Aldred[35]	Lake Cooper 64 miles long[36]	Lake Wilson: 166 miles long; 15,500 acres[37]; 50,500 acre-feet[38]	High pool: 110 sq. mi. (70,400 acres) in Passamaquoddy Bay (USA & Canada). Low pool: 40 sq. mi. (25,600 acres) in Cobscook Bay (USA)[39]	Dnipro Reservoir, 87 miles long; area of 123 sq. mi. and contains almost 53 billion cubic feet[40]
Dam dimensions	Wing dam 785' long, up to 27' high,[41] Head of 165' to 180'[42]	2,392' to 2,700' wide;[43] 55' to 95' high[44]	Powerhouse & Lock was nearly 9,000' (1.7 miles) long; 32' high[45]	137' high[46] by 4,541' long[47]	7 miles of dams; longest 6,000';[48] deepest 180';[49] 16–28' head	200' high, 2,600' long[50]
Other structures			Navigation lock 400' × 110', and lift 40'[51]	Main lock 110' × 600' with max lift of 100'—highest single lift lock east of the Rockies. Average 3,700 vessels pass through locks each year.	Pumped storage auxiliary power plant: 2 units @ 22,500 kW at Haycock Harbor;[52] 300-mile transmission line to Boston[53]	3 chamber navigation lock[54]
Turbines & generators	11 generators, each sitting atop a 180' vertical shaft[55]	The original Kingsbury bearings, in use since 1912, are an IHME engineering landmark. Previous roller bearings would last about 2 months under the generator's 220-ton load.[56]	When built, the largest turbines ever made. Each of the 15 turbines was more than 16' wide, generated 10,000 hp. Each turbine-generator weighed a million pounds.[57]	Initially 8 units @ 32,500 hp;[58] 21 generating units; 663 MW as of 2019[59]		9 turbines by Newport News Shipbuilding and Drydock Company; 5 generators from General Electric Company, 4 from Siemens of Germany[60]
Total cost: per kW, kWh/yr. MM = millions			$25 MM[61] to $30 MM[62]	$37[63] to $47[64] MM (dam); $90[65] to $150 MM[66] (total); $200 to $740 per kW[67]	$75 to $100 MM;[68] $.0075 per kWh[69]	$100 MM[70] to $110 MM;[71] $400 MM w/ metal mfg.[72]

	St. Lawrence/ Niagara River: Niagara Falls, Ontario, Toronto Generating Station Canada	Susquehanna River: McCall Ferry, Pennsylvania, Holtwood Dam USA	Mississippi River: Keokuk, Iowa USA	Tennessee River: Muscle Shoals, Tennessee, Wilson Dam USA	Passamaquoddy Bay: Maine and New Brunswick, International Plan USA & Canada	Dnieper River: Ukraine Soviet Russia
	When built, the largest hydraulic tunnel in the world, for the tailrace: 33' tall and 19' wide. 2,200' long, exiting *under* Horseshoe Falls.[73] 142-mile transmission line to Buffalo, NY—which was 90 miles or 3 *times farther* than anyone had ever transmitted electric power.[74] National Historic Site[75]	Pioneering design using triangular cross-section monolith that would resist overturning due to water pressure from the dam—primary characteristic for main stem river dams for next 75 years. Pioneering use of temporary crib dams while building permanent dams.[76]	When built, the largest concrete dam; largest concrete monolith;[77] largest single powerhouse electric-generating plant in the world.[78] Power transmitted to St. Louis, 110 miles away (3 times farther than had been done),[79] on Hugh Cooper-designed 110,000 volt system, which is the model for most modern high-power transmission systems.[80]	First major public power project. Originally designed by US Army Corps of Engineers, then redesigned by Army colonel Hugh Cooper with construction management by Dexter Cooper. National Historic Site	If built, Quoddy would have been the most powerful hydroelectric dam in the world and the first commercial-scale, tidal-hydroelectric power plant.	When built, the most powerful hydroelectric dam in the world. Hugh Cooper was awarded the Order of the Red Star, the first foreigner to receive the highest civilian honor bestowed by the Soviet government.[81] Still the largest dam in Europe.
Notes: See ch. 9, "Anticipation" for a comparison of Quoddy to private power dams in Maine and public power dams elsewhere.						
Book chapter number and section title	3, "Political Calculus"	Not covered in text	1, "The Engineers"	3, "Waterpower and Electricity"; 8, "Communism versus Capitalism"; 8, "The Commander in Chief"	3, "Lessons from World War I"— through 14	6, "American Technology"; 7, "The Campaigns"; 8, "Communism versus Capitalism"; 8, "The Commander in Chief"; 12, "United They Stood"

APPENDIX 4

Vital Statistics of Quoddy Tidal Power Project, Army Plan D, 1935–1936

PROJECT DATES

Federal approval, May 16, 1935[1]
President Franklin Roosevelt, via PWA

Opening ceremonies, July 4, 1935
Eastport, ME

Construction ceased, August 16, 1936[2]
Dam repairs continued until March 1937

Official end, October 31, 1936[3]
USACE "Quoddy District" closed

Quoddy Village, etc., sold, October 7, 1949[4]
Grossmans Surplus primary buyer

BUDGET, FUNDING, AND USE OF FUNDS

Original budget	$30,125,000.00 (ERA funds)[5]
Initial allocation, June 7, 1936[6]	$10,000,000.00
Allocation cut $5 million, October 21, 1935[7]	−$5,000,000.00
Reinstated $2 million, January 11, 1935[8]	*+$2,000,000.00*
Net federal funding	$7,000.000.00
Receipts from sale of village and equipment	*$1,192,191.80[9]*
Total revenue	$8,192,191.80
Gross expenditures	*$7,386,889.36[10]*
Cash balance as of October 28, 1936	$805,302.44[11]* (retained by US Army)

Item	Constr. Cost[12]*	% Complete[13]*	Notes (see Key)
District overhead[14]*	$617,226.44[15]	n/a	A; 15 months
Surveys	$833,000.00[16]	90	Land-A; sea-C
Land acquisition	$131,932.25.[17]	70	A: total of 308 acres[18]
Preliminary engineering	$155,283.80[19]	90	A: village, dams, reservoirs, etc.
Detailed engineering	$445,310.00[20]	40	A: Village, dams, reservoirs, etc.
2 labor camps	$235,850.00[21]	100	A; T; estimate $100,000[22]
Administration building	$230,315.56[23]	100	C; T; estimate $132,814[24]
2 apartment houses: 80 units	$285,015.80[25]	100	C; P
1 dormitory: 85 rooms	$199,317.60[26]	100	C; P
120 houses Quoddy Village	$745,102.37[27]	100	A; T
9 houses on "Snob" Hill	$245,232.33[28]	100	C; P; estimate $102,088[29]
Diesel power plant	$ 62,302.01[30]	20	A; P; building and 1 of 5 generators[31]
Groups 3 buildings	$258,352.49[32]	100	A; P; hospital, fire, police, exhibit, etc.
Village heating plant	$276,104.52[33]	100	See chapter 11, "Limbo"
Shops, warehouses, etc.	$180,900.00[34]	100	
Roads, water, sewer, electricity	$316,971.05[35]	100	
Pleasant Point Dam	$167,096.60[36]	70	A; estimate $150,000; repair 3/37[37]
Carlow Island Dam	$163,414.32[38]	70	A; estimate $100,000; repair 3/37[39]
Treat-Dudley Dam	$ 47,733.69[40]	18	A; washed out & abandoned
Tidal Power Station	$ 14,178.44[41]	10	A; site work only
Navigation lock	$ 64,919.89[42]	5	A; site work only
Filling gates	*$ 94,313.70[43]*	*5*	*A; site work only*

Key: A = army; C = contractor; P = permanent building on foundation; T = temporary building on piers

Employment (as of October 23, 1936)
Maximum employed: 5,430
    (11/30/35)[45]
Total payroll: $3,119,873.00[47]

Total worker hours: 5,851,479[44]
Average cost per worker-year:
    $2,011.28[46]
Fatalities: none

Gertrude Sturgis Cooper

June 20, 1889–December 18, 1977

As superintendent of the Vanderbilt Mansion in Hyde Park, New York, Gertrude hosted her daughter Nancy's wedding reception there in 1941. After her tenure at the mansion ended in 1945, Gertrude became assistant director and then director of Greenwich House, the famous immigrant settlement and community development organization in New York City.[1]* She retired in 1948.[2]

Some years later, Gertrude moved to Essex, Connecticut, to live near her youngest daughter Elizabeth ("Boo"). Elizabeth ran a small graphic production service called Brushmill Books in the neighboring town of Chester. Her largest client was across the street, Pequot Press—later Globe Pequot Press, now the parent company of Down East Books, the publisher of this book.

Gertrude died at Aaron Manor, a nursing home a few hundred yards away from the Brushmill Books office and about a mile from where I have lived for the last thirty years. Aaron Manor is the same facility where my mother-in-law passed away in 2016.

Gertrude is buried next to her husband and their daughter Elizabeth at the cemetery of St. Anne's Church, their spiritual home on Campobello.

Roscoe Conklin Emery

March 28, 1886–December 16, 1969

Roscoe remained active in the real estate business and civic organizations until his death. The year before his death, the *Saturday Review* published an article, "Portrait of a Declining Town: Eastport, Maine," to illustrate the plight of small towns that had lost their industry. In it, Roscoe was described as "spry, at eighty-two. . . . He has changed little from a 1930s picture . . . except that that his flattop crewcut has turned white and he had replaced his wire spectacles with thick black hornrims concealing a hearing aid."[1]

Roscoe's daughter, Joyce Emery Kinney (1924–2016), was the founder of the Border Historical Society in Eastport, which maintains the Quoddy Model Museum at 72 Water Street.

Roscoe's second cousin once removed,[2] Theodore C. Holmes (1920–2015), an Eastport native and graduate student in history at the University of Maine in 1955, wrote his master's thesis on Quoddy. Much to my surprise, I discovered that Theodore's sister, Olive Katherine Holmes (1917–1998), married Charles Roosevelt Borton (1913–1978). Charles was not a Roosevelt by blood; his middle name was apparently given to him in honor of former president Theodore Roosevelt. As there are few Bortons, it seemed likely Charles was a distant cousin of mine, but a search of the Eastport phone book yielded no leads to living descendants.

While on Campobello, I struck up a conversation with Laurel Storm at the tourist information center. She was intrigued by my project and offered to connect me to a possible descendant of Moses Pike. Within an hour I received a call from Davis Pike, a nephew, who in turn connected me to Anne Pike Rugh, Moses's daughter and a trove of documents. The next day I received a call from Laurel, who said, "I hope you don't mind, but I broke protocol. I gave your telephone number to my best friend when I remembered that her maiden name is Borton." The following morning, I got a call from Kate, Theodore Holmes's niece and a cousin I didn't know I had. We share John "the First" Borton, who arrived in New Jersey in 1687, as our mutual progenitor, ten generations back. Kate and I met in Eastport and had a riotously good time telling each other about heretofore unknown family history.

The US Army Corps of Engineers

Philip B. Fleming (1887–1955): Promoted to brigadier general (1941); major general (1942); recipient of Distinguished Service Medal (1946).[1] "Known for his friendliness and wit," Fleming "attended weekly White House cabinet meetings from the time of the Pearl Harbor attack to 1949."[2] He was with President Harry S. Truman when the latter announced the end of World War II in Europe on May 8, 1945.[3] Fleming headed the Federal Works Administration (1941–1949); he was under secretary of commerce (1950–1951) and ambassador to Costa Rica (1951–1953).[4]

Royal B. Lord (1900[5]–1963): Lord's building of Quoddy Village is proudly recounted in his official obituary, along with the tale of the private rail cars attached to the "milk train." After Quoddy, Lord headed construction for the New Deal's Farm Security Administration. During World War II, he was instrumental in the "Red-Ball Express" of rapidly built roads and bridges, which sped supplies to Allied forces as they pushed the Nazis back across France to ultimate defeat. He repeatedly worked for Philip Fleming and retired in 1946 as a major general.[6]

Hugh J. Casey (1898–1981): Post-Quoddy, Casey worked on flood control on the Connecticut River. In 1941, he helped design the Pentagon in Washington, D.C. That fall he went to the Philippines and was promoted to brigadier general in January 1942. He escaped from Corregidor with General Douglas MacArthur via PT boat in March 1942 and was thereafter awarded the Distinguished Service Medal. In an interview in the 1970s, and apparently contradicting his statements in *The March of Time* (1938), he said the US Army's Plan D was not financially justified because of low fuel costs in the 1930s, but the International Plan might have been justified.[7]

Donald Leehey (1897–1968): Leehey became district engineer for Portland, Oregon (1940–1943), and worked on the Bonneville Power Authority and the International Boundary Commission. He retired from the army in 1946 as a colonel and joined the Atomic Energy Commission, with responsibilities for developing, testing, and storing nuclear weapons. Quoddy received half a sentence in his West Point obituary.[8]

Samuel D. Sturgis Jr. (1897–1964): Sturgis worked on Mississippi River projects after Quoddy. During World War II, he served in New Guinea and the Philippines. After the war he focused on the Missouri River. Sturgis was chief of the engineers (1953–1956). His official biography on the Army Corps's website[9] doesn't mention Quoddy, despite his role during its construction (1935–1936) and his having headed the IJC's "full-dress" survey (1956–1958).

APPENDIX 8

The Passamaquoddy Reservation

In February 1964, Joseph Stevens found a man cutting down trees in the Passamaquoddy Reservation. When Joseph protested, the man said he had legally acquired the land (in a card game). Through dozens of such deals over almost two hundred years, 6,000 acres and all the islands had been chiseled from the original 23,000 acres specified in the Treaty of 1794.[1] The tribe appealed to Maine's governor John Reed, who scoffed at their claims, so members of the tribe peacefully blockaded the disputed property and were promptly arrested. The tribe then convinced the courageous (or too new to know better) attorney Don Gellers of Eastport to represent them.[2] Gellers documented how the tribe had lost their land and, with the aid of Tom Tureen, a law student, began building a case against Massachusetts and Maine for violations of the treaty. On March 8, 1968, Gellers filed a suit against Massachusetts for $150 million as compensation for past mismanagement of the tribe's trust fund. Three days later, Gellers was arrested for possession of marijuana and subsequently run out of town.[3]*

The following summer, Tom Tureen moved to Maine to continue the case. Digging in, Tureen discovered the central issue was sovereignty. The Indian Nonintercourse Act of 1790 had been passed by Congress to protect Indian lands by prohibiting their "alienation" without congressional approval. Massachusetts and the twelve other original American states believed the act didn't apply to them.[4] Tureen thought it did. Federal power over the Indians had been used by President Andrew Jackson and others to destroy Indian sovereignty. But in 1934, the Indian Reorganization Act began reestablishing Indian sovereignty.

Realizing what he was up against legally and politically, Tureen assembled nonprofit funders and A-team attorneys to prosecute the case so he wouldn't suffer illegal potshots like Gellers. Rather than seeking restitution of the tribe's lands as defined in the treaty, the team decided to lay claim to their aboriginal territory, roughly the St. Croix River Valley, or about one million acres, on the grounds that the Treaty of 1794 violated the Nonintercourse Act of 1790 because it had not been approved by Congress.[5] They argued that George Washington's promise in 1776 to protect the Passamaquoddies' sovereign rights to their traditional hunting lands defined the scope of the tribe's land claim.

As almost two centuries had elapsed since the treaty was signed, Tureen feared some statute of limitation, laches (failure to act), or adverse possession (squatter's rights) might nullify the claim; it didn't. Only an explicit limitation set by Congress could do so. As it happened, just such a limitation had been quietly enacted on July 18, 1966, and would extinguish all aboriginal land claims nationwide on July 18, 1972, *then just eight months away.* Thus, Tureen and his team would have to convince the federal government to sue the states on behalf of the tribe in just eight months' time.[6]

The secretary of the interior, a Mohawk Indian, was sympathetic, but Associate Solicitor for Indian Affairs William Gershuny wasn't solicitous: "[I]t was high time the Indians accepted the facts of life. . . . [N]o court had ever ordered the Federal government to file a lawsuit on behalf of anyone, much less a multi-million-dollar lawsuit on behalf of a powerless and virtually penniless Indians tribe."[7]

Maine's new governor, Kenneth Curtis, was sympathetic, saying every American deserved their day in court. Maine's senators Margaret Chase Smith and Edmund Muskie and many in Maine felt similarly— partly because they didn't think the Indians would win. Nonetheless, the federal government didn't like being asked to sue a state. It let the clock tick. So the tribe sued the federal government to sue Maine.

The court in *Passamaquoddy v. Morton* decided the Nonintercourse Act was applicable and had established a trust relationship between the United States and the tribe. It was appealed and upheld and became law on March 22, 1976.[8] Consequently, on January 11, 1977, the Interior Department's draft report to the Justice Department recommended ejectment actions be filed on behalf of the Passamaquoddy and Penobscot Tribes against the 350,000 people living within the 12.5-million-acre region claimed by the Indians, and against the politically powerful timber and paper companies that possessed the largest privately owned forest in the world. Public sentiment turned vehemently against the Indians, causing Maine's political leaders to backtrack.[9]

Neither the tribes nor the federal or state governments wanted decades of litigation or the enormous expense and risk of a trial. A jury could, after all, decide for or against the Indians. So they began settlement negotiations. On October 10, 1980, President Jimmy Carter signed the Maine Indian Claims Settlement Act with a feather pen.[10] It provided $81.5 million to the Passamaquoddy tribe to buy 300,000 acres of land (from willing sellers, at fair market rates) and a $27 million trust fund.[11] It was far and away the largest Indian victory of its kind in the history of the United States.

Chief Neptune's "rumpus" was over. The Passamaquoddy Tribe once again had control of their land, its resources, and financial assets with which to earn a living.

Campobello, New Brunswick

No one is sure what happened to the Sturgis cottage on Campobello, but it seems to have been lost to fire or torn down sometime after 1931.[1] There are no photographs of it known to exist other than a few remote images.[2] While searching for photos, I came across the "Murdoch" map of the island from 1918, which showed the names of the property owners. The Sturgis cottage was directly across Narrows Road from the two Roosevelt cottages. To the north of the Sturgis cottage was the Cope lot, and north of that was a lot labeled "J. Shipley Newlin." The name sounded familiar. I contacted my sister, who within the hour informed me that Joseph Shipley Newlin was our great-great-grandfather.

Newlin offered the lot for sale in 1934.[3] Despite promoting Eastport as "The Summer Home of the President—Why Not Yours?" as a slogan on the front page of the *Eastport Sentinel* for three years;[4] despite the slogan appearing in *The March of Time* newsreel about Quoddy in theaters all over the English-speaking world in 1936; and despite offering the purchaser the opportunity to be the next-door neighbor of the president of the United States, the Newlin lot sat on the market unsold until 1941,[5] when it was sold by the Sentinel Agency for $150.

Eleanor sold the big red cottage on Campobello to her son Elliot in 1946.[6] He put it up for sale in 1950. It was purchased in 1952 for $12,000 by Armand Hammer,[7] the American industrialist, philanthropist, art dealer and forger, and self-proclaimed confidant of Joseph Stalin. Hammer had a fondness for Maine, having made his first real fortune from "blended whiskey" made mostly from Maine potatoes. The right to make alcohol for drinking was a valuable wartime concession granted to him by FDR in return for an unusual service to his country.[8]* Hammer invited Eleanor to use the cottage whenever she wanted.[9] Despite his claims that she was there more often, Eleanor used it only once,[10] for the August 13, 1962, opening of the Franklin D. Roosevelt International Memorial Bridge connecting Campobello to Lubec. She stayed in the red cottage but was too ill to attend the event. She died at Hyde Park, New York, on November 7, 1962.

In May 1963, Hammer "donated" the cottage to the US government (for a sizable tax deduction) to help create the Roosevelt Campobello International Park, which opened in 1966, the first in the world to be operated jointly by two countries.

APPENDIX 10

Eastport, Maine

When you drive into Eastport from Perry, you travel over two dams built for the Quoddy Tidal Power Project in 1936 that now form the Route 190 causeways from Pleasant Point to Carlow Island to Moose Island. Turning east on Vanessa Road will bring you into Quoddy Village. Grossman's tried hard to attract businesses and residents to Quoddy Village by selling pieces to individual owners, but with little success. While there is a large cinderblock drill hall and several Quonset huts from the Seabee's tenure in the 1940s, all the large wooden structures built for the tidal power project in 1935–1936 fell victim to fire, the last being the Kendall Avenue Apartments, on November 8, 2017. The original "action model" of the Quoddy Tidal Power Project, constructed by the Army Engineers, resided in the Exhibition Hall at Quoddy Village until December 5, 1997. The eight-ton, 14-by-16-foot concrete model was then trucked to downtown Eastport, where it was inserted through a gaping hole cut into the front of 72 Water Street and has ever since been the main attraction of the Quoddy Model Museum.[1] It still works.

The Holmes Packing Company sardine factory, owned by Moses Pike (1897–1989), the last sardine cannery operating in Eastport, closed in 1983 and was destroyed by fire on February 3, 1993.[2] Roscoe Emery's office as mayor of Eastport was on the third floor of the Eastport Savings Bank at 43 Water Street, now the Tides Institute & Museum of Art.

It is ironic that the town on the East Coast of the United States with the most hydroelectric power potential turned to imported crude oil for its economic salvation. In 1968, Eastport granted an option to the Pittston Company to build an oil refinery on the 254-acre site of Eastport's airport, with ship docking facilities in Deep Cove. While the tidal power project had enjoyed almost unanimous popular support in Washington County and still had overwhelming support in 1968,[3] the oil refinery was highly controversial and set off a fifteen-year fight.[4] For fear of environmental damage to the bay, and despite the energy crisis of the 1970s, Canada refused to grant permission for the more than 1,000-foot-long oil tankers to transit Canada's tricky tidal waters on their way to Eastport, thus effectively strangling the refinery plan. The fact that the largest oil refinery in Canada is located in St. John, New Brunswick, just 50 miles away, may also have had something to do with Canada's decision.

Tidal Power Development, 1950-2020

The tides that slosh into and out of the Bay of Fundy twice per day are 2,000 times the discharge of the St. Lawrence River. This flow makes Fundy's *theoretical* tidal power potential an astronomical 209 million hp—or about equal to 250 nuclear power plants, four times as many as were operating in the United States in 2019.[1]*

In the late 1950s, Nova Scotia Light and Power considered building a tidal power plant in Minas Basin, the southeastern fin of Fundy's fishtail. With an average tidal range exceeding 22 feet (and up to 50 feet), it could theoretically generate 5 million hp. While appealing, the $3 billion cost was too much to bear. In the 1960s, the Canadians again looked at but rejected tidal power.[2] Elsewhere in the world, Soviet Russia completed its tidal-electric plant in the Arctic in 1968. It now produces a modest 1.7 megawatts (2,270 hp).[3] The proposed tidal power barrages across England's Severn and Mersey Rivers were not built.

Fueled by the oil embargo engineered by the Organization of the Petroleum Exporting Countries in 1973, tidal energy's fortunes started to rise. In 1976, the Passamaquoddy tribe proposed a small-scale tidal power project in Half Moon Cove with a dam at the old Toll Bridge. It, as well as projects at Pennamaquan River, Straight Bay, and others, was actively considered through the mid-1980s but not completed due to economic and environmental concerns.[4] The nuclear disasters at Three Mile Island in Pennsylvania (1979), Chernobyl in Soviet Russia (1986), and Fukushima in Japan (2011) spurred efforts to develop cleaner and renewable power, including tidal.

In the 1970s, Canadians investigated a large power plant in Cumberland Basin, the northern fin of Fundy's fishtail. The plan called for a three-quarter-mile-long barrage to be constructed across the basin at Joggins, Nova Scotia. It would have generated 1,085 megawatts, roughly equal to Dexter's International Plan, but it was rejected.[5]

A massive project was proposed in 1982 for Cobequid Bay in Minas Basin. Promoters' mouths watered over the potential to generate 4,560 megawatts, 90% of which was to be exported to the United States. It was to have a four-mile barrage across the bay, ironically, at Economy Point. The barrage was to be made from 100 office-building-sized concrete caissons, which would be cast on shore, floated into location, filled with rock and sunk, and then connected to form a continuous barrier. During construction, Fundy's tides would surge around them with ever-increasing velocity as each additional caisson forced the current into an ever-narrower channel. Designed as a "single effect" system, it would generate power only for two six-hour periods per bay during ebb tide. One unintended consequence would be to shorten Fundy's tidal basin by 28 miles, thereby shortening the Gulf of Maine's period of oscillation, thereby coinciding more closely with the moon's period of oscillation, thereby increasing the bay's harmonic resonance and the tide's height by six inches—in Boston, 390 miles away. Bostonians threatened to rise in revolt at such a foreign invasion, but they didn't have to, as the $25 billion price tag put the foreigners under water.[6]

The tidal power dream finally came true for Canada in 1984 when Nova Scotia Power built an experimental project at Annapolis, close to where the de Mons–Champlain tidal grist mill was built in 1605. The site was chosen because a dam had been built across the Annapolis River in the 1950s, and thus some construction costs could be avoided and the environmental impact of damming the estuary had already occurred. The dam had been installed as a cheap alternative to maintaining several dozen miles of agricultural dykes (*aboiteaux*) built along the river centuries before by the French Acadians.

Much of Nova Scotia and New Brunswick is covered with of soft sandstone, which is more easily eroded than Maine's granite. The erosion creates the region's famous "chocolate rivers" and mammoth mudflats. The dam across the Annapolis River slowed the tidal currents, which dropped their sediment load, and the estuary quickly filled with silt, significantly reducing the aquatic habitat and species diversity.[7]

The Annapolis Tidal Power Station was built as a full-scale single turbine prototype for the 120-turbine project proposed for Minas Basin. The 25-foot tidal range at Annapolis flowing through the 24-foot diameter turbine generated up to 20 megawatts for 4.5 hours during the ebb tide. The prototype was massively overbuilt (and therefore not cost efficient) for the Annapolis site but would provide data on the actual equipment planned for the Cobequid Dam (just as Dexter Cooper had proposed for Quoddy in 1925).[8] Lots was learned. The full project was never built due to high costs and environmental concerns, especially siltation. Since Annapolis generated only intermittent power, it was operated alternately with a freshwater hydroelectric station, thereby eliminating the need for an expensive pumped storage facility (just as Colonel Fleming suggested in 1936).[9] After 35 years of service, it was shut down in 2019, due to the cost of a major overhaul, its future unknown.

The largest tidal power station in the world is now on Incheon Bay in South Korea. Completed in 2011, it can produce 254 megawatts from an eighteen-foot tidal range and a twelve-square-mile basin.[10] It and all the existing tidal power projects are "single-pool" configurations and hence cannot produce continuous power as Dexter's plans would have done.

The high cost of tidal barrages and their negative environmental impacts have refocused most development efforts on "in-stream" or "hydro-kinetic" generators utilizing the horizontal flow of the water (tidal currents) rather than vertical drop (tides). The idea is to put a relatively small propeller-like device in the flowing water; some resemble "airplane propellers," others "eggbeaters," and old-fashioned push-style helical "lawnmowers." Typically, only a few percentage points (or less) of the potential energy in the stream is captured, but ideally at less cost, and therefore more cost efficiently. In-stream devices can be constructed and installed relatively quickly and easily; are largely invisible, thus avoiding the NIMBY[11]* eyesore issue; and can be set in deep water to allow ships to pass safely over them. In addition, their slow-turning blades[12]* allow fish and marine mammals to swim through them unharmed. These smaller projects can potentially find niche markets where transportation makes fossil fuel costs high, rather than the low-priced mass markets formerly targeted by the mega projects.[13]

In the United States, Maine still leads tidal research. In 2012, ORPC of Portland and Eastport set a 98-foot-wide, 30-foot-tall helical generator on the bottom of Cobscook Bay and, from the 4–6 knot tidal currents, generated up to 150 kW for a month, delivering it to Bangor Hydro-Electric's grid.[14] Having learned a lot, they have since installed a second-generation floating generator in a river in Alaska, and sometime in the future they hope to install a third-generation neutrally buoyant generator (tethered midway in the water column, where the current is strongest, but where there is less sediment) at "the motherlode" of tidal currents in Passamaquoddy Bay, between Eastport and the Old Sow whirlpool, where tidal currents can reach 11 knots.[15]

In the Piscataqua River, at Portsmouth, the University of New Hampshire bolted a six-foot-tall "eggbeater" style vertical shaft turbine onto the Memorial Bridge in 2017.[16] With a 5.5 knot current, it can theoretically generate up to 25 kW. In New York City's East River, Verdant Power had six five-meter-diameter "airplane propeller" turbines generating 35 kW in 2007–2009 and installed three new, fifth-generation turbines in 2020.[17]

Canada set up the Fundy Ocean Research Center for Energy (FORCE) in Parrsboro, Nova Scotia, as a test facility for in-stream tidal technology in 2008. FORCE experienced some of Eastport's tidal power heartbreak. The first turbine to be tested—a 36-foot-diameter model that had been used in rivers successfully for years—was knocked out of commission within three weeks.[18] While the turbine was designed for use in 8.5 knot currents, Fundy's 10.5 knot currents produced almost twice as much force.[19]* Trying again in 2016, a 50-foot-diameter, 1,000-ton turbine was installed. It worked fine for six months and was then removed for analysis and maintenance. The day after a modified turbine was installed in 2018, the company went bankrupt.[20] It has been a painful process, but the test facility concept is proving its worth by sharing the lessons learned and the infrastructure cost distributed over many experiments. Stay tuned.

A century ago, Samuel Insull's idea was to electrify everything.[21] He was right. We did. Franklin Roosevelt was also right: "[T]here cannot be enough electric energy, provided it is cheap,"[22] and he believed that usage would increase by 10% to 15% per year. It did.

It is ironic that Boston, home of Quoddy's most vocal critics, has ever since been bemoaning the high price and limited supply of electricity. For at least 40 years there have been proposals to transmit electricity from Hydro-Quebec to sate Boston's thirst for power. Proposals for cutting rights-of-way through Vermont and New Hampshire were blocked by conservation groups. Another proposal was to lay a 375-mile-long subsea cable from St. John, New Brunswick, to Pilgrim Nuclear Power plant (decommissioned in 2019 after 47 years) at Plymouth, Massachusetts.[23] In 2019, an agreement was reached to cut a 145-mile transmission line through Maine to bring 1.2 gigawatts (the same amount Dexter's International Plan could have been producing for the intervening century) to Central Maine Power's Lewiston station. The cost for *just* the transmission line is projected to be $950 million.[24] Meanwhile, New England still has some of the highest electric utility rates in the continental United States.[25]

And so the "fight for power" continues in Passamaquoddy Bay, and in Augusta, and in Boston, and in Washington, and in *your* legislature.

In the latest ironic twist, to protect it from sea level rise due to two centuries of burning fossil fuels, Boston is now thinking of building tidal barriers across its outer harbor. Some have proposed they be used to generate emission-free tidal-electric power.[26]

Oh, how the tide turns!

APPENDIX 12

Abbreviations and Document Sources

Abbreviation	Item
BDN	*Bangor Daily News* (1889–), Bangor, ME
BH	*Boston Herald* (1846–), Boston, MA
BHE	Bangor Hydro-Electric
BHEN	*Bangor Hydro-Electric News*, Bangor, ME
BHS	Border Historical Society, Eastport, ME (merged into TI&MA in 2021)
CMP	Central Maine Power Company, Augusta, ME
Cooper-26	"Some Aspects of the Passamaquoddy Tidal Power Project." In *Papers and Transactions for 1926* and *Proceedings of the 42nd Annual Meeting*, The Connecticut Society of Civil Engineers. New Haven, CT: Tuttle, Morehouse and Taylor, 1926.
Cooper-FPC-28	*Passamaquoddy Power Development*, Dexter P. Cooper, Incorporated, Eastport, ME, with handwritten annotation: "Project No 463, As submitted originally Oct 19- 1928, International Scheme." Collection of Dexter Cooper III.
Cooper-NPPC-36	"National Power Policy Committee, Passamaquoddy Project, Eastport, Maine, by Dexter P. Cooper, Advisor, 18 June 1936." Collection of Dexter Cooper III.
CPL	Campobello Public Library, Campobello, NB

DPC	Dexter Parshall Cooper (10 July 1880–2 February 1938)
Durand-36	*Passamaquoddy Tidal Power Project: Report of Board of Engineers on Cost Estimates of Project*, May 1936 (aka PWA's Independent Review or Reclamation-Durand Report).
EMERY	Roscoe C. Emery Papers, Fogler Library, University of Maine, Orono, ME
EPL	Eastport Public Library, Eastport, ME
ER-My-Day	"My Day" nationally syndicated newspaper column by Eleanor Roosevelt. From *The Eleanor Roosevelt Papers Digital Edition* (2017), https://www2.gwu.edu/~erpapers/myday/.
ES	*Eastport Sentinel* (1818–1953), Eastport, ME. Most originals are available at EPL; some originals and digital copies are found at Washington County Courthouse, Machias, ME; microfilm copies of most editions are found at UM, Orono, ME.
FDR	Franklin Delano Roosevelt (30 January 1882–12 April 1945)
FDRL	Franklin Delano Roosevelt Library, Hyde Park, NY. Files of interest include

OF: Official Files
    025n: Army Engineers
    0284: Power
    0788: National Emergency Council
    0982: Passamaquoddy Project
PPF: Personal Papers Files
    0072: DELANO, Frederic A.
    1411: BRANN, Louis, J.
    1560: CORCORAN, Thomas G.
    1766: COOPER, Hugh L., Dexter P., and Gertrude S.
    3529: VANDENBERG, Arthur
    7861: BREWSTER, Ralph O.
           FLEMING, Philip B.
           OLDS, Leland

ROOSEVELT, James
WILLIAMS, Aubrey
PSF: President's Secretary's Files, 1933–1945
Example: FDRL-0982: 1935-11-05_Brann to FDR_Brann's position
Annotations: Year-Month-Day_Sender to Recipient_Subject

FDR-Day-by-Day	Franklin Delano Roosevelt: Day-by-Day. On-line calendar and search. http://www.fdrlibrary.marist.edu/daybyday/.
Final-36	*Final Report: Passamaquoddy Tidal Power Development*, 23 October 1936. Army Corps of Engineers. Available at *Quoddy Tides*, Eastport, ME; appendices at NA-MA.
Fleming-49	Memorandum of Conversation between General Philip B. Fleming and Sidney Hyman, August 19, 1949. FDRL-FLEMING: IV.A4_1949-08-19.
FPC	Federal Power Commission
FPC-34	*Passamaquoddy Bay Tidal Power Commission Report*, 30 March 1934. FDRL-OF-0982-1.
FPC-41	Passamaquoddy Tidal Power Project: Letter from the Chairman of the Federal Power Commission: In Response to Senate Resolution 62, 76th Congress, February 1939. https://lccn.loc.gov/41050507. Call number: TK1425.P3 A5 1941d. [Same as:] US Congress, Senate. *Report on the Passamaquoddy Tidal Power Project, Maine.* S. Doc. 41, 77th Cong., 1st Sess., 1941.
FTC	Federal Trade Commission
GE	General Electric
GSC	Gertrude Sturgis Cooper (20 June 1889–18 December 1977), "Passamaquoddy" (1970) (collection of Dexter Parshall Cooper III; and FDRL-Small Collections, Misc. Docs: 1970-

05_Gertrude Sturgis Cooper Monograph_Passamaquoddy Tidal Power Project); "Hyde Park" (ca. 1970) (collection of Dexter Parshall Cooper III); "Early Years" (ca. 1970) (collection of Dexter Parshall Cooper III).

HLC
Hugh Lincoln Cooper (28 April 1865–24 June 1937)

IJC-Eng-50
Report to International Joint Commission on Scope and Cost of an Investigation of Passamaquoddy Tidal Power Project, Washington, DC: International Passamaquoddy Engineering Board, March 1950. FDRL-OLDS: Passamaquoddy Project: 1950-03_IJC International Passamaquoddy Engineering Board_Estimate for Full Survey_intro.

IJC-Eng-59
Passamaquoddy Engineering Investigations: International Passamaquoddy Engineering Board Report to International Joint Commission, October 1959. https://legacyfiles.ijc.org/publica tions/ID308.pdf.

IJC-Eng-59-B
Investigation of the International Passamaquoddy Tidal Power Project: Report to the International Joint Commission by the International Passamaquoddy Engineering Board, October 1959. https://ijc.org/sites/default/files/AA13 .pdf. 20-page brochure.

IJC-50
Report of the International Joint Commission on an International Passamaquoddy Tidal Power Project, October 20, 1950. https://ijc.org/ sites/default/files/Docket%2060%20Passama-quoddy%20Final%20Report%201950-10-20 .pdf (cover letter only).

IJC-Fish-34
Report of the International Commission Appointed to Investigate the Probable Effect of the Damming of Passamaquoddy and Cobscook Bays on the Fisheries of That Region. Ottawa: King's Printer, 1934. FDRL-OF-0982: 1034-04-03_ Department of State to FDR.

	[Same as]
	US Congress, House. *Report by the International Passamaquoddy Fisheries Commission.* H. Doc. 300, 73rd Cong., 2nd Sess., 1934 (discussed in chapter 6, "Courting the Canadians").
IJC-Fish-59	*Passamaquoddy Fisheries Investigations, International Passamaquoddy Fisheries Board Report to International Joint Commission,* October 1959. https://legacyfiles.ijc.org/publications/ID308.pdf.
IJC-61	*Investigation of the International Passamaquoddy Tidal Power Project: Report of the International Joint Commission,* Docket 72, Investigations of the International Passamaquoddy Engineering and Fisheries Boards, April 1961. https://legacyfiles.ijc.org/publications/ID308.pdf.
Improper-36	*Report on Alleged Improper Official Activities in Connection with the Passamaquoddy Project at Eastport, Maine,* 19 February 1936. Communicated by Louis Glavis to Secretary Ickes, March 4, 1936. FDRL-OF-0982.
LEJ	*Lewiston Evening Journal* (1866–1979), Lewiston, ME
LH	*Lubec Herald* (188?–1958), Lubec, ME
LHS	Lubec Historical Society, Lubec, ME
McInnis	Collection of Ruth McInnis, Eastport, ME
MEIGS	Notes on postcard (aerial view of Lubec) by Peveril Meigs (1903–1979), apparently based on conversation with C. A. Bangs, undated (probably 1970s). Tide Mill Institute, Meigs Collection, https://www.tidemillinstitute.org.
MHS	Maine Historical Society, Portland, ME
MSPB-34	*Report on Passamaquoddy Tidal Power Project by the Maine State Planning Board,* December 1934 (sometimes listed as January 1935). Maine State Library, Augusta, ME. https://digitalmaine.com/spo_docs/96/.
NA-DC	US National Archives, Washington, DC

NA-MA	US National Archives, Woburn, MA
RG-77:	Army Corps of Engineers
NA-MD	US National Archives, Silver Spring, MD
RG-69:	Works Progress Administration
RG-71:	Bureau of Yards and Docks
RG-119:	National Youth Administration
RG-59:	Department of State
NA-NY	US National Archives, Hyde Park, NY (FDRL)
NYT	*New York Times* (1851–), New York, NY
RCE	Roscoe Conklin Emery (28 March 1886–16 December 1969)
RCIP	Roosevelt-Campobello International Park, Campobello, NB
Reel-38	*Confidential Report of Investigation Of Quoddy Project*, by Frank Reel. FDRL-ROOSEVELT, James: 1938-03-31_WPA Niles to James Roosevelt_confidential report on Quoddy.
PEN	*Portland Evening News* (1927–?), Portland, ME
PPH	*Portland Press-Herald* (1921–), Portland, ME
PWA	Public Works Administration
PWA-Hunt-35	*Report of the Passamaquoddy Bay Tidal Power Commission*, 17 January 1935. FDRL-0982: _1935-01-17_Hunt Report.
Ruby Black-33	FDRL-0284: 1933-11-02_Ruby A. Black to Eleanor Roosevelt
SCC	*St. Croix Courier* (1865–), St. Stephens, NB
SILLS	Kenneth C. M. Sills (1879–1954), president of Bowdoin College (1918–1952). Sills Letter Files on Passamaquoddy Power Project at Bowdoin College Library, George J. Mitchell Department of Special Collections & Archives, Brunswick, ME.
Sills-35	Sills Commission Report, January 4, 1935. SILLS_1935-01-04: Sills to Brann_Sills Commission report.
TI&MA	Tides Institute & Museum of Art, Eastport, ME, https://www.tidesinstitute.org
TMI	Tide Mill Institute, https://www.tidemillinstitute.org

TVA   Tennessee Valley Authority
Udall-63   *The International Passamaquoddy Tidal Power*
                       Project and Upper Saint John River Hydroelec-
                       tric Power Development. Summary Report to
                       President John F. Kennedy. Stewart L. Udall,
                       secretary, Department of the Interior, July 1963.
Udall-65   Report to President Lyndon B. Johnson: *Conser-*
                       vation of Natural Resources of New England:
                       The Passamaquoddy Tidal Power Project and
                       Upper Saint John River Hydroelectric Develop-
                       ment. US Department of the Interior, July 9,
                       1965 (8-page summary, plus support letter from
                       governors at TI&MA).
UM   University of Maine, Fogler Library, Orono, ME
WPA   Works Progress Administration

All references to the current value of historic prices are based on "CPI Inflation Calculator," US Bureau of Labor Statistics, https://www.bls.gov/data/inflation_calculator.htm.

In the appendixes and endnotes, cross-references to other chapters in this book are noted by chapter and subheading (for example, chapter 1, "The Champion").

Notes

PROLOGUE: WHO KILLED QUODDY?

1. Wayne Curtis, "Eastport, Maine Tidal Project—Trapping the Moon," *Yankee Magazine*, 25 August 2015, https://newengland.com/yankee-magazine /travel/maine/eastport-maine-tidal-project/. The crowd was estimated at thirteen thousand according to Holmes, 54, citing *Bangor Commercial*, 3 July 1935. Mayor Emery estimated fifteen thousand according to Holt, 42.

2. "President Roosevelt is expected to discharge the explosive by touching a key in the White House," *Boston Herald*, 29 June 1935, 22. President Roosevelt was in the White House all day on 3–5 July 1935 (FDR-Day-by-Day).

3. *Baltimore Sun*, 27 August 1947.

CHAPTER 1. THE PROTAGONISTS

1. Lt. William Halsey Jr. was posted to the USS *Flusser* on 16 August 1912, and detached on 7 February 1913, then reposted on 23 February 1913, and detached on 5 September 1913. *Logbook*, USS *Flusser*, NA-DC.

2. Halsey, 18–19, says this took place in 1913, whereas Cross, 42–43, citing Frank Freidel, *Franklin D. Roosevelt: The Apprenticeship* (Boston: Little, Brown, 1952), 165, says it took place in 1916. But as the *Flusser*'s logbooks prove (see note 1), the event must have taken place in 1913. (An asterisk following an endnote referent indicates that the endnote includes information that might be of interest to the general reader in addition to the source citation.)

3. Franklin D. Roosevelt was assistant secretary of the US Navy from 17 March 1913 to 26 August 1920.

4. RCIP, *Roosevelts at Campobello*, 47.

5. Halsey, 17–18.

6. For a discussion of steam turbines, see Klein, *Power Makers*, 372–75.

7. Halsey, 17–18.

8. A knot, or one nautical mile (historically, one minute of latitude; now defined as 1,852 meters) per hour, is equal to 1.15 statute miles per hour.

9. Wells, *Campobello: An Historical Sketch*, 71.

10. A similar event took place in 1917 aboard the USS *Cherokee* (Klein, *Beloved Island*, 81–85).

11. Russell Sturgis (1805–1887) went on to become head of Baring Brothers Bank in London. Samuel Russell (1789–1862), founder of Russell & Co. and Russell & Sturgis, lived in Middletown, Connecticut. His house is now owned by Wesleyan University and serves as the offices of the Philosophy Department—a happy coincidence, as the philosopher George Santayana (1863–1952) was the grandson of Russell Sturgis. Santayana's most famous and relevant quote was "Those who cannot remember the past are condemned to repeat it." A further coincidence—I spent several months reading microfilms of the *Eastport Sentinel*, shipped to me by the University of Maine, at the Russell Library in Middletown, Connecticut, two blocks from the Russell house.

12. Cross, 25.

13. Klein, *Beloved Island*, 5, citing Geoffrey C. Ward, *Before the Trumpet: Young Franklin Roosevelt* (New York: Harper, 1985), 66.

14. Halsey, 17–18; and Cross, 42–43, citing Freidel, *Franklin D. Roosevelt*, 165.

15. "The question of damming the Mississippi River has for a long time been used as a metaphor for the impossible, and some of the objections to the project on the part of the world's big financiers have been, to say the least, amusing." Chief Engineer [Hugh] Cooper to the People of Keokuk, Keokuk *Daily Constitution Democrat*, 12 January 1911, 2, as quoted in Hallwas, 41.

16. Hallwas, 72, 75.

17. Hallwas, 76, citing letter to Gladstone Califf, 11 December 1963, 6.

18. Zebulon M. Pike (1779–1813) was a US Army brigadier general and explorer for whom Pikes Peak in Colorado was named. He led expeditions through the newly acquired Louisiana Purchase territory in 1805–1806 and 1806–1807. He served in the War of 1812 until he was killed during the Battle of York, in April 1813, near Toronto, Ontario. His family was from New Jersey and Maine (Hallwas, 9, 34). The town of Keokuk was named in 1834 for Sauk Chief Keokuk (1780–1848).

19. Zachary Taylor (1784–1850) was the twelfth president of the United States (1849–1850).

20. Wikipedia, s.v. "Keokuk, Iowa," accessed 7 February 2019, https://en.wikipedia.org/wiki/Keokuk%2C_Iowa, citing Jerry Sloat, "Lee County,

Iowa"; Caleb Atwater, *Remarks made on a tour to Prairie du Chien: Thence to Washington City, in 1829* (Columbus, OH: Issac Whiting, 1831), 58–59; and "The Half-Breed Tract" (archived 2 February 2008 at the Wayback Machine, Lee County History). See also Wikipedia, s.v. "Half-Breed Tract," https://en.wikipedia.org/wiki/Half-Breed_Tract, citing "Croton, Iowa" (archived 20 February 2008 at the Wayback Machine); and "Indian Land Cessions in the United States, 1784 to 1894," in *A Century of Lawmaking for a New Nation: U.S. Congressional Documents and Debates, 1774–1875,* Library of Congress American Memory.

21. Hallwas, 10–13.

22. Hallwas, 12.

23. Martin, "Keokuk Energy Center."

24. The lock was 310 feet long, 80 feet wide, and 6 feet deep.

25. Hallwas, 9, 23, 8.

26. "In the 1903 veto of private construction of a dam and power stations on the Tennessee River at Muscle Shoals, Alabama, [Theodore] Roosevelt protected the site for later government development, but he also helped to establish the principle of national ownership of resources previously considered only of local value. In this particular case, Roosevelt recommended using revenue from power production to finance navigation improvements." Billington, Jackson, and Melosi, 37.

27. Hallwas, 38.

28. *Brooklyn (NY) Daily Eagle*, 11 January 1914, 4; and Kiwanis Club of Stamford, 368.

29. EMERY_5.8.13, news clipping dated 14 December 1927, otherwise not labeled; *ES*, 25 March 1925; "An Interesting Person: The Man Who Harnessed Niagara Falls and the Mississippi," *American Magazine*, December 1913, 4, 47; and *Brooklyn (NY) Daily Eagle*, 14 January 1914, 4. Sir William Thomson, 1st Baron Kelvin (1824–1907), was a Scots Irish mathematical physicist and engineer who formulated the first and second laws of thermodynamics. For his work on the transatlantic telegraph project, he was knighted in 1866 by Queen Victoria. The quote, "university of pick and shovel" is from the *American Magazine*, January 1904, 45–47.

30. EMERY_5.8.13, unlabeled news clipping dated 14 December 1927; *ES*, 25 March 1925; *ES*, 16 March 1938, 1; and FDRL-0982: 1938-12-27_Delano to FDR_CONFIDENTIAL history of Quoddy.

31. Hallwas, 39. It took Hugh Cooper six years to find the money, according to M. Meigs, 192–204.

32. Hallwas, 44.

33. Nehf, 775–93.

34. Billington, Jackson, and Melosi, 116, citing Hugh L. Cooper, "The Water Power Development of the Mississippi River Power Company at Keokuk Iowa," *Journal of the Western Society of Engineers* 17 (January–December 1912): 213. This article came from the paper presented on 18 October 1911.

35. Hallwas, 39; "An Interesting Person," *American Magazine*, 4, 47; and *Brooklyn (NY) Daily Eagle*, 4.

36. Martin, "Keokuk Energy Center."

37. Hallwas, 56.

38. EMERY, Roscoe_box 6.20: Stanley Cotton, "The Man Behind Quoddy," *Summer School Review* 1, no. 1 (August 1935): 3–4, Washington State Normal School, Machias, Maine.

39. Cotton, "Man Behind Quoddy," 4.

40. GSC, "Keokuk."

41. *New York Herald*, 12 March 1912, 12. Coincidently, "[a]round 1827, John Jacob Astor established a post of his American Fur Company at the foot of the bluff [at Keokuk]. Five buildings were erected to house workers and the business. This area became known as the 'Rat Row.'" Wikipedia, s.v. "Keokuk, Iowa."

42. *Harvard Alumni Bulletin* 22, no. 5, 13 October 1919, 109, https://www.google.com/books/edition/Harvard_Alumni_Bulletin/vZEBAAAAYAAJ?hl=en&gbpv=1&bsq=Russell. Russell Sturgis was a class of 1902 (FDR graduated in 1903). Russell married Louise Brady on 24 April 1915 in Keokuk, Iowa. *University of Chicago Alumni Magazine* 7, no. 7 (May 1915), 237, https://campub.lib.uchicago.edu/view/?docId=mvol-0002-0007-0007#page/1/mode/1up; and *Descendants of Nath'l Sturgis*, 29.

43. This is supposition on my part, as I found no direct evidence of this, but as will be seen, circumstantial evidence makes this very likely.

44. GSC, "Keokuk."

45. Dexter Parshall Cooper and Gertrude Sturgis were married on 12 April 1912 at Emanuel Church in Boston.

46. Hallwas, 83, citing *Macomb Journal*, 22 July 1912, 5.

47. Hallwas, 106, citing Keokuk *Daily Gate City,* 1 June 1913, 1. See also *Chicago Sunday Tribune*, 1 June 1913, 4. Gertrude unwittingly contributed more to the ceremony than just her presence:

One of my favorite wedding presents was a small brass box of Russian origin, I believe, in which I kept buttons. One day I needed a button and went to the spot where I kept my box. It was not there. I searched high and low, aided by my faithful (I thought) little maid. We couldn't find a trace of it, not even a button.

A week or so later my husband called me from his office and said that he was sending a car up for me. He often did this when there was something interesting going on down on the job. I didn't even bother to change my dress but grabbed a hat and drove down to the dam. To my horror I saw that the whole place was swarming with people—Movie cameras, Reporters and general public, all dressed to the nines. Suddenly it dawned on me that this was the day when the last concrete was to be poured. Dexter met me and escorted me to the place of honour [*sic*] and one of the foremen stepped up and handed me the box full of coins which it is customary to bury in the last shovelful of concrete. It was my little brass button box, all shined up by my maid who had pretended to help me look for it. Dexter had made her promise that she would not tell me where it had gone and she had kept her promise.

Naturally, with cameras popping and people standing all around me watching the dumping of the last concrete I couldn't do a thing about it, so I bade it a silent goodbye, and popped it in. (GSC, "Keokuk")

48. Hallwas, 104, citing Keokuk *Daily Gate City*, 27 May 1913, 5.

49. Hallwas, 49, 74.

50. Martin, "Keokuk Energy Center."

51. Hallwas, 39, citing Keokuk *Daily Gate City*, 8 January 1910, 1.

52. Keokuk had a population of 14,008 in 1910 and 10,780 in 2010. The population of Chicago was 2,185,283 in 1910 and 2,698,000 in 2010. The population of St. Louis was 687,029 in 1910 and 319,294 in 2010. Wikipedia, s.v. "Keokuk, Iowa," citing https://en.wikipedia.org/wiki/Demographics_of_Chicago.

53. Hallwas, 38. Theodore Roosevelt's second trip to Keokuk was in 1908, during which his ride from Keokuk to Memphis was acclaimed as the "largest steamboat parade in history." Billington, Jackson, and Melosi, 115, citing Roald D. Tweet, *A History of the Rock Island District U.S. Army Corps of Engineers: 1866–1983* (Rock Island, IL: US Army Engineer District, Rock Island, 1984), 244.

54. Meigs was the son of Montgomery C. Meigs, quartermaster general of the US Army during the American Civil War. The elder Meigs was an assistant to Robert E. Lee during earlier efforts to improve navigation at Keokuk. The younger Meigs was president of the Keokuk Canal Company and served with the US Army Engineers for 52 years. Hallwas, 22.

55. Martin, "Keokuk Energy Center."

56. Ancestry.com, accessed 9 April 2019.

57. *True: The Magazine for Men*, January 1961, 79.

58. *ES*, 19 April 1933, 1; and *ES*, 28 June 1933, 6. In a speech from the steps of the Eastport Public Library, President Taft announced his plan for trade reciprocity (free trade) with Canada.

59. *ES*, 29 October 1924. The man was Thomas McDonald, from Patterson, New Jersey.

60. *ES*, 16 October 1912, 2; *ES*, 17 April 1912; and Holt, 66.

61. *ES*, 17 April 1912. Svante August Arrhenius (1859–1927) of Sweden is often referred to as one of the founders of the science of physical chemistry. He received the Nobel Prize for Chemistry in 1903 and became the director of the Nobel Institute (1905–1927). Nikola Tesla replicated Arrhenius's experiment with developmentally retarded children in New York City, without success. In one of the scandals of science, Tesla, the inventor of most fundamental alternating current (AC) technology and radio technology, was never awarded a Nobel prize, whereas many who followed him were. Seifer, 357–58, citing "Electrified Schoolroom to Brighten Dull Pupils," *NYT* 18 August 1912, 5, 1; and "Tesla Predicts More Wonders," *NYT*, 7 April 1912, 5.

62. *ES*, 10 January 1912, 5; *ES*, 24 July 1912, 3; *ES*, 2 July 1913, 8; *ES*, 17 April 1911, 8; *ES*, 3 January 1912, 3; *ES*, 10 January 1912, 7; and *ES*, 3 January 1912, 5.

63. *ES*, 24 July 1912, 1. Whether Roscoe's reporting on the subject was coincident, prescient, or instigated by an outbreak of typhoid in the Eastport area in the summer of 1912, we do not know—but Franklin Roosevelt did come down with the disease while at Campobello in July. Klein, *Beloved Island*, 74.

64. *Eastport Directory 1935*, 208; *ES*, 24 April 1912, 2; and *ES*, 3 January 1912, 2.

65. *ES*, 1 July 1914, 2, 4; a total of about 15,000 residents.

66. *ES*, 5 August 1914, 2; this continued throughout the rest of the year.

67. Unsigned editorial by retiring editor of the *Eastport Sentinel*, 14 October 1914, 2.

68. Unsigned editorial by the new editor of the *Eastport Sentinel*, 21 October 1914, 2.

CHAPTER 2. ONE BAY–THREE ISLANDS–THREE NATIONS

1. See appendix 2, "The Astronomy That Creates the Ocean Tides."

2. Broduer, 74–75.

3. In 1850, Nicola Tenesles, a Passamaquoddy Indian, with the help of several citizens of Middletown, Connecticut, including C. H. Pelton, published a book on the Passamaquoddy language, then known as Etchemin. Soctomah, *Save the Land*, 144. Etchemin had been so little used that almost no one could speak it by 1950, and it had to be largely recreated based on the written accounts of European explorers and traders from the seventeenth and eighteenth centuries. Broduer. Although Passamaquoddy was not originally a written language, since 1976 the Passamaquoddy Bi-Lingual Project has devised a phonetic alphabet and published a dictionary and many books

in the Passamaquoddy language. Lester, 18. The effort to record and hence save the words and deeds of the Passamaquoddy began with two men, one of whom was Tomah Joseph. Lester.

4. Miss A. Chadbourne, Professor of Education, "An Excellent Paper on Maine Indian Names," *ES*, 11 August 1920, 5.

5. *IJC-Eng-59-B*, 17.

6. Broduer, 74–75.

7. *National Geographic* (date uncertain), as quoted in *ES*, 12 March 1925.

8. Ungava Bay in arctic Canada is now believed to equal or better Fundy's record. White, 191–218.

9. "Some Historic Spots in Washington County," from "a paper by Miss Grace Donworth, read by her at the October meeting of the D.A.R. at Machias: Dochet Island," *ES*, 12 November 1919, 9.

10. National Park Service, US Department of the Interior, International Historic Site: Maine, "Saint Croix Island" (illustrated brochure, undated, ca. 2018).

11. Grace Donworth, paper presented at October meeting of the D.A.R. at Machias: *ES*, 12 November 1919, 9; *ES*, 12 November 1919, 9; and National Park Service brochure, "Saint Croix Island," GPO 2015-388-437/30561.

12. National Park Service, US Department of the Interior, "Archeology Program: The French Along the Northeast Coast—1604-1607," accessed 6 February 2019, https://www.nps.gov/Archeology/SITES/npSites/champlain. htm. They built the continent's first watermill at Beaver Brook, just to the east of St. Croix Island. "Isle de sainte Croix," Osher Map Library, https://osher maps.org/browse-maps?id=104955.

13. The remnants of this grant can be seen in the northern boundaries of Vermont and New York.

14. Donworth, paper presented at D.A.R. meeting: *ES*, 12 November 1919, 9.

15. Soctomah, *Save the Land*, 17.

16. Kilby, 44.

17. Kilby, 43; and Nowlan, 13–14.

18. Nowlan, 15–16.

19. Wells, "Quoddy Hermit."

20. In 1755, what is now New Brunswick was claimed by the British as part of Nova Scotia. It became a separate colony in 1784 following the mass immigration of Loyalist refugees from the American Revolutionary War. In May 1786 alone, 18,000 refugees landed at the mouth of the St. John River in New Brunswick and founded the city of the same name. By 1834, St. John had grown to 119,457, and by 1857 there were over 200,000 residents. Burrows, 214, 219.

21. Nowlan, 19.

22. Muskie, 7; and Nowlan, 20.

23. Nowlan, 8, 10, 29; and Muskie, 7.

24. Nowlan, 26, 42; and Klein, *Beloved Island*, 19–20. To the desperately poor "rabble" or "swinish" of England, an indenture—what amounted to semi-slavery of four to seven years, but with the possibility of a modest land grant at the end—was better than their prospects in England's hopelessly rigid feudal system.

25. Nowlan, 21. At the Campobello Historical Society, the Francis W. Dean, C.E. map of 1883 says "AREA, 10180 ACRES"; the Gilbert G. Mardock, D.L.S. map of 1918 says "Approximate area—10,180 acres."

26. "Campobello," advertisement from 1880s, courtesy of CPL. Manhattan is currently 14,611 acres. In this ad the size of Campobello is stated as "fully ten thousand acres," whereas in Nowlan it is said to be 12,000 acres. 12,000 is 82% of 14,611.

27. Wells, *Campobello*, 14, 47.

28. Nowlan, 33–34.

29. Nowlan, 29–38; Wells, *Campobello*; and Klein, *Beloved Island*, 20.

30. Nowlan, 50.

31. Nowlan, 61.

32. Burrows, 223, citing *The Quoddy Hermit; or, Conversations at Fairfield on Religion and Superstition* (Boston: privately printed 1841). Owen's family was upset by the public revelation of his illegitimate birth and tried on several occasions to erase proof of his birth, going so far as to lay a false trail.

33. Nowlan, 37.

34. Nowlan, 63.

35. Wells, *Campobello*; and Burrows, 220. See also large map "Campobello Island, New Brunswick," compiled by Gilbert. S. Mardock, D.L.S., 1918, at CPL, in which "Mill Road" runs eastward from Dunns Beach north of Deer Point to Harbor de L' Outre (alternately, "Havre de Lutre"), where a dam and mill building are shown.

36. Nowlan, 64; and Wells, *Campobello*.

37. "The cannon still remained as sentinels, till someone on board the brig *Sam French*, which was going to California for gold, stole them and carried them round Cape Horn. When the brig reached San Francisco it fired a salute; but as the Admiral had forewarned the Southern authorities of the capture of his guns, the timely or untimely salute betrayed their presence, and the guns were seized and returned to Campobello. After the removal of the Owen family to England, one of the guns, which had been bought from them by Mr. Best, an Island resident at that time, was given by him to General Cleaves, who placed it on one of the islands in Portland harbor, where two or three years ago [circa 1890] it exploded and was shattered to pieces. The other gun was bought by George Batson, Esq., and was placed in his store on the Island, where it

became an object of wonder to all newcomers." Wells, *Campobello*. See also Wells, "Brass Cannon."

38. "A raised veranda at the front door was L-shaped and provided a covered walk which Captain Owen frequented. His local detractors likened it to the quarter deck of a ship." Burrows, 221.

39. Nowlan, 66.

40. Nowlan, 66; and Burrows, 230, citing Wells, "Quoddy Hermit."

41. Burrows, 219–20, citing Owen Papers, New Brunswick Museum, St. John, NB; and Nowlan, 68.

42. Wells, *Campobello*.

43. Burrows, 236, citing *Journal and Proceedings of the House of Assembly of Nova Scotia for 1844*, appendix 19.

44. Burrows, 227–29, citing J. G. Boulton, "Admiral Bayfield," *Transactions of the Literary and Historical Society of Quebec*, n.s., no. 28 (1910): 27–95.

45. Burrows, 230, citing *New Brunswick Courier* (St. John, N.B.), 7 November 1857.

46. Daniel Webster (1782–1852) was secretary of state under Presidents William Henry Harrison, John Tyler, and Millard Fillmore. He is most famous for the Nullification Crisis debate of 1830. He argued against then vice president John C. Calhoun's political theory of "states' rights" and that South Carolina did not have the right to nullify federal laws, as claimed by Calhoun in response to unpopular tariffs on imported manufactured goods and later extended to the issue of slavery.

47. Klein, *Beloved Island*, 18; and Muskie, 5, citing William F. Gagnon, *Royal Society of Canada, Proceedings and Transactions*, 7 (May 1901): 294–95.; cf. *ES*, 22 July 1931, 12. See also Gunther, *Inside U.S.A.*, 86.

48. *ES*, 22 July 1931, 12.

49. Klein, *Beloved Island*, 20, citing "Spell of Campobello, Maine's Neighbor Isle, Summer Home of President Roosevelt," *Lewiston Journal*, 8 July 1933; and Lura Jackson, "Benedict Arnold and the St. Croix Valley," *Calais Advertiser*, 20 April 2017.

50. Burrows, 231, citing Owen Papers, New Brunswick Museum, St. John, N.B; and Nowlan, 81.

51. Nowlan, 81.

52. "Campobello," in *The Way We Were: 1908—Calais, St. Stephen, Woodland, Eastport, Campobello, St. Andrews*, courtesy of CPL.

53. Klein, *Beloved Island*, 23.

54. Barber. The house was being built for the Porter family, one of the owners of the Campobello Company. Wells, *Campobello*, "David Owen" section.

55. Nowlan, 93.

56. Klein, *Beloved Island*, 25.

57. Cross, 27–28.

58. Klein, *Beloved Island*, 31.

59. Klein, *Beloved Island*, 30.

60. Klein, *Beloved Island*, 115.

61. Klein, *Beloved Island*, 7, citing Mrs. James (Sara) Roosevelt, *My Boy, Franklin Roosevelt* (New York: Ray Lang and Richard Smith, 1933), 13.

62. Klein, *Beloved Island*, 31, quoting Rexford G. Tugwell, *FDR: Architect of an Era* (New York: Macmillan, 1967), 10. Coincidently, the tides are strongest during the full and new moon phases of the lunar cycle.

63. Klein, *Beloved Island*, 39.

64. Riggs, 47, citing Frank Freidel, *Franklin D. Roosevelt: The Apprenticeship* (Boston: Little, Brown, 1952), 61f. Ripley's book, *Wall Street and Main Street* (Boston: Little, Brown, 1927), would likewise have a significant effect on Franklin Roosevelt and the nation.

65. Seifer, 270; and Berton, 262.

66. Display at Roosevelt-Campobello International Park.

67. Klein, *Beloved Island*, 34.

68. Klein, *Beloved Island*, 4, citing *Down East Magazine*, August 1958, 28–29.

69. Klein, *Beloved Island*, 43; and Goodwin, *No Ordinary Time*, 77.

70. Klein, *Beloved Island*, 37. Franklin's father had been vice president of the Delaware & Hudson Railway. It was on a D&H train that Theodore Roosevelt was sworn in as president of the United States on 14 September 1901, at North Creek, New York, following the assassination of President William McKinley.

71. Klein, *Beloved Island*, 37, 235.

72. Klein, *Beloved Island*, 37, citing Geoffrey C. Ward, *Before the Trumpet: Young Franklin Roosevelt* (New York: Harper, 1985), 254.

73. Klein, *Beloved Island*, 38, 235. Eleanor's mother, Anna Hall, died in December 1892 of diphtheria when Eleanor was eight years old. Her younger brother, Elliot Jr., died four months later of scarlet fever. Her father, Elliot B. Roosevelt, Theodore's younger brother, who had recently run away with the maid (Catherine Mann) and their illegitimate child (Elliot Roosevelt Mann), but whom Eleanor nonetheless loved, died in a fall or committed suicide. Thus, Eleanor lost her mother, a brother, and her father within 20 months. Klein, *Beloved Island*, 13; and Ken Burns, *The Roosevelts*, episode 1, "1858–1901: Get Action," September 14, 2014, https://www.pbs.org/kenburns/the-roosevelts/film-credits.

74. Klein, *Beloved Island*, 38, 42, 43, 235, citing Ward, *Before the Trumpet*, 340. Goodwin, *Leadership*, 53–55, says the announcement was made at Thanksgiving.

75. Holt, 99–100; and Kilby, 67.

76. Holt, 112–13: *Plan of Moose Island in the Province of New Brunswick: Copied from a Plan Drawn by Lieutenant Brendreth, Royal Engineers, 1815*, from the Public Archives of Canada.

77. Zimmerman, 37.

78. Holt, 99–100; "disinclined" borrowed from *Pirates of the Caribbean* (Walt Disney, 2003).

79. Kilby, 131–33; and Holt, 99–100. Moose Island was the last piece of American territory to be returned by the British. "The War of 1812 Historic Sites," BHS, 2012.

80. *ES*, 16 December 1829, 1 (emphasis in original).

81. *ES*, 12 January 1921, 1, quoting E. H. "Quoddy" Kilby, "Eastport Sentinel: A Historical Sketch," 1893.

82. Warren, 54, quoting Fannie Hardy Ekstrom, *Indian Place Names*, 225.

83. Holt, 17.

84. Holt, 99–100.

85. Holt, 17.

86. Holt, 66; and Bangor Hydro-Electric Company, "Bangor Hydro-Electric Company Chronological Charts" (1990), *Bangor Hydro Electric News*, n.d., 63. https://digicom.bpl.lib.me.us/bangorhydro_news/63. This indicates that Eastport Electric Light Co. began on 2 July 1901.

87. For good accounts of the Saxby Gale, see Holt, 57–59; and Lockett, 157–163 (Eastport).

88. Holt, 17–18.

89. Holt, 99–100.

90. Pesha, 187.

91. Pesha, 171, quoting *LH*, 9 February 1950 (Bangs no. 61); and Johnson, 86.

92. However, "George [Comstock] and my father ran the grist mill before the gold from sea water incident. . . . I have no information as to the source of the electricity for the gold project." P. Meigs: C. A. Bangs, 29 March 1970.

93. Pesha, 101, quoting *LH*, 6 January 1949 (Bangs no. 4), in turn citing the *Hartford Courant* (undated).

94. Pesha, 32, quoting *LH*, 4 August 1949 (Bangs no. 38).

95. Pesha, 25–26.

96. The US Bureau of Labor Statistics Consumer Price Index goes back only to 1913—hence equivalent prices listed for dates listed prior to 1913 are listed with a "+" as the exact amount is not determined.

97. Pesha, 35, quoting Railton, Arthur R. "Jared Jernegen's Second Family," *Dukes County Intelligencer* 28, no. 2 (Nov 1986), published by what is now the Martha's Vineyard Museum; and Johnson, 86.

98. Johnson, 18, citing Myron Avery, "A History of North Lubec," 1931.

99. Johnson, 85.

100. Pesha, 58, quoting *LH*, 19 January 1950 (Bangs no. 58); and Johnson, 87.

101. Pesha, 137–38, quoting *LH*, 5 May 1949 (Bangs no. 21).

102. Pesha, 169, quoting *LH*, 9 February 1950 (Bangs no. 61), in turn quoting *BH* 31 July 1898.

103. Pesha, 139–40.

104. Pesha, 152, quoting *LH*, 2 June 1949 (Bangs no. 25), quoting *BH*, 2 August 1898.

105. Pesha, 153–54, quoting *LH*, 19 June 1949 (Bangs no. 26), quoting an unidentified paper.

106. Pesha, 154, quoting *LH*, 19 June 1949 (Bangs no. 26), quoting an unidentified paper; and Johnson, 88.

107. Pesha, 108, quoting *LH*, 27 January 1949 (Bangs no. 7).

108. Johnson, 88.

109. Pesha, 135, quoting *LH*, 15 December 1949 (Bangs no. 53).

110. Pesha, 182, apparently quoting article in *LH*, "What Happened to Jernegan?" and apparently by Bangs, because it follows the same typographic style as others, but it is not labeled.

111. For a curious sequel, see *ES*, 4 April 1934, 4. There is a nice collection of artifacts from the Jernegan Gold Swindle at the Lubec Historical Society Museum. A terrifically well-told version of this story by Winston "Dick" Chapman is included in Johnson, 85–89.

112. *ES*, 4 May 1932, 8, citing *Portland Sunday Telegram*: "In the herring season it [Lubec] lies shrouded 'neath a blanket of dense brown smoke which emanates from the chimneys of the curing bays . . . [creating] Pungent Zephyrs. . . . By no means disagreeable is this pungence from the curing fish. Many actually like it! When however, the vagaries of the zephyrs intermingle the characteristically peculiar odors with those of the sardine and fertilizer factories of Lubec, not so good!"

113. Holt, 88; and *ES*, 9 July 1913.

114. Klein, *Beloved Island*, 65. The Owen Hotel was torn down in about 1900. The 60-room Tyn-y-Coed Hotel (Welsh for "house in the woods") was built in 1882; the Tyn-y-Maes Hotel ("house in the field") was built in 1885. They were closed after the financial panic of 1907 and demolished a few years later.

115. Francis Joseph Neptune (?–1834). Soctomah, *Save the Land*, intro.

116. Wikipedia, s.v. "Francis Joseph Neptune," accessed 12 February 2019, https://en.wikipedia.org/wiki/Francis_Joseph_Neptune#cite_note-Tribe's_ Role_In_Margaretta_Battle-7, citing Katherine Cassidy, "Tribe's Role in

Margaretta Battle Honored at Machias Celebration," *BDN*, June 10, 2005. The victim is identified only as "the officer" in Donworth, paper presented at D.A.R. meeting.

117. Nowlan, 13–14.

118. Donworth, paper presented at D.A.R. meeting; and Wells, *Campobello*.

119. Broduer, 76–78.

120. Broduer, 76.

121. "Treat Island History," Maine Coast Heritage Trust, ca. 2015.

122. "Enclosure: A Memorandum for Indian Eastern Department, 28 May 1783," Founders Online, https://founders.archives.gov/documents/Hamilton/01-26-02-0002-0085-0002. (Original spelling and capitalization retained.)

123. Soctomah, *Save the Land*, 12; and Broduer, 76–78.

124. Broduer, 76–78. Broduer's source is not provided.

125. Wikipedia, s.v. "Massachusetts Banishment Act," accessed 18 February 2019, https://en.wikipedia.org/wiki/Massachusetts_Banishment_Act.

126. David S. Macmillan and Roger Nason, "Pagan, Robert," in *Dictionary of Canadian Biography*, vol. 6 (University of Toronto/Université Laval, 2003), accessed 19 February 2019, http://www.biographi.ca/en/bio/pagan_robert_6E.html.

127. "Welcome to the St. Croix Historical Society," accessed 6 March 2019, http://stcroixhistorical.com. *Skutik*, alternately *Scootic*, means "burn over place." Spring settlements would ideally be located in low-lying areas with no trees or shrubs. The Passamaquoddy would typically burn a site to clear bugs before placing their village.

128. Broduer, 82.

129. Soctomah, *Save the Land*, 33. In Chief Francis Joseph Neptune's petition to Governor John Brooks of Massachusetts on 20 November 1818, "Monsr. Romange" is presumably the same person. The Pleasant Point Reservation having been stripped of firewood, the petition requests Massachusetts to purchase 300 acres of woodland on which they were accustomed to camp in winter—from the former agent, Theodore Lincoln.

130. Soctomah, *Save the Land*, 6.

131. Soctomah, *Save the Land*, 8, 16, 38. By state law enacted in 1821, new agents were prohibited from granting leases of greater than one year.

132. Soctomah, *Save the Land*, 88, 44, 86, 86 and 120, 118, 12, 89, 96.

133. Soctomah, *Save the Land*, 34.

134. Lester, 4, citing Alberta Francis, *Maine Indians: The People of the Early Dawn* (Perry, ME: Waponahki Museum and Resource Center, 1979), 11–14. See also Soctomah, *Save the Land*, 37–38.

135. *ES*, 16 December 1829, 2 (emphasis in original).

136. *ES*, 23 December 1829.

137. *ES*, 30 December 1829, 3.

138. Soctomah, *Save the Land*, 79.

139. *ES*, 14 July 1830, 1.

140. *ES*, 14 July 1830, 1, quoting the *Philadelphia Morning Journal*.

141. *Cherokee Nation v. Georgia* (1831) and *Worcester v. Georgia* (1832).

142. White birch, *betula papyrifera*. Lester, 7, citing Eva L. Butler and Wendell S. Hadlock, "Uses of Birch-bark in the Northeast," Robert Abbe Museum *Bulletin* 7 (1957): 4.

143. Old Town Canoe Co., home page, accessed 1 April 2019, https://www.oldtowncanoe.com/heritage.

144. Lester, 7. There was a Passamaquoddy camp at Gooseberry Point on Campobello. Wyllie, 271.

145. Lester, 16.

146. *NYT,* 20 August 1933, SM10-11, in an article by Franklin's friend, Stephen Chalmers, and quoting H. Morton Merriman, a summer resident of Campobello.

147. James Roosevelt took his son Franklin (age five) to meet President Grover Cleveland at the White House in 1887, a visit during which the president said, "I am making a strange wish for you, little man, a wish I suppose no one else would make—but I wish for you that you may never be president of the United States." Irwin, 29, citing Don Wharton, *Roosevelt Omnibus* (New York: Alfred A. Knopf, 1934), 39. And of course, Franklin Roosevelt met his cousin, President Theodore Roosevelt, on numerous occasions and would have met William H. Taft, T. R.'s vice president, at the inauguration in 1905.

CHAPTER 3. CONVERGING CURRENTS

1. For a good discussion of the difference between waterwheels and turbines, see Hughes, 262–63.

2. One hp = 750 watts; 1,000 watts = 1 kilowatt (kW); 1 hp = 1.341 kW. Compared to a horse, a human is a decimal point—we can sustain only about 0.1 hp, although trained athletes can briefly peak at about 3.0 hp. Water weighs 62.5 pounds per cubic foot. There are 27,874,400 square feet in a square mile.

3. Marmer, 216: 1—Bay of Fundy, Canada, 44'; 2—Port Gallegos, Argentina, 36'; 3—Frobisher Bay, Davis Strait, 35'; 4—England, West Coast, 33'; 5—Koksoak River, Hudson Strait, 30'; 6—Cook Inlet, Alaska, 30'; 7—Magellan Strait, Argentina, 30'; 8—France, North Coast, 28'; 9—Ashe Inlet, Hudson Bay, 23'; 10—Collier Bay, Western Australia, 23'; 11—Arabian Sea, India, 23'; 12—Maraca Island, Brazil, 23'; 13—Colorado River, Mexico, 22'; 14—Chemulpo, Korea, 22'; 15—St. Croix River, Maine, 20'.

4. Though the flatrod system, *Stangenkunst*, invented in the sixteenth century in Sweden for pumps and lifts, could transmit power up to four kilometres. Some were in use into the 1920s. Wikipedia, s.v. "Flatrod System," https://en.wikipedia.org/wiki/Flatrod_system, citing Terry S. Reynold, *Stronger Than a Hundred Men: A History of the Vertical Water Wheel* (1983).

5. Smith, *Power Policy of Maine*, 5, 10, citing Clive Day, "Capitalist and Socialist Tendencies in the Puritan Colonies," in *Annual Report, 1920* (Washington, DC: American Historical Association, 1925), 110. See also J. T. Trowbridge, *Tinkham Brothers' Tide Mill* (1882; Arlington, MA: Arlington Historical Society, 1999).

6. Smith, *Power Policy of Maine*, 1, 7–8. The Great Pond Ordinance applied to all ponds over ten acres and gave citizens the right to access all such ponds by foot trespass if no other public access was provided. It also established that on the seacoast, the private property line was at the low-water mark, not the high-water mark, as is more typical for the rest of the United States.

7. Smith, *Power Policy of Maine*, 19, citing the *Auburn v. Union Water Power Company* (1887).

8. Warren, 2, 12, 27.

9. Tide mills in Passamaquoddy Bay in Maine included (1) Bell's Tide Mill Farm in Edmunds; (2) at the canal in North Lubec; (3) at Mill Creek in North Lubec; (4) on the end of the breakwater in Lubec Narrows in downtown Lubec (a "boat mill" to grind salt, driven by the horizontal movement of the tidal current rather than the vertical fall of the tidal waters); and (5) at Dennysville. Tide mills in Passamaquoddy Bay in New Brunswick included (1) at Harbor de L' Outre on Campobello, (2) at Mill Cove on Campobello, (3) at Mascarene, and (4) at Mill Cove on Frye Island at L'Etete Harbor.

10. There are believed to have been at least 1,500 tide mills worldwide: Eastern North America, 445; Europe, 903; South America, 135; Persian Gulf, 2; Oceania, 7; Asia, 45; Africa (none known); and Western North America, 3. Sonnic. There were at least 65,000 water-powered mills in the Eastern United States in 1860, according to the US Census. Elliott et al. Riverine mills were also much more powerful, further increasing their economic impact relative to tide mills.

11. The first was built in 1633 in Dorchester. Several other tide mills were built in surrounding communities. Meigs, "Energy in Early Boston," 84. In 1693, a floating tide mill, or "boat mill"—a barge with a paddle wheel anchored in a tidal stream—was located in what was Shirley Gut separating Deer Island from the mainland. The gut was filled in by eroded material during the hurricane of 1938, thus connecting the island to the mainland. Earl Taylor, historian, TMI, and cofounder, treasurer, and president of Dorchester Historical Society, "Pinpointing Boston's Early Tide Mills" (paper presented at Tide Mill Institute

conference, 2019); and National Park Service, "Boston Harbor Islands: Deer Island," https://www.nps.gov/boha/learn/historyculture/facts-deer.htm.

12. Cotting's definition of "millpower" was unique to each mill. The term was later standardized in Lowell, Massachusetts, where it became a larger unit of power. Patrick Malone and Bob Gordo, personal correspondence with the author, 19 April 2020 and 12 June 2021.

13. Meigs, "Energy in Early Boston," 88.

14. Gordon and Malone, 3–29. The six mills used a total of 16 millpower, of the 81 millpower the company expected to sell.

15. The use of Boston's mill ponds and streams as dumping grounds started early. In 1656, Boston specified that the North Bridge over Mill Creek was the only place into which butchers might cast "their beasts entrails and gardidg [*sic*]" without being fined. Meigs, "Energy in Early Boston," 87–88.

16. For all its power, a steam train could not project its power even six inches to the side of the railroad tracks. Electricity, however, could spark action remote from the power plant. Also, electricity must be used when generated, since batteries have been woefully inefficient at commercial scale, whereas fuels can be stored indefinitely.

17. Engineering and Technology History Wiki (ETHW), s.v. "Pearl Street Station," https://ethw.org/Pearl_Street_Station; and Wasik, 20.

18. Weil, 4; and US Energy Information Administration, "Electric Power Monthly," https://www.eia.gov/electricity/monthly/epm_table_grapher.php ?t=epmt_5_6_a.

19. Engineering and Technology History Wiki (ETHW), s.v. "Milestones: Vulcan Street Plant, 1882," https://ethw.org/Milestones:Vulcan_Street _Plant,_1882. The Rogers House, complete with its original 1882 electric wiring and fixtures, still stands and is open for tours. In 1886, inventor and entrepreneur Charles D. Van Depoele installed the first permanent electric streetcar systems in Appleton, Wisconsin, and New Orleans, Louisiana. Electric street cars, or trolleys operated by "traction" companies, were critical to the development of the electric industry because they used large amounts of electric power and did so during the day, when power was not being used for lighting. McDonald, 36–37.

20. See Nye.

21. Berton, 44–45.

22. Kiwanis Club of Stamford, 361–62.

23. Berton, 204. About $2 million in 2020 dollars. "The History of Power Development in Niagara," https://www.niagarafallsinfo.com/niagara-falls -history/niagara-falls-power-development/the-history-of-power-development -in-niagara/. Ball, 18, says there was an additional payment of $8,000 to Horace Day, a former developer who also lost a fortune.

24. "Niagara Falls: History of Power," www.NiagaraFrontier.com/power. html. Kiwanis Club of Stamford, 362, says that "by 1882" they also were supplying seven mills with waterpower, and the Brush Electric Light and Power Company did the lighting. A second 2,100-hp-generating station was built on the Canadian side in 1893 to power a tourist train using Westinghouse's AC equipment. Kiwanis Club of Stamford, 295, 364. The railway continued in operation until 1932. Ball, 23 and 29, says the development of the tourist railroad was driven in part by the desire to put the unscrupulous coachmen or "hacks" out of business, as well as to generate royalty revenue to the Ontario Parks Commission.

25. The Niagara River Hydraulic Tunnel Power & Sewer Company. See also Seifer, 132–37; Klein, *Power Makers*, 330–44; and "The History of Power Development in Niagara: The Hydraulic Tunnels of Niagara."

26. Berton, 206; and Klein, *Power Makers*, 331.

27. Kiwanis Club of Stamford, 363.

28. "The History of Power Development in Niagara."

29. *ES*, 2 February 1927, 1.

30. Klein, *Power Makers*, 334.

31. Horsepower is reported variously: 10,000 (Ball, 11); 5,500 (Ball, 75, and a reprint of Niagara Falls Power Company brochure, "Information for Visitors," 1913).

32. "The History of Power Development in Niagara: The Hydraulic Tunnels of Niagara."

33. Ball, 75. Seifer, 171–72, says it was enough electricity to power nearly one-quarter of the continent.

34. The International Commission included one representative each from England, France, Switzerland, Germany, and the United States (but not Canada). Porter, 274.

35. Kiwanis Club of Stamford, 363.

36. Kiwanis Club of Stamford, 363. In 1890, Niagara Falls had a population of about 5,000, and Buffalo about 250,000. Klein, *Power Makers*, 332. In 1900, Buffalo had a population of 352,387. *Population of the 100 Largest Cities and Other Urban Places in the United States: 1790 to 1990*, US Census Bureau, 1998, https://www.census.gov/library/working-papers/1998/demo /POP-twps0027.html.

37. Berton, 209; Seifer, 132; and Klein, *Power Makers*, 334–42.

38. *ES*, 5 December 1934, 2.

39. Berton, 209.

40. Kiwanis Club of Stamford, 363, 365; and Ball, 23–24. Sebastian Ziani Ferranti created the Depford Power Station in London, England, in 1890 and was an investor in one of the potential developers of hydroelectric power at

Niagara, which was eventually taken over by Edward Dean Adams in March 1892. In an 1891 experiment, electricity was successfully transmitted 108 miles in Germany using AC. Riggs, 15. See also Klein, *Power Makers*, 335–36.

41. See Seifer, 22–23.

42. Patents including those by Nikola Tesla, Galileo Ferarris, Lucian Gaulard, and George Gibbs. Employees including William Stanley, Olive Shallenberger, and Henry Byllesby. Seifer, 48–55.

43. Weil, 9.

44. "[P]rivate investigators reported that GE had arranged for a draftsman to be hired at the Westinghouse plant, paid a monthly fee by GE, provided with a fund to bribe others for confidential papers, and given a code to telegraph information to GE. . . . The incident resulted in a hung jury." Riggs, 15–16; see also Klein, *Power Makers*, 342–44.

45. Wikipedia, s.v. "War of the Currents," https://en.wikipedia.org/wiki/War_of_the_currents, citing Craig Brandon, *The Electric Chair: An Unnatural American History* (Jefferson, NC: McFarland, 1999), 82.

46. Klein, *Power Makers*, 341.

47. Berton, 218.

48. The Westinghouse-Tesla AC generators at the Columbian Exhibition produced three times as much power as all of the power plants serving Chicago. GE refused to sell its (Edison's) patented lightbulbs to Westinghouse, forcing Westinghouse to manufacture his own, albeit inferior, bulbs. While Edison lightbulbs were only about 5% efficient (95% of the energy was lost as heat), they were about five times as efficient as gas lights. Seifer, 115–16, 119; and Klein, *Power Makers*, 300–323. The original contract called for only 92,622 lamps. Wikipedia, s.v. "World's Columbian Exposition," https://en.wikipedia.org/wiki/World's_Columbian_Exposition.

49. Seifer, 111–15, 120, reports that Tesla subjected himself to electricity of up to 300,000 volts at various times. Klein, *Power Makers*, 315, says 100,000 volts.

50. Seifer, 56–57.

51. "Niagara Falls: History of Power." The contract signed in October 1895 included the building of transformers that could handle 1,250 hp and the stringing of overhead wires capable of transmitting 11,000 volts. "The History of Power Development in Niagara: Powerhouse #1 & #2." See also Kiwanis Club of Stamford, 363–74. However, McDonald makes no mention of Nikola Tesla but states on page 37, "In 1886 George Westinghouse and William Stanley perfected a device—the transformer—that made it possible to transmit electricity by means of alternating current." Seifer says this was a deliberate and widespread campaign by his competitors to deny Tesla's pioneering achievements, circumvent his patents, and avoid paying royalties.

52. Klein, *Power Makers*, 349.

53. Klein, *Power Makers*, 350; and Berton, 222.

54. "Alcoa got the utility company to agree to never sell power to any other aluminum company. It did the same with the rivers feeding into the rapids of the St. Lawrence River in Quebec. For years it resisted efforts by the Roosevelt administration and others to develop more use of the St. Lawrence for power." Klein, *Call to Arms*, 160, citing *Fortune*, May 1941, 67–68; and Stone, 61–62.

55. Klein, *Power Makers*, 351; and Kiwanis Club of Stamford, 364.

56. Klein, *Power Makers*, 351.

57. Klein, *Power Makers*, 350.

58. Klein, *Power Makers*, 375. The growth was from the time NFPC opened for business in 1895.

59. Klein, *Power Makers*, 395.

60. Berton, 69–71, quoting Anna Jameson.

61. The American park was to be funded by the State of New York, while the Canadian was to be self-supporting. Kiwanis Club of Stamford, 266–67, 291, 293.

62. Kiwanis Club of Stamford, 291–93, 364. Queen Victoria Park opened on Victoria Day, 24 May 1888. The Canadian government did in fact own the "Queen's Chain," a 66-foot wide strip along the river that had been reserved for military purposes. Berton, 53.

63. Starting in 1892, leases were granted to the Canadian Power Company, later the Ontario Power Company; Canadian Niagara Power Company, later a subsidiary of the Niagara Power Company; and a syndicate composed of William MacKenzie, a famous Canadian railroad builder, Henry M. Pellatt, a Toronto investment broker, and Frederick Nichols, one of the founders of the Toronto Electric Light Company, which later became the Canadian General Electric Company. Kiwanis Club of Stamford, 364; 295; and "Niagara Falls: History of Power."

64. Berton, *Niagara*, 286–96; and "The History of Power Development in Niagara: Ontario Hydro & Sir Adam Beck." See also Kiwanis Club of Stamford, 296. However, according to the Public Power Association, "The first public power utility was born on the evening of March 31, 1880, in the farm community of Wabash, Indiana. Shortly after 8 pm that evening, mechanics hitched a threshing machine engine to the west wall of the Wabash County Courthouse and sent motive power to a generator in the basement. Within minutes, lights atop the courthouse bathed downtown Wabash in brilliant light. One eyewitness account described the scene in Wabash that night as follows: 'People stood overwhelmed with awe, as if in the presence of the supernatural. The strange, weird, light, exceeded in power only by the sun, rendered the square as light as midday. Men fell on their knees; groans were uttered at the

sight and many were dumb with amazement. We contemplated the new wonder in science as lightning brought down from the heavens.'" "Public Power: A Rich History, a Bright Future," *Community Engagement* (blog), American Public Power Association, 15 February 2018, https://www.publicpower.org/blog /public-power-rich-history-bright-future, citing Museum of Electricity, "Brush Arc Lamps, Wabash, Ind. 1880," citing an excerpt from a newspaper account in *Men and Volts: The Story of General Electric*, http://electricmuseum.com /brush-arc-lamps/.

65. "The History of Power Development in Niagara: Ontario Hydro & Sir Adam Beck." See also Berton, 308. George Westinghouse died on 12 March 1914.

66. Electrical Development Company secured its franchise (charter) in January 1903. Kiwanis Club of Stamford, 295.

67. The wing dam was 785 feet long and up to 27 feet high. Kiwanis Club of Stamford, 368.

68. Kiwanis Club of Stamford, 368. The eleven generators of 11,000 hp each, for a total of 91,800 kW.

69. In *ES*, 2 February 1927, 1, Dexter described the scene to Roscoe, who recounted it in the *Sentinel*. A second version of events is told in Kiwanis, *Niagara*, 368–69 but does not provide the name of the "engineer in charge," nor does it cite sources. A third version of events is told in Berton, 284–85:

Construction began on the lower end of the tunnel in 1903. A shaft 150 feet deep was sunk in the bank at the edge of the Horseshoe Falls, and a second construction tunnel driven at right angles out to the very brink, 700 feet of it under the water, to a point where the excavation for the main tunnel would commence. To save both money and time, the contractor, Anthony C. Douglass, an American from Niagara Falls, New York, decided that the debris from both tunnels would not be hauled back to shore and up the connecting shaft. Instead it would be dumped into the chamber the river had gnawed out between the falling sheet of water and the limestone face of the cliff over which that water tumbled.

But to accomplish this, he needed to rip a hole in the wall of the escarpment. A small opening was made near the ceiling of the construction tunnel, but thick clouds of spray from the cataract burst in. Obviously a larger opening was needed through which the tailings could be removed and the water allowed to drain out, leaving a clear passageway.

Douglass then had 18 holes drilled in the rock face and loaded with ten cases of dynamite. The blast that resulted tore a jagged hole in the cliff, but still it wasn't large enough. Meanwhile, the tunnel, open to the spray, was filling up with water.

[Anthony C.] Douglass had a flat bottomed boat lowered down the shaft in the river bank. The tunnel was now so full of water that the boat couldn't clear the roof and had to be weighted down. Three miners with several boxes of dynamite and coils of copper wire then boarded the boat and started off down the tunnel, lying on their backs and propelling the craft with their hands and feet.

When they reached the opening, they placed the dynamite around it, attached the copper wire, and headed back to the shaft. Just as they reached it, their boat sank under them. They climbed to safety, and a moment later a tremendous explosion rocked the gorge. But the hole in the cliff face still wasn't big enough.

The only solution was to apply the dynamite to the face of the cliff behind the wall of water, a dangerous enterprise. The company's chief engineer, Hugh L. Cooper, and its resident engineer, Beverly R. Value, donned rubber suits and roped themselves together like mountain climbers. Starting from the Scenic Tunnel, long a tourist attraction, the pair headed for the opening that had previously been blasted. They scrambled precariously along the cliff face, blinded by the intensity of the spray and buffeted by the force of the wind created by the intense pressure of the falling water. Soaking wet in spite of their precautions, chilled to the bone, and thoroughly miserable, they finally reached the opening in the tunnel wall.

Here they were battered by two forces of water—the backlash churning up from the base of the falls and the powerful jets of spray coming at them from every side. Again and again they made this journey, risking their lives each time, until they had secured four tons of dynamite around the opening, chaining the boxes into position to prevent them from being torn away by the incessant blast of water.

This effort worked. The obstructions were at last removed, and the water ran out of the tunnel, which, as Nichols later told the Empire Club dinner in Toronto, "is as dry and pleasant as this room."

70. *ES*, 29 October 1924.

71. In 1912, a German engineer by the name of Pein proposed a tidal power plant on the North Sea coast. "Power is the Key," 50 (ca. 1957, publication name unknown). The Petitcodiac River in Moncton is nicknamed the "Chocolate River" because of the high sediment load it carries.

72. Halifax Bloggers, "The Origins of Tidal Power Discovered in Our Bathroom," *Noticed in Nova Scotia* (blog), 9 June 2016, https://halifaxbloggers .ca/noticedinnovascotia/2016/06/the-origins-of-tidal-power-discovered-in -our-bathroom/.

73. Campobello trumpeted its being hayfever free and celebrated "St. Hayfever Day" on 15 August, presided over by summer resident W. M. Patterson, vice president of the US Hayfever Association, known affectionately as "The Big Sneeze." *ES*, 19 August 1914, 3.

74. Theodore Roosevelt had suffered from asthma as a child, and when he graduated from Harvard his doctor told him he had a weak heart and should choose a desk job and quiet life. He didn't follow that advice. Theodore Roosevelt's daughter, Alice Lee Roosevelt, was born on 12 February 1884. The following day Theodore's mother, Mittie, died of typhoid fever. The next day

Theodore's wife, Alice, died of kidney disease and complications from child-birth. "The Strenuous Life" was the name given to a speech given by Theodore Roosevelt on 10 April 1899 in Chicago, in which he extolls the virtues of manliness in physical effort and the imperative to do good for one's family and country. At the time he was governor of New York, and it came to character-ize Roosevelt's pugnacious optimism. Wikisource, s.v. "The Strenuous Life," https://en.wikisource.org/wiki/The_Strenuous_Life.

75. Klein, *Beloved Island*, 77, citing Geoffrey C. War, *Before the Trumpet: Young Franklin Roosevelt* (New York: Harper, 1985), 313.

76. Klein, *Beloved Island*, 78–80, citing J. William T. Youngs, *Eleanor Roosevelt: A Personal and Public Life* (Boston: Little, Brown, 1985), 108.

77. Holt, 99–100; and Klein, *Beloved Island*, 78–80, citing Morgan, 172–74.

78. Cross, 36, 42; and Klein, *Beloved Island*, 78–80, citing Morgan, 172–74. See also *ES*, 4 October 1916.

79. Seifer, 377, 392–93. More than 25 years later, in 1943, Lighthouse Board documents that Franklin provided won for Nikola Tesla and the US government patent infringement cases at the US Supreme Court brought by Guglielmo Marconi. *Marconi Wireless Tel. Co. v. United States,* 320 U.S. 1 (1943).

80. Eastport's ship-burning business began in 1901 when the USS *Minnesota* (built 1854) was burned, and the USS *Vermont* (built 1818–1825) was burned in 1902. There was then a hiatus of a decade, recommencing under Roosevelt's auspices as assistant secretary of the navy when three Civil War vintage ships were burned for salvage. The USS *Wabash* (built 1854–1855) was burned on 26 June 1913; the USS *Franklin* (built 1854–1864) on 2 October 1916; and the USS *Richmond* (built 1858–1860) on 13 May 1920.

81. All information about the ship burning is from the historical markers at Cony Beach, Eastport, Maine. Similar salvage work was still being done in Harpswell, Maine, in the 1960s. Todd Griset, email to author, 24 August 2020. Before setting the ships ablaze, anything of value was stripped off, especially the ornamental woodwork, windows, and cabinetry from the officers' quarters. These salvaged bits of naval history were then built into houses all around the Bay, including, probably, Roscoe Emery's house on Elm Street.

82. McDonald, 163.

83. McDonald, 165–66, 169–87.

84. McDonald, 169–187. $1.3 billion in 1917 would be close to $30 billion in 2020.

85. McDonald, 169–87.

86. McDonald, 169–87; Hughes, 291–92.

87. Klein, *Beloved Island*, 85, citing Roosevelt and Rosenau, *F.D.R.: His Personal Letters, 1905–1928*, 351. After the war, a bomb literally hit home on 2 June 1919. It destroyed the Washington, D.C., home of Attorney General

A. Mitchell Palmer, just across the street from Franklin's home. The blast blew out Franklin's windows and sent bits of the bomber onto his doorstep— a threshold that he and Eleanor had crossed just minutes before. Seifer, 409; and Wikipedia, s.v. "1919 United States Anarchist Bombings," https://en.wikipedia.org/wiki/1919_United_States_anarchist_bombings#cite_note -Brands-11, citing Brands, 134.

88. Klein, *Beloved Island*, 48.

89. Lora Whalen, "WWI Veterans from Passamaquoddy Tribe Honored in Maine," *Quoddy Tides* (blog), US World War One Centennial Commission, https://www.worldwar1centennial.org/index.php/communicate/press-media /wwi-centennial-news/1264-passamaquoddy-tribe-ww1-veterans-honored.html.

90. Funigiello, xiv, 226–31, citing Charles Keller, ed., *The Power Situation During the War* (Washington, DC: Government Printing Office, 1921), 1; and Hughes, 285–323.

91. Billington, Jackson, and Melosi, 37; and Dorn, 27, citing Joseph Sierra Ransmeier, *The Tennessee Valley Authority* (Nashville, 1942), 41–44; and Preston J. Hubbard, *Origins of the TVA* (Nashville, 1961), 2–3.

The US federal government's first experience with hydroelectric power development was the Salt River Dam (later Theodore Roosevelt Dam) in 1903–1911. Built outside of Phoenix, Arizona, for irrigation, it was the Reclamation Service's first dam and included a 9,500 kW hydroelectric plant for use during construction, which was then repurposed to provide power to nearby ranches, thus putting the federal government in the power-producing and -distribution businesses. The idea was officially sanctioned in 1906 with an amendment to the Reclamation Act. The project took at least three times as long as projected and cost five times the original estimate. It was completed nine days before Arizona became a state. Billington, Jackson, and Melosi, 98–107, citing Arthur P. David, *Irrigation Works Constructed by the U.S. Government* (New York: John Wiley & Sons, 1917), 18–19. See also Funigiello, 33; and Riggs, 29.

92. Following his pioneering success at Niagara Falls, Hugh had been hired by the McCall Ferry Power Company to build (1905–1910) a dam and hydroelectric powerplant on the Susquehanna River at Holtwood, Pennsylvania. According to the army, at McCall Ferry, "Cooper established a structural type that became the characteristic of main stem dams over the next 75 years." In essence, it was a concrete dam with a roughly triangular cross-section that was heavy and stable enough to prevent it from being tipped over by the horizontal pressure of the water above the dam. Equally innovative, and far more technically challenging, was the construction process Hugh conceived and Dexter executed. Based on their experiences with the Niagara wing dam, they blocked half the river with a cofferdam to allow construction there while the river ran through the open half. Thereafter, the river was redirected through the mostly

completed first half of the dam while the second half was built within a second cofferdam. Hugh and Dexter used the same design and process for the seven times bigger flow of the Mississippi River at Keokuk, and for the nineteen times more powerful flow of the Tennessee River at Muscle Shoals. Wikipedia, s.v. "William Murray Black," https://en.wikipedia.org/wiki/William_Murray _Black; and Billington, Jackson, and Melosi, 114–17, citing Valentine Ketcham and M. C. Tyler, "Hugh Lincoln Cooper Memoir," *Transactions, American Society of Civil Engineers* 103 (1938): 1772–77.

93. Hughes, 331, citing Kellogg, "Conowingo Hydroelectric Project," 7, 9.

94. Billington, Jackson, and Melosi, 114–19, citing Ketcham and Tyler, "Hugh Lincoln Cooper Memoir." See also Dorn, 327, citing "Conflict in House Reports on Col. Cooper's Muscle Shoals Work," *Engineering News-Record* 84 (1920): 1082.

95. *ES*, 1 January 1919, 8.

96. *ES*, 10 September 1919, 2.

97. EMERY_box 11, folder 15 (Family History), Wedding invitation, 23 October 1919.

98. *ES*, 10 September 1919, 2.

99. *ES*, 10 September 1919, 10.

100. *ES*, 2 February 1921, 2.

101. *ES*, 2 February 1921, 2.

102. The company's first hydroelectric plant was built in 1906 at the site of the Pembroke Iron Works (destroyed by fire in 1879) on the Pembroke River. In 1921, a new powerhouse and dam were erected at the outlet of Pennamaquan Lake. *ES*, 16 September 1925, 1. See also Bangor Hydro-Electric Company, "Bangor Hydro-Electric Company Chronological Charts" (1990), *Bangor Hydro Electric News*, 63. https://digicom.bpl.lib.me.us/bangorhydro_ news/63. This indicates Pennamaquan Power Co. began on 4 November 1908.

103. *ES*, 28 December 1921, 1.

104. Roscoe Emery wrote an elegant eulogy for President Theodore Roosevelt. Thirty years later, his words could have applied to President Franklin Roosevelt (*ES*, 8 January 1919):

With the passing of Theodore Roosevelt, there has gone out not only a figure great in the world's affairs by reason of his tremendous accomplishments and vital personality, but a guide, a light and an inspiration to hundreds of thousands of people, who saw in him their ideal of manhood and patriotism. For this reason, his influence like all great leaders, will go on and on, and its affect on the national life and character will never be effaced. In this fact lies his greatest claim on the affections of his countrymen and on the esteem of the historian,—that he quickened, as none other could, the conscience of America; that by his unerring power to see and his undaunted will to do, the right, he renewed the moral force of a nation for the problems of peace and war alike. He needs no memorial other than that which will endure forever in the hearts of Americans, and the history of their country.

While he was denied any active part in the prosecution of the war, more than any other single individual or agency he was responsible for the unity of purpose and thought of America with reference to it. He perceived the German menace while others were talking about "neutrality in thought and deed" and "peace without victory," and his fearless denunciation of Germany and German ideas early after the war broke out saved the country from the spineless apathy, into which German propaganda sought to throw, and otherwise might have succeeded in throwing it. His voice, raised strongly for unity, justice and victory, was a mighty factor in awakening and maintaining a national sentiment that stopped pro-Germanism before it had fairly started. His work with voice and pen on the home front was worth that of many divisions on the firing line. He was forbidden to participate in the struggle as an official. As an individual and citizen he made himself a tower of strength to the common cause.

In Roosevelt spoke the voice of America, for he represented, as no other American, Lincoln alone excepted, the courage, the honesty, the wisdom and the ideal of our people.

His death, coming at this time, is a disaster to his party, to his country, and in a larger sense to the world, for his vision, courage and wise statesmanship were never needed more than they are now, amid the problems following the end of the war and renewal of peace.

105. Goodwin, *Bully Pulpit*, 606.

106. Goodwin, *Bully Pulpit*, 606.

107. Percival Proctor Baxter (1876–1969) was governor of Maine in 1921–1925.

108. Bert Manfred Fernald (1858–1926) was governor of Maine in 1909–1911 and a US senator in 1916–1926.

109. Richard Judd, "Walter Wyman and River Power," Maine History Online, https://www.mainememory.net/sitebuilder/site/815/page/1225/display.

110. The first state-level departments of public utility control were established in 1907 by Wisconsin, New York, and Massachusetts. McDonald, 114, 120–21.

111. Weil, 15–16.

112. Cantelon, 61–63. As early as 1906, federal law provided that municipal utilities, as public entities, would have a preference in buying power from federally financed generators. Under the Federal Water Power Act of 1920, Congress extended the preference to all dams built on navigable waters. Lawyers joked that "navigable" is anything that can float a Supreme Court decision. With the advent of large-scale federal water power projects, this greatly worried the private power companies. Weil, 184–85.

113. GSC, "Passamaquoddy," 3.

114. *NYT*, 27 September 1925; *NYT*, 4 February 1938.

115. Holmes, 6, citing statement by Mrs. Cooper, 13 August 1954.

116. As of 2020, Central Maine Power (CMP) is a subsidiary of Avangrid, a Spanish-owned conglomerate that also owns gas and electric utilities in Massachusetts, Connecticut, and New York.

117. *ES*, 13 October 1920, 4.

118. Full-page (15" × 21") ads for CMP appeared in the *Sentinel* in 1920 on 13, 20, and 27 October and 4 November, followed by a one-third-page ad on 10 November seeking a securities salesman.

119. CMP ads appeared in the *Sentinel* in 1921 on 2, 9, 23, and 30 March; 6, 13, 22, and 29 April; 11 and 25 May; 1, 8, 15, and 22 June; 13 and 27 July; and 3 and 17 August.

120. *ES*, 9 March 1921, 4; *ES*, 8 June 1921 (unpaginated); and *ES*, 27 March 1921. CMP also had a dividend reinvestment plan with 993 participants. *ES*, 1 June 1921.

121. *ES*, 2 March 1920, 2. *ES*, 4 April 1920 says "Over 5,900 people share in earnings of C.M.P. Co." *ES*, 11 May 1921 says 6,000 Maine stockholders. As of 3 August 1921, CMP had "over 6,500 profit-sharing friends." As of 17 August, the number was up to 7,000.

122. CMP was raising capital for construction of a hydroelectric dam on the Kennebec River at Skowhegan, Maine.

123. *ES*, 9 March 1920, 4.

124. *ES*, 9 March 1920, 4. See also *ES*, 8 June 1921, "Does Maine's Future Mean Anything to You?"

125. *ES*, 23 March 1920 (unpaginated); and *ES*, 27 July 1921 (unpaginated).

126. *ES*, 29 December 1920, 1; and *ES*, 5 January 1921, 1–2.

127. Holmes, 6, citing statement by Mrs. Cooper, 13 August 1954. See also GSC, "Passamaquoddy," 5.

128. *ES*, 4 April 1934, 1, 7. See also GSC, "Passamaquoddy," 5–6.

129. GSC, "Passamaquoddy," 6.

130. Jim Jollotta, "In Retrospect," *Quoddy Tides*, undated, in the collection of Dexter P. Cooper III.

131. Wikipedia, s.v. "1920 Democratic National Convention," accessed 4 March 2019, https://en.wikipedia.org/wiki/1920_Democratic_National _Convention#cite_note-7, citing Staff writer(s), "Roosevelt Given Second Place; Convention Ends," *San Francisco Chronicle*, 6 July 1920; and Klein, *Beloved Island*, 94, 237.

132. Theodore Roosevelt: NY State Assembly, 21st District in Manhattan (1882–1884); assistant secretary of the US Navy (1897–1898); governor of New York (1899–1900); vice president of the United States (4 March 1901–14 September 1901); and 26th president of the United States (1901–1909). Frank-

lin D. Roosevelt: New York State Senate, 28th District (1911–1913); and assistant secretary of the US Navy (1913–1920).

133. Running Lubec Narrows at a speed of 18 knots would have been 5–10 knots faster than normal for a ship of the *Hatfield*'s size. Interestingly, both the speed and the fact "Assistant Secretary of the Navy conning" are noted in the ship's official log. Lt. Commander Max B. DeMott was in charge of the ship. Navy logbooks, USS *Hatfield*, NA-DC.

134. *ES*, 28 July 1920, 2.

135. Klein, *Beloved Island*, 94, 237.

136. *ES*, 14 July 1920, 1–2.

137. *ES*, 28 July 1920, 1.

138. *ES*, 4 August 1920, 1; and *ES*, 28 July 1920, 1.

139. Holmes, 2, quoting from an interview with Roscoe Emery, 22 December 1954. See also *ES*, 16 March 1938.

CHAPTER 4. INTERNATIONAL ACTION (1921–1924)

1. Klein, *Beloved Island*, 100.

2. Klein, *Beloved Island*, 100, 104–5.

3. "It [the Federal Emergency Relief Administration] was this highly imaginative white-collar program that drew the most heated criticism, especially after an aldermanic investigation in New York revealed that a FERA official was teaching 150 men to make 'boondoggles,' such items as woven belts or linoleum-block printing. Ever after, 'boondoggling'—the official claimed the term went back to pioneer days, but others insisted it had been invented by a fifteen-year-old Eagle Scout seven years before—served as the main epithet for critics of federal relief." Leuchtenburg, 123.

4. Klein, *Beloved Island*, 107.

5. Klein, *Beloved Island*, 85, citing Lash, *Eleanor and Franklin*, 224; and Klein, *Beloved Island*, 107.

6. Klein, *Beloved Island*, 107–8.

7. Klein, *Beloved Island*, 108–11.

8. Klein, *Beloved Island*, 114–15; Johnson, 134–35; and ER-My-Day, 11 September 1944.

9. Klein, *Beloved Island*, 115.

10. Klein, *Beloved Island*, 107–15; and *ES*, 21 September 1921, 1.

11. *ES*, 7 September 1921, 1.

12. "Plymouth Tercentenary Photographs," Plymouth Public Library, https://www.digitalcommonwealth.org/collections/commonwealth:8049gm70r.

13. *ES*, 7 September 1921, 1.

14. *ES*, 7 September 1921, 1.

15. *ES*, 30 January 1924, 1.

16. Written by F. Scott Fitzgerald, *The Great Gatsby* was published in 1925.

17. Wasik, 130.

18. New Brunswick Electric Power Commission estimated that demand was growing by 15% per year from 1924 to 1928. *Cooper-FPC-28*, I-10.

19. Guy Eastman Tripp (1865–1927) was chairman of the Westinghouse Electric Corporation board of directors (1912–1927). During World War I he was an assistant to the US Army's chief of ordnance and was promoted to brigadier general in 1918. President Woodrow Wilson awarded him the Distinguished Service Medal in recognition of his efforts to convert America's production capacity to wartime materiel. He died of complications after intestinal surgery. Wikipedia, s.v. "Chief of Ordnance of the United States Army," https://en.wikipedia.org/wiki/Chief_of_Ordnance_of_the_United_States_Army.

20. Tripp, 14, 19.

21. Tripp, 45. Actually, Nikola Tesla was the inventor of AC; Westinghouse was the licensor and promoter and coinventor of the transformer, which increased the voltage for easier transmission and then decreased it for easier distribution and use.

22. Goodwin, *Leadership*, 166.

23. Klein, *Beloved Island*, 123–24; and Goodwin, *Leadership*, 167.

24. Goodwin, *Leadership*, 167. See also Klein, *Beloved Island*, 118, citing: Gunther, *Roosevelts in Retrospect*, 229. The most detailed account is in Chase, *FDR on His Houseboat*.

25. *ES*, 16 March 1938, Special Cooper Memorial Section, citing *ES*, 29 October 1924. In March 1925, *Sentinel* readers learned, "During the World War he [Dexter Cooper] was in the service of the Council of National Defense, and was one of that body's most active and useful officials, on whom devolved practically all of the plans for organizing the power resources of the country. He was then employed by the Canadian government for six months, with a corps of fifty assistant engineers, to investigate and report on power plans to be put into effect at Niagara and other points on the St. Lawrence waterway. From there he went to Muscle Shoals to act as first assistant to his brother, Col. Hugh Cooper, who was the engineer in charge. His work there terminated last April and he then came direct to Quoddy." *ES*, 25 March 1925, 1.

26. EMERY_5.8.13, *Bangor Daily News*, 14 December 1927. Nancy Parshall Cooper was born 27 November 1913 in Keokuk, Iowa. Elizabeth ("Boo") J. Cooper was born 12 January 1917 in Connecticut. Dexter ("Decks") Parshall Cooper Jr. was born 12 April 1921 in Buffalo, New York. *Descendants of Nath'l Sturgis*, 65.

27. *NYT*, 28 April 1937, 20.

28. Holmes, 11, citing *Legislative Record of the 82nd Legislature of the State of Maine*, Augusta, 1925, 125; *BDN*, 14 March 1925; and Dexter P. Cooper, "Some Aspects of the Passamaquoddy Tidal Power Project," in *Papers and Transactions for 1926 and Proceedings of the 42nd Annual Meeting*, The Connecticut Society of Civil Engineers (New Haven, CT: Tuttle, Morehouse and Taylor, 1926), 29.

29. Holmes, 8, citing Federal Power Commission, *Fourth Annual Report* (Washington, DC: Government Printing Office, 1924), 166. The preliminary permit was issued on 28 May 1926. Federal Power Commission, *Sixth Annual Report* (Washington, DC: Government Printing Office, 1926), 148.

30. See also Bangor Hydro-Electric Company, "Bangor Hydro-Electric Company Chronological Charts" (1990), *Bangor Hydro Electric News*, 63, https://digicom.bpl.lib.me.us/bangorhydro_news/63. BHE was formed on 2 June 1924.

31. *ES*, 25 March 1925, 1.

32. Pesha, 183, quoting *Hartford Courant*, 17 January 1926.

33. *ES*, 25 March 1925, 1; and Smith, *Power Policy of Maine*, 233.

34. *ES*, 30 January 1924, 1.

35. *National Geographic* article reprinted in *ES*, 12 March 1924, 1, 4, 7.

36. The first appearance (of hundreds) of the Passamaquoddy Bay Tidal Power Project in the *New York Times* was on 6 April 1924, X19.

37. *ES*, 16 March 1938; and *ES*, 12 January 1938, 1.

CHAPTER 5. AMERICAN POWER (1925)

1. *ES*, 25 March 1925, 1.

2. Wasik, 122.

3. Attributed to William Ford Gibson (1948–).

4. *ES*, 2 February 1927, 1.

5. *ES*, 15 December 1926, quoting *New York World*.

6. *ES*, 6 February 1935, 4, quoting *PEN*.

7. Smith, *Power Policy*, 71–72.

8. Funigiello, 4–6, citing 66 Cong. Rec. 1101–1107 (22 January 1925); and US Congress, Joint Resolution No. 329, 68th Cong., 2nd Sess. (Washington, DC, 1925).

9. SILLS: flyer dated 21 November 1925, and letter dated 1926. Knowlton's US patents: Tide Motor, #1,209,97526, December 1916; Hydraulic Air-Motor #1,342,682, 8 June 1920; and Hydraulic Motor #1,496,470, 3 June 1924. Power was to be stored as compressed air. Another inventor, Alexander Wilson of Bar

Harbor, Maine, presented a plan to MPUC proposing to generate energy from the tides by compressing air. *BDN*, 13 February 1925 (Holmes, iin).

10. *ES*, 30 May 1928, 6.

11. *Cambridge Tribune* 64, no. 1, 5 March 1921, 13, https://cambridge .dlconsulting.com/cgi-bin/cambridge?a=d&d=Tribune19210305-01&e=-------en-20--1--txt-txIN-------.

12. *ES*, 30 May 1928, 6.

13. *ES*, 25 February 1925, 8.

14. *ES*, 25 February 1925, 8.

15. The Sentinel Real Estate Agency was in business at least as early as January 1919; *ES*, 1 January 1919, 4.

16. *ES*, 6 May 1925, 6.

17. *ES*, 15 February 1925, 8; and Smith, *Power Policy*, 235.

18. Holmes, 12, citing *BDN*, 14 March 1925; and *ES*, 18 March 1925, 1.

19. Charles Blanchard Carter (1880–1927) was general counsel to the Great Northern Paper Company and counsel for the Maine Central Railroad. In 1925, Carter was elected as a Republican to the Maine Senate representing Androscoggin County. Carter was a leader in the effort to prevent hydroelectric companies from exporting surplus power out of Maine. He died at age 47 of a heart attack.

20. Holmes, 12, citing *BDN*, 14 March 1925.

21. Frederick Wheeler Hinckley (1878–1933), a Republican, was mayor of South Portland, Maine, in 1919 and served in the Maine House of Representatives from 1919 to 1922.

22. Holmes, 14, citing *Maine Legislative Record*, 1925, 979–80.

23. Holmes, 14, citing *Maine Legislative Record*, 1925, 1251–52.

24. Bert Manfred Fernald (1858–1926) was governor of Maine, 1909–1911 and a US senator, 1916–1926.

25. *ES*, 18 March 1925, 8.

26. *ES*, 18 March 1925, 8.

27. Smith, *Power Policy*, 276.

28. *ES*, 30 November 1932, 3. The date of the deaths of the Norwood boys is not reported in the article, but they were the uncles of Robert Norwood Jr., who narrowly escaped a similar fate with his friend Hubert Chaffey in 1932.

29. *ES*, 30 November 1932, 3.

30. *ES*, 1 April 1925, 1.

31. *ES*, 8 April 1925, 6.

32. *ES*, 15 April 1925, 4.

33. *ES*, 8 April 1925, 1; and *ES*, 15 April 1925, 1.

34. *ES*, 15 April 1925, 1; and Holmes, 16, citing *Portland Sunday Telegram*, 13 April 1925.

35. *NYT*, 27 September 1925, XX4.

36. EMERY_5.7c.1-36. See also *ES*, 3 June 1925, 1.

37. *NYT*, 19 August 1925, 21.

38. Why the MPUC didn't do the same to Knowlton is unknown.

39. *ES*, 15 April 1925, 1.

40. *ES*, 10 June 1925, 7; *ES*, 8 August 1925; *ES*, 1 July 1925, 1; and *ES*, 22 July 1925, 1.

41. *ES*, 1 July 1925, 1.

42. *ES*, 13 July 1925, 1.

43. *ES*, 22 July 1925, 1.

44. *ES*, 22 July 1925, 1.

45. *ES*, 9 September 1925, 1.

46. *ES*, 11 June 1924; and *NYT*, 27 September 1925, XX4.

47. *ES*, 16 September 1925, 1.

48. *ES*, 16 September 1925, 1.

49. Robert L. Duffus, "Fundy's Giant Tides are to be Harnessed," *NYT*, 27 September 1925, XX4.

50. Dexter P. Cooper, Inc. (DPC, Inc.) was formally incorporated on 17 October 1925. The officers of the corporation were President—Dexter P. Cooper; Treasurer—Gertrude S. Cooper; Clerk—C. L. Andrews; Directors: Dexter P. Cooper, Gertrude S. Cooper, F. A. Havey, and E. H. Bennett. The issuance of stock in DPC, Inc. was approved by the Maine Public Utilities Commission on 1 October 1926. *Final-36*: appendix A-3, narrative p. 3.

51. *ES*, 16 September 1925, 1. See also Bangor Hydro-Electric Company, "Bangor Hydro-Electric Company Chronological Charts" (1990), *Bangor Hydro Electric News*, 63, https://digicom.bpl.lib.me.us/bangorhydro_news/63. This suggests Pennamaquan Power Co. was acquired by BHE on 1 January 1927.

52. *ES*, 23 September 1925, 1.

53. *Ruby Black-33*.

54. *ES*, 30 September 1925, 2.

55. Hughes, 296–97; and Funigiello, 227–28, citing W. S. Murray, *A Superpower System for the Region Between Boston and Washington* (June 1921); and McDonald, *Let There Be Light: The Electric Utility Industry in Wisconsin, 1881–1955* (Madison, 1957), 182–85. The superpower survey called for private development of large-scale thermal and hydroelectric power plants (60,000 to 300,000 kw) and transmission systems. It was instigated by the World War I power shortage crisis and a belief in the need for preparedness for future emergencies. "Giant Power" was a competing concept with a focus on public benefits as espoused by Gifford Pinchot and Morris Cooke. See Hughes, 297–313.

56. The Murray & Flood analysis of the Quoddy Project cost $24,550. *Final-36*: appendix A-3, narrative p. 7. Emphasis added.

57. *Reel-38*, 2.

58. *ES*, 18 November 1925, 1–2.

59. *ES*, 18 November 1925, 1–2.

60. US Congress, House, Debate over Brewster's charge that Thomas G. Corcoran had blackmailed him, 79 Cong. Rec. pt. 10, 10,660 (2 July 1935). See also *Ruby Black-33*.

61. Klein, *Power Makers*, 396.

62. McDonald, 18; and Wasik, 3–4, 10, 12, 23. Insull was also present at the creation of another industry: In 1903, Edison invited his energetic protégé to accompany him to dinner one evening with a former employee, Henry Ford, at which Ford showed his former boss his plans for mass producing automobiles. Wasik, 88.

63. McDonald, 58–59, 98; Wasik, 79–80, 88; and Riggs, 20. Insull's phrase "massing of production" was abbreviated as "mass production" and often attributed to Ford.

64. Wasik, 29, 45, 50.

65. Wasik, 79–80, 105; McDonald, 177; and Riggs, 22–24. Insull first proposed state regulation in 1898, but it was not well received by the industry. Progressive politicians picked up the idea, and in 1907 New York governor Charles Evans Hughes and Wisconsin governor Robert M. LaFollette organized the first two state departments of utility control. Originally oriented to the public interest, the state agencies were quickly corrupted by the industry, leading Milwaukee mayor Daniel Hoan to fume, "No shrewder piece of political humbuggery and downright fraud has ever been placed upon the statute books. It's supposed to be legislation for the people. In fact, it's legislation for the power oligarchy." Riggs, 23–24, citing Richard Rudolph and Scott Ridley, *Power Struggle: The Hundred Year War over Electricity* (New York: Harper and Row, 1986), 40. In the "regulatory bargain," states would give territorial monopoly rights to utilities, which would agree to have their rates set by the regulators; thus regulation became the "surrogate for competition." One problem with this model is that the "public interest" is not defined as that of the customers (ratepayers) but as that of the customers *and* the utility company investors, who were entitled to a "fair return" on their investment. Weil, 11–13.

66. McDonald, 181–87.

67. Riggs, 45, citing Insull, "Economics of the Public Utility Business," *Electrical World*, 25 March 1922; McDonald, 203–5; Wasik, 109.

68. McDonald, 172, 185.

69. Wasik, 81. The first attempt to prosecute a manufacturing company under the Sherman Antitrust Act (1890) was against a holding company that con-

trolled 98% of sugar refining. The court ruled in favor of the holding company on the illogical premise that the holding company itself did not operate across state lines, even though it owned operations in many states. General Electric created the first utility holding company, the Electric Bond & Share Company, in 1905. Riggs, 25, 32, citing United States v. E.C. Knight Co., https://supreme.justia.com/cases/federal/us/156/1/case.html; Charles R. Geisst, *Monopolies in America: Empire Builders and Their Enemies, from Jay Gould to Bill Gates* (New York: Oxford University Press, 2000), 57–58; and William Z. Ripley, *Main Street and Wall Street* (Boston: Little, Brown, 1927), 87.

70. McDonald, 58–59; and Riggs, 74–80, citing FTC, 72-A, 37.

71. *ES*, 25 November 1925, 1.

72. *Ruby Black-33*; and *ES*, 25 November 1925, 1.

73. Samuel Insull was a close friend of and an inactive partner of H. M. Byllesby in his namesake company from 1902 to 1912. Byllesby helped Edison create the Pearl Street Station in New York in 1882 and investigated and recommended Tesla's AC system for Westinghouse in 1886 and for Niagara in 1894. Along with doing contract engineering, Byllesby owned several small electric utilities. He died in 1924. McDonald, 125, 132, 269; and Seifer, 46, 134.

74. *ES*, 6 December 1925.

75. *Final-36*: appendix A-3, narrative p. 5; and *Reel-38*, 2.

CHAPTER 6. CROSSING THE BORDER (1926–1929)

1. *ES*, 6 May 1925, 1; Holmes, 19, citing *Evening Times Star*, 21 April 1925; and Smith, *Power Policy of Maine*, 235.

2. Holmes, 19n, 20, citing *SCC*, 1 Apr 1926.

3. Holmes, 20: "L.R. Day, Scientific Assistant to the Director, St. Andrews Biological Station, St. Andrews, N.B., to the writer, Jan. 14, 1955. In Washington County, Maine, there were 4,400 people employed in the fisheries in 1930, and the value of the manufactured fish products was three million dollars." In Charlotte County, New Brunswick, more than two thousand people are employed in the fisheries, which are worth an average of $1.5 million per year. *IJC-Fish-34*, 3.

4. Holmes, 20, citing *SCC*, 16 April 1925.

5. Holmes, 21, citing *SCC*, 29 April 1926; and Holmes, 21n.

6. *ES*, 17 February 1926, 1, quoting the *St. John Telegraph*.

7. *NYT*, 15 May 1926, 5; *Final-36*: appendix A-3, narrative p. 2; and *ES*, 16 March 1938, 9.

8. *ES*, 19 May 1926, 1; and *ES*, 26 May 1926, 1.

9. *ES*, 2 June 1926, 1.

10. *NYT*, 20 May 1926, 37. See also Holmes, 23, citing *Acts of the Parliament of the Dominion of Canada* (Ottawa: King's Printer, 1926), ch. 23, 13.

11. For damage in the Lower Pool, Cooper budgeted $49,400 for wharves and misc., and $70,550 for losses to herring and salmon fisheries. For the Upper Pool, Cooper budgeted $32,000 for the bridge to Minister's Island; $9,800 for St. Andrews' sewer; $7,200 for St. Stephen's sewer; $5,800 for Calais's sewer; $13,000 for foundations at St. Stephen and Calais; $12,000 for the power plant above St. Stephen; and $3,500 per herring weir, with 15 "Total Loss" and 100 "Possible Loss," for a grand total of $425,000. *Cooper-FPC-28*, N-7-8.

12. *ES*, 12 January 1927, 1.

13. EMERY_5.7.11-12. See also *ES*, 13 March 1928, 9.

14. Holmes, 21n, citing *SCC*, 24 June 1926.

15. *ES*, 8 December 1926, 4.

16. *ES*, 8 December 1926, 4.

17. Gruening, 216–17; and *ES*, 3 November 1926, 4.

18. *ES*, 3 November 1926, 4.

19. "Re 'Murray Report' on Electric Utilities; Refutation of Unjust Statements Contained in a Report Published by the National Electric Light Association Entitled, 'Government Owned and Controlled Compared with Privately Owned and Regulated Electric Utilities in Canada and the United States' Respecting the Hydro-electric Power Commission of Ontario," 1922, Internet Archive, https://archive.org/details/remurrayreporton00ontarich /page/n3/mode/2up.

20. McDonald, 158–60, 170–72, 182–84.

21. Gruening, 144–46.

22. Virtually identical language appears in *Manufacturers News* 31 (1927): 16, https://www.google.com/books/edition/Manufacturers_News/p_zmAAA AMAAJ?hl=en&gbpv=1&bsq=we,—everybody%E2%80%94finance+their +building+and+buy+their+products+from+ourselves+as+owners+.

23. The "notable change" in the decline in patent medicine advertising was the result of the Pure Food and Drug Act of 1906, which required accurate information on labels about a product's ingredients (many patent medicines were either just sugar water or, alternatively, deadly or addictive) and forbade additives used to bulk up a product (such as chalk added to milk or flour) or to make fouled food appear edible. The Pure Food and Drug Act was instigated by "muckraking" investigative journalists and Upton Sinclair's novel, *The Jungle*, about Chicago's meat-packing industry. It was championed by then president Theodore Roosevelt. See Goodwin, *Bully Pulpit*, 364–66.

24. *ES*, 17 November 1926, 4.

25. Smith, *Power Policy of Maine*, 4–5.

26. *ES*, 17 November 1926, 4.

27. *ES*, 8 December 1926, 1. For Dexter's financial position at this time, see *Final-36*: appendix A-3, narrative p. 3.

28. *ES*, 29 December 1926, 1.

29. *ES*, 29 December 1926, 4.

30. Bangor Hydro-Electric Company, "Bangor Hydro-Electric Company Chronological Charts" (1990). *Bangor Hydro Electric News*, 63, https://digi com.bpl.lib.me.us/bangorhydro_news/63.

31. *ES*, 30 March 1927, 1.

32. SILLS: Sills Letter File, Rosco Emery to Sills, 24 September 1935. He worked for Shredded Wheat.

33. *ES*, 2 February 1927, 1.

34. *ES*, 20 July 1927, 1.

35. FDRL-OLDS: 1950s_history of Quoddy_p.2: Murray and Flood Investigation, Report, 15 April 1927, to Gerard Swope, Pres, Gen. Elec. Co., on water power resources.

36. *ES*, 12 January 1927, 1.

37. *ES*, 27 April 1927, 4. This for-profit power pool was created in reaction to the Governor Pinchot's proposal for a publicly owned, managed, and regulated "Giant Power" pool. William S. Murray called Pinchot's proposal "communistic." Hughes, 297–313, 324–34.

38. *ES*, 27 April 1927, 4.

39. *Final-36*: appendix A-3, narrative pp. 5–6. (Boston Edison, GE, and Middle West Utilities become investors in Quoddy.)

40. McDonald, 263, says the total exceeded $160,000. Riggs, 59, reports the total was $500,000.

41. Julius Rosenwald, head of Sears, Roebuck & Company. McDonald, 263.

42. McDonald, 154, 254–56, 262–65. McDonald (1962 ed.) is generally favorable to Insull, as opposed to later authors Wasik and Riggs.

43. Riggs, 59.

44. Senator James Reed (D-MO) was co-chair with progressive senator Robert M. LaFollette Jr. (R-WI). McDonald, 262–66.

45. Wasik, 146–47; McDonald, 267; and Riggs, 59–61.

46. The dam across the Colorado River separating Nevada and Arizona was built in 1931–1936 just downstream at Black Canyon and called Boulder Dam from 1933 to 1947; then it was renamed the Hoover Dam.

47. McDonald, 267–68.

48. *ES*, 27 April 1927, 4. Possibly to counteract the bad publicity, in 1927 Central Maine Power (owned by Insull's Middle West Utilities since 1924) opened a speakers' bureau. CMP archives at MHS, Collection 2115, https://www.mainehistory.org/PDF/findingaids/Coll_2115.pdf.

49. Wasik, 170; and Marvin, 5, citing 79 Cong. Rec. pt. 10, 10,660 (2 July 1935). See also comments by Hon. John Wilson of Bangor, Republican candidate for governor in 1930, *ES*, 4 June 1930, 1.

50. FDRL-COOPER: 1934_09-03_Gertrude Cooper to FDR_Brann's election.

51. *The Octopus* (1905) by Frank Norris was a novel about the conflict between wheat growers and the railroad trusts and corrupt government. Pinchot referred to the network holding companies as an evil spider's web. Hughes, 305.

52. *ES*, 29 June 1927, 6.

53. *ES*, 28 September 1927, 1; *ES*, 1 May 1929, 1; *ES*, 19 October 1927, 1; *ES*, 19 October 1927, 1; *ES*, 16 March 1938, 9; and *ES*, 16 March 1938, 9: The two engineers were Richard Field and E. S. Hedstrom. Paul Jaenicke was added to the staff on 18 January 1927.

54. *Final-36*: appendix A-3, narrative pp. 7–8.

55. *ES*, 28 September 1927, 3; Cantelon, 62–64. The FPC was not able to hire its own employees until 1928. The FPC suffered from 500% executive turnover during the 1920s.

56. *Institutes of Justinian*, 2.1.1.

57. Billington, Jackson, and Melosi, 3, citing Ben Moreell, *Our Nation's Water Resources—Policies and Politics* (Chicago: Law School, University of Chicago, 1956), 15–16. Coincidently, Ben Moreell was "Father of the Sea Bees," who honored him by naming the drill hall at Quoddy Village for him.

58. Gruening, 87.

59. In *Charles River Bridge v. Warren Bridge*, 36 U.S. (11 Pet.) 420 (1837), the US Supreme Court set the foundation for more modern "public utility" law under which private entities own and operate infrastructure deemed an essential public good, as a regulated monopoly, until the public good calls for something different: "While the rights of private property are sacredly guarded, we must not forget that the community also have rights, and that the happiness and well-being of every citizen depends on their faithful preservation."

60. McDonald, 68, 97–100, 103–4, 137–38.

61. Highly touted but tiny "rate reductions" were strategically timed to mask the enormous profit margins the utilities enjoyed. Rate schedules charged base fees plus a declining usage fee based on volume—thus encouraging more volume. Weil, 19.

62. Nehf, 775–93.

63. Weil, 12–13.

64. Nehf, 775–93.

65. "The Right of People to Rule" (address delivered at Carnegie Hall, New York City, 20 March 1912). See Theodore Roosevelt, "Social Justice and

Popular Rule: Essays, Addresses, and Public Statements Relating to the Progressive Movement (1910–1916)," Library of Congress, https://www.loc.gov/collections/theodore-roosevelt-films/articles-and-essays/sound-recordings-of-theodore-roosevelts-voice/.

66. Funigiello, 6–7, citing US Congress, Senate, *Electric Power Industry, Control of Power Companies*, 69th Cong., 2nd Sess. (1927), Doc. No. 213; and Thomas K. McCaw, *TVA and the Power Fight, 1933–1939* (Philadelphia, 1971), 4–9; and "Politics Discovers a Power Trust," *Literary Digest* 93 (2 April 1927), 12.

In response to the critics of the FTC report, in June 1927 the National Electric Light Association (NELA) organized the Joint Committee of National Utility Associations, representing the electric, gas, and electric railway companies and became the instrument for executing their propaganda campaigns. At the same time, General Electric formed the United Corporation, a "super-holding company," which dominated nearly all others except a few, including those associated with Insull. Funigiello, 10–13, citing King, "The Genesis of the TVA," ch. 16, p. 15; McDonald, *Insull*, 249; and Vincent P. Carosso, *Investment Banking in America: A History* (Cambridge, MA, 1970), 259.

67. Gruening, 90–91. The phrase "watering the stock" originated in the nineteenth century with the cattle drover Daniel Drew. He allegedly kept his cows thirsty for a couple of days, and then, on the sale date, he allowed them to gluttonously drink their fill. When Drew became a railroad speculator, he issued and sold extra shares and pocketed the proceeds, likening his oversold stock to his overwatered cattle. Riggs, 34, citing Frederick Lewis Allen, *Lords of Creation* (New York: Harper & Brothers, 1935), 12.

68. Gruening, 3–4.

69. Gruening, 5, 7–8.

70. Gruening, 8–9; Hughes, 300 (re: Morris Cooke's view of professional societies); Riggs, 47–48, 63, 143 (re: co-op in Massachusetts); Funigiello, 154–55, 146–48, 175–80 (re: Bonneville, 1936); and Springer, 30–31, 148 (re: Tennessee Valley Authority in Maryland court), citing John A. Schwarz, *The New Dealers: Power Politics in the Age of Roosevelt* (New York: A.A. Knopf, 1993), 150.

"Walsh declared that FTC commissioner William E. Humphrey worshiped at the temple of big business. The larger the corporation the more sacrosanct it was to him." Funigiello, 17, citing 69 Cong. Rec. 2892ff. (13 February 1928); and 2953–54 (14 February 1928). See also Thompson, 5; and Lief, 307. "Humphrey had previously accused the FTC of being 'an instrument of oppression and disturbance' to big business. During his chairmanship he instituted a new set of procedures for the FTC to follow, including closed hearings, settlement of cases by 'stipulation' rather than by prosecution, and the end to sweeping economic studies." Funigiello, 17n62, citing Collum, "Transformation of the FTC," 445–49; "Portrait" [Humphrey], *Review of Reviews* 71 (June 1925): 70; and Pendelton Herring, "The Federal Trade Commissioners," *George Washington Law Review* 8 (January–February 1940): 356.

71. Holmes, 23–24, citing A. G. Huntsman, *Passamaquoddy Bay Power Project and Its Effect on the Fisheries* (St. John, 1928), 45.

72. *ES*, 7 March 1928, 7.

73. *ES*, 15 February 1928, 1; and *ES*, 7 March 1928, 7.

74. *ES*, 7 March 1928, 7; and *ES*, 16 March 1938, 9.

75. Holmes, 24, citing *Evening Times Globe*, 31 March 1928.

76. Holmes, 25n, and citing *PPH*, 10 March 1928.

77. *ES*, 4 April 1928, 1.

78. The 490-acre Minister's Island was owned by the heirs of the former president of the Canadian Pacific Railroad, Sir William Van Horne, who led the construction of Canada's transcontinental railroad and the development of the famous resorts along it, including the Algonquin House in St. Andrews (1889) and the Banff Spring Hotel in Alberta (1888). At low tide, cars can drive across a gravel bar to Minister's Island. At high tide the bar is covered by at least fourteen feet of water.

79. *ES*, 4 April 1928, 1. See also Holmes, 6, citing *Acts of the Legislative Assembly of New Brunswick* (Fredericton, 1928), ch. 75, 164–76.

80. *ES*, 4 April 1928, 1.

81. *ES*, 23 May 1928, 5.

82. *ES*, 29 August 1928, 1, and *ES*, 16 March 1938, 9; and Smith, *Power Policy of Maine*, 235.

83. *ES*, 31 October 1928, 1.

84. *ES*, 12 December 1928, 1.

85. *ES*, 26 December 1928, 1.

86. Wikipedia, s.v. "Dust Bowl," https://en.wikipedia.org/wiki/Dust_Bowl; and Wikipedia, s.v. "Black Sunday (Storm)," https://en.wikipedia.org/wiki /Black_Sunday_(storm), citing *Black Blizzard*, History Channel, originally aired 12 October 2008.

87. *ES*, 2 May 1928, 6.

88. *ES*, 4 July 1928, 1.

89. *ES*, 15 March 1939, 4.

90. *ES*, 8 August 1928, 10; and *ES*, 17 October 1928, 1. The proposed dam landing sites were at Eastport and Deer Island. For a list of options purchased, see *Cooper-FPC-28*, F-4.

91. *Cooper-FPC-28*.

92. Lyrics from "Moon River" by Henry Mancini and Johnny Mercer. It was sung by Audrey Hepburn in the 1961 film, *Breakfast at Tiffany's*. Huckleberries can refer to many species, including blueberries and cranberries.

93. Franklin Roosevelt was elected to the New York State Senate in 1911 at age 29, just as Theodore Roosevelt had been at age 24; he was assistant secretary of the navy at age 31, as T. R. had been at 39; and he was nominated for vice president of the United States (at 38; T. R. had been at 42). Klein, *Beloved Island*, 67, citing Geoffrey C. Ward, *Before the Trumpet: Young Franklin Roosevelt*.

94. *ES*, 10 October 1928, 1. Al Smith was governor of New York 1918–1920, and 1922–1924.

95. Riggs, 5–6, 50, 74:

> He [Smith] considered himself to be following in the footsteps of two illustrious Republican predecessors. As governors, Theodore Roosevelt had warned against giving "waterpower barons" a monopoly on the state's natural resources and Charles Evans Hughes had declared that New York's water power "should be preserved and held for the benefit of the people and should not be surrendered to private interests."* By the time of Smith's election, however, the Republican party had swung solidly behind private power, and the Republican-controlled Assembly buried his bill in committee. He returned the favor by vetoing their bill to lease the sites to private utilities.**
>
> Republican Nathan Miller defeated Smith in 1920 and won passage of a law intended to place the hydropower sites in the hands of private developers. But before he could implement it, Smith ousted him two years later in a rematch charged with rhetoric that would echo down the years. Miller claimed that public power would be "wasteful and socialistic" and Smith argued that the rivers of the state should not be handed over to the "power barons."***
>
> [*] New York Power Authority, https://nypa.gov/about/timeline. [**] Eldot, Paula, *Governor Alfred E. Smith: The Politician as Reformer.* New York: Garland, 1983, 236–37. [***] Josephson, Matthew and Hannah Josephson, *Al Smith: Hero of the Cities.* Boston: Houghton Mifflin, 1969, 276–77.

96. Watkins, 46.

97. *ES*, 6 February 1929, 1.

98. *ES*, 19 June 1929, 1; *ES*, 23 January 1929, 1, 7; *ES*, 30 January 1929, 6; and *ES*, 6 February 1929, 1, 2, 4. The number of companies to be acquired by MSW&EC was also reported as sixteen and eighteen. The book value of the plant was reported as $245,000.

99. *ES*, 6 February 1929, 1, 4; and *ES*, 23 January 1929, 7.

100. *ES*, 19 June 1929, 1.

101. *ES*, 9 January 1929, 1; and *ES*, 20 February 1929, 1.

102. *ES*, 19 February 1930, 8. This event was not published in *Sentinel* at the time. This summary is quoted from a letter to FDR from the US State Department. FDRL-0982: 1934-04-13_IJC Fisheries Report.

103. *ES*, 10 April 1929, 1. According to *ES*, 16 March 1938, the petition was filed on 6 March 1929. Letter from Cooper to Emery published in *ES*, 17 April 1929, 1.

104. *ES*, 13 March 1929, 1.

105. *ES*, 10 April 1929, 1.

106. Holmes, 30, citing *SCC* 11 and 18 April 1929. See also *ES*, 10 April 1929, 1.

107. Holmes, 30, citing *SCC* 11 and 18 April 1929. See also *ES*, 10 April 1929.

108. Holmes, 30, citing *Evening Times Globe*, 13 April 1929. These numbers are about twice that reported in Holmes, 20, citing L. R. Day, Scientific Assistant to the Director, St. Andrews Biological Station, and *IJC-Fisheries-34*, 3.

109. *ES*, 17 April 1929, 1–2, 4.

110. *ES*, 17 April 1929, 2. Holmes, 31, citing *Evening Times Globe*, 17 April 1929, says vote was 10 in favor, 22 opposed.

111. *ES*, 24 April 1929, 4.

112. *ES*, 19 June 1929, 3; *ES*, 24 April 1929, 4.

113. Holmes, 33, citing *Lewiston Daily Sun*, 3 May 1929; *ES*, 15 May 1929, 1; Holmes, 33, citing *PPH*, 10 May 1929; and *ES*, 16 Mar 1938, 9.

114. *ES*, 17 April 1929.

115. Holmes, 32, citing *SCC*, 18 April 1929.

116. *ES*, 24 April 1929, 4.

117. *ES*, 24 April 1929, 4.

118. *ES*, 24 April 1929, 4.

CHAPTER 7. BACK IN THE USA (1929–1931)

1. *ES*, 16 July 1930, 1.

2. The first pumped storage hydroelectric facility in the United States was built by Connecticut Light and Power Company in 1926–1928 on the Housatonic River, forming Candlewood Lake.

3. *ES*, 29 May 1929, 4. The FPC sent letters to Machias town officials asking whether there were any objections to Dexter's plan to flood part of the town; there were none. *ES*, 26 June 1929, 1. The pumped storage reservoir was reported as being thirteen square miles in area with a volume of nineteen billion cubic feet. *ES*, 24 July 1929, 1.

4. One-third or more of the potential energy stored in the reservoir is consumed by the effort to pump the water uphill into the reservoir. *FPC-41*, 7, 16.

5. *ES*, 29 May 1929, 4.

6. *ES*, 16 July 1930, 4. EMERY_Scrapbook: Quoddy 1933-6: *BDN*, 3 October 1933: "TIDAL POWER PROJECT" illustrated map; and *FPC-41*, 8–9.

7. *ES*, 17 July 1929, 1; *ES*, 24 July 1929; *ES*, 16 July 1930; and *ES*, 6 August 1930. In May 1930, a patent was issued to Cooper for his plan. Dexter also

filed an application with the FPC in May 1929 for an "unconditional" permit with license to sell power in and out of Maine and in New Brunswick, with initial installation to be 464,000 hp and ultimately 1,087,000 hp (the largest number he had ever quoted). *ES*, 22 May 1929, 1.

8. *ES*, 15 May 1929, 1. Attending the meeting at the US State Department in Washington, D.C., were Secretary of State Henry L. Simpson; William Philips, minister to Canada; Assistant Secretary Castle; Governor Gardiner of Maine; Senators Hale and Gould of Maine; and Congressman Nelson.

9. *ES*, 1 May 1929, 4.

10. *ES*, 17 July 1929, 3; and *ES*, 31 July 1929, 6.

11. *ES*, 14 August 1929, 6; *ES*, 28 August 1929, 7; and *ES*, 4 September 1929, 2, 7.

12. *Ruby Black-33*. This was mistakenly dated by Black as 1927.

13. *ES*, 11 September 1929, 1.

14. The Electric Bond and Share Company (EB&S), in which the Morgan interests had large holdings, controlled the American Gas and Electric Company, the American Power and Light Company, and the National Power and Light Company. EB&S owned plants extending from the Atlantic to the Pacific coasts. In January 1927, the US Supreme Court decided *McGrain v. Daugherty*, approving and defining broad inquisitional powers for Congress. McDonald, 266.

15. *ES*, 4 September 1929, 1, quoting *New York World*, 19 August 1929. See also McDonald, 251–52.

16. *ES*, 30 January 1929, 6.

17. In October 1929 Insull created Corporation Securities Company of Chicago, or "Corp" for short. Riggs, 78, citing Samuel Insull, *The Memoirs of Samuel Insull* (Polo, IL: Transportation Trails, 1992), 188–90; and Frederick Lewis Allen, *Lords of Creation* (New York: Harper & Brothers, 1935), 283–84.

18. Wasik, 180–81.

19. Wasik, 170, 180–81; and Cantelon, 67.

20. Gruening, 189–91, 211–29, 236. See also *Summary Report of the Federal Trade Commission to the Senate of the United States on Efforts by Associations and Agencies of Electric and Gas Utilities to Influence Public Opinion*, S. Doc. 92, pt. 71a (1934).

21. Wasik, 108–9; Gruening, 18, says the campaign began in 1918.

22. Gruening, 6 (not dated).

23. *ES*, 20 October 1920, 5; *ES*, 27 October 1920, 4; and Gruening, 110.

24. Gruening, 146–47, 237.

25. Gruening, 173–74.

26. Gruening, 173–74, 216–17, 242–43; and *ES*, 22 April 1924, 2.

27. Gruening, 197–203.

28. "Music: In Chicago," *Time*, 4 November 1929, http://content.time.com/time/subscriber/article/0,33009,787523,00.html.

29. McDonald, 112.

30. "Music: In Chicago."

31. Wasik, 188; and McDonald, 284–85.

32. Wasik, 149–50. "Herbert Hoover, an able engineer, a man who had studied this subject of electricity and its future market, made a speech before the World Power Congress in London in 1924, ten years ago, in which he said: 'The factor that has been primarily responsible for the tremendous changes in the last century, and without which civilization could not exist, is mechanical power. The form in which in increasing degree such power is beling applied, is electrical energy—the greatest tool that has ever come into the hands of man. The degree to which we utilize this tool will largely determine the rate of our industrial progress.'" Joseph F. Preston, *ES*, 5 December 1934, 2.

33. Cantelon, 65–66. See also *United States v. Smith*, 286 U.S. 6 (1932). For more on regulatory capture in the 1920s, see Riggs, 45–46, 72.

34. *ES*, 19 February 1930, 1 and 8, quoting H. Doc. No. 275 (4 February 1930). See also Smith, *Tidal Power in Maine*, 241.

35. *ES*, 12 March 1930, 4; *ES*, 27 February 1929, 6; *ES*, 3 June 1931, 1; *ES*, 19 June 1929, 3; and *ES*, 30 March 1930, 4.

36. *NYT*, 30 April 1930, 16; and *ES*, 11 June 1930, 1.

37. Holmes, 35, citing North American Council on Fishery Investigations, *Proceedings 1921–1930*, 32; *PPH*, 26 March 1930 and 30 July 1931; and *IJC-Fish-34*, 5.

38. *ES*, 6 August 1930, 1. Application filed 7 December 1925, Serial No. 73,595. Patent issued in late May 1930.

39. Rueben "Rube" Goldberg (1883–1970) was an American cartoonist most famous for his cartoons of complicated gadgets performing simple tasks in convoluted ways.

40. *ES*, 13 August 1930, 1. See also Marvin, 6, citing State of Maine, Legislature, An Act to Amend Chapter 111 of Private and Special Laws of 1925, 84th Legislature, special session, 5–6 August 1930.

41. Emails to author from Anne Pike Rugh, 10 May 2021, 18 May 2021, and 17 June 2021.

42. *ES*, 12 February 1930, 1. The Boston News Bureau reported that GE and Westinghouse engineers had been seen at Quoddy in February 1930, though the *Sentinel* discounted it as possibly old news from 1928.

43. *ES*, 26 November 1930, 1. Genevieve Roches married Alexander MacNichol of Eastport. Mr. MacNichol later became a commissioner of the Roosevelt-Campobello International Park. L'Hommedieu. See also *Final-36*:

appendix A-3, narrative pp. 1, 7. (Westinghouse funding ceased as of 29 May 1929 and GE as of 16 February 1931.)

44. *ES*, 11 June 1930, 4; and *ES*, 2 July 1930, 3.

45. EMERY_5.7.15-16: letter from FPC, dated 24 January 1931.

46. FDRL-284: Power, 1933–1934; and Wasik, 191. See Riggs, 81: Private utilities claimed Ontario had more access to hydropower and forced industrial and commercial consumers to subsidize household rates. Emmons (884) argues that although (Louis) Howe's comparisons were crude, more sophisticated analyses also supported Roosevelt's conclusions.

47. Wasik, 191; and Riggs, 81, citing *NYT*, 8 July 1929.

48. Wasik, 171, 191; and Riggs, xxviii, 80–83. The New York State Power Authority was established by Roosevelt on 27 April 1931.

49. *ES*, 11 January 1933, 4. See also *ES*, 6 June 1934, 1, 4.

50. Wikipedia, s.v. "John G. Utterback," https://en.wikipedia.org/wiki /John_G._Utterback#cite_note-1, says 324 votes, citing *Lewiston Daily Sun*, 10 November 1932, 1. *ES*, 11 January 1933, 4, quoting *Current Events*, says 346 votes, and a majority of 20 for Brewster or 800 for Utterback, as later reported in *ES*, 3 January 1934, 1. Utterback was defeated in his reelection effort by Brewster by 2-to-1 margin. *ES*, 24 January 1934, 4. See also *ES*, 23 May 1934, 4.

51. *ES*, 11 January 1933, 4.

52. Regulatory capture was a continuing problem, as was basing current retail prices on future equipment replacement costs. Riggs, 80, citing Davis, *New York Years*, 76–77; and citing Burns, *Roosevelt: Lion & Fox*, 113–14. Tax advantages are described in *ES*, 9 August 1933, 3; *ES*, 27 September 1933, 4; and *ES*, 1 November 1933, 4. "Public utilities" also had the advantage in raising capital compared with other for-profit companies in that the dividends were "exempt from Taxation in Maine . . . and from the Federal Income Tax." *ES*, 27 July 1931, unpaginated.

53. Seifer, 421.

54. *ES*, 12 August 1931, 1; *ES*, 19 August 1931, 1; *ES*, 28 October 1931, 9; *ES*, 7 October 1931, 7; and *ES*, 9 December 1931, 1. Competition from Portland area fishermen was also a factor, it being a day's travel closer to Boston markets.

55. *ES*, 23 December 1931, 1.

56. *ES*, 5 August 1931, 1.

57. *ES*, 16 December 1931, 5.

58. Smith, *Power Policy of Maine*, 74–79. More altruistically, CMP maintained it was dependent on the welfare of the communities they served, and thus it made sense for them to invest in their own communities.

59. *ES*, 17 July 1929, 3; and *ES*, 16 December 1931, 5. For a discussion of the lopsided ownership and voting control of the Insulls' holding company, the New England Public Service Company, see Smith, "Regulation of Some Holding Companies," 293–94.

60. Riggs, xxiv, 5, citing Federal Trade Commission, *Control of Power Companies*, 22 February 1927. By coincidence, one of Eastport's native sons, Ray P. Stevens (University of Maine, 1898), was a principal in some of the largest public utility holding companies. "Mr. Stevens, ever since graduation, has been identified with public utilities, chiefly electrical, being a pioneer in some phases of the industry. He is now chairman of the Board of Stevens & Wood, Inc.; president of American Electric Power Corporation; Pennsylvania Gas & Electric Corporation, Allegheny Gas Company, and Southeastern Ice Utility Company. He is also a Director of several other companies, including Allied Power and Light Corporation, American Super Power Corporation and Niagara Hudson Power Corporation." *ES*, 12 November 1930, 6; and Maine Alumnus 12, no. 3 (December 1930), DigitalCommons@UMaine, https://digitalcommons.library.umaine.edu/cgi/viewcontent.cgi?article=1198&context=alumni_magazines.

61. *ES*, 6 April 1932, 4. McDonald, 274–304, describes the financial maneuverings behind the collapse of the Insull empire in detail, laying much of the blame on speculators who bid up the price of the stock to absurd levels and on short sellers, notably Insull's commercial rival, the House of Morgan, which made a killing from Insull's collapse. Insull was forced out of all his companies on 6 June 1932. The bad blood between Insull and Morgan (and, more broadly, all New York financiers) originated in 1884 in a fight for control of the Edison Electric Light Company, the formation of General Electric in 1889, and Insull being forced out of GE in 1892. McDonald, 31–32, 38–41, 49–52.

62. Wasik, 203.

63. McDonald, 309–11; and *ES*, 18 November 1925, 1–2.

64. Wasik, 201. When it collapsed in 2001, Enron Energy became the largest bankruptcy in US history, with $11 billion in shareholder losses. Enron used a fraudulent accounting practice similar to "watered stock," which it called "marking to market," in which it immediately claimed the future profits of an entity at their current value. Likewise, Enron hid operating and trading losses in layered holding companies and deceitful intercompany transactions.

65. *ES*, 6 April 1932, 4. The version of events included in the finding aid for the CMP collection (no. 2115) at the Maine Historical Society is as follows: "When the Insull System collapsed following the 1929 stock market crash and the major bank failures in 1931–32, Wyman recognized that Middle West's control of Nepsco also meant control of Maine operating subsidiaries employing thousands of workers. To prevent liquidation and to keep the Maine

firms out of the hands of speculators, Wyman organized friendly investors [such] as the Northern New England Company to purchase Nepsco stock with its control of CMP at greatly depreciated prices, thus insuring jobs for Maine workers. Nepsco and the Northern New England Company liquidated on April 14, 1953." Chuck Rand, *Guide to the Central Maine Power Company Archival Collection, 1853–2001* (Maine Historical Society, December 7, 2011), 7, https://www.mainehistory.org/PDF/findingaids/Coll_2115.pdf. See also Smith, *Power Policy*, 74–79; and Smith, "Regulation of Some Holding Companies."

66. *Cooper-NPPC-36*, 3, and appendix B. See note 67.

67. Smith, *Power Policy*, 78, quoting comments made in 1945 by Securities and Exchange Commissioner Sumner Pike, a native of Lubec, Maine (and the brother of Moses Pike). As GE's president Owen D. Young had testified in February 1933, even Insull had been unable to comprehend the crazy patchwork of pyramid companies that was Middle West. Funigiello, 24–25, citing *Hearings Pursuant to S. Res. 84, Stock Exchange Practices, U.S. Congress, Senate*, 72nd Cong., 2nd Sess., 6 pts. (1932–1933), pt. 5, 1516–19; Davis, 21–29; and Landis, 28. See also Smith, "Regulation of Some Holding Companies."

Smith defends the Insulls, Wyman, and CMP on the basis that they brought millions of investment dollars into Maine, which helped to keep Maine's textile and paper mills in operation (and some 4,200 residents employed) at a time when they were in deep financial trouble. Specifically, after the Insull meltdown, a group of investors led by Walter Wyman, and including Stone & Webster and General Electric, purchased (at deeply discounted prices) most of the non-utility assets acquired by the Insulls in Maine and then ran them profitably for a number of years, helped in part by the improving wartime economy. These companies were also purchased to ensure a market for CMP's power and to gain control of the waterpower rights owned by the companies. From the discussion in Smith, one gets an impression of the benevolence and skill of the Wyman-led rescuers of Eastern Manufacturing Company (EMC), but Smith appears unaware that EMC's profitability was at the expense of the public and EMC's competitors, for CMP was selling power to EMC at one-half to one-third the price it was charging everyone else. Bangor Hydro-Electric did the same for manufacturing companies it controlled. *Cooper-NPPC-36*, 6, and appendix B.

68. *ES*, 16 March 1938, 9.

69. Dexter expected to be in Russia for six months. Gertrude and the children would stay in Concord, Massachusetts.

70. *ES*, 10 February 1932, 1.

71. McDonald, 272–73, 309–11.

72. Chicago's population had increased from about 1.1 million (1890) when Insull arrived in 1892, to 3.4 million (303%) in 1930. In the same period, the

total population of Maine had increased only 21% from 661,086 to 797,432. McDonald, 55; "Population Since 1741," *Maine: An Encyclopedia*, https://maineanencyclopedia.com/population-since-1741/; and Wikipedia, s.v. "Demographics of Chicago," https://en.wikipedia.org/wiki/Demographics_of_Chicago.

73. Franklin D. Roosevelt, "Commonwealth Club Address," 23 September 1932, San Francisco, CA, https://teachingamericanhistory.org/library/document/commonwealth-club-address-2/.

74. Wikipedia, s.v. "Theodore Roosevelt Dam," accessed 16 June 2020, https://en.wikipedia.org/wiki/Theodore_Roosevelt_Dam ; and Goodwin, *Bully Pulpit*, 352–53. On a similar transcontinental train trip in 1903, then president Theodore Roosevelt, a Republican, began framing his reformist vision for a "Square Deal." Within its righteous angles, the Square Deal sought fairness "for every man, great or small, rich or poor," Black, White, or Red, and would "give every man his due and not more than his due." That slogan encapsulated T. R.'s effort to address the great problem of his day: the stunning consolidation of dozens of industries into overly powerful "trusts." These trusts controlled what he called "invisible empires" or "combinations" of corporations and industries that corrupted public officials, resulting in economic injustice and the impoverishment of the people.

It was also during this trip that T. R. visited the Grand Canyon with John Burroughs and the great redwood forests of Yosemite with John Muir—strengthening his resolve to protect America's great natural resources: "Keep it for your children, your children's children and for all who come after you." T. R.'s motivation was "not to preserve forests because they are beautiful—though that is good in itself—not to preserve them because they are refuges for the wild creatures of the wilderness—though that too is good in itself . . . but to guarantee a steady supply of timber, grass, and above all, water."

T. R. began turning his resource conservation ideas into laws when he signed the Reclamation Act of 1902, putting the federal government into the irrigation business, and four years later, the electricity generating business. The legislation authorized the Army Corps of Engineers to build its first sizable dam, on the Salt River, in the not-yet state of Arizona. At a breathtaking 280 feet high, it was rivaled only by the dam being built on the Nile River in Egypt. In 1913, Hugh Cooper was hired by the Egyptian government to retroactively design a hydroelectric power plant for the Nile dam. American Society of Civil Engineers, "Hugh Lincoln Cooper," accessed 4 August 2022, https://www.asce.org/about-civil-engineering/history-and-heritage/notable-civil-engineers/hugh-lincoln-cooper.

75. "Power: Protection of the Public Interest, Governor Franklin D. Roosevelt's Speech at Portland, Oregon, September 21, 1932" (issued by the Demo-

cratic National Committee, Hotel Baltimore, New York City), 3, http://www
.fdrlibrary.marist.edu/_resources/images/msf/msf00530.

76. "Power: Protection of the Public Interest."

77. Riggs, 47, citing FTC, *Supply of Electrical Equipment*, xviii; and *Public Utility Commission of Rhode Island v. Attleboro Steam & Electric Co.*, https://supreme.justia.com/cases/federal/us/273/83/. See also Funigiello, 24n93: "Legal authorities, economists, and a number of utility commissioners were less confident [than Power Trust advocates] that state regulation had been effective. They noted that the Court in *Smyth v. Ames*, 169 U.S. 466 (1899), and in the 1927 Attleboro decision had cast doubt on whether a holding company was a public utility and had placed the interstate transmission and sale of energy beyond the reach of public-service commissions. They also observed that the absence of arm's-length transactions between the parent company and its affiliate made regulation of banking transactions difficult; that accounting practices were neither uniform nor public; that only one half of the state commissions were empowered to regulate security issues; that service contracts were, in the main, unregulated; and that only four states even admitted the relationship between depreciation and dividends."

78. "Power: Protection of the Public Interest."

79. McDonald, 117. "The number of supporters of outright nationalization [in the United States] was negligible." Funigiello, 21, 62, in reference to the uproar in 1929–1931 about FTC's revelations of utility holding company corruption and what to do about it.

80. In 1931, the Army Corps of Engineers had issued a plan and sought funding for ten dams and navigational locks on the Columbia River but no power plants, in keeping with President Hoover's policy of private power development rather than public. Riggs, 111. On 31 March 1931, Hoover vetoed Senate Joint Resolution 49, which called for the federal government to operate Muscle Shoals as a public power and fertilizer production facility. "Veto of the Muscle Shoals Resolution," 3 March 1931, The American Presidency Project, https://www.presidency.ucsb.edu/documents/veto-the-muscle-shoals-resolution. Likewise, President Calvin Coolidge used a pocket veto to kill Senator Norris's bill for public operation of Muscle Shoals in May 1928. Funigiello, 35n12; and Riggs, 64.

81. Franklin Roosevelt's use of the term "yardstick" echoes the "measuring stick" phrase of muckraking journalist Ida Tarbell, who, in a 1905 exposé on collusion between railroads and the Standard Oil monopolies, endorsed an idea of Ohio governor Herrick to build a government-owned oil refinery to break the refining monopoly and to create a "measuring stick" to help determine the real cost of refining. Tarbell also recommended pipelines be

regulated as "common carriers" in a way similar to what would emerge as the electric transmission grid. Goodwin, *Bully Pulpit*, 438–39. Likewise, the phrase "People will love him for the enemies he has made" was used by the *New York World* in reference to the trusts that hated Theodore Roosevelt, "the trust-buster." Goodwin, *Bully Pulpit*, 400.

82. Riggs, 45, citing Insull, "Economics of the Public Utility Business"; McDonald, 203–5; Wasik, 109; and Munson, 52.

83. Samuel Insull left for France on 14 June 1932. McDonald, 307.

84. Wasik, 208–9. See McDonald, 315: "[O]n October 9, 1932, when the President [Hoover] of the United States asked Premier Mussolini of Italy to seize Insull when he passed through that country."

85. Funigiello, 31.

CHAPTER 8. DESPAIR AND HOPE (1929–1934)

1. Heymann, 15. See also Wikipedia, s.v. "Vladimir Lenin," https://en.wikipedia.org/wiki/Vladimir_Lenin#cite_note-FOOTNOTEFischer1964423,_582Sandle1999107White2001165Read2005230-367, citing Louis Fischer, *The Life of Lenin* (London: Weidenfeld and Nicolson, 1964), 423, 582; Mark Sandle, *A Short History of Soviet Socialism* (London: UCL Press, 1999), 107, doi:10.4324/9780203500279; James D. White, *Lenin: The Practice and Theory of Revolution* (Basingstoke: Palgrave, 2001), 165; Christopher Read, *Lenin: A Revolutionary Life* (London: Routledge, 2005), 230; and Vladimir Lenin, *Our Foreign and Domestic Position and Party Tasks* (Moscow, 1920). See also Dorn, "Hugh Lincoln Cooper," 322: "Soviet power plus electrification of the whole country, since industry cannot be developed without electrification."

2. Wikipedia, s.v. "Russian Famine of 1921–22," accessed 19 March 2019, https://en.wikipedia.org/wiki/Russian_famine_of_1921–22; and Wikipedia, s.v. "Vladimir Lenin," https://en.wikipedia.org/wiki/Vladimir_Lenin#CITEREFRead2005, citing Louis Fischer, *The Life of Lenin* (London: Weidenfeld and Nicolson, 1964), 507–8; Christopher Rice, *Lenin: Portrait of a Professional Revolutionary* (London: Cassell, 1990), 185–86; and James Ryan, *Lenin's Terror: The Ideological Origins of Early Soviet State Violence* (London: Routledge, 2012), 164.

3. Major General Lasing Hoskins Beach (1860–1945) was chief of the Army Engineers (1920–1924). Dorn, 327–28, citing *Muscle Shoals Propositions, House Comm. on Military Affairs*, 67th Cong., 2nd Sess. (8 February–13 March 1922), 409. See also Billington, Jackson, and Melosi, 118. "Just before Armistice Day, on November 9, 1918, construction resumed on

the dam while the Corps made more subsurface tests and, following Cooper's recommendations, began to make design sketches. Cooper, meanwhile, had gone back into private practice and then, on May 21, 1920, the Corps signed a contract with Hugh L. Cooper & Co. that put the company in charge of design, construction, and inspection of the entire project. Cooper began work, and by election day 1920 he had completed numerous drawings laying out the dam and powerhouse. From them until 1924, Cooper produced drawings and oversaw construction." Billington, Jackson, and Melosi, 118, citing Wegmann, *Design and Construction of Dams*, 8th ed. See, for example, drawings 3450–3452, dated 1 November 1920, and 3453, dated 23 October 1920. The Corps continued to make drawings into October 1920, but Cooper's work had by then become the official record.

4. Dorn, 330; and Watkins, 150.

5. Gruening, 204, quoted in a letter of 23 March 1929 from Mr. S. E. Boney, NELA's Carolina Committee on Public Utility Information, complaining to Mr. Rogers that his statement was "wholly lacking in truth." William Penn Adair Rogers (1879–1935) was a Cherokee citizen born in the Cherokee Nation, Indian Territory, now part of Oklahoma. He traveled around the world three times, made 50 silent films and 21 "talkies" and wrote more than 4,000 nationally syndicated newspaper columns. He died in 1935 with aviator Wiley Post (the first to fly solo around the world) when their experimental airplane crashed in northern Alaska.

6. Funigiello, 34, citing George W. Norris, *Fighting Liberal: The Autobiography of George W. Norris* (New York, 1945), 249–59; and John D. Hickes, *Republican Ascendancy, 1921–1933* (New York, 1960), 62–63. The fundamental reason no one obtained the Shoals in the 1920s, apart from Norris's dogged opposition, was that none of the offers from private interests satisfied Congress. Also, southern farm leaders voted against legislation that failed to guarantee cheap fertilizer. Private power interests fought the chemical industry for control of the site, while, together with their Wall Street affiliates, both attacked Ford's offer. See also Wengert, 145ff.; and Funigielllo, 34n9.

7. Dorn, 330. "Senator Norris replied that according to Ford's proposed method of paying debts the $236 million of public property that Ford would acquire in the transactions would, after a century, be worth $14.5 billion. Sward, *Legend of Ford*, 128–30.

8. Hughes, 293–95, citing Preston J. Hubbard, *Origins of the TVA: The Muscle Shoals Controversy, 1920–1932* (Nashville, TN: Vanderbilt University Press, 1961), 28–146; and "Wilson Dam and Reservoir," accessed October 2020, http://www.encyclopediaofalabama.org/article/h-3268.

9. Dorn, 331, citing Norris, 250–51, 261. See also Billington, Jackson, and Melosi, 114.

10. Dorn, 330, citing Allan A. Benson, *The New Henry Ford* (New York, 1923), 212.

11. Newton D. Baker was secretary of war in the Wilson administration (1916–1921), overlapping with Franklin Roosevelt's tenure as assistant secretary of the navy (1913–1920).

12. Watkins, 149. Quote from 1924.

13. Dorn, 330–31, citing *Muscle Shoals Propositions* [see note 3], 409–29, esp. 410 and 424. "H. L. Cooper Opposed Ford Proposal for Muscle Shoals," *Engineering News-Record* 88 (1922): 335; and *NYT*, 18 February 1922, 12.

14. Dorn, 331, citing *NYT*, 2 October 1925, 32.

15. Riggs, 53, citing Preston J. Hubbard, *Origin of the TVA: The Muscle Shoals Controversy, 1920–1932* (Nashville, TN: Vanderbilt University Press, 1961), 12–16, 21–23.

16. Hughes, 287, citing Preston J. Hubbard, *Origin of the TVA: The Muscle Shoals Controversy, 1920–1932* (Nashville, TN: Vanderbilt University Press, 1961), 7: "The Muscle Shoals nitrate plants was completed just before the war ended in November 1918. Close to $70 million had been invested in it, but no nitrates from it had been used during the war. A transmission line had been erected to bring power from a steam power station because the dam and hydroelectric station were not yet completed. The government continued construction of these until 1925, by which time an additional $45 million had been invested."

17. *ES*, 12 January 1938, 1. Brewster: "Mr. Ford was at one time deeply interested in, and thoroly [*sic*] informed about it. In fact, he gave out the very first interview concerning Mr. Cooper's plan,—before Mr. Cooper himself had announced it. That was just after Cooper and the Ford engineers had been in contact at Muscle Shoals, and just before he withdrew his offer to buy Muscle Shoals from the Government."

18. Holmes, 8, citing *BDN*, 21 November 1924.

19. The contract was claimed to be worth $40 million. Wilson, 162; and Siegel, 106.

20. L'Hommedieu, "MacNichol, David."

21. Dorn, 335, citing *NYT*, 15 February 1927, 6.

22. The Dnieper River is 1,400 miles long and the fourth largest river in Europe. Dorn, 333, citing *Russian Review* [later *Soviet Union Review*] 1 (1923): 104; Isaac Deutscher, *The Prophet Unarmed* (London, 1959), 210–11; and Leon Trotsky, *My Life* (New York, [1930] 1970), 519. Dorn: "What role, if any, Trotsky played in expediting the project remains uncertain. He was removed from the electrotechnical board late in 1926." *NYT*, 3 December 1926, 4. Lenin died in January 1924.

23. Dorn, 335–36, citing H. R. Knickerbocker, *The Red Trade Menace* (New York, 1931), 179.

24. Cooper, "Soviet Russia."

25. Cooper, "Soviet Russia"; Siegel, 86; Dorn, 337, citing American-Russian Chamber of Commerce, comps., *Handbook of the Soviet Union* (New York, 1936), 202. Heymann, 18, says the four turbines were built by the Soviets.

26. "According to a 1933 State Department description, Hugh Cooper was "hard boiled, clear headed, a veteran engineer, strongly anticommunist but eager to make money and achieve professionally," who probably had been allowed to see only the best aspects of life in Russia. Wilson, 90–91, citing Sussdorff (Riga) to State Department, 26 September 1928, US Department of State, Records Relating to International Affairs of Russia and the Soviet Union, 10 January 1929, Record Group 59, files 861.6463/31, 861.6463/34; *Annals of American Academy of Political and Social Sciences* 138 (July 1928); 119; *Hearings on H.R. 15597, H.R. 15927, H.R. 16517, House Comm. on Ways and Means*, 95; *NYT*, 11 July 1933, 16; *Complete Presidential Press Conferences*, 1:143–44; and J. G. Rogers (Assistant Secretary of State), memorandum, May 5, 1931, US Department of State, Political Relations Between the U.S. and Russia and the Soviet Union, Record Group 59, file 711.61/233.

27. Wikipedia, s.v. "GOELRO," https://en.wikipedia.org/wiki/GOELRO, citing Lenin, *Collected Works*, 30: 335. GOELRO is the transliteration of the Russian abbreviation for State Commission for Electrification of Russia. "Collectivization" of farms, usually forced, was also part of the first Soviet Five-Year Plan, particularly in the Ukraine, "the breadbasket of Europe." The results were disastrous. Farm production decreased immediately, and eventually to less than half, causing widespread famine and the death of seven to fourteen million people. Industrialization exacerbated food shortages as labor shifted away from food production. Some view the collectivization, forced migrations, and famine as Soviet genocide of Ukrainians.

28. An article in *Pravda* (Moscow), 28 January 1933, described the economic devastation in the capitalist countries and then crowed, "The victorious fulfillment of the Five-Year Plan in the U.S.S.R. in four years, in juxtaposition with the unprecedented crisis in the capitalist countries, proves that only a *socialist system*, which is not afraid of any crisis, can assure a constant progress of national economic life and a systematic improvement of the standard of living of the laboring class and of the toiling masses of the villages." NA-DC: State Department: 1249-6: 1933-01-28_First year of Second 5-Year Plan.

29. Heymann, 51–52. According to Wilson, 162, Ford supplied 85% of the Russian market.

30. Heymann, 54–55. Among the other major American companies doing business in Russia at this time was the DuPont chemical company. Siegel, 131.

31. Siegel, 133–34. This included 97% of the hydroelectric turbine exports.

32. Wilson, 162–63.

33. Cooper, "Soviet Russia."

34. "The uncommon business argument was that communism resembled capitalism because the Communist state was a 'gigantic corporation', in which all working citizens were considered stockholders. Communism was also sometimes described by U.S. businessmen as capitalism 'extremely magnified,' the 'most extreme type' of capitalism in which the government ended up being the only capitalist. Comparisons of monopoly capitalism with state socialism have become more common in recent years." Siegel, 12.

35. Dorn, 339, citing *NYT*, 28 February 1929, 5, and 6 July 1931, 1–2. For Stalin's speech outlining his program, see *Soviet Union Review* 9 (1931): 146–54.

36. Dorn, 340, citing *NYT*, 10 July 1931, 5.

37. Dorn, 340, citing *Christian Science Monitor*, 31 December 1931, 5.

38. Dorn, 337, citing American-Russian Chamber of Commerce, comps., *Handbook of the Soviet Union* (New York, 1936), 202.

39. Dorn, 337, citing *Soviet Union Review* 10 (1932): 212; *NYT*, 27 August 1932, 2; and *NYT*, 10 October 1932, 1.

40. Klein, *Beloved Island*, 143–44, citing *LH*, 29 June 1933, 1.

41. See Klein, *Beloved Island*, 147, for a telling story about FDR's emotional return to Campobello.

42. Klein, *Beloved Island*, 141, citing *LH*, 22 June 1933, 1.

43. Klein, *Beloved Island*, 143, citing *LH*, 29 June 1933, 8.

44. Amy Davidson Sorkin, "The F.D.R. *New Yorker* Cover That Never Ran," *New Yorker*, 5 May 2012.

45. *ES*, 5 July 1933, 1. Full repeal of the Twenty-first Amendment came on 5 December 1933.

46. Holmes, 37, citing *ES*, 18 June 1933.

47. Watkins, 113.

48. Watkins, 150.

49. More precisely 73% of the private utilities, which amounted to 97% of the entire electric generating industry. Wikipedia, s.v. "Public Utility Holding Company Act of 1935," accessed 19 September 2019, https://en.wikipedia .org/wiki/Public_Utility_Holding_Company_Act_of_1935#cite_note-4, citing Leonard S. Hyman, *America's Electric Utilities: Past, Present and Future* (Public Utility Reports, 1988), 74.

50. Watkins, 150–51.

51. It was also the "Giant Power" vision that Gifford Pinchot and Morris Cooke had for Pennsylvania. Hughes, 297–313.

52. Riggs, 97.

53. Watkins, 114–15.

54. *ES*, 8 March 1933, 4. "The popularity of President Roosevelt and the hope that the 'New Deal' might in some way improve conditions in Quoddy played a powerful but wholly unforeseen part in the municipal elections here Monday and swept into the office of Mayor of Eastport a young man who had sought neither nomination nor election and is probably not entirely pleased with the result. In winning, Mr. Reilly polled the largest vote ever accorded a candidate for mayor here." *ES*, 8 March 1933, 1.

55. *ES*, 8 March 1933, 4.

56. *ES*, 22 March 1933, 4.

57. *ES*, 5 April 1933, 4. See also *ES*, 4 June 1930. Donald Randall Richberg (1881–1960) was instrumental in creating the New Deal as coexecutive director of the National Recovery Administration and coauthor of the Railway Labor Act, the Norris-LaGuardia Act (anti-injunction), and the Taft-Hartley Act (labor-management relations). He was also Harold Ickes's law partner in Illinois when they sued (1913–1916) Insull over excessive electric rates. McDonald, 181, 259.

58. McDonald, 313, 315.

59. Wasik, 212. The deadline for departure was 1 January 1934.

60. Wasik, 213.

61. Wasik, 209–10.

62. McDonald, 317, described this as "kidnapping."

63. Wasik, 210.

64. Wasik, 218.

65. Wasik, 221.

66. Wasik, 227.

67. Wasik, 221.

68. McDonald, 332.

69. Cole Porter's musical comedy *Anything Goes* debuted in 1934. Or, as John P. Coughlan of Pacific Gas and Electric put it, corporate morals in the twenties were no worse than individual morals. *Or at least some individuals' morals.* Funigiello, 26, citing John P. Coughlan, "Public Utilities and the Public," *National Electric Light Association Bulletin* 19 (December 1932): 712.

70. Weil, 22.

71. Wikipedia, s.v. "Fireside Chats," https://en.wikipedia.org/wiki/Fireside_chats.

72. Display at Roosevelt-Campobello International Park, September 2018.

73. Watkins, 126.

74. Watkins, 126–27.

75. FDRL-0284: 1933-04_Frederick Hale to FDR_suggestion Quoddy for work relief. Hale enclosed a favorable report by British authorities about the

Severn River tidal power project. FDRL-COOPER: Frederick Hale to Franklin Roosevelt, 10 April 1933. FDR responded positively on 10 April 1933, saying that he would investigate. Hale also forwarded to FDR a letter from Hugh Cooper on 21 April 1933.

76. FDRL-Day-by-Day: 1933-04-14, also when FDR announced his sailing trip to Campobello.

77. FDRL-COOPER: On 31 March 1933, Senator Norris recommended Mr. McIntyre meet Hugh Cooper in relation to the question of diplomatic recognition of Russia. McIntyre sent for Cooper on 6 April. On 15 April, Robert Anderson Pope also recommended Hugh Cooper to President Roosevelt as the best person to talk with about recognition of Russia. See also FDRL-COOPER: 1933-04-16_McIntyre to Hugh Cooper_Norris suggests meeting president.

78. Dorn, 344, citing *NYT*, 11 July 1933, 16, and 21 October 1933, 2. See also Ickes, *Secret Diaries*, 111, for 21 October 1933. Norris was apparently instrumental in arranging Cooper's meeting with Roosevelt: "I was so impressed with Colonel Cooper that later I made arrangements through which he talked with the President; and I think that the several years of personal contact with the Russians not only was extremely valuable to him, but turned out to be valuable to American officials called upon to deal with the Russian government in later years." Norris, 370. Roosevelt publicly denied that Cooper had advocated recognition, insisting that their discussion was confined to commercial prospects in connection with American-Russian trade. *Complete Presidential Press Conferences*, 1:143–44. "Cooper's influence on the new administration was probably greater than it had been under the previous Republican governments." Wilson, *Ideology and Economics*, 104–5, 127–8.

79. Dorn, 345, citing *NYT*, 18 November 1933, 2.

80. Vyacheslav Mikhailovich Molotov (1890–1986) was chairman of the Council of People's Commissars (1930–1941) and minister of foreign affairs (1939–1949, and 1953–1956). Molotov was the principal Soviet signatory of the German-Soviet non-aggression pact of 1939 (the Molotov-Ribbentrop Pact), whose secret provisions called for the invasion of Poland and partition of its territory between Nazi Germany and the Soviet Union.

81. The reports were made to the US consul in Berlin, Germany, on Hugh's return trip to the United States.

82. Maxim Litvinoff (1876–1951) (also "Litvinov") was Russia's commissar for foreign affairs (1930–1939) and ambassador to the United States (1941–1943).

83. Leuchtenburg, 206–7, citing William Philips, *Adventures in Diplomacy* (Boston, 1952), 157.

84. FDRL-COOPER: 1933-11-21_Telegram from Hugh Cooper to FDR inviting to celebration at Waldorf Hotel in NYC.

85. Funigiello, 256, citing Rosenman, comp., *The Public Papers and Addresses of Franklin D. Roosevelt*, 13 vols. (New York, 1938–1950), 1, 639–46.

86. Riggs, 98, citing Ellsworth Barnard, *Wendell Willkie: Fighter for Freedom* (Marquette: Northern Michigan University Press, 1966), 43; and Kenneth S. Davis, *FDR: New Deal Years, 1933–1937* (New York: Random House, 1986), 93.

87. Watkins, 151.

88. Watkins, 154; and Riggs, 106. Willkie feared low retail rates, selling lots of electrical appliances to increase demand, and federal ownership of the distribution network.

89. This might be the meeting described in the *Sentinel*: "Col. Cooper was a guest of the President at the White House some weeks ago, and the Quoddy project was the subject under discussion on that occasion. Whether Col. Cooper was called to Washington or whether he went there for the purpose of urging upon him the availability of his brother's project has not been announced. It is generally believed here that the former was the case. At any event, Col. Cooper advised Dexter that the President was deeply interested, and Dexter's immediate arrival in this country, a month before his wife and children could leave England, was the next step." *ES*, 26 July 1933, 1.

90. *ES*, 26 July 1933, 1; and *ES*, 16 March 1938, 9. Cooper returned from Washington on 7 August 1933.

91. *ES*, 23 August 1933, 1.

92. *ES*, 26 July 1933, 1. An increase from the 300,000 hp reported in *ES*, 16 July 1930, 1.

93. *ES*, 26 July 1933, 1.

94. FDRL-COOPER: 1933-04-19_Hugh Cooper to Hale_views on Quoddy. However, Hugh Cooper did lend $5,000 to Dexter P. Cooper, Inc., at some point before the PWA's audit in May 1935. *Final-36*: appendix A-3, schedule 2.

95. *ES*, 12 April 1933, 1; and *ES*, 19 April 1933, 1.

96. *ES*, 19 April 1933, 1. See also Cross, 1–24.

97. *ES*, 28 June 1933, 1.

98. Riggs, xxix.

99. Klein, *Beloved Island*, 143–44, citing *LH*, 29 June 1933, 1.

100. FDRL, FDR Speech File: "Informal Extemporaneous Remarks of the President on the Occasion of His Visit to Campobello Island, New Brunswick, June 29, 1933," file msf00658. A slightly different version (noted in brackets) is in *ES*, 5 July 1933, 8.

101. Klein, *Beloved Island*, 144, citing Lash, *Eleanor Roosevelt on Campobello*, Campobello, Roosevelt Campobello International Park (no date), 9.

102. *ES*, 26 July 1933, 1. However, "Cooper then went to Russia where he worked with his brother, Hugh Cooper, on the Dnieperstroy Dam. After that

he went to England. While in England, he received a letter from Senator Hale of Maine urging him to come to Washington and submit his Quoddy plan as a P.W.A. project. Cooper went to Washington in July 1933." *Reel-38,* 2.

103. *ES,* 16 March 1938, 9.

104. *ES,* 26 July 1933, 1, citing radio reports that Dexter "and his brother Hugh had been in consultation with the President." FDR-Day-by-Day does not list any meetings between FDR and Dexter Cooper in July or August 1933, but Harold Ickes appears on the calendar three times on 26 July, so it is possible that Dexter participated in one of those meetings.

105. *ES,* 26 July 1933, 1.

106. Edgerton was the district engineer of the Rock Island District (in which the Keokuk Dam is located) in 1930–1933. *Gateways to Commerce,* 29, 162, 163.

107. *Reel-38,* 2–3.

108. *Final-36*: appendix A-3, narrative p. 7.

109. *Final-36*: appendix A-3, narrative p. 7.

110. GSC, "Passamaquoddy," 8.

111. *ES,* 16 March 1938, 9. Cooper returned from Washington on 7 August 1933. Holmes, 37, citing *LEJ,* 29 August 1933.

112. *ES,* 9 August 1933, 1.

113. *ES,* 9 August 1933, 1.

114. *Final-36*: appendix A-3, exhibit I, 1–6.

115. *ES,* 30 August 1933, 1. To ensure accurate placement of dam fill, Dexter planned to use bottom-dump barges, which would be positioned with the help of "pilot" barges anchored on either side acting as guides. *Cooper-FPC-28,* M-9.

116. *ES,* 13 September 1933, 1. This was quoted by Roscoe in *American Magazine,* December 1935, 159.

117. Ralph Waldo Emerson, "Civilization," 1870, https://www.rwe.org /chapter-ii-civilization/.

118. *ES,* 6 December 1933, 1; *ES,* 13 December 1933, 1, 3; and *ES,* 20 December 1933, 1. See also US Congress, Senate, *Report on the Passamaquoddy Tidal Power Project, Maine,* S. Doc. No. 41, at 3 (1941).

119. *ES,* 22 November 1933, 7.

120. EMERY_6.17: undated.

121. The power trusts' influence over Maine's Public Utilities Commission was a continuing concern. Several towns were at the time preparing to challenge Bangor Hydro-Electric's rates based on information made public by the FTC investigation. See *ES,* 1 November 1933, 1; *ES,* 15 November 1933, 1, 7; *ES,* 22 November 1933, 4; and *ES,* 29 November 1933, 4.

122. EMERY_6.21: letter from Roscoe Emery to Dexter Cooper, 5 Oct 1933.

123. LHS_1936-07-08_The stock certificate book for the Canadian Dexter P. Cooper Company. Numbered certificates are as follows:

1. 6 shares to Dexter P. Cooper, Welshpool, Campobello, New Brunswick, 14 May 1927.
2. 1 share to Gertrude S. Cooper, Welshpool, Campobello, New Brunswick, 14 May 1927.
3. 1 share to Mehille N. Cockburn (?), St. Stephen, New Brunswick, 14 May 1927.
4. 1 share to George H. I. Cockburn, St. Stephen, New Brunswick, 14 May 1927.
5. 1 share to E. Vincent Sullivan, St. Stephen, New Brunswick, 14 May 1927.
6. 1 share to John E. Scovil, St. Stephen, New Brunswick (Director), 15 May 1928.
7. 1 share to Harold H. Murchie, Calais, Maine (Attorney), 15 May 1928.
8. 9 shares to Dexter P. Cooper, Inc., Eastport, Maine, 15 May 1928. "Cancelled" "Note: The word 'cancelled' above was written by mistake for No. 9. (80 shares) on the following stub, D.P. Cooper, 12/28/35."
9. 80 shares to Dexter P. Cooper, Inc., Eastport, Maine, 15 May 1929. "Cancelled—[unclear]—Issued by Mistake, DPC." See also *Final-36*: appendix A-3, narrative p. 8.
10. [no issuing information—stock certificate missing from book]
11. [no issuing information—stock certificate missing from book, but located in private collection in Campobello]

The rest of the certificates are unissued and still in the stock book.

At the time of the PWA's audit in May 1935, DPC, Inc., owed Muchie $4,263.87. *Final-36*: appendix A-3, schedule 4.

124. EMERY_6.21.19: letter from Dexter Cooper to Roscoe Emery, 7 October 1933.

125. EMERY_6.21.18: letter from Roscoe Emery to Dexter Cooper, 28 October 1933.

126. FDRL-ROOSEVELT, James: 1937-09-16_James Roosevelt to Dexter Cooper_SEC and financing concerns.

127. In 1942, the Reconstruction Finance Corporation loaned Henry J. Kaiser $110 million to build the Fontana steel mill near Los Angeles, California, despite repeated attempts by U.S. Steel and others in "Big Steel" to protect

their monopoly. Kaiser accused RFC officials of deliberately impeding his application and eventual operation, even during World War II, in order to favor U.S. Steel. In 1946, Kaiser called for a Senate investigation thereof, as those same companies were trying to "freeze him out of the steel and other commodities he needs desperately for his new venture in automobiles." Kaiser also helped to build the Bouder, Bonneville, and Grand Coulee Dams, for which he recived $110 million in contracts. All told, Kaiser's various companies received close to $2 billion from the US government over a twenty-year period. The most famous of his enterprises at the time was his shipyard, where he was launching up to thirty-two ships per month during World War II. Gunther, *Inside U.S.A.*, 65–67, 73, and citing *NYT*, 8 August 1946.

128. *Report of the Board of Review to Col. H. M. Waite, Deputy Administrator*, reprinted in appendix C, "Passamaquoddy Bay Tidal Power Commission Report," 30 March 1934, p. 1. FDRL-OF_982, box 1.

129. Marvin, 10, citing US Congress, Senate, *Report on the Passamaquoddy Tidal Power Project, Maine*, S. Doc. No. 41, at 4–5 (1941).

130. *ES*, 16 March 1938, 9. Cooper returned from Washington on 7 August 1933.

131. Holmes, 37, citing *LEJ*, 31 October 1933.

132. *ES*, 15 November 1933.

133. *ES*, 15 November 1933, 1; and *ES*, 16 March 1938, 9.

134. *ES*, 8 November 1933, 4.

135. See chapter 7, "Blame."

136. Ruby Aurora Black (1896–1957) became a White House correspondent for United Press in 1933 and covered Eleanor Roosevelt's activities for the next seven years. She wrote *Eleanor Roosevelt: A Biography*, the first biography of the First Lady, in 1940. "Ruby Aurora Black (1896–1957)," Texas State Historical Association, accessed 21 August 2019, https://tshaonline.org/handbook/online/articles/fbl61.

137. Henry M. Kannee, one of FDR's secretaries.

138. Ernest Henry Gruening (1887–1974) was a graduate of Harvard Medical School but switched to journalism. He started with the *Portland Evening News* in 1927 and then worked as editor of the *New York Post* in 1934 before joining Roosevelt's New Deal team in the Department of the Interior. He was appointed governor of the Alaska Territory (1939–1953) and was elected US senator from the state of Alaska (1959–1969).

139. FDRL-0284: 1933-11-02_Ruby A. Black to Eleanor Roosevelt. (Hereafter cited as *Ruby Black-33*.)

140. The finding aid for the CMP Collection (no. 2115) at MHS says, "Financing growth and expansion forced Wyman into discussions with the Insull brothers, Samuel and Martin, who held financial resources in their Middle

West Utilities holding company. On June 26, 1925 Wyman reported that Middle West was offering $140 for each share of CMP's common stock. This offer was accepted almost unanimously by the stockholders. The New England Public Service Company (Nepsco) organized as a subsidiary of Middle West provided the money for plant expansion as it was needed. One of Middle West's 14 constituent companies was the National Electric Power Company which in turn owned the New England Public Service Company. Nepsco was led by Wyman as president which in turn owned Central Maine Power, Cumberland County Power and Light, Central Vermont Public Service, Public Service of New Hampshire, and National Light Heat and Power—which in turn owned Twin State Gas and Electric." Chuck Rand, *Guide to the Central Maine Power Company Archival Collection, 1853–2001* (Maine Historical Society, December 7, 2011), https://www.mainehistory.org/PDF/findingaids/Coll_2115.pdf.

141. The *Portland Press Herald* was formed in a merger of the *Portland Daily Express* (owned by Frederic Hale) and the *Portland Herald* in 1921. Wikipedia, s.v. "*Portland Press Herald*," https://en.wikipedia.org/wiki/Port land_Press_Herald, citing "Press and Herald Join in Portland," *Editor & Publisher, 21 November 1904.*

142. *Ruby Black-33.*

143. The finding aid for the CMP Collection (no. 2115) at MHS says, "Because it was anticipated that the Wyman Station would generate surplus power, a larger market for this surplus power had to be developed. When efforts to obtain authorization to sell this power outside the state failed in 1929 (the state electorate defeated the Smith-Carlton Bill in September 1929 that would have allowed the export of power across state boundaries, thus maintaining the Fernald Law and the Export Ban Act), Wyman proposed the construction of a paper mill in Bucksport as a customer for this energy. The Maine Seaboard Paper Mill was established." Rand, *Guide to the Central Maine Power Company Archival Collection.*

144. Blaine Spooner Viles (1879–1943), a Republican, was at times mayor of Augusta, a Maine state representative and senator, and a member of the governor's council.

145. *Ruby Black-33.*

146. *Ruby Black-33.*

147. *Ruby Black-33.*

148. FDR-Day-by-Day, 16 November 1933.

149. EMERY_6.21: Roscoe Emery to Dexter Cooper, 20 November 1933.

150. EMERY_6.21c.2: 21 November 1933 letter from Emery to Baxter. See also *ES*, 22 November 1933, 1; and *ES*, 29 November 1933, 4.

151. *ES*, 20 December 1933, 1.

152. *ES*, 10 January 1934, 1. See also Holmes, 38, citing letter from the FPC to Federal Emergency Administrator of Public Works, 3 January 1934.

153. "Load factor" is essentially the average utilization.

154. US Congress, Senate, *Report on the Passamaquoddy Tidal Power Project, Maine*, S. Doc. No. 41, at 1 (1941).

155. *ES*, 10 January 1934, 1.

156. Smith, *Power Policy of Maine*, 237, citing *Report on the Passamaquoddy Tidal Power Project*, S. Doc. No. 41, at 1, 3.

157. FDRL-0284: 1934-02-07_E.C. Moran to FDR.

158. FDRL-0284: 1934-02-07_E.C. Moran to FDR. Roscoe theorized the power trust wanted the federal government to pay for the studies to develop more hydroelectric power in Maine and to purchase the land that would be flooded as a result, thus relieving the power trusts of those financial burdens. *ES*, 10 January 1934, 4. See also *Ruby Black-33* for more on Moran versus the power trust.

159. *Reel-38*, 3. See also FDRL-COOPER: 1934-01-24_Dexter via James Roosevelt asks for meeting with FDR. See also FDRL-PPF-3: 1934-01-24_Dexter telegram to White House saying delayed due to snow (Dexter appears on FDR's appointment calendar for 18 January 1935 at 11:00). FDRL-Day-by-Day.

160. Waite resigned on 23 January 1936 because of a power struggle with his boss, Secretary Ickes. Ickes, 130, 141–42, 193–94, 333.

161. *ES*, 21 February 1934, 4. Senators Frederic Hale and Wallace H. White were also involved in the effort (though apparently not in the meeting with the president), as was Representative Edward C. Moran. Interestingly, Maine's chief justice Pattangall may also have been present. See also *ES*, 28 February 1934, 3; and *ES*, 16 March 1938, 9.

162. *ES*, 21 March 1934, 1.

163. *ES*, 21 March 1934, 1, 4; and *ES*, 16 March 1938, 9. This report (apparently) was paid for by twelve private power companies. *BH*, 8 April 1935, 10.

164. EMERY_6.17: letter to Joseph P. Preston, Webber Lumber & Supply Co., Fitchburg, Mass., 19 March 1934.

165. Except Carroll L. Beedy (1880–1947), a US representative (1921–1935). *ES*, 4 April 1934, 1, 7.

166. *ES*, 4 April 1934, 1, 4, 7: "Cooper said he expected to begin with a production of 154,000 kilowatts, increasing to 234,000, and that this would be stepped up when the international project is built to more than 600,000 kilowatts ultimately reaching a maximum of 2½ billion kilowatt hours." Dexter Cooper's strategy to use Quoddy's surplus power by making metals was similar to Walter Wyman's strategy for using surplus power from Wyman Dam for making paper. See note 143.

167. Dexter also explored the possibility of a $30 million loan request to the PWA, which was still under consideration. *BH*, 8 April 1935, 10.

168. *ES*, 4 April 1934, 1, 7.

169. *ES*, 4 April 1934, 1, 7. *Final-36*: appendix A-3, exhibit I. See also *Reel-38*, 9.

170. See the correspondence between Dexter and his investors in *Final-36*: appendix A-3, exhibit I.

171. *ES*, 23 May 1934, 1; and SILLS: Dexter Cooper to Kenneth Sills, 19 Sep 1934. See also FDRL-0982: 1934-06-06_Dexter Cooper to FDR_market for power.

172. *ES*, 4 April 1934, 1, 7.

173. *ES*, 4 April 1934, 1, 7; and Klein, *Call to Arms*, 160. Alcoa was controlled by Andrew W. Mellon (1855–1937). Born into a wealthy family in Pittsburgh, Pennsylvania, he made his first fortune in banking and then invested in emerging industries including steel, aluminum, oil, and electricity. He served Presidents Harding, Coolidge, and Hoover (all Republican) as secretary of the treasury (1921–1932).

174. *Cooper-NPPC-36*, 6. To solve the problem, Dexter proposed the transmission system be government owned and operated as a common carrier, where preferential pricing was not allowed.

175. *ES*, 4 April 1934, 1, 7.

176. Frederic A. Delano was a founding member of the Federal Reserve Board of Governors (1914–1918) and vice chair (1914–1916).

177. *ES*, 4 April 1934, 1, 7.

178. SILLS: "A Brief upon the 'Quoddy Project': An Outline of the Case" (undated, but 1934); and FDRL-DELANO: 1934-05-18_Delano to FDR_ Quoddy Brief.

179. FDRL-DELANO: A.E. Morgan to F. A. Delano, 19 May 1934.

180. EMERY_8.33: Roscoe Emery to M. A. LeHand, 5 May 1934.

181. EMERY_8.33: M. A. LeHand to Roscoe Emery, 8 May 1934.

182. EMERY_8.33: Roscoe Emery to M. A. LeHand, 14 May 1934.

183. EMERY_8.33: M. A. LeHand to Roscoe Emery, 18 May 1934.

184. Signed on 18 June 1934; aka Wheeler-Howard Act.

185. Watkins, 134.

186. Watkins, 134–36.

187. In early May 1934, the *BDN* quipped, "The only chance we see for Engineer Cooper is to take that Quoddy scheme under his arm and carry it down to Tennessee or Alabama. Down there, or even in Colorado, it might sprout." A mere two weeks later, a group of engineers from Chicago applied to the PWA for a $20 million loan with which to build an aluminum plant at

Muscle Shoals—specifically to "break Mellon's hold" and "make aluminum cheaper." *ES,* 23 May 1934, 1, 4.

188. Marvin, 15, citing *LEJ,* 3 May 1934. Moran also said, "The 'New Deal' calls for breaking the aluminum trust, not succumbing to it," and "newspapers have delivered an obituary [for Quoddy]. But in my opinion now is the time to begin to fight, and carry the fight to the court of last resort. This is a call to arms to the people of Maine." *ES*, 9 May 1934, 1.

189. *ES*, 25 November 1925, 1.

190. Harold L. Ickes (1874–1952) started as a newspaper reporter in Chicago and then switched to law and reform politics. During World War I, he worked under Samuel Insull and Bernard Mullaney on the Illinois Committee for Public Information. He got the moniker "Honest Harold" as a result of his subsequent legal fights with the Insulls and others over utility rates and political corruption in Chicago in the 1910s and 1920s. To the Insull crowd, he was known as "Itches." He was appointed by President Roosevelt as secretary of the interior in 1933 and served in that position until 1946, making him the second-longest-serving cabinet member. McDonald, 171, 259.

191. *ES*, 13 June 1934, 1. See also Holmes, 42, citing *PPH*, 17 May 1934 and 8 June 1934.

192. FDRL-0982: 1934-06-09_Utterback to FDR_Eastport is desperate.

193. FDRL-0982: 1934-06-14:_FDR to Mac_I would like to see Dexter.

194. *Reel-38*, 3–4, 9. Regarding Howe's interest, see Marvin, 20, citing Alfred B. Rollins Jr., *Roosevelt and Howe* (New York: Alfred Knopf, 2018), 429.

195. SILLS: Dexter Cooper to President Roosevelt, 29 June 1934. See also FDRL-0982-1; and FDRL-COOPER: 1934-06-29_Dexter Cooper to Franklin Roosevelt.

196. *ES*, 18 July 1934, 3.

197. *ES*, 25 July 1934, 2, includes a transcription of Governor Brann's radio broadcast to State of Maine, 23 Jul 1934, 7:45 p.m. See also FDRL-0982: 1934-07-01_FDR to Brann.

198. *ES*, 25 July 1934, 1.

199. *ES*, 15 August 1934, 1.

200. *ES*, 29 August 1934, 1; and *ES*, 16 March 1938, 9. Raymond Moley was a member of FDR's "brain trust," assistant secretary of state in 1933, and then editor of *Today* from 1933 to 1937. Michael Straus, PWA director of information, joined them on Monday. Ickes, 189–93.

201. Ickes, 189–93.

202. *ES*, 16 March 1938, 9; and Ickes, 191.

203. *ES*, 22 August 1934, 5.

204. *ES*, 22 August 1934, 1. See also Brown, 302.

205. Ickes, 189, a transcription of which is in EMERY_6.18.

206. Ickes, 189–90.

207. FDRL-0982: 1934-09-23_FDR to Sills_No Funds currently available for Quoddy. See also Ickes, 189–91.

208. "Philip B. Fleming, 1911," West Point Association of Graduates, https://www.westpointaog.org/memorial-article?id=e7274805-eaaa-4b1c-8ecf-1796066f4de5.

209. "Philip B. Fleming, 1911," West Point Association of Graduates. See also *Reel-38*, 6.

210. FDRL-FLEMING: IV.A4_1949-08-19_"Memorandum of Conversation between General Philip B. Fleming and Sidney Hyman, August 19, 1949," 1.

211. *Reel-38*, 4.

212. Also in attendance at the 23 August meeting were Dexter Cooper, Governor Brann, Philip M. Benton, PWA financial director, and K. F. Wingfield, PWA industrial engineer. *ES*, 5 September 1934, 1. Wingfield, it turned out, had a few undisclosed conflicts of interest: "According to Mrs. Cooper, this Committee did not meet until it was pestered by her husband. She said that when the Committee did meet, Colonel Fleming was present and that he had with him a young man named Wingfield, who had made the Murray & Flood report. According to Mrs. Cooper, Wingfield had a grudge against the Coopers because Mrs. Cooper had broken up an affair between him and the governess of the Cooper children." *Reel-38*, 5.

213. Railroad electrification was also considered, but its potential was diminishing as the old spark-spewing, coal-fired locomotives were being replaced with new diesel engines. Chicago, Burlington and Quincy Railroad (of which Frederic Delano had been general manager, and where Russell Sturgis had also worked) was at the time pioneering high-speed stainless steel trains powered by diesel engines. *MSPB-34*, 28–29.

214. *ES*, 5 September 1934, 1.

215. Brown, 298–99.

216. *ES*, 5 September 1934, 1.

217. *Reel-38*, 4.

218. *ES*, 5 September 1934, 1. See also EMERY_Scrapbook-Quoddy 1933-6: unidentified clipping, possibly early September 1934, probably *Bangor Daily News*; and SILLS: Roosevelt to Sills, 18 Sep 1934. As published in a "A Message to you about Quoddy," by the Maine Quoddy Association in December: "Other press matters seems to follow the same trend. Newspaper reports of Secretary Ickes' visit to Eastport could hardly be called encouraging. At that time one state paper ran an alleged interview with Major Fleming, in which Major Fleming appears distinctly dubious, stating that in any case PWA was without funds. His attention called to it, Major Fleming wrote Mr. Cooper that he had not seen the article in question, but states, 'If the paper attributed

to me statements which are disagreeable to you, I feel certain that I must have been misquoted, for from the very beginning I have had the greatest admiration for you and your project and have always looked at your project with a most sympathetic eye.'" FDRL-0982: 1935_Maine Quoddy Association_A Message to you about the Quoddy Project.

219. *ES*, 5 September 1934, 1. See also EMERY_Scrapbook-Quoddy 1933-6: unidentified clipping, possibly early September 1934, probably *Bangor Daily News*.

220. SILLS: K. S. Wingate to K. C. Sills, 31 August 1934.

221. In 1930, the populations of Eastport, Maine, and Las Vegas, Nevada, were both about 3,100. In 2020, the Las Vegas metro area's population was 2,777,000. During construction of the Boulder Dam, approximately $70 million in federal funds went into Las Vegas. Power from the dam was sold at cost (not market rates) to a specific group of private companies, notably Pacific Gas & Electric, for 50 years, thus providing hundreds of millions in additional subsidy. In 1984, when the contract was renewed for 30 years, the price differential was 20 to 1 ($0.003 vs. $0.055 per kWh, a 95% discount). "Hoover Dam's Impact on Las Vegas," *Online Nevada Encyclopedia*, https://www.onlinenevada.org/articles/hoover-dams-impact-las-vegas; and *NYT*, 4 May 1984.

222. FDRL-0982: 1936-02-05_Dexter's Plan to Market Power. Irrigation "made the desert bloom"—or, more prosaically, it made it habitable. Food and cities sprouted everywhere water became available, and thus the southwestern states' human population grew with concomitant economic power. The most extreme example is the Colorado River, which was so often dammed and diverted for agricultural, industrial, and residential uses that there ceased to be a Colorado River just north of the US-Mexico border. All that is left for northern Mexico is a dry ditch.

223. Riggs, 119–20, citing *NYT*, 26 November 1934, 1–2; and Richard A. Colignon, *Power Plays: Critical Events in the Institutionalization of the Tennessee Valley Administration* (Albany: State University of New York Press, 1997), 185. There was speculation that Wendell Willkie was behind the lawsuit, as he was president of C&S, Alabama Power Company's (APC) owner, and a director of the Edison Electric Institute. APC also launched a lawsuit. In an effort to sway the results of a referendum in Chattanooga, Tennessee, calling for the city to switch from C&S's subsidiary, TEPCo, to the TVA. TEPCo had bought two empty lots to use as the bogus residence address for 120 nonresident employees. Riggs, 120, citing Thomas K. McCaw, *TVA and the Power Fight: 1933–1939* (Philadelphia: Lippincott, 1971), 128.

224. *ES*, 10 October 1934, 1.

225. *ES*, 17 October 1934, 4; *ES*, 31 October 1934, 4; and *ES*, 12 December 1934, 1.

226. SILLS: 17 November 1934.

227. Brown, 300.

228. FDRL-0982: 1934-06-09_John Utterback to Roosevelt; Utterback to Farley, 8 June 1934; and Rollins, *Roosevelt and Howe*, 429.

229. Marvin, 29, citing Ernest Greenwood, *You, Utilities, and the Government* (New York: D. Appleton Century, 1935), 149–50.

230. EMERY_SCRAPBOOK: Quoddy 1933-6 (pages 13–19); *BH*, 16 August 1934, 26.

231. EMERY_Scrapbook-Quoddy 1933-6: *BDN*, 24 August 1934; and FDRL-0982: 1935_Maine Quoddy Association_A Message to you about the Quoddy Project.

232. *NYT*, 9 September 1934, N1.

233. *NYT*, 24 October 1934, 9; and *ES*, 31 October 1934, 2.

234. *NYT*, 9 September 1934, N1.

235. Holmes, 46, citing *LEJ*, 20 August 1934. See also *ES*, 29 August 1934, 1, for discussion of similar charges in *Collier's* magazine. For the derogatory word "Quoddyize" see *NYT*, 24 October 1934, 9; and *ES*, 31 October 1934, 2. For Ickes's rebuttal see *ES*, 14 November 1934, 1. In the midst of all this, BHE announced a 15% rate cut for residential customers. *ES*, 26 September 1936, 1.

236. FDRL-COOPER: 1934_09-03_Gertrude Cooper to FDR_Brann's election.

237. FDRL-COOPER: 1934-09-10_FDR to Gertrude Cooper_difficulties of Maine politics.

238. *ES*, 26 June 1943.

239. *ES*, 10 September 1934.

240. *ES*, 17 October 1934, 1.

241. EMERY_Scrapbook-Quoddy 1933-6: unlabeled clipping by Ruby A. Black, [*Portland Evening*] *News* Washington Correspondent, "Ickes Denies He 'Bought' Maine: Brands as Lies Charge Quoddy Revived for Political Purposes."

242. *ES*, 14 November 1934.

243. *ES*, 14 November 1934, 1; EMERY_Scrapbook-Quoddy 1933-6: Unlabeled clipping by Ruby A. Black, [*Portland Evening*] *News* Washington Correspondent, "Ickes Denies He 'Bought' Maine: Brands as Lies Charge Quoddy Revived for Political Purposes"; and *ES*, 16 March 1938, 9.

244. FDRL-COOPER: 1934-09-20_Dexter Cooper to FDR_financially exhausted.

245. SILLS: 1934-10-06_Beale to Sills_Dexter illness. The exact date of Dexter's snap is not clear.

246. SILLS: Gertrude Cooper to Sills, 31 October 1934.

247. EMERY_7.32.9-10: letter to Hon. Ralph O. Brewster, 9 October 1934. See also *ES*, 16 March 1938, 9.

248. FDRL-COOPER: 1934-10-09_Dr. Bennett to FDR_Dexter's mental condition.

249. *ES*, 17 October 1934, 1.

250. Frederic A. Delano was a founding member of the Federal Reserve Board of Governors (1914–1918) and vice chair (1914–1916).

251. FDRL-COOPER: 1934-10-19_Delano to FDR_Hugh Cooper views on Quoddy.

252. FDRL-COOPER: 1934-10-19_Delano to FDR_Hugh Cooper views on Quoddy.

253. FDRL-COOPER: 1934-10-10_Frederic A. Delano to President Roosevelt. See also FDRL-0284: 1934-10-19_Delano to FDR_Had consulted with Hugh Cooper believes Quoddy Sound.

254. FDRL-COOPER: 1934-10-10_Telegram from Franklin Roosevelt to Dexter Cooper. See also letter from "Sister" to James Roosevelt, transmitting letter from Dexter Cooper, dated 17 October 1934. James requested FDR's attention.

255. FDRL-COOPER: 1934-11-01_Franklin Roosevelt to Dexter Cooper.

256. *ES*, 30 January 1935, 4. Dexter was confined to his home for several months.

257. FDR met "Mrs. Cooper" on 8 November 1934 at 5:15 p.m. The next appointment was at 6:30. FDRL-Day-by-Day. See also *Reel-38*, 5.

258. *ES*, 21 November 1934, 1. See also FDRL-COOPER: 1934-09-10_ FDR to Gertrude Cooper_difficulties of Maine politics.

259. *ES*, 28 November 1934, 1; and *ES*, 16 March 1938, 9. "The Passamaquoddy Bay Tidal Power Commission was chaired by Cooper." Parkman, 222–23. See also FDRL-COOPER: 1934-11-09_Memo to FDR_ talk with Ickes about Quoddy.

260. *ES*, 21 November 1934, 1; FDR-Day-by-Day: 8 November 1934, Mrs. Cooper at 5:15 p.m. The next item was scheduled at 6:30 p.m. "To Apartments."

261. *ES*, 12 December 1934, 1. See also *Reel-38*, 5–6: "The President then offered Mr. Cooper, through his wife, any other Federal Engineering job in the country that he wanted, but Mrs. Cooper said that her husband felt that the people of Maine were depending on him and the Quoddy job. Mrs. Cooper complained to the President about Brann's Committee saying that it was packed with people favorable to the Utilities. She told the President that Chairman Sills was probably honest but helpless and that she wanted a real investigating Committee. She had been instructed by her husband to ask for a Reclamation Service Engineer. The President promised to send such an

Engineer to Maine. The President sent two men to Maine, one, Henry T. Hund [Hunt], a P.W.A. legal advisor, and H.T. Corey, a Reclamation Service Consulting Engineer. Those two men, together with Dexter Cooper and his assistant, Moses Pike constituted the new committee." The Reclamation Service was the federal agency responsible for irrigation projects, and the army's chief rival in dam construction. Bonneville was built by the US Army and Grand Coulee by the Bureau of Reclamation.

262. SILLS: Sills to Wingate Cram and Harry Crawford, 26 November 1934; *ES*, 26 December 1934, 1, citing *PPH*, 24 December 1934; *ES*, 23 January 1935, 1; and Brown, 307.

263. FDRL-0284: 1934-12-18_Thompson to FDR_summary_see OF-42. Samuel Huston Thompson Jr. (1875–1966) was assistant US attorney general (1913–1918) and on the FTC (1919–1927). These discussions apparently led to a request by a group of private power company executives to ask to meet with FDR on 11 January 1935. FDRL-0284: 1935-01-11_[unknown] to FDR_Couch 1–5.

264. SILLS: Sills to Wallace White, 14 December 1934.

265. *ES*, 12 December 1934, 1.

266. EMERY_6.17: 13 November 1934: Emery to Arthur E. Sewall, Chairman of State Republican Committee.

267. EMERY_6.17: 23 November 1934: Arthur E. Sewall, Chairman of State Republican Committee, to Roscoe Emery.

268. EMERY_6.17: 23 November 1934: Arthur E. Sewall, Chairman of State Republican Committee, to Roscoe Emery.

269. EMERY_6.17: 23 November 1934: Arthur E. Sewall, Chairman of State Republican Committee, to Roscoe Emery.

270. EMERY_6.17: 23 November 1934: Arthur E. Sewall, Chairman of State Republican Committee, to Roscoe Emery.

271. For the increasing demand for power, see *ES*, 17 April 1929, 9; for CMP's planned investments in expanded facilities, see *ES*, 17 July 1929, 3; for CMP ads stressing economic development, see *ES*, 24 July 1929, 4; for Deer Island, NB power, see *ES*, 28 August 1929, 3.

272. Funigiello, 51–52, citing Lilienthal, *Journals*, 1, 40–41; Ickes, *Secret Diaries*, 1, 224–27; Robert Winsmore, "Public Utility Concerns May Become Aggressive in Self-Defense," *Literary Digest* 118 (1 December 1934): 36; "Public Utility Concerns Plan Court Tests of Power Issue," *Literary Digest* 118 (15 December 1934): 5; and "Proposals to Curb Power Holding Companies," *Literary Digest* 118 (8 December 1934), 36.; and *ES*, 1 January 1935.

273. Sugrue, 158.

274. Sugrue, 158.

275. EMERY_1.8: letter to John A. McDonough, State Administrator, F.E.R.A., Augusta, Maine, 20 December 1934.

276. EMERY_1.8: telegrams from John A. McDonough to Roscoe Emery, 20 November 1934 and 4 December 1934, and letter to John A. McDonough, State Administrator, F.E.R.A., Augusta, Maine, 20 December 1934; and EMERY_2.19: undated form letters from Mayor Emery titled "Sewing Machines" sent to multiple women in Eastport.

277. EMERY_1.8: telegrams from John A. McDonough to Roscoe Emery, 20 November 1934 and 4 December 1934, and letter to John A. McDonough, State Administrator, F.E.R.A., Augusta, Maine, 20 December 1934; and EMERY_2.19: undated form letters from Mayor Emery titled "Sewing Machines" sent to multiple women in Eastport.

CHAPTER 9. GREEN LIGHTS (1935)

1. *MSPB-34*, 248–49.

2. FDRL-0982: 1935-02-23_Kimball to FDR via Lehand.

3. *MSPB-34*, 154.

4. *MSPB-34*, 150–51. Compare the costs per kWh noted here versus those quoted by Major Fleming in chapter 8, "Liquidity." *ES*, 5 September 1934, 1.

5. *MSPB-34*, 149–60. The projected cost, utilization rates, and power for the manufacturing plants were aluminum @ $2,500,000, 25%, 81,000,000 kWh; stainless steel @ $2,000,000, 25%, 10,500,000 kWh; nitrogen fixation (fertilizer) @ $2,100,000, 50%, 45,675,000 kWh; other metallurgical plants @ $1,400,000, utilization and electrical consumption not listed. *MSPB-34*, 199–202.

6. FDRL-0982: 1935-01-17_Hunt Report (*PWA-Hunt-35*). The report was presented to President Roosevelt on 21 January 1935. See also *Reel-38*, 6.

7. SILLS: 1935-01-14_Brann to Sills_requests changes; and SILLS: 1935-01-04_Sills to Brann_submits report_p2.

8. SILLS: 1935-01-11_Campbell to Sills_approve report, and similar.

9. SILLS: 1935-01-04_Sills to Brann_submits report_p2.

10. *MSPB-34*, 152–53. See also Marvin, 48. Congress has not maintained any consistent policy for the marketing of power from federal projects. The federal government required that all the energy from Boulder Dam (authorized in 1928 and completed in 1936) be disposed of before construction started. Although Congress did not establish the same requirements for the Bonneville Project (authorized in 1935), contracts for all available power from the development were obtained as rapidly as facilities could be provided. Twentieth Century Fund Power Committee, *Electric Power and Government Policy*, U96-

33U. Most of the power projects that have been financed by the federal government have been multiple-purpose developments, involving flood control, irrigation, navigation, water supply, conservation, and so forth. The sale of power has helped pay for construction costs on such projects, although electric power frequently has not been a major objective. Twentieth Century Fund Power Committee, *Electric Power and Government Policy*, 678; and Pritchett, 33.

11. SILLS: 1935-01-04_Sills to Brann_submits report_p. 4–7. See also Holmes, 50–51, citing *BDN*, 19 January 1935.

12. *ES*, 16 January 1935, 1; and *ES*, 16 March 1938, 10. See also Dello-Russo, 45, who says that Dexter was hired in May by the PWA at an annual salary of $15,000. No sources are listed.

13. *NYT*, 19 January 1935, 11; and FDRL-Day-by-Day_18 January 1935.

14. *ES*, 23 January 1935, 1. The group included Governor Brann, the congressional delegation, and Commissioner Sills.

15. *ES*, 23 January 1935, 1.

16. *ES*, 23 January 1935, 1.

17. *NYT*, 10 February 1935, E6. This also obviated the problem that Quoddy would cost more than Maine's self-imposed constitutional limit on the amount it could borrow.

18. *ES*, 23 January 1935, 1.

19. *ES*, 23 January 1935, 1.

20. *ES*, 23 January 1935, 1.

21. *ES*, 23 January 1935, 4.

22. *ES*, 6 February 1935, 4, quoting the *PEN*.

23. Smith, *Power Policy of Maine*, 24, 72, 80–82, 180–81, 276–90.

24. FDRL-0284: 1935-01-11_[Unknown] to FDR_Couch rural electrification. "List of companies willing to confer with a desire to reach an agreement are: Electric Bond and Share and its associated companies; North America and its associated companies; Billsby Company of Chicago; United Gas & Electric Company; Stone & Webster & its associates; Commonwealth & Southern; American Gas & Electric; Consolidated Gas; Niagara & Hudson; Hartford Electric Company; American Electric & Water Co.; Public Service of New Jersey." Funigiello, 59, says the memo was created by Louis Howe.

25. FDRL-Day-by-Day_11 January 1935.

26. FDRL-0284: 1935-01-11_[Unknown] to FDR_Couch rural electrification. See note 24.

27. FDRL-0284: 1935-01-11_[Unknown] to FDR_Couch rural electrification. See note 24.

28. Riggs, *High Tension*, 135–37.

29. FDRL-Day-by-Day_24 January 1935; and Riggs, 123.

30. Funigiello, 60, citing David P. Lilienthal, *The Journals of David P. Lilienthal*, vol. 1, *The TVA Year* (New York: Harper & Row, 1964), 1, 46–47; and Riggs, 124.

31. In 1933, Maine's electric utilities sold 166,448,989 kWh of electricity to 179,020 customers, earning gross revenue of $6,989,358, or an average consumption of 650 kWh and $39.05 per customer. *MSPB-34*, 166. Per capita income in Maine in 1930 was $570. "Personal Income," in *Maine: An Encyclopedia*, https://maineanencyclopedia.com/personal-income/. The population of Maine in 1930 was 800,000. Total electric utility revenue in Maine in 1931 was $13.6 million. *MSPB-34*, 128.

32. *ES*, 6 February 1935, 1.

33. *ES*, 6 February 1935, 1.

34. Edward E. Chase (?–1953) was a Maine state representative (1940). He was a Maine senator in 1953 when he died in office in an airplane crash, along with former governor William Tudor Gardiner, in Schnecksville, Pennsylvania. "Index to Politicians: Chase," PoliticalGraveyard.com, http://politicalgraveyard.com/bio/chase.html.

35. *ES*, 6 February 1935, 1.

36. The $47 million figure was attributed to Governor Brann.

37. FDRL-0982: 1935-02-14_*Boston Herald*_President, Pain and Quoddy.

38. Frederik Harold Dubord (1891–1964), a Democrat, was mayor of Waterville (1928–1932). In 1934, he challenged Frederick Hale for his US Senate seat but lost by 1%. In 1936, Dubord was the Democratic nominee for governor but lost to Republican Lewis O. Barrows. In 1955, Dubord was appointed to the Maine Superior Court and then served on the Maine Supreme Judicial Court (1956–1962). See also *Ruby Black-33* for his less savory activities.

39. EMERY_6.17.14: Letter from Mayor Emery to Harold Dubord, 9 February 1935.

40. *NYT*, 10 February 1935, E6, by F. Luriston Bullard.

41. *PEN* (undated), quoted in *ES*, 13 February 1935, 1.

42. *ES*, 16 March 1938, 10.

43. *ES*, 27 February 1935, 1.

44. *ES*, 27 February 1935, 1.

45. *ES*, 16 March 1938, 10.

46. *ES*, 20 February 1935, 1, 3.

47. *ES*, 6 March 1935, 1, 8.

48. *ES*, 6 March 1935, 1. At the time, the mayor of Eastport had to stand for election every year. Roscoe Emery was mayor from 1928 to 1932 and again from 1934 to 1935.

49. An editorial critical of Quoddy in *BH*, 19 May 1935, approvingly referred to William R. Pattangall as "one of the most delightfully injudicious

figures in the state" for his political opinions expressed before his tenure on the bench. *BH* then published a series of articles that were apparently intended to enhance Pattangall's political profile, each of which was preceded by his name: first in the subhead, again in the byline, and then twice more in a bio preceding the article; for example, "This is the 17th of a series of articles by William R. Pattangall, who retired recently as chief justice of the supreme court of Maine. In these articles Mr. Pattangall, long Maine's outstanding Democrat, will discuss a wide range of national and local issues." *BH*, 19 May 1935, 54 (Maine's Good Luck–level mountains); and *BH*, 12 August 1935, 22 (Pattangall on Canada treaty).

50. EMERY_SCRAPBOOK: Quoddy 1933–1936, 13–19; Smith, *Power Policy of Maine*, 242, citing Elizabeth M. Craig, in *PPH*, 8 March 1933. See also *NYT*, 7 April 1935, 1.

51. FDRL-Day-by-Day: 11 March 1935: Dexter Cooper at 11:15 a.m.; the next item on the schedule is Sec. George H. Dern at 11:45 a.m.

52. *Reel-38*, 6.

53. Smith, *Power Policy of Maine*, 246, citing *Kennebec Journal*, 18 May 1935.

54. *ES*, 27 March 1935, 1.

55. Eastport also issued scrip in 1933. *ES*, 12 April 1933, 1, 4. At least part of the scrip was ordered burned by the town council in 1934. *ES*, 21 February 1934, 1.

56. Holmes, 51, citing *PEN*, 4 April 1935. The communication between Dexter and Franklin was apparently a phone call, as DPC does not show up on FDRL-Day-by-Day_4 Apr 1933.

57. *BH*, 4 April 1935, 10.

58. *BH*, 4 April 1935, 10.

59. *NYT*, 11 April 1935, 20.

60. *ES*, 1 May 1935, 1.

61. *ES*, 15 May 1935, 1.

62. FDRL-0982: 1935-04-25_Jerome A. Petitti to Louis Howe. It was referred to the Department of the Navy for investigation. Another patent claim was made on 18 May 1935 in a letter to FDR from Dr. G. van Antwerp Clarke, of New York City. He claimed to own the "rights of invention to all methods of harnessing tidal water dating from 1905, and is writing in formal protest against this procedure; states if the development actually starts he will immediately start injunction proceedings in the Fed. Courts to stop the work." FDRL-0982: 1935-05-21_McIntyre to Markham_Clarke patent claim. Clarke's claim was forwarded to the Army Engineers, who replied, "If your patent dates from 1905, it has now expired and no further rights would accrue thereto." FDRL-0982: 1935-06-05_Army to Clarke_patent claim. See also *BH*, 27 September 1935, 1, 22.

63. Marvin, 46–48.

64. *ES*, 9 September 1936, 4.

65. *NYT*, 17 May 1935, 1. See also *Reel-38*, 6; and *ES*, 9 October 1935, 1.

66. *BH*, 17 May 1935, 27_Senator White approves.

67. *ES*, 22 May 1935, 1.

68. A tidal bore (*mascaret* in French) is a rare phenomenon that occurs in about 80 rivers around the world, including the Petitcodiac River at Moncton, New Brunswick; the Shubenacadie River at South Maitland; and the Salmon River in Truro, Nova Scotia. Bores form in river estuaries where there is (a) a large tidal range, (b) a funnel-shaped estuary, and (c) a sandbar or similar shoaling at the mouth of the estuary. The combination of the out-flowing river current and the sandbar keep the incoming tide from advancing steadily, and instead the tide gets "stuck" on the sandbar—not advancing upriver, but nonetheless increasing in height. Eventually the incoming tide overcomes the resistance of the outgoing river current and jumps over the sandbar, racing upriver at speeds up to 25 miles per hour, as a vertical wall of water. At Moncton, this wall of water can be up to three feet high at the full and new moons; and about two feet at "Shubie"—providing plenty of water to give surfers an exciting ride for many miles as the wave travels upstream. The largest bores in the world are on the Qiantang River in China and the Amazon River in Brazil, where they can reach 30 feet in height and travel dozens, even hundreds, of miles upriver.

69. *ES*, 22 May 1935, 1.

70. *ES*, 22 May 1935, 1.

71. *ES*, 22 May 1935, 1. See also EMERY_Scrapbook-Quoddy 1933–36b; and "Special Dispatch to Sunday Telegram," dateline "Lubec, May 18."

72. *ES*, 22 May 1935, 1.

73. *ES*, 22 May 1935, 4.

74. *ES*, 22 May 1935, 8.

75. *ES*, 5 June 1935, 4.

76. FDRL-0982: 1935-06-06_Brann to Howe: Republicans appropriate "Our Child" Quoddy.

77. FDRL-OF-25n: War Dept., Chief of the Engineer Abstracts; and FDRL-OF-25n: 1935-06-06_Brann to FDR_REPUBS credit for Quoddy.

78. *ES*, 22 May 1935, 1.

79. *ES*, 22 May 1935, 1, 4; and *NYT*, 17 May 1935, 1. See Holmes, 54n4: "Army engineers arrived in Eastport on June 27, and began making arrange-ments to start work. There was delay in starting because of a difference of opinion between Cooper and the Army about who was to be in charge. Roo-sevelt had written Cooper promising to put him in charge and reimburse him for the money already spent. Statement by Moses Pike, December 28, 1954."

80. *ES*, 29 May 1935, 1; and *ES*, 16 March 1938, 10. See also *BH*, 24 May 1935, 32_Funding delay.

81. *Reel-38*, 6.

82. FDRL-Day-by-Day: 26 May 1936.

83. *Reel-38*, 6.

84. *ES*, 29 May 1935, 1. As of 2019, the document and the pen were on display at the Quoddy Model Museum in Eastport. Comptroller-General J. R. McCarl approved the allocation on 5 June 1935. *ES*, 16 March 1938, 10.

85. GSC, "Passamaquoddy," 9, 11–13. See also *Reel-38*, 3.

86. *ES*, 26 June 1935, 1. See also *ES*, 19 June 1935, 1; and EMERY_Scrapbook-Quoddy 1933–36b: Elisabeth M. Craig, "Washington Proceeds with Plans for Quoddy Project," *[Portland] Sunday Telegram*, 19 May 1935.

87. EMERY_Scrapbook-Quoddy 1933–36b: Elisabeth M. Craig, "Washington Proceeds with Plans for Quoddy Project," *[Portland] Sunday Telegram*, 19 May 1935.

88. *ES*, 29 May 1935, 1.

89. *ES*, 12 June 1935, 1.

90. *ES*, 5 June 1935, 1.

91. *ES*, 5 June 1935, 1.

92. *ES*, 3 July 1935, 1.

93. *ES*, 26 June 1935, 1.

94. *ES*, 3 July 1935, 2.

95. *ES*, 16 March 1938, 10.

96. *ES*, 17 July 1935, 1.

97. *PPH*, 18 May 1935, 1.

98. *ES*, 28 August 1935, 1.

99. *ES*, 31 July 1935, 8.

100. *ES*, 7 August 1935, 1; and *ES*, 18 September 1935, 1.

101. *ES*, 25 September 1935, 1.

102. *ES*, 4 September 1935, 1.

103. *ES*, 2 October 1935, 1.

104. GSC, "Passamaquoddy," 9–10.

105. *ES*, 12 June 1935, 1.

106. *ES*, 12 June 1935, 1. $58,000 in 1935 is roughly equivalent to $1.1 million in 2020. The REA was a tremendous success. When private power companies refused to build lines even when offered low-cost government loans, REA sponsored the creation of nonprofit cooperatives. In response, private power companies occasionally built new "snake lines"—cursed as "spite lines"—to make the REA's area-wide electrification efforts more expensive. Nonetheless, the promise of REA's light drew farmers, "who voted by the light of kerosene lamps, to borrow hundreds of thousands, even millions of dollars, from the

government to string power lines into the countryside. Finally, the great moment would come: farmers, their wives and children, would gather at night on a hillside in the Great Smokies, in a field in the Upper Michigan peninsula, or on a slope of the Continental Divide, and, when the switch was pulled on a giant generator, see their homes, their barns, their schools, their churches, burst forth in dazzling light. Many of them would be seeing electric light for the first time in their lives." In 1935, 11% of farms were electrified; 40% were by 1941; and 90% were by 1950. Leuchtenburg, 158, citing Morris Cooke, "Early Days of Rural Electrification," *American Political Science Review* 62 (1948): 431–37; *Time* 34 (4 July 1938), 12; and Marquis Childs, *The Farmer Takes a Hand* (Garden City, NY, 1952). Riggs, 138–45.

107. *ES*, 19 June 1935, 7.

108. *ES*, 19 June 1935, 7.

109. *ES*, 3 July 1935, 1.

110. *ES*, 19 June 1935, 7.

111. *BH*, 17 May 1935, 27_Senator White approves.

112. *MSPB-34*, 150–56.

113. FDRL-0982: 1935-02-14_*Boston Herald*_President, Pain and Quoddy; and EMERY_6.17: 23 Nov 1934: Arthur E. Sewall, Chairman of State Republican Committee to Roscoe Emery.

114. *NYT*, 11 April 1935, 20.

115. *ES*, 19 June 1935, 7.

116. Captain Sturgis is profiled in *ES*, 18 September 1935, 1; Captain Leehey in *ES*, 25 September 1935, 1; Captain Lord in *ES*, 2 October 1935, 1; and Captain Casey in *ES*, 4 September 1935, 1.

117. *ES*, 3 July 1935, 1.

118. FDRL- COOPER: 1935-06-19_Gertrude Cooper to James Roosevelt.

119. FDRL-0982: 1935-10-16_Dexter to James Roosevelt_Quoddy sale negotiations. See also Parkman, 224–25: "A preliminary agreement was signed on June 26 (a final agreement was reached on October 18, [1935]), according to which . . . Cooper was appointed Advisor to the National Power Policy Committee to find a market for Quoddy power. The next day, June 27, the active prosecution of Quoddy began." George Bigelow Pillsbury (1876–1951) served in France during World War I as Corps Engineer, 2nd Army Corps, from October 1918 to January 1919. "George Bigelow Pillsbury," Hall of Valor Project, https://valor.militarytimes.com/hero/18045; and "Obituary for George B. Pillsbury (Aged 74)," Newspapers.com, https://www.newspapers .com/clip/39703388/obituary-for-george-b-pillsbury-aged/.

120. *ES*, 6 June 1935, 1.

121. *ES*, 26 June 1935, 1. See also GSC, "Passamaquoddy," 8. "The last act was the signing by Secretary Ickes of the Interior Department, of a commission

to Cooper as special advisor to the national power policy committee. The commission will run from July 1, 1935 to January 1, 1938, the period contemplated as the construction period of the huge tidal power project. He was specifically instructed to conduct investigations in connection with the development of the right to use the power and start its distribution." *BH*, 28 June 1935, 17.

122. *ES*, 3 July 1935, 1. See also *NYT*, 29 June 1935, 31.

123. Dexter's "Five-Party Agreement" (DPC, GE, Westinghouse, BE, and MWU) stipulated that Dexter would be given the contract to manage construction (if undertaken) for a 3% fee. *Final-36*: appendix A-3, narrative p. 6.

124. *Reel-38*, 6.

125. *Reel-38*, 6. See also FDRL-COOPER: 1935-06-07_FDR to Dexter Cooper_negotiations sale of Quoddy rights. See also FDRL-COOPER: 1935-06-28_FDR to Corcoran_Delano letter.

126. Marvin, 70–71, citing FDRL-PPF-1766_1935-06-19: Gertrude Cooper to James Roosevelt. (Note: I did not find the original of this letter in the National Archives, only the following references to it.) "Fleming felt that the $65,000 then being offered to Dexter Cooper was ridiculous. Mrs. Cooper said of the offer, 'He has personally borrowed $85,000 so if he took the $65,000 they are now offering he would be left morally and legally owing $20,000 more—naturally the security was Quoddy so if he sells out for less he would be responsible for the rest and left penniless—a fine reward for fifteen years of work.' Cooper, of course, finally settled for only $60,000." See also FDRL-0982: 1935-06-19_Getrude to FDR_Fleming is opposed to Quoddy.

127. The unaccounted-for debt may have related to development expenses (estimated at $35,000) Dexter paid out of pocket between 1914 and 1924, the latter being the date on which the army audit commenced. *Final-36*: appendix A-3, narrative p. 2.

128. FDRL-COOPER: 1935-06-28_FDR to Corcoran_Delano letter. See also *Reel-38*, 6–7; and Parkman, 224–25.

129. See chapter 10, "Resistance."

130. FDRL-COOPER: 1935-07-11_McIntyre to Comptroller General_Dexter Cooper claim #A-61970: "Colonel Dexter Cooper called at the White House and asked that we look into the matter of his Claim #A-61970 sent to your Department on April 29, 1935, in the amount of seven hundred some odd dollars. He says that this has been unreasonably delayed. What may we advise Colonel Cooper[?]"

131. See chapter 10, "Resistance."

132. All financial decisions regarding Quoddy appear to have been made by the army, even though the funds came from the PWA. To what degree PWA secretary Ickes was aware of this issue or agreed with the army's position is unknown.

133. *Final-36*: appendix A-3, exhibit G, 3: "Five-Party Agreement," 31 October 1927.

134. *Final-36*: appendix A-3, narrative p. 7.

135. FDRL-0982: 1935-07-01_Brann to FDR_Mayor Maybee in Quoddy Commission[?].

136. FDRL-0982: 1935-07-01_Brann to Edward Brown[?] The *Bangor Commercial* sent a letter to the *Sentinel*: "[T]he *Bangor Daily News* Saturday took large and almost exclusive credit for the success of the Quoddy project." EMERY, Roscoe_box 5.12e_misc correspondence 1934–50s.

137. FDRL-0982: 1935-07-02_Morgenthau to FDR_Brewster taking credit.

138. Holmes, 4, citing *Bangor Commercial*, 3 July 1935. Mayor Emery estimated 15,000, according to Holt, *Island City*, 42.

139. *ES*, 10 July 1935, ? (illegible).

140. GSC, "Passamaquoddy," 11. Perhaps it was a foul-up, as the intent to invite Dexter was clearly stated in *ES*, 26 June 26, 1935, 1.

141. Inadvertently departing from his written speech, FDR omitted the word "features," thus making his intent appear even more draconian to utility holding companies. Funigiello, 57; *NYT*, 5 January 1935; Schlesinger, *Politics of Upheaval*, 305. See also Riggs, 123, citing Steve Neal, *Dark Horse: A Biography of Wendell Willkie* (Lawrence: University Press of Kansas, 1984), 30. Wikipedia, s.v. "Franklin Delano Roosevelt's Second State of the Union Address," https://en.wikisource.org/wiki/Franklin_Delano_Roosevelt%27s_Second_State_of_the_Union_Address.

142. Riggs, 125–26, citing Funigiello, 63–65, and others.

143. Riggs, 125, citing Schlesinger, *Politics of Upheaval*, 305–6; Funigiello, 626–33.

144. Wikipedia, s.v. "Public Utility Holding Company Act of 1935," https://en.wikipedia.org/wiki/Public_Utility_Holding_Company_Act_of_1935#cite_note-15, citing Associated Press, "Borah Utilities Bill Change Hit," *Washington [D.C.] Evening Star*, 12 June 1935.

145. For example, *Washington Evening Star* cartoon on 3 July 1935.

146. *NYT*, 3 July 1935, 1.

147. Funigiello, 113, quoting Representative Richard Dunn (D-MI).

148. *NYT*, July 12, 1935, 1; and Marvin, 54, citing Cong. Rec. (2 July 1935). John E. Rankin (1882–1960) (D-MS), US Representative (1921–1953), and coauthor (with George Norris, R-NB), of the TVA bill.

149. Hugo Lafayette Black (1886–1971) was a senator from Alabama (1927–1937), home of the Muscle Shoals Dam, and later appointed by FDR as associate justice of the US Supreme Court (1937–1971). He resigned eight days before his death at age eighty-five. Riggs, *High Tension*, 171; and Wikipedia, s.v. "Hugo Black," https://en.wikipedia.org/wiki/Hugo_Black.

150. John Joseph O'Connor (1885–1960) was a Democratic representative from New York (1923–1939).

151. *NYT*, 3 July 1935, 1; and Funigiello, 99–100.

152. *NYT*, 3 July 1935, 16.

153. "A brilliant graduate of Harvard Law School, once a private secretary to Oliver Wendell Holmes, a young man flaming in his beliefs and almost ascetic in his appetites, an intense social economic crusader—such is Mr. Corcoran. Although he has always been one of the members of the Brain Trust who enjoyed the President's highest esteem he has, until recently, managed to keep well out of the limelight. This was good for him. He was gaining, rather than losing, in avoiding the calcium glare so essential to the happiness of Rexford G. Tugwell. He was mounting high in the President's opinion. Partly because of a cultivated and agreeable social quality—partly because he is a fine ballad singer and accompanies himself pleasantly on the piano or accordion—Mr. Corcoran has become more and frequently in demand at very small, very intimate White House gatherings. An able, brilliant young Harvard man, who thinks closely along the same lines of another Harvard man who can adorn hours of relaxation as well as hours of toil, can go a long way with his fellow-Cantab when that one happens to be the President of the United States." *NYT*, 3 July 1935, 16.

154. *Time*, 22 July 1935; Funigiello, 99, citing Charles A. Beard, "The New Deal's Rough Road," *Current History* 42 (September 1935): 630.

155. *Time*, 22 July 1935.

156. Caro, 87–88.

157. Ernest Gruening ("greening") was director of the Division of Territories and Island Possessions of the Department of the Interior (1934–1939) and administrator of the Puerto Rico Reconstruction Administration (1935–1937). Gruening also wrote *The Public Pays: A Study of Power Propaganda*, published in 1931. Gruening was editor-publisher of the *Portland (ME) Evening News* (1927–1932). His departure presumably led to the position being offered to Roscoe Emery. EMERY_6.21.18_Roscoe Emery to Dexter Cooper, 28 October 1933; and "The Congress: Boomerang & Blackjack," *Time*, 22 July 1935.

158. *Time*, 22 July 1935.

159. *ES*, 24 July 1935, 4.

160. *Time*, 22 July 1935; Funigiello, 99, citing Beard, "New Deal's Rough Road," 630.

161. Holmes, 57 citing *LEJ*, 26 July 1935.

162. Holmes, 7–8, citing *LEJ*, 15 July 1935.

163. *Boston Globe*, date unknown, from Roscoe Emery's Scrapbook, "Quoddy 1935–6."

164. *ES*, 17 July 1935, 4.

165. Lilienthal's interest in regulation had been sparked at Harvard Law School in the public utilities class taught by future Supreme Court justice Felix Frankfurter. Frank Walsh, whom FDR appointed as chair of the New York Power Authority, then introduced Lilienthal to Donald Richberg, a lawyer who represented the City of Chicago in suits against the Insulls and Consolidated Edison. Lilienthal also helped Wisconsin senator Robert A. Follette's Progressive Party campaign for US president in 1924. Riggs, 101–2.

166. *ES*, 10 July 1935, 1.

167. *ES*, 10 July 1935, 1.

168. For a profile of Wendell Willkie and his relations with the TVA see Riggs, 89–109.

169. For a detailed discussion of the private power lobbying and propaganda campaign, as exposed by Senator Black's investigation, see Funigiello, 98–121.

170. Funigiello, 107–14.

171. Funigiello, 112, 116–18.

172. For an excellent summary of Senator Hugo Black's investigation, see Gregory and Strickland, "Hugo Black's Congressional Investigation," 543.

173. Wikipedia, s.v. "Howard C. Hopson," accessed 17 September 2019, https://en.wikipedia.org/wiki/Howard_C._Hopson#cite_note-23, citing "Utility-Baron in Eliza Role," *Literary Digest*, August 17, 1935. See also "The Congress: Hobson Hunt," *Time*, 19 August 1935, http://content.time.com/time/subscriber/article/0,33009,848115,00.html.

174. Basil O'Connor met Franklin D. Roosevelt in 1920 when the latter was running for vice president, and became his legal adviser. In 1924, the two men associated in their own law firm, which existed until Roosevelt's first presidential inauguration in 1933. In 1927, he and Roosevelt and a group of friends created the Georgia Warm Springs Foundation, in which O'Connor served first as treasurer and later as president. O'Connor was a member of the "Brain Trust" that advised Roosevelt on political strategy during his 1932 presidential campaign. Roosevelt appointed O'Connor to the American Red Cross, which he served as chairman (1944–1947) and president (1947–1949). Wikipedia, s.v. "Basil O'Connor," https://en.wikipedia.org/wiki/Basil_O%27Connor, citing Alden Whitman, "Basil O'Connor, Polio Crusader, Dies," *NYT*, 10 March 1972.

175. The quote is attributed to Arthur Krock (undated, presumably from the *New York Times*). Wikipedia, s.v. Howard C. Hopson; Wikipedia, s.v. "Public Utility Holding Company Act of 1935"; and Gregory and Strickland, 552, 562–64. PUHCA was signed into law on 26 August 1935. "Congress: Hobson Hunt," *Time*, 19 August 1935.

176. The House ignored the lobbying by the power trusts. Wikipedia, s.v. "Public Utility Holding Company Act of 1935." See also Funigiello, 114, citing *Hearings Pursuant to S. Res. 165, and S. Res. 184, Senate Special Comm. to Investigate Lobbying Activities*, 74th Cong., 1st Sess. (1935), 61–64, 92, 23; and Gregory and Strickland, 552, 562–64. PUHCA was signed into law on 26 August 1935. Riggs, 131–32, citing Roger K. Newman, *Hugo Black: A Biography* (New York: Pantheon, 1984), 183–84. See also Paul W. White, "The Public Utility Company Act of 1935," http://paulwwhite.com/the-public-utility-company-act-of-1935_316.html, citing *NYT*, 1 August 1935 and 13 August 1935.

177. *ES*, 17 July 1935, 4.

178. Holmes, 56, citing *LEJ*, 26 July 1936.

179. *Final Report-36*, 32.

180. *ES*, 4 September 1935, 1; and *ES*, 11 September 1935, 1, 3. A full-page co-op ad in the *Sentinel*—for cement, lumber, plumbing, paint, wallpaper, and furniture—boosted the campaign.

181. *ES*, 6 November 1935, 1.

182. *ES*, 18 September 1935, 1; *ES*, 6 November 1935, 1; and *American Magazine*, December 1935, 159.

183. *Reel-38*, 11–12.

184. *ES*, 18 September 1935, 5, 8.

185. *Reel-38*, 15.

186. *ES*, 14 August 1935, 1. See also FDRL-0982: 1935-08-21_Emery to Simkhovich_Water Company.

187. *ES*, 14 August 1935, 1.

188. Mary Simkhovitch, who with Major Fleming and Dexter Cooper was on the Eastport Planning and Zoning Commission, wrote to FDR, "I enclose a letter I received this morning from the Mayor of Eastport. He is intelligent & progressive. The water company ought not to get away with it. If there is any way by which federal funds can be advanced to the City of Eastport for the purpose, it would be a great gain. A bigger Eastport ought to be free from the high rates the water company has imposed on this community for a long time. But Eastport can borrow from no other source than the Federal Government." FDRL-0982: 1935-08-22_Simkhovitch to FDR_Eastport Water Company monopoly. See also FDRL-0982: 1935-09-11_FDR to Simkhovitch_PWA will consider loan to Eastport for water company.

189. Unidentified newspaper clipping, BHS collection at TI&MA.

190. *ES*, 24 July 1935, 1. For Roscoe Emery's personal perspective, see *Reel-38*, 11.

191. *Boston American*, 20 July 1935, BHS collection at TI&MA.

192. *NYT*, 21 July 1935. See also *BH*, 21 July 1935, 1.

193. *NYT*, 22 September 1935, E7.

194. *ES*, 15 November 1933, 2: "My interest in the study of the tides, which had its beginning over twenty years ago, was the means of attracting my interest also to the Cooper Tidal Power Development Project, I having the honor of having been, on the fifth day of September 1925, the first person, aside from Mr. Cooper and his backers, to invest money on the strength of belief in the practical advantages of the enterprise." While Dexter's financial records are now quite incomplete, I could not find any corroborating evidence for Stevens's claim.
Stevens had also agreed to sell his land to the government at the original site for Quoddy Village at Deep Cove. *ES*, 24 July 1935, 1. See also *Final Report-36*, appendix. Stevens agreed to sell his 3.27 acres (parcel PR-5-A) to the army for $220, but "Offer void at death of Mr. Stevens."

195. *ES*, 31 July 1935, 8.

196. EMERY_5.11: Fleming Correspondence 1934–1940. See also *Final Report-36*, appendix; and NA-MD, RG-71, NA Identifier 1110318: Naval Property Case Files, 1941–1958: Maine: 20, Quoddy Village: C20-9-QV box 538.

197. Richard O. Boyer, "Maine Power Concerns Will Fight Quoddy Dam in Courts, Legislature," *BH*, 24 July 1935, 1–2.

198. Boyer, "Power Concerns Will Fight," 1–2.

199. Boyer, "Power Concerns Will Fight," 1–2.

200. Richard O. Boyer, "Seek 'Killing,'" *BH*, 25 July 1935, 1, 10.

201. David MacNichol was married to Dexter's longtime secretary, Genevieve Roche. The MacNichol sardine factory on Eastport's Deep Cove burned in 1935. L'Hommedieu, "MacNichol."

202. Boyer, "Seek 'Killing.'"

203. *BH*, 26 July 1935, 14.

204. EMERY, Roscoe_box 5.13_*Boston Herald* 1935: Emery to Richard O. Boyer, 26 July 1935. The letter in Roscoe's files is a carbon copy with numerous handwritten corrections. At the bottom is the notation "RCE. VL," suggesting that it was typed by his wife, Vera Leonard. As typical of a carbon copy, it is not signed. Hopefully Vera prevailed over Roscoe's anger. Roscoe also sent a long letter to the editor, which was so long they refused to print it.

205. EMERY_box 5.3: 1935-07-29_Emery to Bridges. For Boyer-Emery disputes, see *ES*, 4 March 1936, 1.

206. Boyer, "Seek 'Killing,'" *BH*, 25 July 1935, 1, 10.

207. *ES*, 31 July 1935, 8; and *ES*, 27 November 1935, 8.

208. *ES*, 24 July 1935, 1. George and Carrol Rice properties purchased for Quoddy Village totaled 56 acres. *Final-36*: appendix.

209. Spring Farm, also known as "McFaul property," was ten acres. *Final-36*: appendix.

210. *ES*, 31 July 1935, 1.

211. *ES*, 31 July 1935, 1.

212. *ES*, 7 August 1935, 1.

213. *ES*, 21 August 1935, 1. The number of buildings shifted slightly over time.

214. *ES*, 28 August 1935, 2.

215. *ES*, 28 August 1935, 4.

216. *ES*, 28 August 1935, 1.

217. *ES*, 28 August 1935, 4.

218. *ES*, 28 August 1935, 4.

219. *ES*, 2 October 1935, 8.

220. *ES*, 11 September 1935, 1. The highest temperature, 90.4°F, was on 17 August 1935.

221. *ES*, 18 September 1935, 1; and *ES*, 27 November 1935, 8.

222. *ES*, 21 August 1935, 1.

223. Compare this 1,000 kW diesel plant with the 15,000 to 20,000 hp tidal power prototype Dexter Cooper proposed to build first to demonstrate feasibility, test equipment, and supply power during construction of the full-scale project. *ES*, 25 February 1925, 8.

224. "Royal B. Lord, 1923," West Point Association of Graduates, https://www.westpointaog.org/memorial-article?id=3885f497-0f59-4b4e-a90f-f745aa044e6c. The actual date of this incident is not certain.

225. *ES*, 2 October 1935, 1.

226. *ES*, 28 August 1935, 2.

227. *ES*, 16 October 1935, 1.

228. "The History of Power Development in Niagara," https://www.niagarafallsinfo.com/niagara-falls-history/niagara-falls-power-development/the-history-of-power-development-in-niagara/.

229. *BHEN*, July 1936, 26, http://digicom.bpl.lib.me.us/bangorhydro_news/36.

230. *Reel-38*, 18.

231. *ES*, 8 April 1936, 7. For more on the borings, see *ES*, 31 July 1935, 1; *ES*, 4 September 1935, 1; *ES*, 25 September 1935, 1; *ES*, 23 October 1935, 1; *ES*, 27 November 1935, 1; *ES*, 1 January 1936, 1; *ES*, 8 January 1936, 4; *ES*, 13 May 1936, 1; and *ES*, 30 December 1936, 1.

232. *ES*, 27 November 1935, 1. "Soil borings, dry land and subaqueous areas, are to have a combined depth of about 2300 feet, while rock core drillings will total 6270 feet." *ES*, 31 July 1935, 1.

233. *BH*, 24 July 1935, 1–2.

234. Roberts, *Trending into Maine*, 283. In shallower water, such as between Pleasant Point and Carlow Island, timber cribs were built, towed into

position, and then sunk with rocks to create stable drilling platforms. For a detailed discussion of Dexter's drilling and blasting plans, see DPC-III: 1928-12-19_O'Rourke drilling report.

235. *ES*, 2 October 1935, 4, quoting the *PEN*.

236. *Reel-38*, 25–26.

237. For dam specs, see *ES*, 19 June 1935, 7. Granite at 168 pounds per cubic foot × 27 cubic feet per cubic yard × 4,100,000 cubic yards to build Eastport-Treat Island Dam / 2,000 pounds per ton = 9.3 million tons.

238. The Eastport-Treat Island dam was to be 3,400 feet long and as deep as 140 feet. The angle of repose or "spread" of the rock fill dams would vary from about 2:1 to 6:1 (as per cross-sectional diagrams) for areas subject to the strongest tidal currents. "Tentative Program for the Construction of Quoddy Power Project, Maine," War Department, Corps of Engineers, Washington, DC, 1935, TI&MA.

239. With a 6:1 spread, the widest part of the dam would be 840 feet, or about one-sixth of a mile, not one-half. New Hampshire's Mount Washington is 6,288 feet high. The volume of a cone (e.g., a mountain) = $\pi R^2 * h/4$; thus, the volume of Mt. Washington is roughly half a million times more.

240. NA-MD: 77-ned-83-r1: B&W 10:31, silent, 16mm movie, *Passamaquoddy Tidal Power Development, Eastport, Maine*; and Corps of Engineers, "Hydraulic Model Tests: Rock Fill Dam, Scale 1:50." See also similar tests in the same collection; reports therefrom are at NA-MA and in the collection of Anne Pike Rugh.

241. *BH*, 19 May 1935, 6.

242. Kenneth Roberts, "Quoddy," *Saturday Evening Post*, September 1936. The Great Pyramid at Giza, Egypt, was originally 481 feet tall and about 756 feet square at its base, for a volume of roughly 92 million cubic feet, weighing about 6 million tons. Wikipedia, s.v. "Great Pyramid of Giza," https://en.wikipedia.org/wiki/Great_Pyramid_of_Giza.

243. As per *PWA-Hunt-35*.

244. *ES*, 13 November 1935, 1. One of the reasons for the increased costs was the army's plan to raise the elevation of the dams surrounding the Haycock Harbor Reservoir by five feet. DPC-3: 1935-09_SUMMARY—LOW POOL DEVELOPMENT_p.4. The reason for this increased elevation is not clear. Most surveys in the Haycock area were based on the 1794 Solomon Cushing survey; hence the 1935 boundaries and ownership were hard to determine. Haycock Harbor reservoir would require 11,560 acres from some 200 deeded parcels. *Final-36*, 14.

245. *ES*, 13 November 1935, 1. *Final-36*, 16. Lake Utopia in St. George, New Brunswick, was also considered.

246. *ES*, 13 November 1935, 1. The army's rationale for changing plans to Cobscook Bay being a high pool instead of a low pool is in *Final-36*, 16.

247. *Reel-38*, 20–21.

248. *Reel-38*, 9–10.

249. *Reel-38*, 12, 28.

250. *ES*, 27 April 1927, 4.

251. Wikipedia, s.v. "Public Utility Holding Company Act of 1935"; and Riggs, 133. At the same time, the Federal Power Act of 1935 became law, giving the FPC the authority to regulate electric transmission rates for the first time. Retail rate regulation was left to the states. Weil, 29–30.

252. FDRL-0284: Power 1935 Aug-Dec: 1935-09-03_Rural Electrification to FDR_PUHCA avoidance; and FDRL-0284: Power 1935 August–December: 1935-09-13_FPC McNinch to FDR_Fergusson Hartford Electric Light. In Fergusson's defense, he had criticized speculators and holding companies during congressional testimony in 1928. Funigiello, 9.

253. *ES*, 4 September 1935, 1; and *ES*, 11 September 1936, 1.

254. The last advertisement for the sale of the *Sentinel* appeared in *ES*, 21 August 1935, 8.

255. *ES*, 21 August 1935, 1.

256. The US Postal Service recorded an average of $2,000 of money orders sent per week. WNAC Radio address by Will Beale for Passamaquoddy Public Relations Association, 10 April 1936, 5 (of 5), BHS collection at TI&MA.

257. *ES*, 14 August 1935, 2.

258. *ES*, 24 July 1935, 1.

259. *ES*, 21 August 1935, 1.

260. *ES*, 4 September 1935, 1.

261. *ES*, 2 October 1935, 2. As of 23 October 1936: purchase orders issued, 11,548, totaling $2,250,000; payments, 14,000 accounts established; and more than 90,000 checks written totaling $6,651,407.51. *Final-36*, 31.

262. *ES*, 2 October 1935, 1.

263. Parkman, 231–32.

264. *Final-36*, 30.

265. *ES*, 8 April 1936, 7.

266. Parkman, 231.

267. *ES*, 18 September 1935, 2; *ES*, 31 July 1935, 1; and *ES*, 21 August 1935, 1.

268. *ES*, 30 October 1935, 4.

269. *ES*, 16 October 1935, 1.

270. *ES*, 16 March 1938, 10.

271. *ES*, 21 August 1935, 1.

272. *ES*, 9 October 1935, 1.

273. *ES*, 2 October 1935, 2.

274. *Portland Sunday Telegram and Sunday Press Herald*, 26 April 1935, B-8.

275. Goodwin, *No Ordinary Time*, 42.

276. GSC, "Passamaquoddy," 10.

277. *ES*, 28 August 1935, 1; and *BH*, 28 August 1935, 24.

278. The nine houses on Redoubt Hill were actually built by John H. Simonds Company of Portland for $150,632. *ES*, 8 April 1936, 7. In a document titled "Failure to Carry Out Plans as Publicly Announced or Contracted for," in Roscoe Emery's files, an item dated 27 August 1935 reads, "Bids opened for nine permanent houses on Redoubt Hill. John H. Simonds Co., Portland awarded contract @ $150,632 (John Osborne, West Medford, Mass., $102,008; Charles A. Mitchell, Eastport, $111,840). Contract called for completion November 4. Only 2 ready for occupancy November 27, remaining 7 at intervals of 4 days each." EMERY_5.11. Smith, *Power Policy*, 250, says the total cost estimates for the nine houses, building the road up to them, and so forth, ranged from $150,000 to $300,000, citing *PPH*, 16 January 1936. According to the army (*Final-36*, 24), total costs for Snob Hill were $245,232,33. See appendix 4, "Vital Statistics," for original estimates and actual expenses.

279. *ES*, 28 August 1935, 1.

CHAPTER 10. YELLOW LIGHTS (1935–1936)

1. *ES*, 4 September 1935, 1.

2. Riggs, 138.

3. *ES*, 25 September 1935, 1.

4. *NYT*, 24 September 1935, 4.

5. *ES*, 2 October 1935, 4, quoting *BDNs*. See also FDRL-0982: 1935-09-06_Sec War Dern to FDR_Quoddy Authority; and FDRL-0982: 1935-09-10_Stanley Reed to FDR_Quoddy Authority.

6. Mary Kingsbury Simkhovitch (1867–1951) was a pioneering social worker and a founder of the Greenwich House in New York City, an immigrant settlement and community development organization.

7. *ES*, 16 October 1935, 3, quoting Mary Kingsbury Simkhovitch, "Passamaquoddy Dam and Eastport," *American City Magazine*, October 1935.

8. FDRL-0982: 1935-09-26: FDR to Markham_Brann visit follow-up. Governor Brann met with FDR on 20 September 1935 at noon in Hyde Park, New York. FDRL-Day-by-Day: 20 September 1935.

9. EMERY_2.19: J.E. Barlow, Portland, Maine, City Manager, to Paul Edwards, Acting E.R.A. Administrator, Augusta, Maine, 1 October 1935; and form letter "To the Mayors of Cities and Selectmen of Towns of Maine," 7 October 1935.

10. EMERY_Scrapbook-Quoddy 1933-6: *PEN*, 8 September 1933. Rates mentioned are for summer 1933.

11. EMERY_2.19: Letter to Arthur F. Minchin, Safety Division, E.R.A., Augusta, Maine, 23 March 1935; letter to Walter Brennan, State Highway Dept., Augusta, Maine, 23 March 1935; and form letter from John A. McDonough, State Administrator, Emergency Relief, "To Local Administrators," 25 August 1934.

12. EMERY_2.19: John Bache-Wiig, Area K. Supervisor, ERA, State of Maine Emergency Relief Administration, Calais, Maine, 29 January 1935.

13. *ES*, 16 March 1938, 10.

14. *NYT*, 23 October 1935.

15. *ES*, 23 October 1935, 1.

16. *ES*, 30 October 1935, 3.

17. *ES*, 23 October 1935, 1.

18. *ES*, 30 October 1935, 4.

19. *ES*, 30 October 1935, 4.

20. FDRL-0982: 1935-10-16_Dexter to James Roosevelt_Quoddy sale negotiations; and *Final-36*: appendix A-3. See also *NYT*, 29 October 1935, 6; and *ES*, 30 October 1935, 1. Westinghouse held the shares because it was the original investor; GE, BE, and MWU joined later.

21. *ES*, 4 April 1934, 1, 7; *Final-36:* appendix A-3, exhibit G. See also *Reel-38*, 9. [Frank Reel:] "I found that there was quite a question of whether or not the Westinghouse company, which had owned 850 shares of Dexter Cooper, Incorporated, stock, had relinquished such ownership."

22. *NYT*, 18 February 1936, 1.

23. *Final-36*: appendix A-3, exhibit G includes letters from and to Dexter from each of the companies, all of which wish him well or encourage him in his efforts to secure government funding for Quoddy, though they may not have known the details of Dexter's "Power & Metals" Plan, for which he was seeking the funds.

24. *ES*, 23 October 1935, 1.

25. Walter Davenport, "High Tension Jitters," *Collier's*, 25 October 1935. A copy is in EMERY_6.21.

26. *NYT*, 3 November 1935, E7.

27. FDRL-0982: 1935-11-05_Brann to FDR_Brann's position.

28. *ES*, 6 November 1935, 1. Samuel Pierpont Langley (1834–1906) was the third secretary of the Smithsonian Institution. It was Langley's test pilot, Charles Manly, who was fished out of the Potomac on 7 October 1903.

29. *ES*, 20 November 1935, 1.

30. *ES*, 21 August 1935, 7.

31. *ES*, 13 November 1935, 1.

32. *ES*, 13 November 1935, 1.

33. *ES*, 20 November 1935, 1.

34. *ES*, 13 November 1935, 1.

35. An Advent calendar counts down to Christmas. The first day of Advent varies between 27 November and 3 December. The calendar is constructed such that a numbered door is opened on each day, revealing a religious image, token, or treat.

36. *ES*, 13 November 1935, 1.

37. *ES*, 16 Mar 1938, 10.

38. *ES*, 15 April 1936, 1. As of 23 October 1936, the total had grown to 1,090 railroad carloads, amounting to 31,000 tons from 22,000 items. *Final-36*, 31.

39. *American Magazine*, December 1935, 24.

40. *ES*, 20 November 1935, 1.

41. *ES*, 20 November 1935, 1.

42. *ES*, 6 February 1935, 4, quoting *PEN*.

43. *ES*, 27 November 1935, 8.

44. *ES*, 11 December 1935, 8.

45. *ES*, 4 December 1935, 1. Fleming will occupy house 1 on the north end of semicircle, Capt. Lord 2, Sturgis 3, and Casey 4. *ES*, 27 November 1935, 8.

46. *ES*, 4 December 1935, 1. This proved to be the highest level of employment on the project.

47. *ES*, 27 November 1935, 5.

48. *ES*, 27 November 1935, 5.

49. *ES*, 4 December 1935, 1.

50. *ES*, 16 March 1938, 10.

51. *ES*, 11 December 1935, 1. House A-14-L was on the southwest corner of the oval on Kendall Street facing south. National Archives_RG-71.

52. Parkman, 232.

53. *ES*, 18 December 1935, 1.

54. The *Sea King* was subsequently sold for $18,500 and left Eastport on 29 January 1936. Apparently it was not a particularly good yacht, bunk house, or tow boat. *ES*, 25 December 1935, 1; and *ES*, 29 January 1936, 1.

55. *ES*, 4 December 1935, 1.

56. FDRL-0982: 1935-12-24_Gen. Pillsbury to McIntyre.

57. FDRL-0982: 1935-12-24_Gen. Pillsbury to McIntyre_$62MM estimate: "[I]nstead of the $35,000,000 estimated by the Special Commission which reported on this project early this year." The army had raised the red flag about probable cost increases as early as 6 September, when Secretary of War Dern wrote to FDR (FDRL-0982:1936-09-05). In Dern's letter, he agrees with the proposal of the attorney general to halt heavy construction work on Quoddy because of the pending constitutional questions (i.e., TVA) and the lack of action on the Quoddy Authority Bill by the Maine Legislature, but argues for continuing engineering work; the completion of Quoddy Village, and the preparation of revised estimates by 1 March 1936.

Dexter made a comparative analysis of his estimate of December 1934, versus the army's plan of September 1935. The first version of this list included $30.3 million for Cooper's estimate and $50 million for the army's, with other big differences being for the Eastport-Lubec Dam: for Dexter $3.5 million, for the army $10.3 million, and +$5 million for housing. DPC-III: 1935-09_SUMMARY—LOW POOL DEVELOPMENT_p.1–3.

58. Roberts, *Trending into Maine*, 372: "President Roosevelt himself, in spite of the gigantic outlay of taxpayers' money which he had cheerfully authorized in every other state, is reported on good authority to have used profanity in condemning the high cost of the Quoddy Project and the extravagance of army engineers."

59. *ES*, 8 January 1936, 1; and *ES*, 15 January 1936, 1.

60. *ES*, 8 January 1936, 1.

61. *ES*, 8 January 1936, 4.

62. *ES*, 8 January 1936, 1.

63. *NYT*, 31 December 1935, 5.

64. FDRL-0982: Governor Bran to President Roosevelt, 4 January 1936.

65. Ickes, 513–14: "The President told George Dern that he ought to fire all his Army Engineers. He went on to tell me General Markham had made a very labored written report to him, and he has had General Markham and Major Fleming in. The $35 million originally estimated as the cost will take care of only one pool and the power produced from that will be more than if it were produced from coal. The President said that, of course, this would never do. He also said that word is quietly going around Boston that the Passamaquoddy project will never be built. I told the President that I always had my fingers crossed with respect to the Army engineers when it came to power projects, and he said, 'That's right.' I also told him that I thought their figures on the cost of the Florida canal were 'cock-eyed.' The President told me that Dexter Cooper is in Washington and that Cooper said he will stake his reputation as

an engineer that the whole project can be built within his original estimate. He told me to send for Cooper tomorrow and have him tell me his story. Then he wants me to send two or three of our best reclamation engineers up to Passamaquoddy to study the project. He told me that I could justify this by saying that since PWA money was being used, I want to check back and see just what the actual cost will be. He said that he would tell Dern that I was doing this on his instructions."

On behalf of Governor Brann, Jack Maloney polled the Maine Legislature's willingness to hold a vote on the Quoddy Authority Act and found 139 in favor, 6 opposed. Whether those legislators would actually vote for the act in the same proportion was a different question entirely. FDRL-0982: 1935-12-31_Maloney to Brann_Q Authority likely vote totals.

66. *NYT*, 14 December 1935, 4.

67. *ES*, 5 January 1936, 4.

68. FDRL-Day-by-Day: 2 January 1936.

69. *ES*, 15 January 1936, 1; and *NYT*, 25 January 1936.

70. FDRL-0982: 1936-01-10_Markham to FDR_cost revision. The last paragraph of this letter, containing the $36.5 million estimate, $2 million of immediate funding to carry on through 1 April, and $5 million to be available thereafter, is marked in pencil, "Ok'd by FDR, Jan 11 / 35." See also FDRL-COOPER: General Markham to FDR, 10 January 1936. See also *ES*, 15 January 1936, 1, 4. *However*, in responding to Senator Vandenberg's query, on 22 September 1936 General Markham told Vandenberg, "The estimated cost of the project with an auxiliary plant of this type [diesel or steam] was materially less than the cost of the pumped storage plant, being approximately $37,000,000." FDRL-0982: 1936-09-22_Markham to Vandenberg_$62 Million Durand estimate. Markham's statement was reinforced by FDR. FDRL-0982: 1936-09-26_FDR to Kannee_cost estimates lowest $37MM.

71. *ES*, 15 January 1936, 1.

72. FDRL-0982: 1936-01-10_Markham to FDR_cost revision; and FDRL-COOPER: 1936-02-04_Dexter Cooper to James Roosevelt_angry at Army Engineers' report. The report, *Passamaquoddy Tidal Power Development Estimates*, was dated 11 January 1936.

73. FDRL-COOPER: 1936-02-04_Dexter Cooper to James Roosevelt_angry at Army Engineers' report.

74. See chapter 10, "Resistance," and *Cooper-NPPC-36*, 6.

75. FDRL-0284: 1935-01-11_[unknown] to FDR_Couch rural electrification.

76. Ickes, 513–14; and *Final-36*, 3. The Reclamation Service investigation of Quoddy was headed by Dr. W. M. Durand of Stanford University and included engineers Charles H. Paul and Joseph Jacobs; hereafter referred to as *Durand-36*. See also Abrams, 263.

77. *ES*, 22 January 1936, 4.

78. *ES*, 1 January 1936, 1.

79. *ES*, 22 January 1936, 4.

80. *ES*, 15 January 1936, 1; and *ES*, 22 January 1936, 1.

81. *ES*, 22 January 1936, 1.

82. *ES*, 22 January 1936, 1. The army had rented twenty separate offices in town. *Final-36*, 28.

83. *ES*, 15 January 1936, 1; and *ES*, 8 April 1936, 7, a reprint of an excellent article by C. H. Vivian in *Compressed Air* magazine, serialized in the *Sentinel* on 1 and 8 April. *ES*, 31 July 1935, 8, says the plan was for 320 feet.

84. *ES*, 5 January 1936, 1.

85. *ES*, 15 January 1936, 1.

86. *ES*, 22 January 1936, 1. From the date it opened until 23 October 1936 (21 months), it received 78,750 pieces of mail and dispatched 81,580. *Final-36*, 31.

87. Marvin, 70, citing "Moon Unharnessed from Recovery," *Literary Digest*, 2 February 1936, 9.

88. *ES*, 19 February 1936, 8.

89. GSC, "Passamaquoddy," 10.

90. Roberts, *Trending into Maine*, 371.

91. See also *Reel-38*, 3–4; Roberts, *Trending into Maine*, 382–83, Smith, *Power Policy*, 250, citing Roberts, *Trending into Maine*; *Reel-38*, 3–4; and *Fleming-49*, 5–7.

92. *ES*, 8 January 1936, 1, 4.

93. *ES*, 8 January 1936, 1.

94. *ES*, 22 January 1936, 8.

95. *NYT*, 18 February 1936, 1.

96. EMERY_box 5, folder 11: "Correspondence, Philip Fleming etc. 1934–1940." See also 3 January 1936: Roscoe agreed to sell 11.8 acres on Carlow Island for dam abutment and/or quarry; this was the basis for a later bitter dispute that went on between 1937 and 1940.

97. *ES*, 12 February 1936, 1.

98. *ES*, 12 February 1936, 1.

99. *ES*, 1 April 1936, 1.

100. *NYT*, 16 February 1936, E11.

101. Total projected costs were variously reported as $42–$44 million for the Quoddy Tidal Power Project in Maine; $138–$150 million for the Florida ship canal, $8.5 million for the Conchas dam in New Mexico; $13 million for the Bluestone reservoir in West Virginia; and $10.5 million for the Sardis reservoir in Mississippi. *ES*, 19 February 1936, 1. See also *NYT*, 22 February 1936, 5.

102. *ES*, 12 February 1936, 1. See also *NYT*, 11 February 1936, 1.

103. *ES*, 12 February 1936, 1, 5, 8.

104. *ES*, 12 February 1936, 1.

105. *ES*, 4 December 1935, 1; and *ES*, 18 December 1935, 1.

106. *ES*, 19 February 1936, 1.

107. The rental rates were calculated based on the assumption of a five-year project. *Final-36*, 34.

108. *ES*, 26 February 1936, 1.

109. *ES*, 9 February 1936, 1.

110. *ES*, 26 February 1936, 1.

111. One diesel generator cost $47,361.79. *Final-36*, Cost Summary sheet 5. The cost of the building sized for five generators is not indicated in *Final-36*. See also *BHNE*, March 1936, 9, http://digicom.bpl.lib.me.us/bangorhydro_news/34, for connecting generator to BHE network.

112. *NYT*, 16 February 1936, E-11.

113. *NYT*, 16 February 1936, E-11.

114. FDRL-0284: 1934-12-18_Thompson to FDR_friendly suit to test constitutionality of the TVA.

115. Riggs, 148.

In *Ashwander v. Tennessee Valley Authority*, 297 U.S. 288 (1936), the US Supreme Court was asked to decide whether the federal government had a constitutional right to generate and sell electric power. The plaintiffs, the preferred shareholders of the Alabama Power Company, argued that the government was undertaking a "coup" that would "open every essential industry and service to direct and permanent governmental competition." The plaintiffs challenged the TVA's "validity in every respect, in its entirety, and in detail, asserting that the program and all of its essentials, the means employed to promote it, its dominant objectives, and its arbitrary methods, do not consist with the letter and spirit of the Constitution." The court decided the government had the right to be in the power generation business, and the plaintiffs' other claims were too broad and hypothetical to warrant further action by the court. Wikipedia, s.v. "*Ashwander v. Tennessee Valley Authority*," https://en.wikipedia.org/wiki/Ashwander_v._Tennessee_Valley_Authority#cite_ref-1.

116. Interestingly, there was no objection when the Maine Public Utilities Commission authorized the Maine Central Railroad to take land for "public use" in the creation of spur lines in Eastport to service Quoddy Village. *ES*, 22 January 1935, 1; and *ES*, 12 February 1935, 1.

117. EMERY_5.7b. March 1943. Exact date uncertain. For a private report on the power trust propaganda effort, see FDRL-0982: 1936-02-24_Frederick Greene to Morgan_Power Trust propaganda.

118. *NYT*, 23 February 1936, E-4. The American Federation of Utility Investors was largely a front for and financially supported by the private power industry, as was the Women Investors of America. Funigiello, 102–3.

119. Riggs, xxx, 152.

120. *ES*, 26 February 1936, 3.

121. *ES*, 26 February 1936, 3.

122. *ES*, 26 February 1936, 3.

123. National Archives Identifier 1110318: RG-71, Records of the Bureau of Yards and Docks, Naval Property Case Files, 1941–1958: Maine, 20, Quoddy Village, C20-9-QV, box 538. The army sought permission from the state of Maine to take a piece of tribal property at Pleasant Point as an additional source of stone for the dams. *ES*, 1 March 1936, 1.

124. *ES*, 11 March 1936, 1.

125. Brands.

126. A provision of the National Industrial Recovery Act (NIRA) was dissolved in 1935, by the court case *Schechter Poultry Corp. v. United States*.

127. Cross, 93.

128. Cross, 93–94, citing Ward Geoffrey, *Closest Companion* (Boston: Houghton Mifflin, 1995), 72. The USS *Potomac* was built in 1934 as a submarine chaser and converted to presidential service in 1936.

129. *ES*, 15 April 1936, 7.

130. The US Army Corps of Engineers' 46-foot launch *Machias* was built by American Car and Foundry Co. It cost $16,800 (*ES*, 31 July 1935, 1) and arrived as a new boat from Wilmington, Delaware, on Monday, 15 July 1935. It was named by Major Fleming "in honor of the Washington County town where occurred the first naval battle of the War of Independence." *ES*, 17 July 1935, 1. President Roosevelt wrote to the chief of the engineers on 16 June 1935, "I hear that bids have been opened for certain floating equipment for the Quoddy project. One of them a boat 45 ft. long for engineer's use. It seems to me that a great deal of floating equipment can be obtained from other Departments. For instance, the Coast Guard has a number of boats of various sizes which are laid up. I feel certain that it is unnecessary to buy any 45 foot boat. There must be a lot of other equipment now owned by the government which could be used thereby keeping down the overhead under the allotment. F.D.R." FDRL-0982: 1935-06-17_FDR to Chief of Engineers. The Treasury answered for the Coast Guard that no such vessel was available. FDRL-0982: 1935-06-22_Treasury to FDR_no yacht like vessel available. The army also said the search was conducted before the requisition was placed. FDRL-0982: 1935-06-19_War Dpt to FDR_previous search for yacht. In 1925, Dexter had built a similarly sized vessel, the *Reconnaissance*, for $5,967.25, a photo of which appears in

Harendy et al., *Campobello Island*, 47. Presumably it would have suited the need. It was listed in the PWA's audit of May 1935 with a fair market value of about $500. *Final-36:* appendix A-3, narrative p. 14.

131. This would be *Durand-36.*

132. Ickes, 542.

133. The Florida Ship Canal (aka Cross Florida Barge Canal) was proposed to follow the route of St. Mary's River south from Jacksonville to Palatka and then cut west across the state for about 107 miles and emerge in the Gulf of Mexico in Citrus County. Construction was started and stopped several times. (It is often confused with Okeechobee Waterway, which was completed in 1937 and can accommodate barges up to 50 feet wide, 250 long, and 10 feet deep. It runs approximately 110 miles from Stuart to Fort Meyers.)

134. *NYT*, 10 March 1936, 9.

135. *Durand-36.*

136. *ES*, 11 March 1936, 1.

137. *ES*, 11 March 1936, 1.

138. Smith, *Power Policy*, 253, citing *PPH*, 1 February 1939.

139. FDRL-COOPER: 1936-01-16_Dexter Cooper to FDR_All Quoddy Power sold.

140. *NYT*, 12 March 1936, 8.

141. *ES*, 18 March 1936, 4.

142. *ES*, 18 March 1936, 4.

143. *ES*, 18 March 1936, 1.

144. *ES*, 18 March 1936, 1.

145. *ES*, 25 March 1936, 1, 5.

146. *NYT*, 1 April 1936, 24.

147. There were at the time 25 Republican and 71 Democratic senators. Maine's junior senator, Wallace White, sat in the third row.

148. *ES*, 25 March 1936, 1.

149. *ES*, 25 March 1936, 1.

150. Irwin, 95, citing CMP *Exciter*, March 1946, 7. Brunswick is on the Androscoggin River, Shawmut is on the Kennebec River, and Union Falls and West Buxton are on the Saco River. See also *BHNE*, March 1936, http://digicom.bpl.lib.me.us/bangorhydro_news/34.

151. G. J. Stewart, J. P. Nielsen, J. M. Caldwell, and A.R. Cloutier, *Water Resources Data: Maine, Water Year 2001* (US Geological Survey, US Department of the Interior, 2002), https://me.water.usgs.gov/reports/Maine01adr.pdf, 126. The flow on Kennebec had ranged from 1,160 cubic feet per second (7 July 1988) to 232,000 cfs (2 April 1987), or a range of 1:200. https://en.wikipedia.org/wiki/Kennebec_River, citing *Stewart et al.,Water Resources Data.*

152. *ES*, 25 March 1936, 6.

153. *ES*, 1 April 1936, 1.

154. *ES*, 8 April 1936, 1.

155. Only 2% of BHE's power was consumed east of Machias. *ES*, 1 April 1936, 1.

156. *ES*, 1 April 1936, 1.

157. *ES*, 18 March 1936, 4.

158. *ES*, 1 April 1936, 1.

159. *ES*, 13 February 1935, 1; *ES*, 1 April 1936, 1; and *ES*, 22 April 1936, 7.

160. WNAC Radio address by Will Beale for Passamaquoddy Public Relations Association, 10 April 1936, 5, BHS.

161. FDRL-0982: 1936-01-11_Pastors to FDR_impropriety. This letter from Reverends Blaney and McKay was declassified on 25 February 1972. Marvin, 74a. See also FDRL-0982:1936-02-04_FDR to PWA_impropriety charges.

162. FDRL-0982: 1936-03-04_Ickes to FDR_undercover investigation of complaints.

163. *Report on Alleged Improper Official Activities in Connection with the Passamaquoddy Project at Eastport, Maine*, 19 February 1936, communicated by Louis Glavis to Secretary Ickes, 4 March 1936, FDRL-0982. The report was considered "confidential," but no date of declassification was given. Marvin, 74a. See also a description of cultural differences between natives and imported workers in *American Magazine*, December 1935, 159.

164. FDRL-COOPER: 1936-04-13_Gertrude to Franklin_Rosoce Emery.

165. FDRL-COOPER: 1936-04-18_FDR to Gertrude Cooper_will see Emery.

166. FDRL-PSF: 1936-04-24_Gertrude Cooper to FDR_clandestine meetings and Roscoe Emery.

167. FDRL-Day-by-Day: 1936.

168. *NYT*, 12 April 1936, 3.

169. *ES*, 22 April 1936, 6. See also *NYT*, 16 April 1936, 2.

170. The fable of the Tar Baby is about a doll made of tar used by the villainous Br'er Fox to entrap Br'er Rabbit. The more that Br'er Rabbit fights the Tar Baby, the more entangled he becomes. Originally a slave tale, it is said to have been first transcribed and published by Robert Roosevelt (uncle of President Theodore Roosevelt) in *Harper's* magazine in the 1870s and, a few years later, more successfully retold by Joel Chandler Harris. The story is criticized for its racist undertones. Wikipedia, s.v. "Robert Roosevelt," https://en.wikipedia.org/wiki/Robert_Roosevelt.

171. *ES*, 22 April 1936, 6.

172. EMERY_6.20.19: 21 April 1936: *Press Herald Bureau*.

173. EMERY_6.20.19: 21 April 1936: *Press Herald Bureau*.

174. *NYT*, 15 April 1936, 3.

175. *NYT*, 19 April 1936, E-10.

176. *NYT*, 19 April 1936, E-10.

177. FDRL-0982: 1936-01-10_Markham to FDR_cost revision. The last paragraph of this letter, containing the $36.5 million estimate, $2 million of immediate funding to carry on through 1 April, and $5 million to be available thereafter, is marked in pencil, "Ok'd by FDR, Jan 11 / 35." See also *ES*, 15 January 1936, 1, 4.

178. *ES*, 29 April 1936, 5.

179. FDRL-0982: 1936-09-17_Vandenberg to General Pillsbury_cost estimates. How and when the Reclamation-Durant report became public is not clear, but its $68 million cost estimate is referenced in a speech in the fall of 1937 by Henry E. Riggs, president of the American Society of Civil Engineers, stating that it "reduced the tidal . . . project to an absurdity." Likewise, the army's $62 million estimate appeared in the society's *Proceedings* in February 1938. Abrams, 263–65.

180. Reprinted in *ES*, 22 April 1936, 4.

181. Reprinted in *ES*, 29 April 1936, 4.

182. *ES*, 5 May 1936, 4.

183. EMERY_6.21.12-16: *Portland Sunday Telegram and Sunday Press Herald*, 26 April 1936.

184. Bainbridge Colby (1869–1950) was a cofounder of the National Progressive Party in 1912, US secretary of state (1920–1921) under President Wilson, and a Democrat thereafter.

185. EMERY_6.20: *Portland Sunday Telegram*, 26 April 1936.

186. *ES*, 22 Apr 1936, 1. "In his message of transmittal of the Giant Power report, [Gifford] Pinchot compared the electrical supply revolution to the steam revolution. 'Steam,' Pinchot wrote, 'might well say of electricity, "One mightier than I cometh, the latchet of whose shoes I am not worth to unloose."'" Hughes, 304.

187. EMERY_ 6.20.19: *Portland Sunday Telegram*, 24 April 1936.

188. *ES*, 29 April 1936, 1.

189. *ES*, 5 May 1936, 1.

190. Westbrook Pegler, "Fair Enough," *Green Bay (WI) Gazette*, 29 July 1936.

191. Pegler, "Fair Enough"; and *ES*, 13 May 1936, 1. The hospital had 24 beds.

192. *ES*, 5 May 1936, 1.

193. *ES*, 13 May 1936, 1.

194. *ES*, 20 May 1936, 1.

195. *ES*, 13 May 1936, 1; and FDRL-0982: 1935-09-26: FDR to Markham_ Brann visit follow-up.

196. *ES*, 13 May 1936, 1.

197. Holmes, 63, citing *Portland Evening Express*, 14 May 1936.

198. Marvin, 78–79, citing *Florida Ship Canal and Passamaquoddy Tidal Power Projects, Hearings on Senate Joint Resolution 266, Senate Comm. on Commerce*, 74th Cong., 2nd Sess. (20 May 1936), 3–10.

199. *ES*, 20 May 1936. 1.

200. *ES*, 20 May 1936, 1.

201. *ES*, 27 May 1936, 1.

202. *ES*, 27 May 1936, 8.

203. *Madera (California) Tribune* 63, no. 247 (28 January 1955), accessed 25 October 2018, https://cdnc.ucr.edu/cgi-bin/cdnc?a=d&d=MT1955 0128.2.34&e=-------en--20--1—txt-txIN--------------1.

204. *Madera (California) Tribune* 63, no. 247.

205. *Madera (California) Tribune* 63, no. 247.

206. *NYT*, 27 May 1936, 33.

207. *Madera (California) Tribune* 63, no. 247.

208. Holmes, 63, citing *PPH*, 27 May 1936.

209. *ES*, 16 September 1936, 1, 6, quoting Kenneth Roberts, "QUODDY," *Saturday Evening Post*, 19 September 1936.

210. *NYT*, 31 May 1936, 1.

211. *NYT*, 30 May 1936, 15.

212. *NYT*, 31 May 1936, 1. See also US Congress, Senate, Debate on Passamaquoddy and Florida Ship Canal Projects, 80 Cong. Rec. pt. 6, 8391, 8313–14 (30 May 1936).

213. *NYT*, 31 May 1936, 1.

214. *NYT*, 31 May 1936, 1.

215. *ES*, 3 June 1936, 1.

216. *NYT*, 7 June 1936, 64.

217. *ES*, 3 June 1936, 1.

CHAPTER 11. RED LIGHTS (1936–1938)

1. *BH*, 5 June 1936, 5.

2. *ES*, 10 June 1936, 1.

3. *ES*, 3 June 1936, 4.

4. *ES*, 10 June 1936, 1.

5. *ES*, 10 June 1936, 1.

6. *ES*, 3 June 1936, 1.

7. *ES*, 3 June 1936, 1.

8. *ES*, 3 June 1936, 1.

9. *ES*, 3 June 1936, 1. Senator Duncan Fletcher (D-FL), a longtime friend of Quoddy, said, "The friends of Quoddy were responsible for its defeat." See also *NYT*, February 22, 1936, 5; *NYT*, 10 March 1936, 9; and *ES*, 29 April 1936, 5.

10. *ES*, 10 June 1936, 1. Daniel Oren Hastings (R-DE) (1928–1937).

11. *ES*, 3 June 1936, 4.

12. *ES*, 27 May 1936, 1.

13. *ES*, 1 July 1936, 4.

14. *ES*, 1 July 1936, 1.

15. *ES*, 29 July 1936, 1. There is an interesting satirical biography of Senator Frederick Hale at the Library of Congress, which includes brief descriptions of his achievements, for instance, "Chapter 3: His Record as a Business Man," followed by a single sentence: "In the course of time Mr. Hale will inherit a large fortune." Likewise, "Chapter 4: His Military Record": "Mr. Hale attained the rank of Colonel by serving on Governor Hill's staff for two years"; and "Chapter 7: His Qualification for the Office of United States Senator": [blank page]. *Frederick Hale: A Biographical Sketch*, https://archive.org/details/frederickhalebio00port/page/n1/mode/2up.

In a letter to the editor of the *Portland Press Herald* on 18 July 1936, Roscoe Emery wrote, "They [Hale and White] had a chance, ready-made and hard to muff, to do something for their constituents when the Quoddy resolution came up in the Senate, but they booted it with the result that thousands of Maine men are being thrown out of work and the hopes of a whole section of the State lies in ruins. . . . They both own luxurious homes in Washington. One at least comes to Maine only in the summer and some years not even then. His representation of Maine is and always has been a standing joke in Washington." EMERY, Roscoe_box 2.9b_Letters to Editor 1930-60s.

16. *ES*, 17 June 1936, 1.

17. *ES*, 24 June 1936, 1. *Final-36*, 31, says maximum 5,430 on 30 November 1935. *ES*, 4 December 1935, 1, says maximum of 5,122.

18. *ES*, 16 September 1936, 1; and *ES*, 5 August 1936, 1.

19. *ES*, 1 July 1936, 1.

20. *NYT*, 29 June 1936, 8.

21. *ES*, 1 July 1936, 1.

22. *ES*, 8 July 1936, 4.

23. Parkman, 234–35, says the shutdown order was issued on 6 July 1935.

24. *NYT*, 12 July 1936, E-12.

25. *BH*, 10 July 1936, 1, 10. Of the $7 million allocated to Quoddy, Colonel Fleming reported on 15 July 1936 the following summary of expenses: $2 million for engineering, $2 million for field construction, $1.5 million for housing for 2,400 laborers, and $1.5 million for materials and equipment. *ES*, 15 July 1935, 1. See appendix 4, "Vital Statistics," for actual expenses.

26. *ES*, 15 July 1936, 1.

27. *ES*, 15 July 1936, 1. These specific statements don't agree with general comments made by Hugh Casey in *The March of Time* (September 1936; generally positive) and that he made in 1993 in an interview in which he said the single-pool project was uneconomical because of the inherent cost of low-head hydroelectric generating stations, the problems of saltwater corrosion of metal and concrete, tidal power's lunar day being fifty minutes longer and hence out of sync with the solar day and regular peaking power demands, particularly the redundant costs of pumped storage. Casey also criticized what he appears to have mistakenly believed to be the plan to sell Quoddy power directly to local consumers through a redundant distribution system. Casey did think that the two-pool International Project could be economically justified but overall believed the support for Quoddy was primarily political, not practical. Hugh J. Casey, ed., *Major General Hugh J. Casey: Engineer Memoirs* (Washington, DC: United States Department of the Army, 1993). Copy courtesy of Wayne Wilcox. Previously available online, it was no longer available as of 8 May 2020.

28. The *Herald*'s sources and construction cost assumptions weren't identified.

29. *BHNE*, July 1936, 26.

30. *ES*, 15 July 1936, 1; *ES*, 22 July 1936, 1; and *ES*, 29 July 1936, 5.

31. *ES*, 8 July 1936, 1.

32. *NYT*, 29 July 1936, 5. During ninety hours of sailing, he had covered 437 miles since setting sail at Pulpit Harbor in Penobscot Bay (Maine), and only once had he set foot onshore, for the picnic at L'Etang Island (New Brunswick).

33. *ES*, 29 July 1936, 5.

34. Smith, *Power Policy*, 255, says 700 people attended the picnic at the Roosevelt Cottage on Campobello on 30 July 1936, citing *PPH*, 30 July 1936.

35. *NYT*, 30 July 1936, 1.

36. *NYT*, 30 July 1936, 1.

37. *NYT*, 30 July 1936, 1. Actually, the Aluminum Corporation of America, aka Alcoa.

38. *NYT*, 30 July 1936, 1.

39. *NYT*, 28 July 1936, 1.

40. *ES*, 29 July 1936, 1.

41. ER-My-Day: 25 July 1936.

42. ER-My-Day: 25 July 1936.

43. *ES*, 29 July 1936, 1.

44. *ES*, 29 July 1936, 1. Prime Minister McKenzie's visit to Eastport was on 22 July 1936.

45. *ES*, 21 July 1936, 1.

46. Parkman, 232–33.

47. *Fleming-49*.

48. "Entrepreneurs, Inventors & Innovators," https://digitalcommons.nyls
.edu/entrepreneurs_inventors_innovators/12/.

49. *ES*, 5 August 1936, 4.

50. *ES*, 12 August 1936, 1.

51. *ES*, 12 August 1936, 4. Aluminum production at Dnieper began in 1933.
NA-MD: Department of State: 1933-01-28_First year of Second 5-Year Plan,
CDF 1930-39: 861.6463/61, microfilm T-129-6, declassified 28 Jan 1980.

52. *NYT*, 11 April 1935, 20.

53. The mocking phrase "drive trolleys and milk cows" was also used in
"Money Flows into the Sea," *NYT*, 11 April 1935, 20; "Tides to Be Harnessed
In Maine," *NYT*, 26 May 1935, E11; and "Economic Moonshine," *NYT*, 22
January 1939, 68.

54. *NYT*, 2 August 1936, E8.

55. FDRL-0982: *BH*, 11 July 1936. The Quoddy Tidal Power Project was
mentioned in the *Boston Herald* more than 140 times in 1935 and as many in
1936—averaging two editorial comments (essentially all negative) and one
article per week.

56. FDRL-0982: *Boston American*, 6 July 1936.

57. EMERY_6.22b. Source and date not identified.

58. *ES*, 19 August 1936, 1.

59. *ES*, 19 August 1936, 1.

60. *ES*, 5 August 1936, 8.

61. FDRL-0982: 1936-08-12_War Dept press release on Quoddy.

62. *Saturday Evening Post*, 19 September 1936; "A History of the Saturday
Evening Post Inside and Out: More Than Just Covers," CollectingOldMaga-
zines.com, https://collectingoldmagazines.com/magazines/saturday-evening
-post/; "History of the Saturday Evening Post," https://www.saturdayevening
post.com/history-saturday-evening-post/. Robert's article was excerpted in *ES*,
16 September 1936.

63. *ES*, 2 September 1936, 6.

64. See also *War in China; Maine's Passamaquoddy*, 1937, https://player
.bfi.org.uk/free/film/watch-war-in-china-maines-passamaquoddy-1937-online.

65. *ES*, 2 September 1936, 7.

66. *ES*, 9 September 1936, 7, 8.

67. *ES*, 16 September 1936, 1.

68. *ES*, 23 September 1936, 4.

69. *ES*, 7 October 1936, 4. See also Ickes, 693.

70. FDRL-0982: 1936-05-18_Vandenberg to FDR_wants copies of new
Quoddy & Florida reports.

71. FDRL-0982: 1936-05-22_FDR to Vandenberg_evasive.

72. FDRL-0982: 1936-09-17_Vandenberg to General Pillsbury_cost estimates.

73. FDRL-0982: 1936-09-22_Clay of Engineers to Kannee_Vandenberg info request.

74. *Final-36*, 4.

75. FDRL-0982: 1936-09-22_War Dept to Vandenberg_$62 Million Durand estimate.

76. Ickes, 513–14; and *Durand-36*, 7. Dexter had allowed 25% for settlement and drift. *Cooper-FPC-28*, M-11.

77. *ES*, 21 October 1936, 1.

78. *ES*, 23 September 1936, 4. Gertrude echoed the charge in her memoir (1970): "Westinghouse and the others were tied in with the power companies and did not want cheap power like the TVA. . . . Representatives of the power companies even informed Dexter that they could stop it from being built by the government, buy it later at a fraction of the cost, and build it with private capital. But Dexter felt the truth, if it were revealed, would hurt the morale of the country, and so he worked doggedly on." GSM, "Passamaquoddy," 11–13.

79. *ES*, 4 November 1936, 4.

80. *ES*, 4 November 1936, 4. Kim Philips-Fein, in *Invisible Hands: The Businessmen's Crusade against the New Deal* (New York: Norton, 2009), attributes this to Roosevelt's campaign manager. See also *NYT*, 5 November 1932, 2.

81. *ES*, 16 December 1829, 1.

82. "Selection bias" (also "confirmation bias") is the tendency to seek, find, and interpret information in a way that reinforces one perspective and discounts alternatives.

83. *BH*, 10 July 1936, 1, 10: Richard Boyer, "Army Engineers Bitter."

84. *ES*, 16 March 1938, 10. There was a "[n]et unobligated balance of $805,302.44 unexpended and available for future work." This amount did not include the costs incurred for work after 23 October 1936, such as repairs to the Carlow Island dams. *Final-36*, 7.

85. *NYT*, 3 November 1935, E7.

86. *ES*, 30 September 1936, 1. The Army Engineers ceased all active operations on 16 August. On 31 October, the Eastport District was discontinued, and operations relating to the final demobilization of the project were placed under the direction of the Boston District. About $5.9 million has been expended on the project. Parkman, 236–37.

87. FDRL-0982: 1936-08-06_Paulsen to FDR_use Quoddy Village for worker training.

88. *Final-36*, 37.

89. *Final-36*, 1.
90. *Final-36*, 2, 8.
91. *Final-36*, 11.
92. *Final-36*, 3 (emphasis added).
93. *Final-36*, 6.
94. *Final-36*, 9. This also contradicts the army's press release of 12 August 1936, issued while Lt. Col. Fleming was in charge.
95. *Final-36*, 10–11.
96. *Cooper-NPPC-36*, 3.
97. See appendix 4, "Vital Statistics."
98. *Final-36*, 13.
99. *Final-36*: appendix A-3, narrative p. 10. In addition, Dexter had invested personally approximately $35,000 in the project between 1914 and 1924. Westinghouse was the original investor in 1925. GE, BE, and MWU joined on 31 October 1927.
100. Some $450,000 annually for Washington County alone. *Reel-38*, 23.
101. *ES*, 30 December 1936, 4.
102. FDRL-COOPER: 1936-12-22_K to McIntyre_Dexter trying to collect; and FDRL-COOPER: 1937-01-04_War Department to FDR Chief of Staff_payment due Dexter. *Final-36*, 3, states, "Under this agreement an initial payment of $50,000 was to be made and a further payment of $10,000 was to be paid upon the fulfillment of any one of three specific conditions set forth in the contract. One of those conditions was based on the event of abandonment of the project by the United States and based on a provisions of this particular clause, a claim was filed in July 1936 by Dexter P. Cooper with this office for the payment of the remaining $10,000, and forwarded to the General Accounting Office for pre-audit."
On page 17 of *Final-36*, the army refuted Dexter's claim of breach of contract caused by the army's switching Cobscook Bay from low pool to high: "No evidence was submitted that such was the case, as the terms of this agreement did not include any specific provisions in regard to whether the pools are maintained at high or low tide levels but state only, 'The Government agrees that it will construct and utilize the dams, structures, and/or power plants in or near Cobscook or Passamaquoddy Bays, to be constructed by it, in such manner as not (unless physical conditions at the site shall otherwise require) unduly to impair the feasibility of the further construction by Cooper, or the Canadian Corporation, or others, of other dams, structures and/or power plants within the Dominion of Canada.'"
103. "The northeast slope [of the dam] resembles a sharply sloping pavement of huge granite blocks, placed so cleverly, one can easily imagine it the work of Paul Bunyon, gripped with the jig-saw fever. The big derrick booms

have handled the granite blocks just as nimbly and accurately as mammoth fingers could have done with intricately shaped pieces of a game. It has been most interesting, in fact perhaps the most interesting job of anything done here at Quoddy since the start of operations." The rock fill came from Quoddy quarries; granite for the armor coat came from quarries in Franklin and Frankfort, Maine. *ES*, 17 February 1937, 1. On 6 October 1936 Captain Sturgis estimated 6,000 cubic yards of rock would be needed, but on 25 November ordered 15,000 cubic yards. Repairs to the Carlow Island dams were completed on 15 March 1937 (EMERY_5.11: "Failure to Carry Out Plans as Publically Announced or Contracted for").

104. *ES*, 30 December 1936, 1.

105. *ES*, 30 December 1936, 1.

106. *ES*, 27 January 1937, 5.

107. *ES*, 6 January 1937, 1. See also *BDN*, 4 January 1937, 1.

108. *ES*, 3 February 1937, 4.

109. *ES*, 3 February 1937, 1. Organized under the auspices of the National Resources Committee for Public Works Planning, the "blue-ribbon" committee was chaired by Harold Ickes of the PWA and included Secretary of War Harry H. Woodring; Secretary of Agriculture Henry A. Wallace; Secretary of Commerce Daniel A. Roper; Secretary of Labor Frances Perkins; WPA head Harry L. Hopkins; Frederic A. Delano and Charles E. Merriam of the NRC; businessman Henry S. Dennison; and Beardsley Ruml, chair of Macy's department stores and head of the New York Federal Reserve. See also *ES*, 6 October 1937, 1.

110. *ES*, 3 February 1937, 4. See also FDRL-COOPER: 1937-11-08_MAD to FDR_Dexter message. "A suggested alternative was for the Bangor Hydro-Electric Company to operate Quoddy as a corporation subsidiary to the state or federal government, thus attempting to combine both public and private power interests." Smith, *Power Policy*, 258, citing *PPH*, 18 January 1939.

111. *ES*, 3 February 1937, 4, citing *PPH*.

112. *ES*, 31 March 1937, 1.

113. One diesel generator cost $47,361.79, and the building to house all five generators cost another $14,941.22. The cost of the electricity purchased from BHE is not listed. *Final-36*: Cost Summary, Sheet 5, 7.

114. *ES*, 23 September 1936, 1.

115. *ES*, 18 November 1936, 1.

116. *ES*, 17 February 1937, 1.

117. *ES*, 24 March 1937, 1.

118. *ES*, 28 October 1936, 2.

119. *ES*, 24 March 1937, 1.

120. *ES*, 3 March 1937, 4.

121. *ES*, 17 March 1937, 2.

122. EMERY_5.7b. March 1943. Also in Roscoe's files is a two-page document titled, "Failure to Carry Out Plans as Publically Announced or Contracted for," which contains a list of nine items dated 27 June 1935 to 23 December 1936. The most serious item would appear to be that the contract to build the nine permanent houses on Redoubt Hill was not awarded to the low bidder but to the highest bidder. It is unclear whether this document is all or part of what Roscoe referred to in his letter to Dexter Cooper.

123. *ES*, 7 April 1937, 1.

124. *ES*, 14 April 1937, 1.

125. *ES*, 7 April 1937, 1.

126. *ES*, 9 June 1937, 1, by Grace B. Carter.

127. *NYT*, 28 April 1937, 20.

128. Author's correspondence with Peter Homans, Dexter P. Cooper's grandson, June 2020.

129. *ES*, 30 June 1937, 1.

130. *ES*, 16 June 1937, 1; and *ES*, 23 June 1937, 1. For dramatic purposes, the author has converted a direct quote from the newspaper report of the event into a quote from the pilot and added emphasis using italics.

131. The party was led by Congressman Ralph O. Brewster (R-ME) and included Senator Theodore F. Green (D-RI) and Congressmen Warren G. Magnusson (D-WA), James W. Mott (R-OR), Melvin Maas (R-MN), Norman R. Hamilton (D-VA), George J. Bates (R-MA), Aime Forand (D-RI), Col. J. C. Fegan of the U.S. Marine Corps, and two pilots, whose names were not reported by the *Sentinel*.

132. *ES*, 16 June 1937, 1; and *ES*, 23 June 1937, 1.

133. *ES*, 16 June 1937, 1; and *ES*, 23 June 1937, 1.

134. *ES*, 6 October 1937, 4.

135. *ES*, 6 October 1937, 1, quoting Elizabeth May Craig in *PPH*, 4 October 1937. See also Riggs, 164; Funigiello, 182–83; and Leuchtenberg, "Roosevelt, Norris and the 'Seven Little TVAs,'" 418–19, citing *Washington (DC) Star*, 5 February 1937.

136. *ES*, 6 October 1937, 4.

137. *ES*, 13 October 1937, 4. See also *Cooper-NPPC-36*, 1–8.

138. Philip H. Gadsden (1867–1945) was a scion of the famous and infamous plantation-owning family from Charlestown, South Carolina, who played significant roles as rebels in the American Revolutionary War, and as Confederates in the American Civil War. P. H. Gadsden led the lobbying against PUHCA in 1934 and was the first witness to be subpoenaed by Senator Hugo Black, thus exposing the tip of the iceberg of the power trusts' propaganda campaign and frauds perpetrated by Hobson. "The Public Utility

Company Act of 1935," http://paulwwhite.com/the-public-utility-company-act-of-1935_316.html, citing *Hearings, Investigation of Lobbying Activities, Senate Special Comm. to Investigate Lobbying Activities*, 74 Cong., 1st Sess. (1935), 611–50.

139. "Bonneville Dam," in *Oregon Encyclopedia*, https://oregonencyclope dia.org/articles/bonneville_dam/#.XvdlyC2z3_8.

140. FDRL-284: Power_A Study in Relative Economy 1937, 7.

141. FDRL-284: Power_A Study in Relative Economy 1937, 1–40. Frederic Delano wrote a very interesting commentary on New York's analysis in response to FDR's query, in which he *partially* defended Potter's statements as having been taken out of context and repeatedly stated that the choice of steam vs. hydro is so site-specific that blanket statements are not appropriate. (Potter's comments were in a report by a committee of the Power Policy Board, of which Delano was vice chair.) Delano also stressed that waterpower was essentially "conservation," whereas steam power was "consumption" and would eventually exhaust the natural resources of coal and oil. Finally, he discussed how the lower initial cost of steam plants better suited private developers, whereas the high initial cost was better able to be handled by governments, which could borrow money at much lower interest rates and thus afford to finance waterpower infrastructure (some of which would last for centuries) over longer periods of time. FDRL-0284: 1937-11-24_Delano to FDR_Gov Hydro vs. Private Steam.

142. Funigiello, 267, citing Kendall K. Hoyt, "Political and Economic Factors in the Slackening of the Federal Power Program," *Annalist* 53 (8 June 1939): 798–800.

143. *ES*, 10 November 1937, 4.

144. *ES*, 25 November 1936, 1–2.

145. Kenneth Roberts, "Quoddy," *Saturday Evening Post*, 19 September 1936, 17.

146. By writing to Aubrey Williams at the PWA, who then engaged the White House, Roscoe Emery was able to increase the number of work relief jobs by 85. FDRL-0982: 1937-12-13_Aubrey Williams to LeHand_Emery letter unemployment.

147. *NYT*, 22 December 1937, 4. The City of Eastport's debts were a mere 17% of the funds allocated to Quoddy but never spent and retained by the Army Engineers. *ES*, 28 October 1936, 1. This amount was $805,302.44 according to *Final-36*, 5. This figure did not include repairs to Carlow Island dams.

148. *Saturday Evening Post*, 19 September 1936, 17.

149. *ES*, 22 December 1937, 1.

150. *ES*, 29 December 1937, 1.

151. *Reel-38*, 23.

152. *ES*, 29 December 1937, 2.

153. *ES*, 2 February 1938, 1.

154. GSC, "Passamaquoddy," 16–17.

155. "Maud Maxwell Vesey," https://dhil.lib.sfu.ca/doceww/person/4458.

156. *St. Croix Courier*, reprinted in *ES*, 16 March 1938, 1.

157. The Brewster-Corcoran debacle of July 1935 was not included.

158. *ES*, 2 February 1938, 1.

159. *ES*, 9 February 1938, 1.

160. *ES*, 16 March 1938.

161. *ES*, 16 March 1938.

162. FDRL-COOPER:1938-03-01_Roscoe Emery to FDR_ask for memorial to Dexter; and FDRL-COOER: 1938-03-04_Early to Emery_regrets no statement on Dexter.

163. *NYT*, 3 February 1938, 23; and *ES*, 23 March 1938, 6. It was not the only embarrassment of the day for the *New York Times*, as described in the *Sentinel*. The Associated Press had also supplied the *New York Times* with the wrong photograph of recently deceased Fairfax Harrison, the one-time president of the Southern Railway, whose image had been mistakenly replaced with that of John Jeremiah Pelley, president of the Association of American Railroads. *Time* magazine reported the errors, noting, "The best newspapers, like the best people, make mistakes. But seldom is any paper so unfortunate as was the exemplary *New York Times* last week."

164. CPL_1935_Dexter's salary record 1929-1935; and CPL_1936_Dexter & Gertrude Cooper personal tax return 1935: See also DelloRusso, 46. No sources are listed for DelloRusso's figures.

165. CPL_1936_Dexter & Gertrude Cooper personal tax return 1935.

166. DelloRusso, 46 (source not given); see the next paragraph in the narrative, the section "Demobilizing," and note 102.

167. Smith, *Power Policy*, 249–50, citing *Lewiston Daily Sun*, 6 March 1936; and *PPH*, 18 January 1939. The erroneous reports of Dexter Cooper's $25,000 salary probably originated with the army's report (*Final-36*, 3), which stated, "A sum of $25,000 was to be paid annually to Mr. Cooper and his staff to investigate the market possibilities for the power to be produced and to make recommendations as to its ultimate disposal." This amount presumably would have also included funds to cover expenses, including other staff. In any case, the total wasn't paid as personal compensation to Dexter.

168. FDRL-ROOSEVELT, James: 1934–1937_Salary payments to Dexter Cooper (undated). See also CPL_1936_Dexter & Gertrude Cooper personal tax return 1935, which shows a gross income of $14,400. Gertrude Cooper could have pressed a claim with President Roosevelt if (a) she knew about the amount still due, (b) it hadn't already been paid (and we are just missing

the documentation), and (c) Dexter hadn't given up on collecting the debt. That such a claim is not included in Frank Reel's report (see chapter 12, "The Turn of Events") would not appear relevant, as he investigated other issues, and this payment issue would have been entirely within the jurisdiction of the Army Engineers.

169. Author's telephone interview with Carla Cooper, Dexter P. Cooper's granddaughter, 21 October 2021.

CHAPTER 12. THE WAR YEARS (1939–1945)

1. A. Frank Reel gained notoriety when he was appointed to defend Japanese general Tomoyuki Yamashita in a war crimes trial. Yamashita, who surrendered in September 1945 to Gen. Douglas MacArthur, was tried in Manila a month later by a military commission of five generals untrained in the law. Accused of failing to prevent atrocities by his troops, he was found guilty in December and hanged the following February. In Reel's book *The Case of General Yamashita* (1949), he asserted that Yamashita had no connection to the atrocities and that the sham trial, including the verdict, had been scripted by MacArthur. "A. Frank Reel; Defended Japanese WWII General," *LA Times*, 11 April 2000, https://www.latimes.com/archives/la-xpm-2000-apr-11-me-18324-story.html.

2. Author's correspondence with Peter Homans, Nancy Cooper Homans's son, 7 July 2020.

3. *Reel-38*, 3–5.

4. FDRL-0982: 1934-06-29_Dexter Cooper to FDR_Feds to take Quoddy. This letter is summarized in chapter 8, "Liquidity." See also FDRL-0982: 1934-06-29_DPC to FDR_proposed to sell Quoddy to Feds. This could have been inserted in the White House files at the time of its writing, or in 1938 when Gertrude Cooper met with FDR and he subsequently initiated Frank Reel's investigation. Given its relatively deteriorated condition compared to other documents in the file, this may be the carbon copy provided by Gertrude in 1938. The two copies of the letter in the White House files were typed separately with different indents, one on plain paper and one on carbon paper. The carbon copy matches the copy in the Sills files.

5. SILLS: 1934-06-29_ DPC to FDR_Feds to take Quoddy.

6. FDRL-0982: 1934-07-05_DPC to Howe_exercise of option.

7. *Reel-38*, 5. Gertrude also charged that the army's Plan D, with Cobscook Bay as the high-level pool, violated Dexter's sales agreement with the US government because it would prevent the eventual development of the International Project which, for technical and political reasons, required Cobscook

Bay to be the low-level pool. Other than claiming the breach of contract, in the documents that remain in existence, no mention is made of what Gertrude thought the remedy for such a breach might be.

8. Smith, *Power Policy*, 250, citing Lewiston *Evening Journal*, 17 July 1937; and Lewiston *Daily Sun*, 4 January 1936: "Numerous Quoddy supporters contended that the Army Engineers 'gummed up' work at Passamaquoddy dam. It was suggested that Army Engineers were less practical than engineers who handle work for industrial concerns, because the former are not trained to handle commercial jobs 'where dollars and cents count,' both in the original cost and in the eventual earning power of the project. Theirs is a military training. You might say they were trained to destruction not to construction."

9. *Reel-38*. See also Smith, *Power Policy*, 250; and Roberts, *Trending into Maine*, 382–83. The cultural divide between Eastport natives and imported workers was alluringly revealed in *American Magazine*, December 1935, 159: "But to see beer sold right out in the open in the state where prohibition grew up and to have women, the wives of engineers and others at work on the project, wearing pants and painting their fingernails . . . well, weren't they inviting some visitation of heavenly wrath? 'I saw,' said a farmer of West Lubec, 'two girls coming across my field when I was a-hayin'. Honest, I thought I would die. They were both wearin' short pants like men's underwear and they was painted red all over their fingernails. 'Twas a wonder my bull didn't go after them, and I tell you, mister, it made me sick . . . They was nice ladies, too.'"

10. This was reported in *Reel-38*, 6–7: "While Cooper was in Washington trying to get more money for his plans, Mrs. Cooper came to Boston to spend a few days at the Hotel Lincolnshire. She claimed that at lunch one day, in the Lincolnshire dining-room she heard a group of men criticizing the President because of the Quoddy proposition. One of the men, a tall tanned fellow, seemed to know quite a bit about Quoddy. Mrs. Cooper wrote on a card, which she had the waiter deliver to this stranger: 'Dexter Cooper has not yet sold his plans to the Government.' On receipt of the card, the tanned gentleman got up and came over to Mrs. Cooper's table and said to her: 'The War Department is trying to ruin Mr. Cooper. They offered him $973.00 so he would get disgusted. Their plan would not work so they realized they would have to change their tactics. They have decided that if they cannot build Quoddy, they will see to it that nobody else does.'" A slightly different version appears in Gertrude's memoir (1970), as quoted in chapter 9, "Celebrations."

11. "Paul Crewes" and "Whitset" according to *Reel-38*, 21. However, Reel seems to have made a mistake. Paul F. Kruse (phonetically "Crewes") was the highest-ranking civilian engineer working at Quoddy. He officially "joined" BHE on 22 July 1936. *BHEN*, July 1936, 26.

12. *Reel-38*, 16. The comment, attributed to an army officer, was in a letter from Richard Field, one of Cooper's surveying engineers who had been on the National Power Policy Committee with Cooper and was then working on the TVA.

13. *Reel-38*, 2. The Murray & Flood report on Quoddy was issued in 1925. Most of the funding from Dexter's investors came thereafter. See chapter 11, "Demobilization," for dates and amounts.

14. *IJC-61*, 3.

15. See, for example, *IJC-61*, 3.

16. FDRL-OLDS: 1950s_history of Quoddy_p2. Reel's reporting of Gertrude's assertions appears to contradict the statement of Walter S. Wyman, president of CMP (formerly an Insull company) in Smith, *Power Policy of Maine*, 235. Smith's statements appear to contradict themselves: "Thus, although it had been supported by four great electrical concerns with $400,000, the project collapsed because of Canadian opposition and the depression"; and "Walter S. Wyman, president of CMP, one of the backers of Cooper's first investigations, maintained in 1940 that the private companies gave up the project when studies revealed that tidal power would be uneconomical; it could not produce power cheap enough and firm enough to compete with other forms of production of energy. See May Craig, Portland *Sunday Telegram*, 22 Dec 1940."

17. *Reel-38*, 2–3. However, the malicious intent of the Murray & Flood reports seems not to have been understood by the Department of the Interior as late as the 1950s, when they were mentioned in a report to Leland Olds without any reference to their origin. Roosevelt appointed Olds to the FPC in June 1939, and he served as chairman from 1940 to 1949. FDRL-OLDS: 1950s_history of Quoddy_p2.

18. *Reel-38*.

19. GSC, "Passamaquoddy," 16.

20. Some of Dexter P. Cooper's (DPC) materials survive in the possession of his grandson, Dexter P. Cooper III (DPC-III) and in a small trunk at the Campobello Public Library. The latter are believed by DPC-III to be a limited set of materials DPC kept in Boston. The trunk may have been lent after DPC's death by Gertrude to an engineer for study and subsequently returned to Campobello, but after her departure therefrom. Author's conversation with DPC-III, fall 2019.

21. *ES*, 22 June 1938, 1. Sentimental attachments were the reason for another pair of visitors that week, the "Nation's Number One Honeymoon Couple," John Aspinwall Roosevelt and Anne Lindsay Clark. The "beloved island" of Campobello had brought the young couple together. He was the youngest son of Franklin and Eleanor Roosevelt, and she was the oldest

daughter of Frances Sturgis Clark (Gertrude Sturgis Cooper's younger sister). *ES*, 22 June 1938, 1; Wikipedia, s.v. "John Aspinwall Roosevelt," https:// en.wikipedia.org/wiki/John_Aspinwall_Roosevelt; and *Descendants of Nath'l Sturgis*, 29, 64–65.

The enmity of the pernicious press toward anything related to the Quoddy project extended even to Dexter Cooper's innocent children, as evidenced by this snide bit of reporting from the International News Service (INS) of 21 June 1938: "The 'Cinderella' of the John Roosevelt-Anne Lindsay Clark wedding today was back at work as a file clerk for the WPA. She is Nancy Cooper, 21, cousin of the wealthy girl who married the youngest son of President Roosevelt. At Nahant, last Saturday, Nancy basked in the sun of society of which she was a member a few years ago. Today she was back at work while Anne and John were honeymooning at Campobello Island, N.B."

22. Another Supreme Court ruling in January 1939, in *TEPCo v. TVA*, re-solved all constitutional issues in favor of the Roosevelt administration. While FDR hadn't succeeded in "packing" the Supreme Court by increasing the number of judges, he did effectively do so in the PUHCA registration require-ment case by appointing Hugo Black to the bench, from which the latter did not recuse himself despite having previously characterized holding companies as "rattlesnakes" and urged their destruction as "networks of chicanery, deceit, fraud or graft." Riggs, 179–89, quoting *NYT*, 29 March 1938. See also *Electric Bond Share Company v. Securities & Exchange Commission*, 303 U.S. 419 (1938), https://supreme.justia.com/cases/federal/us/303/419/, and *Tennessee Elec. Power Co. v. TVA*, 306 U.S. 118 (1939), https://supreme.justia.com /cases/federal/us/306/118/.

23. *ES*, 17 August 1938, 1. In January, Brewster had made a secret trip to Dearborn, Michigan, to meet with Henry Ford, the first person to publicly comment on and endorse Dexter's tidal power plans back in 1924, and still a potential private funder of the tidal power project. *ES*, 12 January 1938, 1.

24. *ES*, 14 September 1938, 1.

25. Madison Electric Works was founded in 1888. In 1993, MEW, which for some years had been buying power from CMP, tried to buy its power from another source. CMP sued to prevent that. MEW won, and its customers en-joyed a 40% savings. Weil, 183.

26. *ES*, 12 October 1938, 1.

27. This would probably have been at Half Moon Cove (aka Bar Harbor), between Pleasant Point and Eastport. FDRL-0982: 1938-12-02_FDR to FPC_ Engineers letter Quoddy revival.

28. *ES*, 18 January 1939, 1.

29. The mocking phrase "drive trolleys and milk cows" was also used in the *New York Times* articles "Money Flows into the Sea," 11 April 1935, 20;

"Tides to be Harnessed in Maine," 26 May 1935, E11; and "The President on Power," 2 August 1936, E8.

30. *NYT*, 22 January 1939, 68.

31. *PPH*, 18 January 1939, quoted in Smith, *Power Policy*, 253.

32. The private power companies in the Connecticut River valley (most of which were owned by International Paper company affiliates) wanted federally funded flood control dams built *without* hydroelectric power plants *unless* the power plants were turned over to private companies. *ES*, 26 January 1938, 1, citing *Boston Post*, 20 January 1938. See also Joseph F. Preston, "Quoddy Project Might Rejuvenate N.E. Railroads," *ES*, 6 October 1937, 1.

33. *ES*, 25 January 1939, 1. Elizabeth May Craig's article continued: "Take the Congressman from New York, Bruce Barton, who is really a rather liberal Republican and a business man . . . [he] said that the New Deal ought to change the course of the Gulf Stream and make it flow north around New York and New England. 'This would be just as practicable as Passamaquoddy and would give Maine and Vermont summer weather, whether the Governor of Vermont wanted it or not. In fact, if you use New Deal bookkeeping you can figure that the increased admissions at the World's Fair would pay for the whole project.' This is typical wisecrack. It shows entire lack of understanding of . . . the power project, but it also is evidence of the hostility toward spending on projects of this sort." See also *ES*, 1 February 1939, 4, for editorials by the *Sentinel* and the *Lubec Herald*.

34. *ES*, 1 February 1939, 1; *ES*, 8 February 1939, 1; *ES*, 15 February 1939, 1. See also *NYT*, 20 January 1939, 5; *NYT*, 3 February 1939, 9; and *NYT*, 16 February 1939, 2. Parkman, 236, seems to have misinterpreted Vandenberg's intent as positive rather than as subterfuge: "Quoddy, however, continued to have its advocates. And, ironically, it was Senator Vandenberg who introduced a resolution in February 1939 requesting the Federal Power Commission to review its reports on Quoddy and bring them up to date."

35. *ES*, 22 March 1939, 1; *ES*, 22 March 1939, 1; *ES*, 26 April 1939, 1; and *ES*, 3 May 1939, 1.

36. *ES*, 9 August 1939, 1, 3.

37. *ES*, 5 April 1939, 1.

38. CPL_1936-07-08_List of Stockholders of the Canadian Dexter P. Cooper Company.

39. *ES*, 15 March 1939, 1 (emphasis added).

40. *ES*, 1 February 1939, 1.

41. *ES*, 29 March 1939, 1, due to competition from trucking.

42. Smith, *Power Policy*, 251, 259.

43. The bill was signed by Governor Barrows on 5 April 1939. *NYT*, 6 April 1939, 18.

44. *ES*, 22 February 1939, 1, 4. According to the financial audit of Dexter P. Cooper, Inc., and the Canadian Dexter P. Cooper, Inc., conducted by the PWA in May 1935, the corporate investors invested cash of $375,250. Interest thereon through 1 January 1932 was $78,802.50 but discontinued thereafter (for reasons unclear) until 24 May 1935, when Dexter Cooper sold all his interests to the US federal government. About $50,000 was also paid by them for the two investigations and reports by Murray & Flood. *Final-36*: appendix A-3; and FDRL-PSF: 1935-05-24_ Cooper_War Dpt to Dexter_Audit Quoddy Sale.

45. *ES*, 15 March 1939, 4. As I could find no other record of these visitors in 1929, it is possible that these events took place on 3–4 July 1928. *ES*, 4 July 1928, 1. Note the slightly different list of attendees. See also chapter 6, "Troubled Waters."

46. *ES*, 26 April 1939, 5.

47. FDRL-COOPER: 1940-04-29_Cobb of National Bank of Calais to FDR_Cooper house foreclosed.

48. FDRL-COOPER: 1940-05-02_FDR to Cobb_Cooper house for sale.

49. As the *Sentinel* noted, York, Maine, is closer to New York City than to Eastport. *ES*, 4 October 1937, 4.

50. *ES*, 16 June 1937, 1.

51. *ES*, 14 June 1939, 1.

52. FDRL-0982: 1940-04-06_FDR to Aubrey Williams_Quoddy model at Worlds Fair; and FDRL-0982: 1940-04-18_Aubrey Williams to FDR_Quoddy Model at World's Fair. FDR had suggested exhibiting the model at the Federal Building, but as there was no room available, it was instead exhibited in what was first the Metals Building and rechristened the Hall of Industry.

53. The location was reported as "a public park on Pennsylvania Avenue," presumably Lafayette Park, across from the White House (*ES*, 18 June 1941, 4), and as a "14th Street and Constitution Avenue," presumably in the Department of Commerce, or possibly the Smithsonian Natural History Museum. FDRL-0982: 1941-10-09_Aubrey Williams to FDR_Quoddy Model at 14th Street in DC.

54. Klein, *Call to Arms*, 7–8; and EMERY_8.33: RCE to FDR, 12 August 1939. The Quoddy model was mentioned briefly in ER-My-Day, 19 August 1940.

55. Wikipedia, s.v. "Sudentenland," https://en.wikipedia.org/wiki/Sudeten land#cite_note-sudetenland1-12, citing Max Domarus and Adolf Hitler, *Hitler: Speeches and Proclamations, 1932–1945; The Chronicle of a Dictatorship* (1990), 1393.

56. Goodwin, *No Ordinary Time*; and Funigiello, 229–30.

57. *ES*, 2 August 1939, 6.

58. *ES*, 9 August 1939, 1.

59. *ES*, 4 October 1939, 1.

60. *ES*, 13 December 1939, 1.

61. Funigiello, 230, 232, citing "Confidential Memorandum on Shortages of Electric Generating Capacity for War-Time Needs," 1 July 1938, George W. Norris Papers, tray 72, box 2, Library of Congress.

62. "Leo Szilard (1898–1964)," Jewish Virtual Library, accessed 28 February 2021, https://www.jewishvirtuallibrary.org/leo-szilard.

63. https://en.wikipedia.org/wiki/Stone_%26_Webster.

64. *ES*, 16 August 1939, 1.

65. *ES*, 30 August 1939, 4.

66. *ES*, 13 September 1939, 4.

67. *ES*, 13 September 1939, 4.

68. Goodwin, *No Ordinary Time*, 63–64.

69. Wikipedia, s.v. "We Shall Fight on the Beaches," https://en.wikipedia.org/wiki/We_shall_fight_on_the_beaches.

70. Goodwin, *No Ordinary Time*, 64.

71. Wikipedia, s.v. "Battle of the Atlantic," https://en.wikipedia.org/wiki/Battle_of_the_Atlantic#cite_note-32, citing Tom Purnell, "The "Happy Time" in *"Canonesa": Convoy HX72 & U-100*, 11 April 2003.

72. Goodwin, *No Ordinary Time*, 42.

73. Goodwin, *No Ordinary Time*, 177.

74. FDRL-COOPER: 1939-07-22_Gertrude Cooper to Catcomb of Calais National Bank.

75. Dexter Cooper III's recollection was that Gertrude worked at Filene's department store in Boston for some period of time. This is quite possible, maybe even probable, as James "Jimmy" Roosevelt was a manager there (Goodwin, *No Ordinary Time*, 178). However, Peter Homans, Gertrude's grandson via Nancy (email to author, 14 October 2020), and Carla Cooper (telephone communication with author, 21 October 2020) both remember Gertrude working for Jay's Lingerie Shop. Carla remembers that Gertrude got the job at Jay's through Fanny Porter, her next-door neighbor at Campobello. For comparison, Dexter's secretary, Genevieve Roche, had a salary of $25 per week. Sue McNichol, Roche's granddaughter, interview with author, 6 April 2021.

76. FDRL-COOPER: 1940-01-29_FDR to Jim Rowe_job for Gertrude.

77. FDRL-COOPER: 1940-06-18_Gertrude Cooper to FDR_thank you for Vanderbilt job. Other relevant correspondence includes FDRL-COOPER: 1940s_01-17_Gertrude Cooper to FDR_jobless; FDRL-COOPER: 1940-01-29_FDR to Jim Rowe_job for Gertrude; "Dear Gertrude," 14 March 1940; James Rowe Jr. to FDR, 14 May 1940; and FDR to "Dear Gertrude," 14 May 1940. FDR also found jobs for Gertrude Cooper's daughter, Elizabeth ("Boo") in the WAACS, and son Dexter Jr. ("Decks") in the navy.

78. FDRL-PSF: Vanderbilt_1940-07-23_Franklin D. Roosevelt to Mrs. James Laurens Van Alen.

79. FDRL-PSF: Vanderbilt_1940-06-29_Attorney General Robert Jackson to The President.

80. Cara Moore, Archives Technician, Reference Department, National Archives, St. Louis, Missouri, "Faces of the National Parks Service" (presentation at Virtual Genealogy Year, 2016), https://www.archives.gov/files/calen dar/genealogy-fair/2016/session-11-moore-presentation-handout.pdf.

81. National Park Service official history of the Vanderbilt National Historic Site, 44, citing Superintendent to Regional Director, 23 January 1942.

82. National Park Service official history of the Vanderbilt National Historic Site, 19–26.

83. FDRL-COOPER: 1940s_Gertrude to FDR: a minute of time when driving.

84. FDRL-COOPER: 1945-03-16_Note of GGT on White House stationery regarding a telephone call from Gertrude.

85. National Park Service official history of the Vanderbilt National Historic Site, 25.

86. GSC, "Hyde Park," 5.

87. *ES*, 3 July 1940, 4.

88. *ES*, 4 December 1940, 4.

89. *ES*, 2 October 1940, 4.

90. *ES*, 4 December 1940, 4.

91. The private power trusts' political influence was a large problem, as these examples indicate:

1. At a conference in Washington of the six New England states called to discuss flood control and the electric power generated as a by-product of the proposed dams and reservoirs, the New England Council was represented by the lobbyists for the private power trusts. *ES*, 26 January 1938, 1.

2. In 1940, the FPC launched another investigation into the accounting and propaganda practices of five large private power companies in the Northwest. Gruening, introduction.

3. There was a "definite conspiracy in Montana to keep industry out of the state," according to John Kinsey Howard, author of *Montana: High, Wide, and Handsome*, in testimony to the US Senate on 22 September 1945. The established companies, chiefly Anaconda Copper, didn't want rivals or organized labor and were also opposed to public power. "What chance have Montanans who want public power development," wondered Howard, "when they have a state public service commission which decides the private power company is overcapitalized by only 19 million

dollars, in the face of a Federal Power Commission order to squeeze 51 million dollars of 'water' out of its accounts?" Gunther, *Inside U.S.A.*, 185 (Gunther was a friend of Eleanor Roosevelt and a frequent visitor to the White House); and Goodwin, *No Ordinary Time*, 200.

4. "[I]n 1940 the commissioners [of the FPC] analyzed the FPC's procedural problems over its first 20 years, concluding that the period witnessed 'increasingly elaborate and protracted procedures devised by . . . private companies to delay or circumvent' the FPC's objectives. 'Having obtained a status substantially free from competition,' the commissioners declared that private utilities 'now seek . . . to regain the arbitrary control of costs and rates which would be theirs under unregulated monopoly.'" Cantelon, 69–70.

92. *ES*, 16 October 1940, 6.

93. Harold Ickes, in his diary, derisively said, "Dewey threw his diaper into the ring." Leuchtenburg, *Franklin D. Roosevelt*, 312, citing Ickes, 3:92.

94. *ES*, 10 July 1935.

95. Riggs, 181–89; and James B. Jones Jr., "TEPCO," *Tennessee Encyclopedia*, https://tennesseeencyclopedia.net/entries/tepco/.

96. Under the terms of the US Johnson Act of 1934, England was not eligible to sell bonds in the United States because it had defaulted on nearly $3.5 billion of debt from World War I. Wikipedia, s.v. "Johnson Act," https://en.wikipedia.org/wiki/Johnson_Act. The idea for the Destroyers-for-Bases deal is usually credited to Navy Secretary Frank Knox and was initially suggested during a cabinet meeting on 2 August 1940. Goodwin, *No Ordinary Time*, 139, citing William M. Goldsmith, *The Growth of Presidential Power*, vol. 3, *Triumph and Reappraisal* (1974), 1754.

However, in Jay Epstein's book *Dossier: The Secret History of Armand Hammer* (1996), Epstein credits Armand Hammer with the idea for the Destroyer-for-Bases deal and Lend-Lease. A professor of history and journalism at UCLA and MIT, Epstein posits that Hammer was acting as an agent of Soviet Russia, which wanted to keep England fighting Germany, thus delaying a likely attack on Russia. Epstein bases his claim on interviews with and documents from Hammer; analysis of intelligence documents in *Codename BARBOSSA* by Barton Whaley; FBI files on Hammer's family; and material from the FDR Library, which confirms FDR communicated with Hammer on the subject. Unfortunately, Epstein does not provide a meaningful chronology. Epstein, 150–55.

Although not referenced on this subject, much of Epstein's book is based on declassified documents from the former Soviet Union. Epstein's book began as a friendly article for *Time* magazine, arranged for by Hammer as he was campaigning to be nominated for the Nobel Peace Prize. But as Epstein dug into the material, he discovered ever more incriminating evidence that Hammer's public persona of a successful businessman and philanthropist was a fabrication hiding a nefarious past. It's interesting reading.

97. The isolationists made a valid point: in 1940 America was grossly unprepared to fight a war. "The United States had only 160 P-40 planes for the 260 pilots waiting for them; it had no anti-aircraft ammunition, and would not for six months. . . . In August 1940, when the Army conducted its greatest peacetime maneuvers in history in St. Lawrence County, New York, the troops were virtually weaponless. Pieces of stove pipe served for antitank guns, beer cans as ammunition, rainpipes as mortars, and broomsticks as machine guns." Leuchtenburg, *Franklin D. Roosevelt*, 303, citing Langer and Gleason, *Challenge to Isolation*, 487–88, 566–69, 654; and Leuchtenburg, *Franklin D. Roosevelt*, 306–7, citing *NYT*, 1–27 August 1940. See also Goodwin, *No Ordinary Time*, 50–51.

98. Goodwin, *No Ordinary Time*, 148.

99. Larson, 78, 148–49, 198, 273–74.

100. Riggs, 206, citing Joseph Barnes, *Willkie* (New York: Simon & Schuster, 1952), 254.

101. Goodwin, *No Ordinary Time*, 141.

102. *ES*, 6 November 1940, 1. Republicans swept Maine's state and national elections. Governor Lewis O. Barrows was reelected, and Ralph O. Brewster relinquished his House seat and won a Senate seat from the retiring Frederick Hale. Roscoe Emery had run for Brewster's US House seat (*ES*, 20 March 1940, 4) but was defeated in the primary by Dr. W. N. Miner, who, the *Sentinel* noted, had "campaigned strenuously and through paid workers." *ES*, 20 June 1940, 1. Arthur Vandenberg wrote to FDR on the day after he was elected president for the third time, "My dear Mr. President: I was wrong. It is not 'Good Bye Mr. Chips.' It is 'Hail to the Chief!' I wish you every good thing for yourself and for the country and I shall hope to be able to cooperate with you whenever possible." FDRL-PPF-VANDENBERG: 1940-11-07_Vandenberg to FDR_Congrats on re-election.

103. *ES*, 6 November 1940, 4.

104. Riggs, 204, citing Steve Neal, *Dark Horse: A Biography of Wendell Willkie* (Lawrence: University of Kansas Press, 1984), 69–71; and Joseph Barnes, *Willkie* (New York: Simon & Schuster, 1952), 164. For skullduggery at the Republican convention, see Riggs, 205, citing David Levering Lewis, *The Improbable Wendell Willkie: The Businessman Who Saved the Republican Party and the Country and Conceived of the New World Order* (New York: W. W. Norton, 2018), 141–42.

105. Franklin D. Roosevelt, "The Great Arsenal of Democracy," radio broadcast, 29 December 1940, AmericanRhetoric.com, https://americanrheto ric.com/speeches/PDFFiles/FDR%20-%20Arsenal%20of%20Democracy.pdf.

106. Funigiello, 232, citing National Association of Railroad and Utility Commissioners, "Report of the Committee on Generation and Distribution of Electric Power," *Proceedings* (Washington, DC, 1938), 656.

107. *MSPB-34*, 150.

108. "In 1939 Alcoa had produced a record high of 327 million pounds of raw aluminum. . . . It had capacity of about 335 million pounds, but its output depended on the reliability of water power, which could be reduced by drought." Klein, *Call to Arms*, 160.

109. Goodwin, *No Ordinary Time*, 44–45, referring to FDR's address to Congress on 16 May 1940, and alluded to in *ES*, 22 January 1941, 1. A collection of 400 planes brought together for America's embarrassing war game in 1940 was "the greatest concentration of combat planes" ever amassed in the United States—and roughly the number of Allied planes destroyed *every day* by the Nazis. Goodwin, *No Ordinary Time*, 50–51.

110. Goodwin, *No Ordinary Time*, 259.

111. Goodwin, *No Ordinary Time*, 53–55, 156–58, 197. To get the cooperation of big business, FDR had to offer significant tax incentives in the form of accelerated depreciation against income, which effectively subsidized factory expansion for war production by reducing near-term taxes.

112. Franklin D. Roosevelt, "January 6, 1941: State of the Union (Four Freedoms) [Speech]," University of Virginia, Miller Center, https://miller center.org/the-presidency/presidential-speeches/january-6-1941-state-union -four-freedoms.

113. Roosevelt, "January 6, 1941: State of the Union."

114. Klein, *Call to Arms*, 160–63, 230–31; and Riggs, 209.

115. Goodwin, *No Ordinary Time*, 259–60, citing *New Republic,* 27 January 1941, 104; Goodwin's interview with Arthur Goldsmith; and *Time,* 7 July 1941, 10.

116. Edward Stettinius was chair of U.S. Steel and a member of the National Defense Advisory Commission.

117. Goodwin, *No Ordinary Time*, 259–60, citing *New Republic*, 27 January 1941, 104; Goodwin's interview with Arthur Goldsmith; and *Time,* 7 July 1941, 10. At least $322 million in RFC loans was made to Reynolds, Alcoa, and Bohn to build aluminum plants. Another $22 million went to Kaiser for magnesium production. Klein, *Call to Arms*, 162–63.

118. *Churchill and the Great Republic* (interactive exhibit, Library of Congress), https://www.loc.gov/exhibits/churchill/interactive/_html/wc0112.html.

119. Henry Wadsworth Longfellow (1807–1882), a native of Portland, Maine, also wrote *Paul Revere's Ride*, *The Song of Hiawatha*, and *Evangeline*.

120. *Churchill and the Great Republic*.

121. Goodwin, *No Ordinary Time*, 193.

122. Goodwin, *No Ordinary Time*, 193–94, 210–15.

123. Goodwin, *No Ordinary Time*, 193–94.

124. *ES*, 8 January 1941, 4.

125. Roosevelt, "January 6, 1941: State of the Union."

126. Quoted in *Life*, 17 February 1941, but otherwise not attributed.

127. *Life*, 17 February 1941, 19–23.

128. *Life*, 24 February 1941, 30.

129. Larson, 378; and Riggs, 214, citing Joseph Barnes, *Willkie* (New York: Simon & Schuster, 1952), 254.

130. Senator Arthur Vandenberg (R-MI) experienced a similar change of heart, but four years later. "On January 10, 1945, Vandenberg, who had been the nation's bitterest isolationist, rose to his feet in the Senate and made a historic speech in which he declared that isolationism was dead and the One World religion was the true religion. The record indicates that from that point on, Vandenberg's entire personality changed. . . . President Roosevelt caused a minor sensation by appointing The New Vandenberg to be one of America's delegates to the United Nations organizing conference." *Collier's,* 19 June 1948, 15, https://archive.org/details/colliers121aprspri_201803/page/n1031 /mode/2up?q=Vandenberg.

131. Walter Lippmann (1889–1974), winner of two Pulitzer Prizes, was famous for critiquing media and democracy in his newspaper column and several books, including *Public Opinion* (1922).

132. Bennett, 263.

133. S. Res. 62 (1 February 1939); and *FPC-41*, 1. Resolution 62 asked the FPC to update its (negative) report from 1934 and to report on "(1) the relative costs of steam-generated or tide-generated power at Passamaquoddy; (2) the relative costs of the power to consumers; and (3) whether there was a local or export market for power thus generated by either method."

134. *FPC-41*, 1, paragraph 2: capital costs of a tidal power plant at Passamaquoddy Bay for 110,000 kW at 60% load-factor, 578,000 kWh annually of firm power.

135. *Durand-36*, 7, 10–11; *FPC-41*, 5, 10–11, 48–51; and *Final-36*, 16–17. These are the FPC's comments on Dexter Cooper's cost estimates: "1) topographic maps later found to be inaccurate; 2) meager information on foundation conditions; 3) construction costs and machinery and equipment cost supposed to obtain at that time." *FPC-41*, 9.

136. *FPC-41*, 36, paragraph 139; and *FPC-41*, 16. Strangely, the FPC used the then-current price of $5.10 per ton for bituminous coal at Eastport. Given the tiny population and industry in Eastport, this would have been a very low-volume and relatively high price. However, a high-volume purchaser of coal, such as a large steam-electric plant, would likely have been able to negotiate

a significantly lower price. For instance, the Grand Lake Generating Station in Minton, New Brunswick (a mere 90 miles distant) was paying C$2.70 per ton for coal in 1936, and the price didn't begin to increase until the end of the decade. "70 Years of Service," a history of the New Brunswick Power Commission, 1990, 23–24, https://www.nbpower.com/media/1489688/seventy -years-of-service.pdf. The US-Canada exchange rate in 1940 was US$1.00 = C$1.10. Thus C$2.70 = US$2.45. "A History of the Canadian Dollar," 53–54, https://www.bankofcanada.ca/wp-content/uploads/2010/07/1939-50.pdf. Had the FPC used $2.45 as the price of coal, it would have made steampower more economical at the time (1941). The average *retail* price of bituminous coal had been relatively stable during the 1930s ($7.71 to $8.61 per ton) but began to rise sharply in mid-1941 ("Retail Prices of Food and Coal, 1941," Bulletin 707, US Department of Labor, 25, https://fraser.stlouisfed.org/files/docs/publications /bls/bls_0707_1942.pdf), reaching $10.29 by July 1944. "Retail Price of Bituminous Coal for United States," FRED Economic Data, https://fred.stlouisfed .org/series/M04047USM238NNBR.

137. *FPC-41*, 34–35. Consideration of larger and smaller projects was not requested by S. Res. 62. The FPC briefly mentioned the possibility of a smaller two-pool project in Cobscook Bay with Whiting and Denny's Bays forming the high-tide pool (dam from Leighton Point to Denbow Neck) and South Bay as the low-tide pool (dam from Denbow Neck to Seward Neck). "But little reliable data are available for estimating the cost of this project, but it appears that with a suitable auxiliary pumped-storage plant, some 250,000 kWh of firm energy could be made available annually, at a cost possibly not greatly in excess of the estimated cost of firm energy from the larger two-pool project." However, no estimates were provided for the larger two-pool project because of lack of what the FPC considered reliable information.

138. *FPC-41*, 15, paragraph 45; FDRL-0982: Passamaquoddy: 1936-01-10_Markham to FDR_cost revision; and FDRL-1766: 1936-02-04_Dexter Cooper to James Roosevelt_angry at Army Engineers' report. The report, *Passamaquoddy Tidal Power Development Estimates*, was dated 11 January 1936.

139. *FPC-41*, 18–24. 540,000 kW or 2.4 billion kWh/yr at 60% load factor, producing power at 1.87 to 3.26 mills per kWh; 2.36 to 3.86 mills per kWh if transmitted to Eastport. The FPC's plan for the development of Maine's river power was in three stages. Stage I was to include five river-power plants on the West Branch of the Penobscot River, with an aggregate installed capacity of 104,000 kW, 482 million kWh/yr, at 60% load-factor, $22 million, 2.36 to 3.86 mills per kWh. *FPC-41*, 1–2, paragraph 2.

140. *FPC-41*, 29.

141. *FPC-41*, 35, paragraphs 130–31.

142. *FPC-41*, 2, 24–31; and Wikipedia, s.v. Minto, New Brunswick," https://en.wikipedia.org/wiki/Minto,_New_Brunswick.

143. *FPC-41*, 29.

144. *FPC-41*, 36, paragraph 138.

145. *FPC-41*, 27. Noted in Smith, *Power Policy*, 264.

146. FPC-41, 36. See also *ES*, 9 April 1941, 1; and Holmes, 72.

147. Coincidently, in 1941 the SEC discovered more evidence about the evil intentions and actions of the utility holding companies when they found the "Murphy Memo," which stated, "I again wish to impress upon you the importance, in my opinion, of scrambling all of these [holding company] reorganizations together so that about the only thing the Pennsylvania Commission will be able to understand will be the result and not how it was reached." The opinion was written in 1927 by a holding company lawyer named Samuel W. Murphy, who by 1941 had wrangled his way to become president of GE's Electric Bond & Share Company, the largest holding company. Riggs, 205–6, citing Karl C. Smeltzer, "Memories from Early Days of the Securities and Exchange Commission," 2004, 3–7, www.sechistorical.org, http://3197d6d14b5f19f2f440 -5e13d29c4c016cf96cbbfd197c579b45.r81.cf1.rackcdn.com/collection/pa pers/2000/2004_0601_Smeltzer_Memories.pdf. This also contains a good summary of the way holding companies "watered the stock" with the result of diluting the equity and earnings of most utility shareholders in favor of the holding company and of inflating the cost basis used for determining rates.

148. *ES*, 4 June 1941, 1.

149. *ES*, 4 June 1941, 1.

150. *ES*, 4 June 1941, 4. By early 1942, the TVA had twelve hydroelectric and one steam-electric plant under construction simultaneously, employing 28,000 workers. Wikipedia, s.v. "Tennessee Valley Authority," https://en.wikipedia.org/wiki/Tennessee_Valley_Authority.

151. Goodwin, *No Ordinary Time*, 265.

152. *ES*, 4 June 1941, 4.

153. Klein, *Call to Arms*, 231; and Riggs, 209. See also Funigiello, 240–43, 249, citing Eliot Janeway, *The Struggle for Survival* (New Haven, CT: Yale University Press, 1961), 234–35; and Ickes to US Chamber of Commerce, 12 December 1939, Interior Department Records, GAF 1-288, pts. 3–4. Funigiello has an extensive discussion of Harold Ickes's role in the development of the national power policy and his acrimonious relationships with the head of the War Department, FPC, and TVA, particularly in chapters 9 and 10. Three weeks after the Japanese attack on Pearl Harbor, Kellogg acknowledged the existence of a power shortage. Funigiello, 253, citing *NYT*, 20 December 1941.

154. McDonald, *Insull*, 169–87.

155. *ES*, 23 July 1941, 1; and Gallant, *Hope and Fear*, 43–44.

156. *ES*, 30 July 1941, 4. As it turned out, recycled aluminum was not suitable for airplanes and was used for other purposes. Goodwin, *No Ordinary Time*, 260–61.

157. Klein, *Call to Arms*, 160–63, 230–33; and Riggs, 209. By the end of the war, the TVA had doubled its prewar capacity and tripled its output to 11.9 billion kWh. And by 1945, the Bonneville and Grand Coulee Dams were kicking out 9 billion kwh. Riggs, 211–13, citing *TVA Handbook*, 1987; and BPA, *1947 Annual Report*.

158. FDRL-0982: 1941-09-23_Dubord Dem Natl Com to FDR_newspaper editorials.

159. "Operation Barbarossa and Germany's Failure in the Soviet Union," Imperial War Museums, https://www.iwm.org.uk/history/operation-barbarossa-and-germanys-failure-in-the-soviet-union. See also Goodwin, *No Ordinary Time*, 253–54.

160. Dorn, 337, citing *Soviet Union Review* 10 (1932): 212; *NYT*, 27 August 1932, 2; *NYT*, 10 October 1932, 1; and Heymann, 22.

161. Wikipedia, s.v. "Dnieper Hydroelectric Station," https://en.wikipedia.org/wiki/Dnieper_Hydroelectric_Station#cite_note-8, citing "Ukrainian Activists Draw Attention to Little-Known WWII Tragedy, https://www.rferl.org/a/european-remembrance-day-ukraine-little-known-ww2-tragedy/25083847.html (the article includes video of a German attack).

162. "Dnieper Dam Blown Up by Retreating Russians," accessed 11 May 2020, https://forum.axishistory.com/viewtopic.php?t=52940, citing *NYT*, 20 August 1941.

163. *NYT*, 21 August 1941, dateline London via United Press.

164. "1931 Pulitzer Prizes," https://www.pulitzer.org/prize-winners-by-year/1931.

165. Wikipedia, s.v. "Dnieper Hydroelectric Station," citing H. R. Knickerbocker, *Is Tomorrow Hitler's? 200 Questions on the Battle of Mankind* (Reynal & Hitchcock, 1941), 107–8.

166. Maxim Litvinoff was the Russian ambassador to the United States at the time of Germany's invasion of Russia. Russia (and Litvinoff) received diplomatic recognition from the United States in 1993 thanks to Hugh Cooper and others (see chapter 8, "Communism versus Capitalism").

167. Gallant, 44–46.

168. *Collier's*, 19 June 1948, 23, 66, https://archive.org/details/colliers121aprspri_201803/page/n1075/mode/2up?q=Vandenberg. See also Goodwin, *No Ordinary Time*, 258.

169. FDRL-0982: 1941-06-07_FDR to Berle_May Craig articles. See also Goodwin, *No Ordinary Time*, 256.

170. *ES*, 25 June 1941, 1.

171. *ES*, 6 August 1941, 1. Eleanor Roosevelt's "My Day" newspaper columns from the summer of 1941 are interesting and relevant because of their mix of world events and Quoddy news, all told in a personal and immediate style. Of particular note are the following:

21 June: German invasion of Russia
23 June: NYA Quoddy: vocational training and democratic self-management
24 June: Quoddy residents sensitive to transportation, power; visit by Gannett and Craig
25 June: Shopping in St. Andrews and Eastport
26 June: Isolationism; Dr. Bennett of Lubec
27 June: Difficulties of small nations in Europe
28 June: US Rep. Hamilton Fish (D-NY) re: shortsighted isolationism
11 July: Extended Roosevelt family; visit by John D. Rockefeller to Campobello
12 July: Long winters and need for cash; self-help programs; Aubrey Williams at Quoddy
14 July: Justice Felix Frankfurter; youth conference; Mrs. Fountain; Quoddy boys
15 July: Henry Morgenthau visit
16 July: Youth conference attendees visit Quoddy Village
30 July: Innovation and employment issues
31 July: Precariousness of fishing and farming; Downeasters exposed to attack by air and sea
1 August: Campobello ferry; taking visitors to see Quoddy Village
16 August: FDR-Churchill's Atlantic Charter; sons' safety; N. Rockefeller, M. Field visits
19 August: FDR's return to Washington; stories from Atlantic Charter meeting

172. *ES*, 6 August 1941, 1.
173. Goodwin, *No Ordinary Time*, 262–63. Princess Märtha of Sweden (1901–1954) became crown princess of Norway with her marriage in 1929 to her first cousin, the future King Olav V of Norway. Following the Nazi invasion of Norway in April 1940, she fled to her native Sweden, and then to the United States, where she was for many weeks a guest at the White House and became one of FDR's closest friends.
174. The fleet included the *Tuscaloosa, Augusta, Madison, Moffett, Sampson, Winslow,* and *McDougal.*
175. Cross, 139–41.

176. Wikipedia, s.v. "Atlantic Charter," https://en.wikipedia.org/wiki/Atlantic_Charter#cite_note-6, citing Harry Gratwick, *Penobscot Bay: People, Ports & Pastimes* (The History Press, 2009).

177. *ES*, 20 August 1941, 1.

178. *ES*, 17 September 1941, 4.

179. *ES*, 1 October 1941, 1.

180. Queen Wilhelmina's government evacuated to London after the Nazi invasion on 10 May 1940.

181. *ES*, 19 November 1941, 4.

182. *ES*, 8 October 1941, 4.

183. *ES*, 15 April 1942, 1.

184. *ES*, 1 April 1942, 1. Eleanor Roosevelt had this to say in the "My Day" column of 14 July 1941: "On Friday night, in Quoddy, I met a Mrs. Fountain who has won the title of 'Mother' through her kindness to all the boys on the project. The girls in various neighbors' houses are always willing to help her out, so she manages to keep the cookie jar filled for the boys and to make little extra things for them to eat every now and then, which make a homesick boy at once feel less keenly the need of his home environment. The boys are all devoted to her and call her 'Mom.'"

185. *ES*, 18 February 1942, 1. See also ER-My-Day, 23 June 1941 and 31 July 1941.

186. The Canadian government purchased a similar portion of its domestic production for the same purpose. *ES*, 7 January 1942, 8; and *ES*, 25 February 1942, 1.

187. *ES*, 8 August 1945, 1.

188. *ES*, 8 April 1942, 1. See also FDRL-0982: 1942-02-06_FDR to Maine Congressional Delegation_Quoddy delayed until after war.

189. *ES*, 15 April 1942, 4.

190. FDRL-0982: 1943-08-15_FDR to Aubrey Williams_Use of Quoddy Village.

191. *NYT*, 1 September 1943, 21.

192. Roberts, *Trending into Maine*, 371.

193. Author's conversation with local historian Wayne Wilcox, based on his notes of conversations with John Grady, one of the firefighters.

194. *ES*, 6 June 1946; and *ES*, 20 June 1946. The Seabees abandoned Camp Lee Stephenson on 31 May 1946.

195. Zipp, 62, 260–63, 488–89, 491.

196. Cross, 185–88.

197. *ES*, 28 March 1945, 1.

198. *ES*, 18 April 1945, 1.

199. *ES*, 18 April 1945, 4.

200. *ES*, 18 April 1945, 4.

201. Superintendent Cooper, Memorandum for the Director (copy to Region One Director), 8 May 1945, ROVA Central Files 1939–1976, Monthly Narrative Reports, September 1940–March 1947, ROVA, Hyde Park.

CHAPTER 13. THE ATOMIC AGE (1946–1969)

1. Gertrude S. Cooper resigned in May 1945, according to Cara Moore, "Faces of the National Parks Service" (presentation at Virtual Genealogy Year, 2016). But "[w]hen FDR died, Mrs. Cooper, as a Presidential appointee without Civil Service status, was let go almost immediately," according to National Park Service, official history of the Vanderbilt National Historic Site, 21, https://www.archives.gov/files/calendar/genealogy-fair/2016/session -11-moore-presentation-handout.pdf.

2. *ES*, 9 May 1945, 7.

3. *ES*, 6 June 1946, 1.

4. *ES*, 1 August 1946, 1.

5. *ES*, 22 May 1947, 1, 4; and *Boston Daily Record*, May 1947.

6. Butler.

7. *NYT*, 28 October 1947, 50.

8. *ES*, 24 June 1948, 4.

9. *ES*, 18 December 1947, 6.

10. *ES*, 22 January 1948, 1.

11. *ES*, 18 December 1947, 6.

12. *ES*, 22 January 1948, 4.

13. *ES*, 15 September 1949, 2.

14. Smith, *Power Policy*, 240, citing Portland *Sunday Telegram*, 4 January 1948; and Irwin, 59.

15. The US Navy had provided "shore power" on numerous occasions, beginning in the 1920s, most notably during a drought in 1930 at Tacoma, Washington, with the aircraft carrier USS *Lexington* (CV-2). *ES*, 8 October 1930, 3. In the 1960s, the navy used nuclear-powered vessels for shore power. Naval Historical Association, https://www.navyhistory.org/?s=CV-2 &submit=Search.

16. *ES*, 13 November 1947, 4.

17. *ES*, 12 August 1948, 1.

18. As of October 1936, the total cost of Quoddy Village, the two labor camps, and associated structures, excluding the functional features of the dams, was $2,413,884.19. *Final-36*, 25.

19. *ES*, 18 August 1949, 1.

20. *ES*, 24 July 1947, 1. See also *NYT*, 15 August 1947, 15.

21. *ES*, 28 Aug 1947, 1. See also ER-My-Day: 26 August 1947; and *NYT*, 16 November 1949, 45. CMP said it was not against Quoddy. EMERY_Scrapbook_Quoddy 1949.

22. *ES*, 3 June 1936, 1, with my paraphrasing of Governor Brann's comments about Senators Hale and White's "betraying kiss" in 1936. *ES*, 3 June 1936, 1.

23. Smith, *Power Policy*, 239, citing *Portland Press Herald*, 11 September 1947.

24. Smith, *Power Policy*, 239, citing *Portland Press Herald*, 6 September 1947.

25. *ES*, 22 August 1946, 4.

26. *NYT*, 8 August 1947, 19.

27. *ES*, 16 March 1938, 9, with my paraphrasing of Colonel Waite's comments about a possible Public Works Administration loan to Dexter in 1933, *ES*, 15 November 1933, 1.

28. *NYT*, 21 August 1947, 1.

29. *NYT*, 24 August 1947, 1.

30. *NYT*, 21 August 1947, 1.

31. *NYT*, 2 December 1947, 59.

32. *ES*, 25 September 1947, 1.

33. *ES*, 25 September 1947, 1; and *BH*, 23 September 1947, 2.

34. *ES*, 15 January 1948, 1.

35. *ES*, 22 January 1948, 1.

36. *ES*, 12 February 1948, 1.

37. *ES*, 19 February 1948, 1.

38. *ES*, 2 September 1948, 1.

39. "Complete Village for Sale" ad, EMERY_Scrapbook_Quoddy-1949; and *NYT*, 8 October 1949, 29.

40. Complete Village for Sale" ad, EMERY_Scrapbook_Quoddy-1949; and *NYT*, 8 October 1949, 29.

41. *NYT*, 8 October 1949, 29; EMERY_Scrapbook_Quoddy 1949; and AP, 7 October 1949.

42. *ES*, 2 March 1950, 2.

43. FDRL-OLDS: 1960-09-06_Margolies to Davidson at Interior_Passamaquoddy summary p4.

44. Smith, *Power Policy*, 239, citing *PPH*, 30 August 1947. See also *ES*, 20 March 1947, 4.

45. "Address by George L. Brady, Chief Editorial Writer of the Hearst Publications, Boston, before the Washington County (Maine) Chamber of Commerce at Whiting, Feb. 4, 1948." Courtesy of Ruth McInnis.

46. *ES*, 18 March 1948, 1. See also Holmes, 74–75, citing C. Prank Keyser, "Passamaquoddy Tidal Power Project," Library of Congress, Legislative Reference Service, Washington, DC, 1 March 1949, 10–11. An identical bill was introduced in the Senate again in 1949.

47. *ES*, 18 March 1948, 1.

48. Commissioners of the Joint International Commission, poster, https://www.ijc.org/sites/default/files/2018-08/IJCCommissionerPoster.pdf.

49. Holmes, 74–75, citing letter to the International Joint Commission from the United States Department of State, 9 November 1948.

50. *St. John Telegraph Journal*, 2 April 1948, as quoted in *ES*, 15 April 1948, and *St. John Telegram Journal*, quoted in *ES*, 23 September 1948, 1.

51. For a view of Roy L. Fernald's politics, see Gallant, 34. Roy L. Fernald was a cousin of Bert M. Fernald, whose name adorned the bill that prohibited the export of hydropower from Maine. Bert M. Fernald (1858–1926) was governor of Maine (1909–1911) and a US senator (1916–1926).

52. "Some of the utility magnates may seem worried and defensive, but they represent what is probably the most keen witted, ruthless, and effective lobby in the United States. There is no other lobby that one meets at every level, city, county, township, state." Gunther, *Inside U.S.A*, quoted in *ES*, 17 July 1947, 4.

53. *ES*, 17 June 1948, 1. See also *ES*, 18 March 1948, 1.

54. *ES*, 23 September 1948, 1, citing *St. John Telegram Journal.*

55. *ES*, 9 September 1948, 4.

56. *ES*, 12 May 1949, 1.

57. *ES*, 23 June 1949, 1. The vote had a higher symbolic value than cash value, as Maine's appropriation would have been considered a "gift" to the federal government, which it couldn't accept, and also an unauthorized practice of foreign policy.

58. One could say that the campaign was successful, or unneeded, as there were no electric company nationalizations in the United States thereafter. In Canada in 1944, however, private companies in the Montreal area were nationalized into Hydro-Quebec, and there were additional takeovers in 1960s. Wikipedia, s.v. "List of Nationalizations by Country," https://en.wikipedia.org/wiki/List_of_nationalizations_by_country.

59. *ES*, 9 June 1949, 1.

60. "Philip B. Fleming, 2011," West Point Association of Graduates, https://www.westpointaog.org/memorial-article?id=e7274805-eaaa-4b1c-8ecf-1796066f4de5.

61. Wikipedia, s.v. "Hugh John Casey," https://en.wikipedia.org/wiki /Hugh_John_Casey#cite_note-41, citing Hugh J. Casey, ed., *Major General Hugh J. Casey, Engineer Memoirs* (Washington, DC: United States Department of the Army, 1993), 270–71.

62. ES, 5 August 1936, 1.

63. *ES*, 6 February 1935, 4, quoting the *PEN*.

64. *Labor*, 3 September 1949, in EMERY_Scrapbook_Quoddy-1949.

65. *ES*, 25 August 1949, 1.

66. *ES*, 25 August 1949, 1.

67. *NYT*, 21 August 1949, 43.

68. *ES*, 27 January 1949, 1.

69. Holmes, 84, citing *Report of the Power Survey Committee of the New England Council: Power in New England* (Boston, 1948), vii.

70. *ES*, 25 August 1949, 1.

71. *ES*, 3 August 1950, 1–2.

72. *ES*, 3 August 1950, 1–2.

73. Weil, 33.

74. In a letter to the editor of the *PPH* on 18 July 1936, EMERY, Roscoe_box 2.9b_Letters to Editor 1930–60s.

75. Parkman, 236–37.

76. *ES*, 18 November 1925, 1–2; and *IJC-50*.

77. *IJC-50*, 18–23. Dexter Cooper thought otherwise, writing in a letter to FDR in January 1936, "This will require a construction and engineering organization of experience and proven ability. Steps should be taken to create such an organization at Quoddy at once. This statement does not imply that there are unusual construction problems involved. There are none." 1936-01-16_Dexter Cooper to FDR_All Quoddy Power sold.

78. *IJC-Eng-50*.

79. Wikipedia, s.v. "Soviet Atomic Bomb Project," https://en.wikipedia.org /wiki/Soviet_atomic_bomb_project.

80. Wikipedia, s.v. "Berlin Blockade," https://en.wikipedia.org/wiki/Berlin_Blockade.

81. *ES*, 3 August 1950, 2.

82. *NYT*, 27 May 1950, 96; this, despite the *Times* report, "Truman Bars Quoddy Project," on 10 March 1950.

83. *ES*, 18 May 1950, 8. See also FDRL-OLDS:1950-06_Let's Get All the Facts about Quoddy, 20.

84. *ES*, 16 September 1936, 1, quoting Kenneth Roberts, "Quoddy," *Saturday Evening Post*, 19 September 1936.

85. *NYT*, 27 May 1950, 96.

86. *ES*, 27 July 1950, 1.

87. *ES*, 3 August 1950, 2; and *IJC-50*. For more power trust propaganda, see *NYT*, 13 December 1950, 65.

88. "Mr. James Morang, Democratic State Chairman, Randolph, Maine; Mr. Roscoe C. Emery, Eastport, Maine; Mr. Arthur Knobskey, Calais, Maine; Rev. Ernest Heywood, Calais, Maine; Mr. Oscar H. Brown, Eastport, Maine; Mr. Arlo T. Bates, Calais, Maine. In November, Mr. Turney Gratz at Democratic National Committee, forwarded to Mr. Connelly a letter addressed to Mr. Boyle by Davis Clark, National Committeeman for Maine, in which he said several members of Washington County, Maine Chamber of Commerce were interested in so-called Quoddy Project, the harnessing of the tide to develop electrical power on an international basis; that this group wished to send to Washington a delegation to discuss this with the President. They asked to come in January. Mr. Connelly checked this with Mr. Charles Murphy before arranging." "Daily Appointments of Harry S. Truman, Wednesday, January 31, 1951," Harry S. Truman Library & Museum, https://www.trumanlibrary.gov/calendar?month=1&day=31&year=7.

89. *ES*, 22 February 1951, 16, citing *Bangor Sunday Commercial*.

90. *ES*, 8 March 1951, 16; and *ES*, 22 March 1951, 1. See also *ES*, 26 January 1950, 1.

91. *ES*, 5 July 1951, 1.

92. *ES*, 27 September 1951, 1.

93. *ES*, 5 July 1951, 16.

94. *ES*, 12 July 1951, 16, citing *Bangor Sunday Commercial*.

95. *ES*, 27 September 1951, 1; *ES*, 15 November 1951, 1; and *Newsweek*, 1 October 1951.

96. *ES*, 15 November 1951, 1, citing *Newsweek*, 1 October 1951.

97. *ES*, 18 October 1951, 1. Governor Payne had an unannounced meeting with President Truman at the White House in April 1951 in which he claimed that he convinced the president to instigate the sonic test. The meeting was not publicized until April 1952, when Payne got into a very nasty fight with Ralph Brewster for his US Senate seat. *ES*, 17 April 1952, 1.

98. *NYT*, 11 November 1951. See also Parkman, 236–37.

99. *ES*, 6 December 1951, 1.

100. Holmes, 76, citing *BDN*, 3 December 1952; and Holmes, 86, citing *BDN*, 3 December 1952, and 18 March 1953. The engineering firm of Thomas Worcester, Inc., of Boston offered to do the survey for $1.5 million according to the specification set forth by the International Joint Commission's engineering board, based in part on actually using the plans and engineering data compiled by Dexter P. Cooper. *ES*, 21 May 1953, 1.

101. Apparently based on the obsolete and erroneous 1941 report by McWhorter of the FPC.

102. A "highly dubious" report was issued in January 1952 that claimed more hydroelectric power could be generated from the St. John River than from the Columbia or St. Lawrence Rivers, and at a tiny fraction of the construction cost. *ES*, 31 January 1952, 1.

103. *ES*, 18 October 1951, 4.

104. *ES*, 27 December 1951, 1.

105. "Figures of the Normandy Landing," D-Day Overlord, accessed 14 May 2020, https://www.dday-overlord.com/en/d-day/figures.

106. Weil, 186, citing Alex Radin, *Public Power—Private Life* (Washington, DC: American Public Power Association, 2003), 139. But by the end of his second term on 17 January 1961, Eisenhower famously warned Americans of the threat to democracy from the overly powerful "military-industrial complex." "Ike's Warning of Military Expansion, 50 Years Later," NPR, 17 January 2011, https://www.npr.org/transcripts/132942244.

107. *ES*, 23 April 1953, 6.

108. *ES*, 21 May 1953, 2. Rail service to Eastport ceased on 13 November 1978. Holt, 14.

109. Maine's congressional delegation called on Eisenhower in January 1954. He said nothing publicly, but in February, "without debate and without dissent," the Senate approved the $3 million allocation for Quoddy's "full-dress" survey. It was also unanimously approved by the House Foreign Affairs Committee but was not brought to the full House for a vote, thus again showing the House's susceptibility to power trust pressure. *NYT*, 11 February 1954, 13; and *NYT*, 27 February, 1954, 7.

110. Appointed by President Truman, Sumner Pike was on the AEC from 1946 to 1951.

111. EMERY_6.32b.13-16: *Newsweek*, 5 April 1954, 90–94. For a profile of Sumner Pike, see L'Hommedieu, "MacNichol, David."

112. *ES*, 24 January 1934, 4; Ickes, 22 August 1934, 189, a transcription of which is in EMERY_6.18.

113. Wikipedia, s.v. "David E. Lilienthal," https://en.wikipedia.org/wiki/David_E._Lilienthal#cite_note-32, citing David E. Lilienthal, *The Journals of David E. Lilienthal*, vol. 2, *The Atomic Energy Years, 1945–1950* (New York: Harper and Row, 1964), 24–25. David Eli Lilienthal (1899–1981) was codirector of TVA from its creation in 1933 until 1941, vice chair (1939–1941), and chair (1941–1946).

114. *NYT*, 17 September 1954.

115. Holmes, 87–88, citing *Courier*, 30 December 1954; Irwin, 60; and "Fernald, Bert M.," *Maine: An Encyclopedia*, https://maineanencyclopedia.com/bert-m-fernald/. In 1955, Quoddy was also given a boost by the report

of the New England–New York Interagency Committee, which recommended that the survey be undertaken. Parkman, 236–37.

116. *Udall-63*, 10; and Parkman, 236–37. See also Emery_box 5, folder 29, correspondence re IJC, 1956.

117. Riggs, 50. The dam has powerhouses on each side of the river: in the United States, Robert Moses Generating Station produces 912 MW; in Canada, R. H. Saunders Generating Station, 1,045 MW. Wikipedia, s.v. "Moses Saunders Power Dam," https://en.wikipedia.org/wiki/Moses-Saunders_Power _Dam#cite_note-Lawrence-6, citing Daniel Macfarlane, *Negotiating a River: Canada, the US, and the Creation of the St. Lawrence Seaway* (Vancouver: University of British Columbia Press, 2014).

118. *IJC-Eng-59*; *Final-36*, 3b; and *NYT*, 17 October 1959, 1. See also Parkman, 238–39.

119. Commissioners of the Joint International Commission, poster.

120. David E. Bell, Director, Office of Budget, to President Kennedy, 30 January 1961, replying to previous request for information, JFKPOF-070-00, Papers of John F. Kennedy, Presidential Papers, President's Office Files, Departments and Agencies, Bureau of the Budget, 1961, January–April.

121. *NYT*, 2 May 1961, 40; and *NYT*, 16 July 1963, 1, 63.

122. *IJC-61*, 30.

123. *NYT*, 20 May 1961, 34; and *Udall-63*, 15.

124. Cantelon, 73–74; McCraw, 154–207, 219, 234; and Weil, 186–88.

125. Special Message to the Congress on Urgent National Needs," WikiSource, https://en.wikisource.org/wiki/Special_Message_to_the_Congress_on _Urgent_National_Needs. See also Weil, 186–88.

126. *Udall-63*, 15; and *IJC-Eng-59*, 6.

127. Parkman, 239; and *NYT*, 16 July 1963, 1. "Our conclusion is that the Tidal Power Project is economically and egineeringly [*sic*] feasible alone or in conjunction with development of the Dickey site on the Upper St. John River for storage and power, and as a reregulating dam at the Lincoln School site. . . . The Department's proposal is feasible from an engineering and economic viewpoint. The benefit-cost ratio is 1.27 to 1.00 based on current United States' project feasibility standards and using an interest rate of 2-7/8 percent with power repayment within 50 years after each unit becomes revenue-producing." *Udall-63*, 2, 23.

128. *Yankee*, September 1963, citing "Load and Resource Study" (Department of Interior, 1962).

129. *NYT*, 30 July 1936, 1.

130. *NYT*, 17 July 1963, 16; *Udall-63*, 6–7; and Parkman, 238–43.

131. *NYT*, 24 August 1963, 9. *Udall-63*, 22, concludes, "France is already constructing a tidal power project and Russia is planning one. It will enhance

the stature of the United States in the electric power field to proceed at once with the Passamaquoddy project. This project will also add to our store of electrical knowledge and possibly lead to economics of construction which will make feasible other tidal power projects in this country. Harnessing the energy of the tides is a big idea and big undertaking. We must think and act big if we are to take full advantage of the opportunities modern technology holds out to us. Generally, the development of the Passamaquoddy Tidal Project and the Upper St. John River could do as much for New England and the Nation as Grand Coulee Dam has done for the Pacific Northwest and the Nation."

132. "Remarks in Response to a Report on the Passamaquoddy Tidal Power Project, 16 July 1963," John F Kennedy Presidential Library and Museum, https://www.jfklibrary.org/asset-viewer/archives/JFKWHA/1963/JFK-WHA-206-002/JFKWHA-206-002.

133. *NYT*, 21 July 1963, 110. See also *NYT*, 3 Aug 1963, 15.

134. *NYT*, 20 July 1963, 20.

135. Office of the Historian, US House of Representatives, Representatives Apportioned to Each State, *1st to 23rd Census, 1790–2010*, http://history.house.gov.

136. Author's interview with Eastport native Susan McNichol, 5 April 2021.

137. This quote has since been tracked back further: "I think we must develop our natural resources. You cannot bring industry into Ohio unless you have clean rivers. I think the greatest asset that has happened to Ohio during the last few years, except for Governor Di Salle's election, was the building of the St. Lawrence Seaway, and I was proud, though I came from Massachusetts, to vote for it, because it is a national asset and a rising tide lifts all boats. If Ohio moves ahead, so will Massachusetts. [Applause.] Good water, power, transportation, those are necessary to develop the economy of the United States in the 1960's." John F. Kennedy, "Remarks of Senator John F. Kennedy," 27 September 1960, Municipal Auditorium, Canton, Ohio, The American Presidency Project, https://www.presidency.ucsb.edu/documents/remarks-senator-john-f-kennedy-municipal-auditorium-canton-ohio.

138. Wikipedia, s.v. "A Rising Tide Lifts All Boats," accessed 27 August 2013, http://en.wikipedia.org/wiki/A_rising_tide_lifts_all_boats, citing Ted Sorensen, *Counselor: A Life at the Edge of History* (New York: HarperCollins, 2008), 227; and Kennedy, "Remarks of Senator John F. Kennedy," 27 September 1960. The privately owned New England Electric System was still adamantly opposed to Quoddy, as evidenced by an issue of its magazine: *Contact* 44, no. 9 (September 1963): front cover; editorial "Questionable Quoddy" on inside front cover; and "The Folly of Quoddy," on page 4.

139. Gruening, 204, quoted in a letter of 23 March 1929 from Mr. S. E. Boney, National Electric Light Association's Carolina Committee on Public Utility Information, complaining to Mr. Rogers that his statement was "wholly lacking in truth."

140. "Syzygy" is the gravitational alignment (linear or perpendicular) of the of the sun, earth, and moon, thus creating unusually large or small tides (spring versus neap, respectively), or more generally (and not really appropriately), any alignment in support of or opposition to something. See appendix 2 for more on the astronomical causes of the tides.

141. A draft of *Udall-65* was completed in August 1964 and circulated to heads of federal agencies and New England governors for their comments, which were then incorporated into the 9 July 1965 final report.

142. The Rankin Rapids project, proposed in 1961, would have required the flooding of a large part of the Allagash River, one of the last great wild rivers east of the Mississippi. An odd alliance was struck between conservationists, hunting and fishing clubs, and the private power interests to defeat the proposal. The Dickey-Lincoln site was just upstream of the junction of the St. John and Allagash Rivers and would therefore not flood the Allagash's esteemed rapids. *Udall-63*, 10; and Parkman, 240–41.

143. *Udall-63*, 3–5. $586 million for dams, gates, locks, and the first of two 500,000 kW powerplants; plus $227 million for Dickey and Lincoln dams; plus $87 million for transmission lines.

144. The federal interest rate charged to such projects was 3% in 1961 and 3.125% in 1963; then it increased to 3.25% on 1 July 1965. *Udall-65* also used updated market value prices based on new, more efficient power plants being built in Maine and Massachusetts, instead of on the average of existing power plants. In combination, the calculated market value of power dropped from 3.0 mills to 2.6 mills. These calculations were based on the then-current (and ultimately proven to be egregiously low) estimates for the cost of nuclear power and fossil fuels. *Udall-65*, 4–5.

145. Parkman, 242; and *Udall-65*, 1, citing the FPC's 1964 *National Power Survey*.

146. Gruening, sample ads following the introduction.

147. Parkman, 239–43.

148. Thurston, *Tidal Life*, 20.

149. Wikipedia, s.v. "Rance Tidal Power Station," https://en.wikipedia .org/wiki/Rance_Tidal_Power_Station#cite_ref-4, citing http://france.edf.com /html/en/decouvertes/voyage/usine/bilan/usine_bilan_d.html.

150. Thurston, *Tidal Life*, 20.

151. *NYT*, 18 August 1963, 9. Russia completed its tidal-hydroelectric power plant in the Arctic in 1968. It now produces a modest 1,700 kW of

power. Wikipedia, s.v. "Kislaya Tidal Power Station," accessed 19 December 2019, https://en.wikipedia.org/wiki/Kislaya_Guba_Tidal_Power_Station.

152. In 1954, CMP took a 9.5% equity stake in Yankee Atomic Electric Company, which then built Massachusetts Yankee in Rowe, which went online on 1 July 1961. Maine Yankee was 50% owned by Maine utilities. CMP was an investor in the Millstone Unit number 3. As of 1999, it had a 2.5% equity interest. Irwin, 60, 66, 71, 109.

153. GSM, "Passamaquoddy," 15–16.

154. Wikipedia, s.v. "George Galatis," 8 December 2019, https://en.wikipedia.org/wiki/George_Galatis; and "Northeast Utilities History," Funding Universe, http://www.fundinguniverse.com/company-histories/north east-utilities-history/.

155. The Maine Yankee Atomic Power Company was formed in 1966. Construction of the $231 million Wiscasset plant began in 1968, and it started generating a maximum of 900,000 kW in 1972. After only 25 years tt was decommissioned in 1997, and it was dismantled by 2005 at significant cost. The plant's nuclear waste remains on site. Wikipedia, s.v. "Maine Yankee Nuclear Power Plant," accessed 20 May 2019, https://en.wikipedia.org/wiki/Maine_Yankee_Nuclear_Power_Plant.

156. Alexis C. Madrigal. "Moondoggle: The Forgotten Opposition to the Apollo Program," *Atlantic*, 12 September 2012, 2http://www.theatlantic.com/technology/archive/2012/09/moondoggle-the-forgotten-opposition-to-the-apollo-program/262254/.

CHAPTER 14. WHO KILLED QUODDY?

1. See chapter 8, "The Commander in Chief," note 123 for the complete list of the stockholders of the Canadian Dexter P. Cooper Company.

2. GSM, "Passamaquoddy," 14. There was a cash balance of $973,000 in Quoddy's account when the US Army left Eastport on 1 November 1936. *ES*, 16 March 1938, 10. See also *ES*, 18 November 1936, 1.

3. *Reel-38,* 13. This wording echoes Ickes's *Secret Diaries* entry for 12 September 1934: "graceful gesture."

4. FDRL-0982: 1935-12-24_Gen. Pillsbury to McIntyre_$62MM estimate (instead of the original $35 million estimate).

5. Ickes, 513–14.

6. FDRL Day-by-Day: 1936. Dexter's meeting with FDR on 26 July 1933 is documented by him and press reports but does not appear on FDR's official calendar. FDR is known to have had private guests visit the White House under assumed names, such as his former lover, Lucy Mercer Rutherford (Goodwin,

No Ordinary Time, 434–35), so it is possible that he had similarly disguised meeting(s) with Dexter, Gertrude, or Roscoe.

7. FDRL-1766_ 1936-01-16_Dexter Cooper to FDR_All Quoddy power sold; and *ES*, 10 March 1936, 1.

8. While Lieutenant Colonel Fleming apparently believed Cooper (*ES*, 11 March 1936, 1, 4), it is doubtful that Cooper disclosed the names of the potential purchasers to Fleming, as there was no "documentary evidence" of their existence according to the US Army. *Final-36*, 10–11. However, Oscar Brown claimed to have known the purchasers' identities but wasn't asked to identify them. *Reel-38*, 23. Given Brown's knowledge of the potential customers, it seems likely that Roscoe also knew. Given that Dexter's letter is dated 16 January 1936—the same day that FDR directed Ickes to meet with him the next day—it seems likely that Secretary Ickes was also told the identities of the prospective purchasers of Quoddy's power. Ickes, 513–14.

9. *MSPB-34*, 159-60; and *ES*, 29 Dec 1937, 1.

10. Some of the names of the prospective power purchasers may have been stated publicly for the first time by Gertrude Cooper and/or Roscoe Emery in 1956 at an event related to the IJC's "full dress" survey. Emery_box 5.29_International Commission 1956, Gertrude Cooper to Roscoe Emery, 17 June 1956.

The St. Croix Paper mill in Houlton has had several owners since the 1930s. It closed for a few years in the early 2000s and reopened in 2010. Monsanto bought Merrimac in 1929. Bayer bought Monsanto in 2018.

The rural electric co-ops being formed as part of the New Deal would have been part of Dexter's plan for local distribution.

11. *ES*, 12 August 1936, 1. The Norris Dam officially opened on 4 March 1936.

12. Because the army's Plan D was not completed or even fully designed (see appendix 4), the price of the electric power and hence the value of the purchase commitments had not yet been determined. If we assume all the power was sold at wholesale price of 3 mills per kWh, then the purchase commitments Dexter secured would have been worth about $1 million annually. If we assume a construction cost of $42 million and a 50-year mortgage at 2% interest ($24.5 million of interest), as in *MSPB-34*, then mortgage costs would be about $1.3 million annually. Thus, the army's Plan D would not be self-liquidating. If Quoddy cost $71 million, as the army predicted, then the deficit would be greater.

13. The two appearances of the word "how" are underscored in a pen different from that used by Delano to sign the letter, suggesting that FDR had underlined the words, whereas further down in this quote the two ap-

pearances of "not" are underlined by a typewriter, thus indicating that it was Delano who did so.

14. FDRL-0982: 1938-12-27_Delano to FDR_CONFIDENTIAL history of Quoddy. Delano was a member of the National Resources Planning Board and its National Power Policy Committee.

15. *ES*, 18 March 1936, 4; and FDRL-0982: 1936-08-12_War Dept press release on Quoddy.

16. Paul Crewes and Mr. Whitse. *Reel-38*, 21.

17. John Sweeney. *Reel-38*, 13, 20–26.

18. Roscoe Emery told Reel that the source of the "porky pies" was a man named "Blodgett." *Reel-38,* 14–15. Remember that Roscoe himself was biased against the army, and Sturgis in particular, because of the soured land deals and Sturgis's request for a "red light district."

19. See chapter 11, "Demobilization," note 102.

20. Roger Bacon McWhorter (1888–1967). *Engineering News-Record* 93, no. 26 (25 December 1924), 1053, https://www.google.com/books/edition/Engineering_News_record/8NJJAQAAIAAJ?hl=en&gbpv=1&dq=obituary+Roger+B.+McWhorter&pg=PA1053&printsec=frontcover.

21. Henry Matson Waite (1869–1944).

22. *ES*, 15 November 1933.

23. *ES*, 16 March 1938, 9. Cooper returned from Washington on 7 August 1933.

24. Ickes, 333. See also chapter 8, "The Commander in Chief," note 102; GSM, "Passamaquoddy," 9; and *Reel-38*, 3: "Colonel Waite was then the Acting Administrator of the P.W.A. He turned down Cooper's project and according to Mrs. Cooper he refused to let her husband point out the errors in his (Waite's) report. Colonel Fleming was very friendly with a man named Hof, who is now dead, and Mrs. Hof was friendly with Mrs. Cooper and frequently told her of conversations between Fleming and Hof. It was when Colonel Waite turned down the Cooper project [in January 1934] that Mrs. Hof told Mrs. Cooper, according to her report to me, that Hof asked his wife to tell Mrs. Cooper somewhat as follows: 'This is dirty Catholic policies.' Mrs. Cooper, incidentally, appears obsessed with the idea that there was some Catholic conspiracy in the matter, because as she claimed, from General Markham, Chief of Engineers, on down, the entire Army organization that later worked on the project, as well as these P.W.A. heads, were Catholics. I might say here, that although I asked about this matter in Eastport, I never at any other time in my investigation ran into any such charge." The date of Gertrude and Alice's meeting is not clear.

25. *Reel-38*, 16. See chapter 12, "The Turn of Events."

26. Riggs, 146–47, citing Funigiello, 178–80.

27. Riggs, 163–64, citing Billington, Jackson, and Melosi, 189–90; Funigiello, 182–84; and Ickes, *Secret Diaries*, 2:50.

28. *ES*, 3 March 1937, 4.

29. *ES*, 3 June 1936, 1.

30. FDRL-0982: 1938-12-03_FDR to Delano_experimental scale Quoddy.

31. *PWA-Hunt-35*, v–vi. According to Ickes, 542, the Army Engineers didn't check Dexter's cost assumptions before accepting them and launching the project on 16 May 1935.

32. Goodwin, *No Ordinary Time*, re: Destroyers-for-Bases, 64; and Francis Perkins and Lend-Lease, 193.

33. EMERY_5.7b. March 1943.

34. *ES*, 6 October 1937, 1, quoting Elizabeth May Craig in *PPH*, 4 October 1937. See also Riggs, 164; Funigiello, 182–83; Leuchtenberg, "Roosevelt, Norris and the 'Seven Little TVAs,'" 418–19, citing *Washington (DC) Star*, 5 February 1937.

35. *ES*, 24 April 1929, 4; *ES*, 19 June 1929, 3; and *Ruby Black-33*, 3–4.

36. *Reel-38*, 2; and *IJC-61*, 3. M&F estimated the International Plan would cost $125 million. *IJC-61*, 3.

37. Riggs, 209; and Klein, *Call to Arms*, 161, 230–31.

38. Goodwin, *No Ordinary Time,* 259–60, citing *New Republic*, 27 January 1941, 104; Goodwin's interview with Arthur Goldsmith; and *Time*, 7 July 1941, 10.

39. Several important newspapers, most dramatically Guy Gannett's *Portland Press Herald* and Hearst's *Boston Herald* (see notes 44 and 46 about George L. Brady), went from Quoddy opponents to advocates. Similarly, the coverage by the *New York Times* ranged from supportive to critical to quite unfair, depending on the writer, often unnamed.

40. Central Maine Power Company Collection, 1883–1965, Special Collections, Raymond H. Fogler Library, University of Maine, DigitalCommons@ UMaine, 2015, https://digitalcommons.library.umaine.edu/cgi/viewcontent.cgi ?article=1042&context=findingaids.

41. The investment in Dexter P. Cooper, Inc., for Quoddy was made by MWU and apparently managed locally through CMP. The exact amount of the investment isn't clear because it flowed through GE and Westinghouse and apparently included cash to Dexter P. Cooper, Inc., as well as payment to Murray & Flood and possibly others. *Final-36*: Appendix A-3; and Smith, *Power Policy*, 235n14.

42. Chuck Rand, *Guide to the Central Maine Power Company Archival Collection, 1853–2001* (Maine Historical Society, December 7, 2011), https:// www.mainehistory.org/PDF/findingaids/Coll_2115.pdf.

43. This appears to have been written by George L. Brady. See quote in chapter 13 from "Address by George L. Brady, Chief Editorial Writer of the

Hearst Publications, Boston, before the Washington County (Maine) Chamber of Commerce at Whiting, Feb. 4, 1948," McInnis Collection.

44. *ES*, 29 May 1947, 4.

45. "Address by George L. Brady, Chief Editorial Writer of the Hearst Publications, Boston, before the Washington County (Maine) Chamber of Commerce at Whiting, Feb. 4, 1948," McInnis Collection.

46. This ending was inspired by three quotes:

- "Affectionately and more power to you. Gertrude." http://www.fdrlibrary .marist.edu/_resources/images/psf/psf000508.pdf, FDRL-COOPER: Subject File "C," box 141. In a letter to "Mr. President" congratulating him on his State of the Union speech of 7 January 1943.
- "The 'New Deal' will go far toward becoming a square deal if it protects us from these two varieties of financial wolves and more power to the President in his efforts in this direction." *ES*, 8 March 1933, 4.
- "Less than [16] years after President Roosevelt stood at Bonneville and said, 'More power to you,' these same omniscient power executives met at Tacoma and resolved: 'A shortage of power cannot be prevented. It has already begun. Failure to provide additional generating capacity will retard development of the Northwest.'" *ES*, 20 May 1948, 4.

APPENDIX 1. AUTHOR'S NOTES AND ACKNOWLEDGMENTS

1. Smith, *Power Policy of Maine*, 290.

2. George R. Marvin's thesis, "Passamaquoddy! The Dream That Won't Die," was published in the *Ellsworth American* in nine weekly installments, starting on 4 April 1974. A copy is in the collection of the Lubec Public Library.

3. Marvin, 71, 93, 100.

4. See chapter 8, "The Commander in Chief."

5. See chapter 9, "Celebrations."

6. Riggs, 75, 79; and, for example, McDonald, 252n22.

7. McDonald, 274–304.

8. McDonald, 271–73. The allegation that Franklin Roosevelt's policies to rein in holding companies and private power trusts was motivated by a grudge against Hobson (from 1928) fails to recognize the influence of Theodore Roosevelt's Progressive trust-busting ideas on Franklin (1901–1919), Franklin's efforts to establish public power on the St. Lawrence River as a New York State senator (1910–1913) and governor (1928–1932), or Franklin's knowledge of the power trusts' actions relative to Quoddy (1919–1945).

9. By 1948, the *Boston Herald* had switched to being pro-Quoddy: "Address by George L. Brady, Chief Editorial Writer of the Hearst Publications,

Boston, before the Washington County (Maine) Chamber of Commerce at Whiting, Feb. 4, 1948," McInnis Collection.

APPENDIX 2. THE ASTRONOMY
THAT CREATES THE OCEAN TIDES

1. "Tides and Water Levels: What Causes Tides?" National Ocean Service, National Oceanographic and Atmospheric Administration, https://oceanservice.noaa.gov/education/tutorial_tides/tides02_cause.html. For simplicity: the sun = 27 million units; the moon = 390^3 units = 59 million units, or about twice the sun. More precisely, the moon's tidal influence is 2.17 times that of the sun, or the sun's is 46% that of the moon. See also *Eldridge Tide and Coast Pilot*, 234.

2. Chaisson and McMillan, 168.

3. McCully, 23. The centrifugal bulge is the result of the water's inertia temporarily exceeding the gravitational force (the water tries to keep going in a straight line while the earth spins). "Tides and Water Levels: Gravity, Inertia, and the Two Bulges," National Ocean Service, National Oceanographic and Atmospheric Administration, https://oceanservice.noaa.gov/education/tutorial_tides/tides03_gravity.html.

4. One high tide per day is called *diurnal*. In some places, like Tahiti, which is near a Pacific Basin tidal node, tides are dominated by the sun rather than the moon, and consequently high tide is on a 24-hour cycle instead of a 24 hour and 50-minute cycle. Thus, high tide at Tahiti is usually at about noon. McCully, 66–67, 110; and Aldersey-Williams, 222.

5. "Moon in Motion," NASA Science: Earth's Moon, https://moon.nasa.gov/moon-in-motion/overview/.

6. Chaisson and McMillan, 19.

7. Koppel, 196.

8. White, 165–66, and personal correspondence, 22 June 2021; and McCully, 90.

APPENDIX 3. EXCEPTIONAL DAMS BUILT
BY HUGH AND DEXTER COOPER

1. Wikipedia, s.v. "Toronto Power Generating Station," accessed 7 October 2020, https://en.wikipedia.org/wiki/Toronto_Power_Generating_Station.

2. Billington, Jackson, and Melosi, 115.

3. Wikipedia, s.v. "Holtwood Dam," accessed 7 October 2020, https://en.wikipedia.org/wiki/Holtwood_Dam.

4. Wikipedia, s.v. "Wilson Dam," accessed 7 October 2020, https://en.wikipedia.org/wiki/Wilson_Dam.

5. Billington, Jackson, and Melosi, 119.

6. Dorn, 337, citing *Soviet Union Review* 10 (1932): 212; and *NYT*, 27 August 1932, 2, and 10 October 1932, 1.

7. As Dexter was employed by Hugh both before and after, it seems likely that Dexter managed the construction of the McCall's Ferry Dam, but I have not come across evidence to prove this contention.

8. Berton, 26.

9. Riggs, 14.

10. Dorn, 325.

11. Billington, Jackson, and Melosi, 115.

12. Hughes, 326: the record high was at the time of the Johnstown Flood (during which an upstream dam burst) in 1889.

13. Billington, Jackson, and Melosi, 115.

14. Hughes, 331, citing Kellogg, "Conowingo Hydroelectric Project," 7, 9; and Billington, Jackson, and Melosi, 117 (flood capacity).

15. Hughes, 331, citing Kellogg, "Conowingo Hydroelectric Project," 7, 9.

16. *ES*, 12 December 1928, 1.

17. *Brooklyn (NY) Daily Eagle,* 11 January 1914, 4. Kiwanis Club of Stamford, 368, says eleven generators, each with about 11,000 horsepower for a total of 91,800 kilowatts.

18. Hughes, 331, citing Kellogg, "Conowingo Hydroelectric Project," 7, 9.

19. Billington, Jackson, and Melosi, 119, citing Cooper, "Water Power Development on the Mississippi River at Keokuk, Iowa," 1923, 5.

20. Hughes, 331, citing Kellogg, "Conowingo Hydroelectric Project," 7, 9.

21. Hughes, 331, citing Kellogg, "Conowingo Hydroelectric Project," 7, 9.

22. EMERY_5.7.18 appears to be Roscoe's draft history of Quoddy. "Cooper estimated that the project would produce more than 600,000 hp., which would run into several billion kilowatt-hours." See also 800,000 hp estimate reported in *ES*, 26 July 1933, 1; and 300,000 horsepower reported in *ES*, 16 July 1930, 1. The 500,000 to 700,000 estimates are in Dexter P. Cooper, Inc., blueprint 1099, dated 23 July 1927 (University of Maine, Fogler Library); EMERY_5.7c.1-36; and *ES*, 3 June 1925, 1.

23. *FPC-41*, 2; and Smith, *Power Policy*, 235: "The Federal Power Commission issued a preliminary permit to Dexter P. Cooper, Inc. on May 8, 1926. On October 22, 1928, an application for a license was filed, contemplating initial and ultimate power installations of 464,000 hp and 1,087,00, respectively." Power reported as 1,000,000 hp according to "Let's Get ALL the Facts about Quoddy," 7, Maine Development Commission, Augusta, June, 1950, https://digicom.bpl.lib.me.us/cgi/viewcontent.

cgi?article=1291&context=books_pubs. EMERY_5.7.18 appears to be Roscoe's draft history of Quoddy. "The cost of the project was estimated at between $75,000,000 and $100,000,000, and electricity could be produced there for three-quarters of a cent per kilowatt-hour. Other estimates were that eleven to twenty billion kilowatt-hours were theoretically available from the international project, but practical considerations would limit it to three billion kilowatt-hours, or nearly one-half million horsepower each year." Smith, *Power Policy*, 233, citing *Maine Legislative Record*, 1925, 975–79; and *Principles and Facts Underlying Quoddy Tidal Power Project*, pamphlet issued by the Eastport District, US Engineers Office, n.d., 3.

24. *ES*, 11 September 1929, 1; "Dnieper Dam Blown Up by Retreating Russians," accessed 11 May 2020, https://forum.axishistory.com/viewtopic .php?t=52940, citing *NYT*, 20 August 1941.

25. *Brooklyn (NY) Daily Eagle*, 11 January 1914, 4. Kiwanis Club of Stamford, 368, says eleven generators, each with about 11,000 horsepower for a total of 91,800 kilowatts.

26. Wikipedia, s.v. "Holtwood Dam."

27. Martin.

28. Dorn, 326, citing "Water Power Development on the Mississippi River at Keokuk, Iowa," *Engineering News Record* 66 (1911): 355–64; see also Mildred Adams, "A Man Who Makes River Do Our Work," *NYT*, 21 April 1929, 13, and Kerwin, *Federal Water-Power Legislation*, 174.

29. *MSPB-34*, 150; and Hughes, 287n3.

30. EMERY 5.7c.1-36: Flyer from Cooper-Quoddy Publicity Organization, September 1925, Office Headquarters at 83 Water Street, Eastport. Officers do not list Emery.

31. Holmes, 10, citing "Legislative Record of the 82nd Legislature of the State of Maine," Augusta, 1925, 125.

32. EMERY 5.7c.1-36: Flyer from Cooper-Quoddy Publicity Organization, September 1925.

33. *ES*, 12 October 1932, 1.

34. Wikipedia, s.v. "Dnieper Hydroelectric Station," accessed 20 March 2019, https://en.wikipedia.org/wiki/Dnieper_Hydroelectric_Station.

35. Wikipedia, s.v. "Holtwood Dam."

36. Hallwas, 2; and Meigs, 200.

37. "Wilson Dam and Reservoir," *Encyclopedia of Alabama*, accessed 7 October 2020, http://www.encyclopediaofalabama.org/article/h-3268.

38. "Wilson [Dam]," Tennessee Valley Authority, https://www.tva.com /energy/our-power-system/hydroelectric/wilson.

39. EMERY_5.7.18 appears to be Roscoe's draft history of Quoddy. Smith, *Power Policy*, 233.

40. "Dnipro Hydroelectric Station," *Internet Encyclopedia of Ukraine*, http://www.encyclopediaofukraine.com/display.asp?linkpath=pages%5CD%5CN%5CDniproHydroelectricStation.htm.

41. Kiwanis Club of Stamford, 368.

42. Kiwanis Club of Stamford, 368. Eleven generators of 11,000 horsepower each, for a total of 91,800 kilowatts.

43. Wikipedia, s.v. "Holtwood Dam"; and Billington, Jackson, and Melosi, 115.

44. Wikipedia, s.v. "Holtwood Dam"; and Billington, Jackson, and Melosi, 118.

45. Dorn, 326; see also Adams, "A Man Who Makes River Do Our Work"; and Kerwin, 174.

46. *MSPB-34*, 151.

47. Wikipedia, s.v. "Wilson Dam."

48. Holmes, 10, citing *Report to International Joint Commission*, 1950, 5.

49. EMERY_5.7.18 appears to be Roscoe's draft history of Quoddy.

50. Wikipedia, s.v. "Dnieper Hydroelectric Station."

51. Hallwas, 94.

52. Smith, *Power Policy*, 241, citing *Passamaquoddy Tidal Power Project*, S. Doc. no. 41, 77th Cong., 1st Sess. (7 April 1941), 2–3.

53. Holmes, 11, citing Legislative Record of the 82nd Legislature of the State of Maine, Augusta, 1925, 125; *BDN*, 14 March 1925; and Dexter P. Cooper, "Some Aspects of the Passamaquoddy Tidal Power Project," in *The Connecticut Society of Civil Engineers, Papers and Transactions for 1926 and Proceedings of the 42nd Annual Meeting* (New Haven, CT: Tuttle, Morehouse and Taylor, 1926), 29.

54. "Dnipro Hydroelectric Station."

55. Kiwanis Club of Stamford, 368, says eleven generators, each with about 11,000 horsepower for a total of 91,800 kilowatts.

56. Wikipedia, s.v. "Holtwood Dam."

57. Hallwas, 93.

58. Hughes, 331, citing Kellogg, "Conowingo Hydroelectric Project," 7, 9.

59. "Wilson [Dam]," Tennessee Valley Authority; and Wikipedia, s.v. "Wilson Dam."

60. Dorn, 337, citing American-Russian Chamber of Commerce, comps., *Handbook of the Soviet Union* (New York, 1936), 202. Heymann, 18, says the other four turbines were built by the Soviets.

61. Hughes, 293n3, citing US Department of Commerce, Bureau of the Census, *Central Electric Light and Power Stations and Street and Electric Railways, with Summary of Electric Industries, 1912* (Washington, DC: GPO, 1915), 123, and foldout facing 132.

62. Hallwas, 39.

63. *MSPB-34*, 150.

64. Hughes, 287n3; and Wikipedia, s.v. "Wilson Dam."

65. Dorn, 330: "Senator Norris replied that according to Ford's proposed method of paying debts the $236 million of public property that Ford would acquire in the transactions would, after a century, be worth $14.5 billion." Sward, 128–30.

66. Gruening, 204, quoted in a letter of 23 March 1929, from Mr. S. E. Boney, the National Electric Light Association's Carolina Committee on Public Utility Information, complaining to Mr. Rogers that his statement was "wholly lacking in truth." "The entire Muscle Shoals project—including the dam—cost $130 million" according to "Wilson Dam and Reservoir," *Encyclopedia of Alabama*.

67. *MSPB-34*, 150.

68. EMERY_5.7.18 appears to be Roscoe's draft history of Quoddy. Smith, *Power Policy*, 233.

69. Holmes, 10, citing *BDN*, 14 March 1925: Cooper stated that his plant could deliver power to industries bordering New Hampshire and Boston for less than one cent per kilowatt hour.

70. Cooper, "Soviet Russia." See also Dorn, 336, citing *NYT*, 21 August 1932, 1.

71. *ES*, 12 October 1932, 1; and "Dnieper Dam Blown Up by Retreating Russians."

72. Dorn, 336.

73. Kiwanis Club of Stamford, 368.

74. Martin.

75. Wikipedia, s.v. "Toronto Power Generating Station."

76. Billington, Jackson, and Melosi, 115.

77. Hallwas, 49, 74.

78. Hughes, 293n16; and Wikipedia, s.v. "Lock and Dam," accessed 7 October 2020, https://en.wikipedia.org/wiki/Lock_and_Dam_No._19.

79. Martin.

80. Martin.

81. Dorn, 337.

APPENDIX 4. VITAL STATISTICS OF QUODDY TIDAL POWER PROJECT, ARMY PLAN D, 1935–1936

1. *Final-36*, 3.

2. *Final-36*, 3.

3. *Final-36*, 3.

4. EMERY_Scrapbook_Quoddy 1949; and AP, 7 October 1949.

5. "The Initial application for funds to the National Emergency Council was made May 15, 1935, by the Chief of Engineers, as previously mentioned. At this time, $20,000,000 was requested. Advice of allotment to the amount of $10,000,000 was given the District Engineer on June 7, 1935, having been approved by the President May 28, 1935, under Appropriation 013028, Emergency Relief, War, River & Harbors, Flood Control, etc., 1935—1937." *Final-36*, 5.

6. *Final-36*, 5.

7. *Final-36*, 5.

8. *Final-36*, 5.

9. *ES*, 28 October 1936, 1.

10. *ES*, 28 October 1936, 1.

11. *Final-36*, 5 (does not include repairs to Carlow Island dams). *ES*, 28 October 1936, 1, reported $973,000.00 was retained.

12. Does *not* include survey or engineering costs. Items listed here are "Estimated Total" from *Final-36*. See notes for each item as actuals were often significantly different.

13. Percent completed as per *Final-36*, 39–40.

14. Army's summary of July 1936: $2 million for engineering; $2 million for field construction; $1.5 million for housing for twenty-four hundred laborers in barracks and six hundred staff and their families at Quoddy Village; $1.5 million for materials and equipment, and land. *ES*, 15 July 1936, 1. This summary obscured the $617,226.44 spent on "District Overhead." *Final-36*, Cost Summary, Sheet 7.

15. *Final-36*, Cost Summary, Sheet 7.

16. *Final-36*, Cost Summary, Sheet 2. Cost as of 23 October 1936 was $693,772.01.

17. *Final-36*, 6. Cost Summary, Sheet 1 shows an estimated total of $165,000, presumably because of ongoing settlement issues.

18. 137 acres for Quoddy Village and Redoubt Hill; 30 acres for warehouse and camps; 141 acres for dams, locks, gates, powerhouse. *ES*, 2 September 1936, 1.

19. *Final-36*, Cost Summary, Sheet 3. The total estimate was $151,000.

20. *Final-36*, Cost Summary, Sheet 3. Cost as of 23 October 1936 was $392,972.31.

21. *Final-36*, 24, and Cost Summary, Sheet 6. Cost as of 23 October 1936 was $185,902.80.

22. *ES*, 28 August 1935, 1; and *BH*, 28 August 1935, 24.

23. *Final-36*, 24, and Cost Summary, Sheet 6. Administration building cost as of 23 October 1936 was $693,674.46.

24. *ES*, 15 January 1936, 1; and *ES*, 22 January 1936, 1.

25. *Final-36*, 24, and Cost Summary, Sheet 6. Cost as of 23 October 1936 was $302,695.97.

26. *Final-36*, 24, and Cost Summary, Sheet 6. Cost as of 23 October 1936 was $190,884.55.

27. *Final-36*, 24, and Cost Summary, Sheet 6. Cost as of 23 October 1936 was $720,568.07.

28. *Final-36*, 24, Cost Summary, Sheet 5, includes houses, utilities, and site work.

29. *ES*, 28 August 1935, 1; and *BH*, 28 August 1935, 24.

30. *Final-36*, 24, Cost Summary, Sheet 5, 7. Original estimate $3 million. *Reel-38*, 12.

31. One diesel generator cost $47,361.79. *Final-36*, Cost Summary, Sheet 5. The cost of the building sized for five generators is not indicated in *Final-36*.

32. *Final-36*, 24, and Cost Summary, Sheet 6. Cost as of 23 October 1936 was $316,971.05.

33. *Final-36*, 24, and Cost Summary, Sheet 6. Cost as of 23 October 1936 was $233,034.95.

34. *Final-36*, 25. Cost as of 23 October 1936 was $133,546.69.

35. *Final-36*, Cost Summary, Sheet 7.

36. Original estimate was $150,000. *ES*, 19 June 1935, 7. *Final-36*, 23, Estimated total for the unfinished Pleasant Point dam was $202,000.00.

37. *ES*, 17 February 1937, 1. Estimated total for the unfinished Carlow Island dam was $161,000.00.

38. Original estimate was $100,000. *ES*, 19 June 1935, 7. *Final-36*, 23.

39. *ES*, 17 February 1937, 1.

40. Original Eastport-Treat-Dudley Dam estimate $2,500,000. *ES*, 19 June 1935, 7. *Final-36*, 23. Cost Summary, Sheet 6. Estimated total for the unfinished dam was to be $213,600.

41. *Final-36*, Cost Summary, Sheet 5. Estimated total for tidal power station was not given in report.

42. *Final-36*, Cost Summary, Sheet 5. Estimated total for the unfinished navigation lock was to be $302,000.00.

43. *Final-36*, Cost Summary, Sheet 5. Estimated total for the unfinished gates was to be $284,000.00.

44. *Final-36*, 31. Total of 6,687,486 man hours was reported in *ES*, 28 October 1936, 5.

45. *Final-36*, 31.

46. *Final-36*, 31. Average cost per man-year of $1,445.67 was reported in *ES*, 28 October 1936, 5.

47. *Final-36*, 31.

APPENDIX 5. GERTRUDE STURGIS COOPER
(JUNE 29, 1889–DECEMBER 18, 1977)

1. Greenwich House was founded by Mary Kingsbury Simkhovitch, a summer resident of Passamaquoddy Bay and a member of the Eastport Development Board with Dexter P. Cooper in 1935–1936. The board of directors of Greenwich House included such luminaries as Gerard Swope, president of General Electric, and Marshall Field of Chicago retail fame.

2. *NYT*, 19 December 1977, obituary.

APPENDIX 6. ROSCOE CONKLIN EMERY
(MARCH 28, 1886–DECEMBER 16, 1969)

1. EMERY_2.17: "Portrait of a Declining Town: Eastport, Maine," *Maine Digest*, 1969, a reprint of an article by David Butwin in the *Saturday Review*, 5 October 1968, 17, https://www.unz.com/PDF/PERIODICAL/SaturdayRev -1968oct05/17-20/.

2. Thomas Loraine Holmes (1812–1897) was Roscoe's granduncle and Theodore's grandfather.

APPENDIX 7. THE US ARMY CORPS OF ENGINEERS

1. "President Harry S. Truman Double Hand Shaking Major General Philip B. Fleming and Major General Ulysses S. Grant III during Award Ceremony, May 1946" (photo). Photo by George Skadding/The LIFE Picture Collection/ Getty Images.

2. "Not Many Have Heard of Philip Bracken Fleming, but He Helped Change America's Roadways," Hawkeye, 15 October 2017, accessed 27 January 2019, https://www.thehawkeye.com/news/20171015/not-many-have-heard -of-philip-bracken-fleming—-but-he-helped-change-americas-roadways.

3. "President Harry Truman as He Reads the Victory Proclamation" (photo), https://media.gettyimages.com/photos/president-harry-truman-as-he-reads -the-victory-proclamation-to-the-picture-id174537389?s=2048x2048.

4. "Fleming, Philip B. (Philip Bracken), 1887–1955, Person Authority Record," National Archives Catalog, accessed 27 January 2019, https://catalog .archives.gov/id/10572659.

5. *ES*, 2 October 1935, 1.

6. "Royal B. Lord, 1923," West Point Association of Graduates, https://www.westpointaog.org/memorial-article?id=3885f497-0f59-4b4e-a90f-f745aa044e6c.

7. USACE Publication #EP 870-1-18: Engineer-Memoirs—Major General Hugh J. Casey, US Army, December 1993, 105–12.

8. "Donald J. Leehey, 1920," West Point Association of Graduates, https://www.westpointaog.org/memorial-article?id=30109044-6883-4cb7-969f-e288edc9ff43.

9. "Lieutenant General Samuel D. Sturgis, Jr.," Internet Archive: Wayback Machine, https://web.archive.org/web/20050306122846/http://www.hq.usace.army.mil/history/coe4.htm. Nor is there any mention in Wikipedia, s.v. "Samuel D. Sturgis Jr.," accessed 1 May 2020, https://en.wikipedia.org/wiki/Samuel_D._Sturgis_Jr.

APPENDIX 8. THE PASSAMAQUODDY RESERVATION

1. Broduer, 70.

2. Broduer, 70–74.

3. Broduer, 81–83. Don Gellers was posthumously pardoned by Governor Janet Mills on 15 January 2020. Erin Slomski-Pritz, "Tribal Attorney in Maine Posthumously Pardoned for 1968 Pot Charge," *NPR*, 15 January 2020, https://www.npr.org/2020/01/15/796540445/tribal-attorney-in-maine-posthumously-pardoned-for-1968-pot-charge.

4. Broduer, 82.

5. Broduer, 85.

6. Broduer, 86–87.

7. Broduer, 89.

8. Broduer, 95. The original decision was made on 20 January 1975.

9. Broduer, 98–99.

10. Brouder, 129.

11. Broduer, 126.

APPENDIX 9. CAMPOBELLO, NEW BRUNSWICK

1. The empty lot on which the Sturgis Cottage stood was sold by Gertrude Sturgis Cooper to Aaron Tilton in 1959. Wyllie, 56–57.

2. Wyllie, 56 and images on 169 (Friar's Bay Vista, ca. 1890, CPL collection) and 170 (Friar's Bay Vista, ca. 1892–1897, CPL collection).

3. *ES*, 15 August 1934, 2.

4. 28 June 1933 through 12 August 1936.

5. *ES*, 6 April 1941, 8.

6. Klein, *Beloved Island*, 201.

7. Klein, *Beloved Island*, 206, 239; and *NYT*, 10 July 1952, 33.

8. See chapter 12, "Gray Days," note 96. For the full and fascinating story, read Epstein.

9. Klein, *Beloved Island*, 206, 239.

10. *NYT*, 12 May 1963; and Epstein.

APPENDIX 10. EASTPORT, MAINE

1. *Quoddy Tides*, 12 December 1997, 17.

2. Holt, 57.

3. Marvin, 12.

4. Holt, 30–31.

APPENDIX 11. TIDAL POWER DEVELOPMENT, 1950–2020

1. Thurston, *Tidal Life*, 19–20; and Marmer, *Tide*, 223. As of October 2019, there were 58 commercially operating nuclear power plants in the United States with 96 nuclear reactors in 31 states. US Energy Information Administrations, "Frequently Asked Questions (FAQs)," http://www.eia.gov/tools/faqs /faq.cfm?id=207&t=3.

2. Thurston, *Tidal Life*, 20.

3. Wikipedia, s.v. "Kislaya Guba Tidal Power Station," accessed 19 December 2019, https://en.wikipedia.org/wiki/Kislaya_Guba_Tidal_Power_Station.

4. Thurston, *Tidal Life*, 20.

5. Thurston, *Tidal Life*, jacket copy.

6. Thurston, *Tidal Life*, 20–22, 151.

7. Information from tour guide at Annapolis Tidal Generating Station, 15 September 2018.

8. *ES*, 25 February 1925, 8.

9. FDRL-COOPER: 1936-02-04_Dexter Cooper to James Roosevelt_angry at Army Engineers' report. The report, *Passamaquoddy Tidal Power Development Estimates*, was dated 11 January 1936.

10. Wikipedia, s.v. "Sihwa Lake Tidal Power Station," accessed 19 December 2019, https://en.wikipedia.org/wiki/Sihwa_Lake_Tidal_Power_Station.

11. NIMBY is an acronym for "Not In My Back Yard."

12. Most helical blades have airfoil-shaped cross sections and can therefore turn faster than the speed of the current. ORPC's foils turn at an average 5 rpms.

13. Water is 832 times denser than air and, when moving at the same speed, will have hundreds of times more power to turn blades. If air were blowing past a windmill at a speed of a typical tidal flow—say, 5 knots—not much would happen. But water traveling at the same speed hits the blades of a turbine with as much force as a hurricane. Manning, "Tidal Power," 43.

14. *Environmental Report*, ORPC, March 2013, accessed summer 2021, https://www.orpc.co/permitting_doc/environmentalreport_Mar2013.pdf; and "Cobscook Bay Tidal Energy Project," TETHYS, https://tethys.pnnl.gov /project-sites/cobscook-bay-tidal-energy-project.

15. Discussions with Robert Lewis of ORPC, 13 September 2019.

16. R. Crowell, "This Bridge Monitors the Environment and Harnesses Tidal Energy," *Eos,* 7 August 2019, https://doi.org/10.1029/2019EO130389, and https://eos.org/articles/this-bridge-monitors-the-environment-and-har nesses-tidal-energy.

17. "Three Verdant Power Tidal Turbines Deployed in New York City's East River," PR Newswire, 23 October 2020, https://www.prnewswire.com /news-releases/three-verdant-power-tidal-turbines-deployed-in-new-york -citys-east-river-301159031.html.

18. "Turbine Damage Stalls Fundy Tidal Power Test," *CBC News*, 11 June 2010, https://www.cbc.ca/news/canada/nova-scotia/turbine-damage-stalls -fundy-tidal-power-test-1.926011; and "Failed Tidal Turbine Explained at Symposium," *CBC News*, 11 July 2011, https://www.cbc.ca/news/canada/nova -scotia/failed-tidal-turbine-explained-at-symposium-1.1075510.

19. Information from tour guide at Annapolis Tidal Generating Station, 15 September 2018. Power is the square of the speed of the current, thus $8.5 \times 8.5 = 72.25$, whereas $10.5 \times 10.5 = 110.25$.

20. Information from tour guide at Annapolis Tidal Generating Station, 15 September 2018.

21. Wasik, *Merchant of Power*, 68.

22. *NYT*, 2 August 1936, E8.

23. "Pilgrim Nuclear Power Station Shut Down Permanently," Entergy, 3 June 2019, https://www.entergynewsroom.com/news/pilgrim-nuclear-power -station-shut-down-permanently/.

24. "What You Need to Know about the New Mass Clean Energy Project Proposals," 3 August 2017, https://climate-xchange.org/2017/08/03/what -you-need-to-know-about-the-new-mass-clean-energy-project-proposals/; Iulia Gheorghiu, "Controversial $1B Canada-US Transmission Line Gets

Maine PUC Approval," Utility Dive, 11 April 2019, https://www.utilitydive
.com/news/controversial-1b-canada-us-transmission-line-gets-nod-from
-maine-puc-staff/551812/; and "Northern Pass," NHPR, https://www.nhpr
.org/topic/northern-pass#stream/0.

25. "Annual Average Electricity Price Comparison by State," Nebraska
Department of Environment and Energy, https://neo.ne.gov/programs/stats
/inf/204.htm.

26. "Boston Outer Harbor Barrier Could Generate Tidal Power," *Boston
Herald*, 30 August 2020, https://www.bostonherald.com/2020/08/30/boston
-outer-harbor-barrier-could-generate-tidal-power/.

Bibliography

See appendix 1, "Author's Notes and Acknowledgments," for a bibliographic essay on the sources. Dozens of newspapers and magazines are quoted; the primary periodicals, such as the *Eastport* (Maine) *Sentinel, Boston Herald*, and *New York Times*, are discussed in appendix 1. See appendix 12, "Abbreviations and Document Sources," for the list of newspaper and documentary sources consulted and the abbreviations used to cite them in the notes.

Abrams, Ernest R. *Power in Transition*. New York: Charles Scribner's Sons, 1940.

Aldersey-Williams, Hugh. *The Tide: The Science and Stories behind the Greatest Force on Earth*. New York: W. W. Norton, 2016.

Atler, Jonathan. *The Defining Moment: FDR's Hundred Days and the Triumph of Hope*. New York: Simon and Schuster, 2007.

Ball, Norman R. *The Canadian Niagara Power Company Story*. Erin, ON: Boston Mills Press, 2005.

Barber, Bette E. *Guide Book to FDR's "Beloved Island."* N.p.: n.p., 1956.

Bennett, William J. *America—The Last Best Hope: From a World at War to the Triumph of Freedom, 1914–1989*. Nashville, TN: Thomas Nelson, 2008.

Berton, Pierre. *Niagara: A History of the Falls*. Toronto: McClelland & Steward, 1992.

Billington, David P., Donald C. Jackson, and Martin V. Melosi. *The History of Large Dams: Planning, Design and Construction*. Denver, CO: US Department of the Interior, Bureau of Reclamation, 2005. https://www.usbr.gov/history/HistoryofLargeDams/LargeFederalDams.pdf.

Blau, Raphael D. "A New Deal for an FDR Dream." *True Magazine*, January 1964.

Boyer, Richard O. "Maine Power Concerns Will Fight Quoddy Dam in Courts, Legislature." *Boston Herald*, 24 July 1935.

Boyer, Richard O. "Seek 'Killing' on Quoddy Land: Maine Owners after Huge Profit Despite Vital Need of Project." *Boston Herald*, 25 July 1935.

Brands, H. W. *Traitor to His Class: The Privileged Life and Radical Presidency of Franklin Delano Roosevelt*. New York: Anchor Books, 2008.

Broduer, Paul, *Restitution: The Land Claims of the Mashpee, Passamaquoddy, and Penobscot Indians of New England*. Boston: Northeastern University Press, 1985.

Brown, Herbert Ross. *Sills of Bowdoin: The Life of Kenneth Charles Morton Sills, 1979–1954*. New York: Columbia University Press, 1964.

Burns, McGregor. *Roosevelt: The Lion and the Fox*. New York: Harcourt Brace, 1956.

Burrows, E. H. *Captain Owen of the African Survey, 1774–1857*. Southampton, UK: Privately published, 1978.

Butler, Joyce. *Wildfire Loose: The Week Maine Burned*. 3rd ed. Lanham, MD: Down East Books, 1979.

Cantelon, Philip L. "The Regulatory Dilemma of the Federal Power Commission, 1920–1977." *Federal History*, no. 4 (2012): 69–70.

Caro, Robert A. *Working: Researching, Interviewing, Writing*. New York: Alfred A. Knopf, 2019.

Chaisson, Eric, and Steve McMillan. *Astronomy Today*. 3rd ed. Upper Saddle River, NJ: Prentice Hall, 1999.

Chase, Karen. *FDR on His Houseboat: The Larooco Log, 1924–1926*. Albany: State University of New York Press, 2016.

Cline, Adam, *The Current War: A Battle Story between Two Electrical Titans, Thomas Edison and George Westinghouse*. Self-published, 2017.

Complete Presidential Press Conferences of Franklin Delano Roosevelt. 12 vols. New York: Da Capo, 1973.

Cooper, Dexter P. "Some Aspects of the Passamaquoddy Tidal Power Project." In *Papers and Transactions for 1926 and Proceedings of the 42nd Annual Meeting*, Connecticut Society of Civil Engineers. New Haven, CT: Tuttle, Morehouse and Taylor, 1926.

Cooper, Hugh L. "Soviet Russia." Address delivered to the Institute of Politics, Williamstown, MA, 1 August 1930.

Cross, Robert F. *Sailor in the White House: The Seafaring Life of FDR*. Annapolis, MD: Naval Institute Press, 2003.

Davis, Ewin L. "The Influence of the Federal Trade Commission's Investigation of Federal Regulation of Interstate Gas and Electric Companies." *George Washington Law Review* 14 (December 1945): 21–29.

DelloRusso, Barbara A. F. "Waves of the Future? Harnessing Tidal Power—A Historical Review of the Passamaquoddy Bay Tidal Project and the La Rance River Project." *Infinite Energy*, no. 33 (2000): 44–47.

Descendants of Nath'l Russell Sturgis, Edition of January 1, 1925. Privately published. Collection of Dexter P. Cooper III.

Dorn, Harold. "Hugh Lincoln Cooper and the First Détente." *Technology and Culture* 20, no. 2 (1979): 322-47. https://doi.org/10.2307/3103869.

Durand, William F. *Passamaquoddy Tidal Power Project: Report of Board of Engineers on Cost Estimates of Project*. 8 May 1936. Referred to in notes as *Durand-36*.

Eldridge Tide and Pilot Book. Annual. 1854–.

Elliott, Sara, Peter Wilf, Robert Walter, and Dorothy Merritts. "Subfossil Leaves Reveal a New Upland Hardwood Component of the Pre-European Piedmont Landscape, Lancaster County, Pennsylvania." *PloS One* 8 (2013): e79317. https://doi.org/10.1371/journal.pone.0079317.

Emmons, William M., III. "Franklin D. Roosevelt, Electric Utilities and the Power of Competition." *Journal of Economic History* 53, no. 4 (December 1993): 880–907.

Epstein, Edward Jay. *Dossier: The Secret History of Armand Hammer*. New York: Random House, 1996.

Final Report: Passamaquoddy Tidal Power Development. Eastport, ME: Army Corps of Engineers, 23 October 1936. Referred to in notes as *Final Report-36*.

Funigiello, Phillip. *Toward a National Power Policy*. Pittsburgh, PA: University of Pittsburgh, 1973. https://digital.library.pitt.edu/islandora/object/pitt%3A31735057894564/viewer#page/1/mode/2up.

Gallant, Gregory P. *Hope and Fear in Margaret Chase Smith's America: A Continuous Tangle*. Lanham, MD: Lexington Books, 2014.

Gateways to Commerce. National Park Service, Rocky Mountain Region, 1992. https://irma.nps.gov/DataStore/DownloadFile/466343.

Goodwin, Doris Kearns. *The Bully Pulpit: Theodore Roosevelt and the Golden Age of Journalism*. New York: Simon & Schuster, 2013.

Goodwin, Doris Kearns. *Leadership in Turbulent Times*. New York: Simon & Schuster, 2019.

Goodwin, Doris Kearns. *No Ordinary Time: Franklin & Eleanor; The Home Front in World War II*. New York: Simon & Schuster, 1994.

Gordon, Robert, and Patrick Malone. "'Perpetual Power' from the Tides in Boston, Massachusetts, USA, 1813–1858." *Water History* 11 (2019): 3–29.

Gregory, William A., and Rennard Strickland. "Hugo Black's Congressional Investigation of Lobbying and the Public Utility Holding Company Act: A Historical View of the Power Trust, New Deal Politics, and Regulatory Propaganda." *Oklahoma Law Review* 29 (1976): 543–76.

Gruening, Ernest. *The Public Pays . . . and Still Pays: A Study of Power Propaganda; New England Edition*. New York: Vanguard Press, 1964. First published 1931.

Gunther, John. *Inside U.S.A.* New York: Harper & Brothers, 1947.

Gunther, John. *Roosevelts in Retrospect: A Profile in History*. New York: Harper & Brothers, 1950.

Hallwas, John E. *Keokuk and the Great Dam*. Chicago: Arcadia, 2001.

Halsey, William F. *Admiral Halsey's Story*. New York: McGraw-Hill, 1947.
http://penelope.uchicago.edu/Thayer/E/Gazetteer/People/William_Halsey
/HALAHS/2*.html#illustration_p34H.

Harendy, Jim, and Jane Diggins Harendy. *Campobello Island*. Charleston, SC:
Arcadia, 2002.

Heymann, Hans. *We Can Do Business with Russia*. Chicago: Ziff Davis, 1945.

Holmes, Theodore C. *A History of the Passamaquoddy Tidal Power Project*.
Orono: University of Maine, 1955.

Holt, John "Terry." *The Island City: A History of Eastport, Moose Island, Maine,
from Its Founding to Present Times*. Eastport, ME: Eastport 200 Committee,
1999.

Hubbard, Preston J. *Origin of the TVA: The Muscle Shoals Controversy, 1920–
1932*. Nashville, TN: Vanderbilt University Press, 1961.

Hughes, Thomas O. *Networks of Power: Electrification in Western Society,
1880–1930*. Baltimore, MD: Johns Hopkins University Press, 1983.

Ickes, Harold. *The Secret Diaries of Harold Ickes*. New York: Simon & Schuster,
1954.

Insull, Samuel. "Economics of the Public Utility Business." *Electrical World*,
25 March 1922.

International Joint Commission (IJC). For reports, see appendix 12.

Irwin, Clark T., Jr. *The Light from the River: Central Maine Power's First Century of Service*. Augusta: Central Maine Power, 1999.

Johnson, Ryerson, ed. *200 Years of Lubec History, 1776–1976*. Lubec, ME: Lubec Historical Society, 1976.

Justinian, Caesar Flavius. *The Institutes of Justinian*. 5th ed. Translated into English by J. B. Moyle. 1913. Released 11 April 2009; updated 6 February 2013.
https://www.gutenberg.org/files/5983/5983-h/5983-h.htm.

Kerwin, Jerome G. *Federal Water-Power Legislation*. New York: Columbia
University Press, 1926.

Kilby, William Henry. *Eastport and Passamaquoddy: A Collection of Historical
and Biographical Sketches*. Eastport, ME: Border Historical Society, 2003.
First published 1888.

Kiwanis Club of Stamford, Niagara Falls, Ontario. *Niagara—River of Fame*.
Condensed ed. Toronto: Kiwanis Club of Stamford, 1968.

Klein, Jonas. *Beloved Island: Franklin & Eleanor and the Legacy of Campobello*. Forest Dale, VT: Paul S. Eriksson, 2000.

Klein, Maury. *A Call to Arms: Mobilizing America for World War II*. New York:
Bloomsbury, 2013.

Klein, Maury. *The Power Makers: Steam, Electricity, and the Men Who Invented
Modern America*. New York: Bloomsbury, 2008.

Koppel, Tom. *Ebb and Flow: Tides and Life on Our Once and Future Planet*.
Toronto: Dundurn Group, 2007.

Landis, James M. "The Legislative History of the Securities Act of 1933." *George Washington Law Review* 28 (October 1959): 29–39.

Larson, Erik. *The Splendid and the Vile: A Saga of Churchill, Family, and Defiance during the Blitz.* New York: Crown, 2020.

Lash, Joseph P. *Eleanor and Franklin.* New York: Signet New American Library, 1973.

Lester, Joan A. *History on Birchbark: The Art of Tomah Joseph, Passamaquoddy.* Bristol, RI: Haffenreffer Museum of Anthropology, Brown University, 1993.

Leuchtenburg, William E. *Franklin D. Roosevelt and the New Deal, 1932–1940.* New York: Harper TorchBooks, 1965. https://archive.org/details/in.ernet.dli .2015.213904/page/n1/mode/2up.

Leuchtenberg, William E. "Roosevelt, Norris and the 'Seven Little TVAs.'" *Journal of Politics* 14, no. 3 (August 1952), https://doi.org/10.2307/2126212.

L'Hommedieu, Andrea. "MacNichol, David Oral History Interview," 2002. Edmund S. Muskie Oral History Collection, 230. http://scarab.bates.edu /muskie_oh/230.

Lief, Alfred. *Democracy's Norris.* New York: Octagon, 1939.

Lockett, Jerry. *The Discovery of Weather: Stephen Saxby, the Tumultuous Birth of Weather Forecasting, and Saxby's Gale of 1869.* Halifax, NS: Formac, 2012.

Macy, Christine. *Dams.* Norton/Library of Congress Visual Sourcebooks in Architecture, Design and Engineering. New York: Norton, 2010.

Maine State Planning Board. *Report on Passamaquoddy Tidal Power Project by the Maine State Planning Board, State House, Augusta, to the Quoddy Hydro-Electric Commission Appointed by Governor Brann.* Maine State Planning Board, Alfred Mullikin. 22 December 1934. https://digitalmaine.com /spo_docs/96/. Referred to in notes as *MSPB-34.*

Manning, Jean. "Tidal Power." *Infinite Energy Magazine* 33 (2000): 43.

Marmer, H. A. *The Tide.* New York: D. Appleton, 1926.

Martin, Kent. "Keokuk Energy Center: Harnessing the Power of the Mississippi." *Hydro Review*, 7 November 2013.

Marvin, George. "Passamaquoddy! The Dream That Won't Die." *Ellsworth American*, 1974. Reprint of the author's honor's thesis, "The Passamaquoddy Tidal Power Project and the New Deal," Bowdoin College, 25 May 1972.

McCraw, Thomas K. *Prophets of Regulation: Charles Francis Adams, Louis D. Brandeis, James M. Landis, Alfred E. Kahn.* Boston: Belknap/Harvard University Press, 1984.

McCully, James Greig. *Beyond the Moon: A Conversational Common Sense Guide to Understanding the Tides.* Singapore: World Scientific Press, 2006.

McDonald, Forrest. *Insull: The Rise and Fall of a Billionaire Utility Tycoon.* Washington, DC: Beard Books, 2004. First published 1962 by University of Chicago Press.

Meigs, M. "Operating Experiences at the Keokuk Lock and Dry Dock." *Military Engineer* 12, no. 62 (March–April 1920): 192–204. https://www.jstor.org /stable/44573794.

Meigs, Peveril. "Energy in Early Boston." *New England Historical and Genealogical Register*, 127 (April 1974): 83–84.

Morgan, Ted. *FDR: A Biography*. New York: Simon & Schuster, 1985.

Munson, Richard, *From Edison to Enron: The Business of Power and What It Means for the Future of Electricity*. Santa Barbara, CA: Praeger/ABC-CLIO, 2005.

Muskie, Stephen O. *Campobello: Roosevelt's Beloved Island*. Rockland, ME: Down East Books, 1982.

Nehf, Harley W. "The Concentration of Water Powers." *Journal of Political Economy* 24, no. 8 (October 1916): 775–93. https://doi.org/10.1086/252872.

Norris, George K. *Fighting Liberal*. New York: Macmillan, 1961. First published 1945.

Nowlan, Alden. *Campobello: The Outer Island*. Toronto: Stoddard, 1993.

Nye, David E. *Consuming Power: A Social History of American Energies*. Cambridge: MIT Press, 2001.

Parkman, Aubrey. *Army Engineers in New England: The Military and Civil Work of the Corps of Engineers in New England, 1775–1975*. Waltham, MA: US Army Corps of Engineers New England Division, 1978. https://digitalcommons .library.umaine.edu/cgi/viewcontent.cgi?article=1006&context=dickey_lincoln.

Passamaquoddy Bay Tidal Power Commission Report. Washington, DC: U.S. Federal Power Commission, 30 March 1934. Referred to in notes as *FPC-34*.

"Passamaquoddy Tidal Power Project: Letter from the Chairman of the Federal Power Commission: In Response to Senate Resolution 62, 76th Congress, February 1939." In *Report on the Passamaquoddy Tidal Power Project, Maine*. US Congress, Senate, S.D. 41, 77th Cong., 1st Sess., 1941. Referred to in notes as *FPC-41*.

Pesha, Ronald, ed. *The Great Gold Swindle of Lubec, Maine*. Charleston, SC: History Press, 2013.

Porter, Peter A., *Guide: Niagara—River—Falls—Frontier*. N.p.: printed by Matthews Northrop Works.

Pritchett, Charles H. *The Tennessee Valley Authority: A Study in Public Administration*. Chapel Hill: University of North Carolina Press, 1943.

Radin, Alex, *Public Power—Private Life.* Washington, DC: American Public Power Association, 2003.

Riggs, John A. *High Tension: FDR's Battle to Power America.* Diversion Books, 2020.

Roberts, Kenneth. "Quoddy." *Saturday Evening Post*, 19 September 1936.

Roberts, Kenneth. *Trending into Maine*. Boston: Little, Brown, 1938.

Rollins, Alfred B., Jr. *Roosevelt and Howe*. New York: Knopf, 1936.

Roosevelts at Campobello. Campobello, NB: Roosevelt Campobello International Park, 2003.

Schlesinger, Arthur M., Jr. *Politics of Upheaval, 1935–1936.* Boston: Houghton Mifflin, 2003.

Seifer, Marc J. *Wizard: The Life and Times of Nikola Tesla, Biography of a Genius.* New York: Citadel Press, 1996.

Siegel, Katherine A. S. *Loans and Legitimacy: The Evolution of Soviet-American Relations, 1919–1933.* Lexington: University Press of Kentucky, 2014.

Smith, Lincoln. *The Power Policy of Maine.* Berkeley: University of California Press, 1951.

Smith, Lincoln. "The Regulation of Some New England Holding Companies." *Land Economics* 23, no. 3 (August 1949): 289–303.

Smith, Lincoln. "Tidal Power in Maine." *Land Economics* 24, no. 3 (August 1948): 239–52.

Soctomah, Donald. *Let Me Live as My Ancestors Had, 1850–1890: Tribal Life and Times in Maine and New Brunswick.* Pleasant Point, ME: Passamaquoddy Cultural Heritage and Resource Center, 2005.

Soctomah, Donald. *Passamaquoddy at the Turn of the Century, 1890–1920: Tribal Life and Times in Maine and New Brunswick.* N.p.: Maine Humanities Council, Passamaquoddy Tribe of Indian Township, 2002.

Soctomah, Donald. *Save the Land for the Children, 1800–1850: Passamaquoddy Tribal Life and Times in Maine and New Brunswick.* Perry, ME: Tribal Historic Preservation Department, Passamaquoddy Tribe, 2009.

Soctomah, Donald, and Jean Flahive. *Remember Me: Tomah Joseph's Gift to Franklin Roosevelt.* Gardiner, ME: Tilbury House, 2009.

Sonnic, Ewan. "A Worldwide Inventory of Tide Mills Reflecting an Essential Western History and Consequent Distribution." Paper presented at Tide Mill Institute Annual Conference, 2019.

Springer, Vera. *Power and the Pacific Northwest: A History of the Bonneville Power Administration.* Tucson: University of Arizona Press, 1986.

Stone, I. F. *Business as Usual: The First Year of Defense.* New York: Modern Age Books, 1941.

Sugrue, Thomas. "Santa Clause Comes to Way Down East." *American Magazine,* December 1935.

Sward, Keith. *The Legend of Ford.* New York: Rinehart, 1948.

Thompson, Carl D. *Confessions of the Power Trust.* New York: Dutton, 1932.

Thurston, Harry. *A Place between the Tides: A Naturalist's Reflection on the Salt Marsh.* Vancouver, BC: Graystone Books, 2004.

Thurston, Harry. *Tidal Life: A Natural History of the Bay of Fundy.* Camden East, ON: Camden House, 1990.

Tripp, Guy E. *Super Power as an Aid to Progress.* New York: Knickerbocker Press/Putnam, 1924.

Twentieth Century Fund Power Committee. *Electric Power and Government Policy*. New York: The Twentieth Century Fund, 1914.

Warren, Earle, "Budd." "The Salt Water Mills of Maine in Context: A Preliminary Exploration." Unpublished manuscript, 2012.

Wasik, John F. *The Merchant of Power: Sam Insull, Thomas Edison, and the Creation of the Modern Metropolis*. New York: Palgrave/Macmillan, 2006.

Watkins, T. H. *The Great Depression: America in the 1930s*. Boston: Little, Brown, 1993.

Weil, Gorgon. *Blackout: How the Electric Industry Exploits America*. New York: Nation Books, 2006.

Wells, Kate Gannett. "The Brass Cannon of Campobello." *New England Magazine* 5 (1891).

Wells, Kate Gannett. *Campobello: An Historical Sketch*. Campobello, NB: privately published, 1894.

Wells, Kate Gannett. "The Quoddy Hermit." *Atlantic*, June 1885, 821–26.

Wells, Walter. *The Water Power of Maine*. Augusta, ME: Sprague, Owen & Nash, 1869.

Wengert, Norman. "Antecedents of the TVA: The Legislative History of Muscle Shoals." *Agricultural History* 26, no. 4 (October 1952): 141–47.

White, Jonathan. *Tides: The Science and Spirit of the Ocean*. San Antonio, TX: Trinity University Press, 2017.

Wilson, Joan Hoff. *Ideology and Economics: U.S. Relations with the Soviet Union, 1918–1933*. Columbia: University of Missouri Press, 1974.

Wyllie, Robin H. *The Franklin Delano Roosevelt Cottage and the Campobello Summer Colony, Roosevelt-Campobello International Park*. Campobello, NB: Roosevelt Campobello International Park, 1992.

Zimmerman, David. *Coastal Fort: A History of Fort Sullivan, Eastport, Maine*. Eastport, ME: Border Historical Society, 1984.

Zipp, Samuel. *The Idealist: Wendell Willkie's Wartime Quest to Build One World*. Boston: Belknap/Harvard University Press, 2020.

Index

skeptics of, 96–97; test drilling, 118; US Army Corps of Engineers and, 230–31, 232–36, 253–60, 274, 277–78, 281–86, 313, 445–46; water management patent, 146; whispering campaign against, 220; workers/worker concerns, 262–65, 276–80, 293–94, 311; as World War II casualty, 388–89
Quoddy Village: closing of, 311, 332, 340–41, 393; construction/design of, 256, 276–77, 281, 301, 302, 318–20, 325, 357; cost of, 233, 283; electric service interruptions, 286; location of, 252–55, 459; sale of, 395–98; US Army and, 249, 291; for vocational training, 339, 342, 367, 389

Rankin, John, 231
Rankine, William Birch, 56–57
Reconstruction Finance Corporation (RFC), 240, 346, 382, 531n127
Redman, Fulton J., 364–65
Reed, James, 121
Reel, Frank, 355–58, 359, 579n1
regulatory bargain, 125
regulatory capture, 127, 428, 517n52
Relief Bill for Maine, 222–23, 224
Rexall Orderlies, 15
riparian rights, 123–24
Ripley, William Z., 31, 127
river rights, 123–24
Roberts, Kenneth, 259, 325
Roches, Genevieve, 147
Rogers, Will, 147, 412–13
Roosevelt, Eleanor, 32–33, 82, 84, 89, 181–82, 318–19, 457, 458, 484n73, 594n171, 595n184
Roosevelt, Franklin, 157; as assistant secretary of the navy, 63–64, 81; Atlantic Charter, 386; atomic bomb concerns, 368; attempted

assassination of, 14; birth and early years, 30–33; Boulder Dam, 198, 267; canoe use, 49; Churchill, Winston and, 386, 390; Cooper, Dexter P. and, 176–77, 184–85, 194, 201–4; Cooper, Hugh and, 172–75, 176; death of, 391–92; decoy charade by, 385–86; election of, 167–69; on electricity demand, 315–16, 323; Emergency Banking Act, 169–70; fireside chats, 171, 376; Fleming, Philip B. and, 319–21; government regulation of monopolies, 153–55; as governor of New York, 147–48; Keokuk Dam and, 8, 9, 13; Lend-Lease program, 379, 385; Lubec Narrows and, 3–5, 165–66; on Nazi Germany, 367, 368, 376, 378–80; negotiations with private power trusts, 215–17; New Deal, ix, 171–72, 200, 527n54; New Deal Public Power Policy, 155; Passamaquoddy Bay and, 6, 166–67; piloting/sailing skills, 3–6, 82, 314–15, 385; polio and, 63, 82–84, 88–89; political beginnings, 81–82; political successes, 132–33; as presidential candidate, 153–55; Public Utility Holding Company Act, 240, 244, 260; Quoddy project and, 175–77, 195–207, 275, 284–85, 290–91, 292, 299, 315–17, 319–21, 424–25; Quoddy shutdown, viii, 310–12, 424–25; reelection of, 330–31; St. Lawrence–Franklin D. Roosevelt Power Project, 408; third term as president, 375–77; tidal-hydroelectric power, 74, 77; TVA lawsuit, 287–88; as vice-presidential candidate, 74–77, 81; Wheeler-Rayburn Act, 240; World War I and, 66